Library Service, Doncaster Royal Infirmary
T017827

# A TEXTBOOK OF CLINICAL OPHTHALMOLOGY

*3rd edition*

## A Practical Guide to Disorders of the Eyes and Their Management

**Ronald Pitts Crick**
FRCS, FRCOphth

*Honorary Consultant Ophthalmologist,*
*King's College Hospital, London*

*Lecturer Emeritus in Ophthalmology,*
*School of Medicine and Dentistry of King's College,*
*University of London*

*President, International Glaucoma Association*

**Peng Tee Khaw**
PhD, FRCP, FRCS, FRCOphth, FIBiol, FRCPath, FMedSci

*Professor of Glaucoma and Wound Healing*
*Consultant Ophthalmic Surgeon,*
*Director, Wound Healing Research Unit*
*Moorfields Eye Hospital and Institute of Ophthalmology,*
*University College, London*

*Adjunct Professor,*
*University of Florida, U.S.A.*

*1st Hong Leong Visiting Professor,*
*National University of Singapore*

*Published by*

World Scientific Publishing Co. Pte. Ltd.

P O Box 128, Farrer Road, Singapore 912805

*USA office:* Suite 1B, 1060 Main Street, River Edge, NJ 07661

*UK office:* 57 Shelton Street, Covent Garden, London WC2H 9HE

**British Library Cataloguing-in-Publication Data**
A catalogue record for this book is available from the British Library.

**A TEXTBOOK OF CLINICAL OPHTHALMOLOGY – (3rd Edition)**



ISBN 981-238-128-7
ISBN 981-238-150-3 (pbk)

Printed by FuIsland Offset Printing (S) Pte Ltd, Singapore

To

Roger and Valerie

Jocelyn and Peggy

**WORLD SCIENTIFIC PUBLISHING**

ACKNOWLEDGE THE SUPPORT GIVEN IN THE PRODUCTION OF THIS BOOK BY

**THE INTERNATIONAL GLAUCOMA ASSOCIATION**

Registered Charity No: 274681

108C Warner Road, London, SE5 9HQ UK

TO WHICH THE AUTHORS' ROYALTIES WILL BE DONATED

---

WORLD SCIENTIFIC PUBLISHING

ACKNOWLEDGE THE SUPPORT GIVEN IN THE PRODUCTION OF THIS BOOK BY

THE INTERNATIONAL GLAUCOMA ASSOCIATION

Registered Charity No. 274681

108c Warner Road, London, SE5 9HQ U.K.

TO WHICH THE AUTHORS' ROYALTIES WILL BE DONATED

The authors and the publisher of this volume have taken care to make certain that the doses of drugs and schedules of treatment are correct and compatible with the standards generally accepted at the time of publication. Nevertheless, as new information becomes available, changes in treatment and in the use of drugs become necessary. The reader is advised to consult carefully the instruction and information material included in the package insert of each drug or therapeutic agent before administration. This advice is especially important when using new or infrequently used drugs. The authors and publisher disclaim any liability, loss, injury or damage incurred as a consequence, directly or indirectly, of the use and application of any of the contents of this volume.

## SUB EDITORS

**David C. Broadway    MD, BSc (Hons), FRCOphth, DO**
*Consultant Ophthalmic Surgeon,*
Norfolk and Norwich University NHS Trust.
Honorary Senior Lecturer, University of East Anglia
Ocular Trauma and Conjunctiva.
Chapters 7, 8, 24

**Peter Constable    MA, MD, FRCOphth**
*Consultant Ophthalmic Surgeon,*
Royal Berkshire Hospital, Reading.
Strabismus and Disorders of Ocular Motility,
Functional Eye Disorders and Therapeutics Pt.1
Chapters 11, 19, 21, 32

**M. Francesca Cordeiro    MB BS, MRCP, FRCOphth**
*Welcome Trust Research Fellow,* Glaucoma Unit and
Wound Healing Research Unit, Department of Pathology,
Institute of Ophthalmology and Moorfields Eye Hospital, London.
Ophthalmic Aspects of Blood Diseases,
Endocrine and Systemic Immune Disorders
Chapters 15, 16, 18

**Martin D. Pitts Crick    MB BS, FRCS, FRCOphth**
*Consultant Ophthalmic Surgeon,*
Royal Bournemouth Hospital.
Vascular, Vitreous and Retinal Disease
Chapters 6, 14, 27, 29

**Alexander J.E. Foss    DM, MA, FRCOphth, MRCP**
*Consultant Ophthalmic Surgeon,*
University Hospital, Queen's Medical Centre, Nottingham.
Uveal Diseases
Chapters 7, 28

**Helena J. Frank    B Med Sci, MB BS, FRCS, FRCOphth, DO,**
*Consultant Ophthalmic Surgeon,*
Royal Bournemouth Hospital.
Ophthalmic Optics and Refraction
Chapter 4

**David S. Gartry    MD, BSc (Hons), FRCS, FRCOphth, DO, FCOptom**
*Consultant Ophthalmic Surgeon, Corneal and External Diseases Service,*
Moorfields Eye Hospital, London.
Diseases of the Cornea, Sclera and Lens
Chapters 6, 7, 25, 26

**Paula D. Gormley MB BCh, FRCS, FRCOphth, DO**
*Consultant Ophthalmic Surgeon,*
University Hospital, Queen's Medical Centre, Nottingham
Diseases of the Uvea
Chapters 7, 28

**Peter L.O. Heyworth MB BS, FRCOphth**
*Consultant Ophthalmic Surgeon,*
Bunbury Regional Hospital, Western Australia
Therapeutics Pt2 and Iatrogenic Diseases
Chapters 33, 34

**Ordan J. Lehmann MA, BM BCh, FRCOphth**
*IGA Research Fellow in Glaucoma Genetics,*
Institute of Ophthalmology and Moorfields Eye Hospital, London.
Development of the Eye, Congenital Anomalies and Genetic Disorders
Chapters 3, 20

**Andrew I. McNaught MD, FRCOphth**
*Consultant Ophthalmologist,*
Cheltenham General Hospital, Gloucester.
*Visiting Professor in Ophthalmology,*
Cranfield University, Gloucester.
Ocular and Visual Examination and Visual Physiology
Chapters 5, 12, 13

**Ian E. Murdoch MSc, MD, FRCOphth**
*Consultant Ophthalmic Surgeon,*
Moorfields Eye Hospital, London.
*Senior Research Fellow,*
Department of Epidemiology and International Eye Health,
Institute of Ophthalmology, London.
Infections and Infestations of the Eye,
Nutritional Deficiencies
Chapter 17

**John F. Pitts MRCP, FRCS, FRCOphth**
*Consultant Ophthalmic Surgeon,*
Harold Wood Hospital, Romford.
Diseases of the Eyelids, Lacrimal Apparatus and Orbit.
Chapters 9, 22, 23, 30

## ACKNOWLEDGEMENTS

The authors would like to thank, in addition to the principal illustrators, Professor Geoffrey Arden for Fig.13.4; Mr. Eric Arnott for Figs. 6.10, 6.11; Mr. David Broadway for Figs. 2.1, 2.16, 2.17; Mr. Robert Cooling for Plates 6.23, 6.24; Mr Peter Choyce for Plates 6.5, 17.10; Ms Sheena Caraway for Figs. 11.16, 11.21; Mr. Geoffrey Davies for Plates 6.22, 16.10, 29.1; Professor Andrew Elkington, Miss Helena Frank and Mr. Michael Greaney for Fig 5.4; Professor Frederick Fitzke for Fig 13.3; Miss Helena Frank for Figs. 4.1, 4.3, 4.6, 4.7; Mr. David Gartry for Figs. 6.9, 6.12, 6.13, 6.14; Mr. Michael Gilkes for Fig. 5.1; Professor Roger Hitchings for Plates 31.14(a), 31.14(b); Mr. Paul Hunter for Plates 17.15, 29.4; Dr. I Johansen and Dr. John Thygeson for Plate 6.4; Mr. Jack Kanski for Plates 7.4, 8.1, 8.5, 8.6, 29.10; Dr Kok-Tee Khaw for Fig. 19.10; Mr. Ordan Lehmann for Plates 7.1(a), 7.11, 16.7, 16.11, 22.7, 23.1, 28.3; Professor N. J. McGhee, Dr. A. D. Brown and Dr. K. H. Weed for Plate 25.14, Fig 25.13; Mr. Andrew McNaught for Figs. 13.1, 13.13, 13.19; Mr Ian Murdoch for Plates 17.12, 17.13 and Figs. 17.1, 17.2, 17.3, 17.8, 17.9; Professor B. C. K Patel for Fig. 2.12; Mr John Pitts for Plates 19.11, 22.2 and Figs. 19.12, 22.5, 23.2; Mr John Talbot for Plate 15.1 and Fig. 15.2; Mr Maurice Tuck for Fig. 31.20; Mr. Ananth Viswanathan and Professor Frederick Fitzke for Plate 31.12; Dr. Janet Voke for Fig. 7.12; Mr. David Wright for Fig. 4.15. We are also greatly indebted to the Ophthalmic Photographer, King's College Hospital, Mr. Colin Clements for much assistance over many years.

In addition we are grateful for illustrations to Clement Clarke International, London for Figs. 4.22, 4.25, 5.2, 11.15, 13.8, 13.9; to Hamblin Instruments, London for Fig 12.4, to Interzeag Medical Technology, Switzerland for Fig 13.11(a); to Keeler Instruments, London for Figs. 5.3, 12.5; to Tinsley Medical Instruments, Croydon, England for Fig. 13.10; to Carl Zeiss-Humphrey Systems, Oberköchen, Germany for Figs. 13.7, 13.11(b).

We are indebted for permission to reproduce illustrations to the Editors of Blackwell Scientific Publications for Fig 5.4; of the British Journal of Ophthalmology for Plate 18.9 and Figs. 7.23, 7.25, 18.7, 18.8; of Eye News for Plate 25.14 and Fig. 25.13; of Glaucoma Forum for Plate 13.12; of Ophthalmic Epidemiology for Fig. 31.20 and of Research and Clinical Forums for Figs. 13.4, 13.5(a), 13.5(b), 13.6.

# CONTENTS

Chapter

## 5: REGIONAL OPHTHALMOLOGY

Chapter

# CHAPTER 1

# INTRODUCTION

At an early stage the authors found that the schematic students' ophthalmic texts were of limited value in the real world, but Henry Stallard in his books always 'seemed to have been there before' and his practical advice gave great relief and encouragement to the reader faced with the needs of his patients, especially if isolated. Accordingly, the authors in turn have attempted what he achieved and this has been the spirit in which they have written this book, sensible always of the debt they owe their teachers and colleagues.

All books of instruction are a compromise and a practical textbook must make concessions to brevity. However, the authors are convinced that statements without attempted explanations merely add to learning without comprehension so that by selection they have tried to invite the medical student, the general practitioner, the optometrist and the ophthalmologist, especially when in training, to consider with them what is essential in ophthalmology. They hope that it will not be necessary to remember what is written, but merely that it will be incorporated in a reasonable, general medical approach to patients who present with eye problems or in whom the eyes share in general disease. The carefully chosen section on Common Ophthalmic Problems will summarise for the medical student what is most important for examination purposes and will indicate to the family doctor how to deal with eye problems, which can be so worrying in practice. This 'book within a book' extends its use by cross references beyond the limitations inevitable in a short student text and the authors hope this will make it a familiar book to which students will refer in their subsequent practice of medicine. It is frustrating when consulting a medical book for information about a disease or treatment to find that it is not even mentioned, accordingly they have tried to ensure by careful cross references and indexing that a reader can rapidly find at least a short description of a wide range of ophthalmic conditions.

The relevance of eye conditions to general medicine is emphasised by the section 'Systemic Ophthalmology', but in addition there were some aspects peculiar to the eyes which required consideration separately as Regional Ophthalmology. Finally therapeutics is presented as a separate section for reference and for examination purposes. The first edition of the book owed much to the work of Roger Trimble MRCP, FRCS, FRCOpth. and although sadly he is no longer with us, his influence is still strong in the present edition and tribute is paid to him as both friend and colleague and in the dedication of this edition which is even more widely collaborative. In view of accelerating advances in ophthalmology the authors are greatly indebted to a number of highly qualified ophthalmologists, who have special research and practical experience in particular aspects of ophthalmic work, for their collaboration as sub-editors, as described in the title pages. This ensures that while retaining the original aims of the book, those sections reflect the most modern concepts and practice. Ever increasing international communication and travel means that a study of medicine must embrace a global range of disease and this we have tried to emphasise, including relevant tropical and deficiency diseases, accepting that medical facilities are scarce in many parts of the world, so that methods of examination and treatment which require less equipment have also been described.

There are two advances regarding the illustrations in this 3rd Edition. Instead of being confined to colour sections they are now distributed throughout the text and there are more of them which results in a well illustrated book. In addition to this we have collaborated with Professor Arthur S Lim FRCS, FRACS, FRACO FROphth and Dr Tien Yin Wong FRCSE, MMed (Ophth) MPH of the National University of Singapore and Professor Ian Constable FRCSE, FRACS, FRACO, of the University of Western Australia by giving LCW cross references by number and page to illustrations in the new 4th edition of their Colour Atlas of Ophthalmology. If the two books are used together, our book is enhanced by many more excellent colour illustrations and their concise text appropriate to an Atlas, can be broadened by rapid reference to a reasonably comprehensive text.

For those facing the prospect of ophthalmology degrees and diplomas, an entirely new book, the Ophthalmological Examinations Review, written by Dr. Tien Yin Wong, would be invaluable for martialling their knowledge for presentation, thus completing a trio of basic volumes for the ophthalmologist in training, as a foundation for their study of subspecialist monographs and papers.

Stereoscopic pairs of certain conditions in the semi-transparent tissues of the eye make an appreciation of their true nature so much clearer that in relevant cases they have been included and we are indebted to Martin Pitts Crick FRCS, FRCOphth for them. We thank David Dorrell FRCS, FRCOphth for most of the informative line drawings of surgical technique reproduced by kind permission from his book 'The Surgery of the Eye' (Blackwell Scientific Publications). The authors also thank Adrian Pitts Crick BA Hons (Fine Arts) for the majority of the other line drawings, Jonathan Pitts Crick PhD FRCP for many of the black and white photographs and Peter Wright FRCS FRCOphth for many of the colour plates of external eye disease. The sources of other illustrations are gratefully acknowledged on page xi .

We are greatly indebted to Scientific Editor, Ms Lim Sook Cheng, and Senior Editor, Mr Steven Patt of World Scientific Publishing Company, and subsequently, Ms Nee Phua, Manager of the London office, for their invariably helpful collaboration. At the International Glaucoma Association our special thanks are due to Jane Eastick, General Manager for re-typesetting the book and to David Wright, Chief Executive for his continued cooperation. We have also appreciated the help of Jennifer Murray, Medical Secretary at the Institute of Ophthalmology for coordinating the Sub-Editors contributions.

Ronald Pitts Crick

Peng Tee Khaw

***Stereophotographs***

To view pairs of stereophotographs place the book flat on a table in good lighting to avoid shadows. View the illustration binocularly through a stereo-viewer or through sph +5.00D trial lenses in a trial frame. To avoid vertical doubling align your eyes horizontally parallel to the pair of pictures. If horizontal doubling occurs move the lenses horizontally together. If it is difficult to achieve the required degree of exotropia, addition of a 4-8 prism dioptres base out prism will help.

# CHAPTER 2

# PRACTICAL ANATOMY AND PHYSIOLOGY OF THE EYE AND ORBIT

## The eye

The eye lies in the front half of the orbit surrounded by fat and connective tissue and is supported by a fascial hammock. The *optic nerve,* which connects the eye with the brain leaves the orbit at its apex through the optic foramen in which it lies close to the ophthalmic artery. Attached to the eye are six *extraocular muscles,* four rectus and two oblique. They take origin posteriorly from the orbital walls or from a tendinous ring which bridges the superior orbital fissure and includes the optic foramen. They are innervated by the 3rd, 4th and 6th cranial nerves which enter the orbit at its apex through the superior orbital fissure. The branches of the ophthalmic division of the 5th cranial nerve also pass through the superior orbital fissure to convey sensory impulses from the eye and the upper part of the face.

The exposed front of the eye (Fig. 2.1 p5) consists of a central transparent convex portion, *the cornea,* surrounded at the corneo-scleral limbus by opaque white *sclera* which is covered by the loose bulbar *conjunctiva* continuous at the fornices with the more adherent palpebral conjunctiva which lines the eyelids.

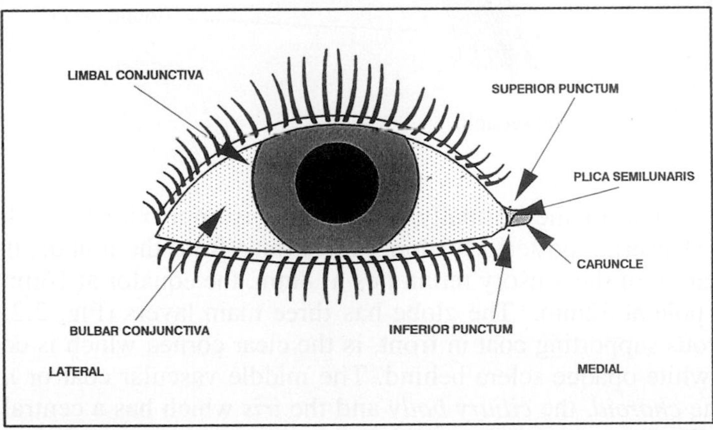


Fig. 2.1 An external view of the eye.

The *lacrimal gland* is situated just behind the upper outer angle of the bony orbit and its ducts discharge tears into the upper fornix. The front of the eye is protected by the eyelids which form an elliptical opening, the *palpebral aperture.* The area of the outer and inner angles of the palpebral aperture are known as the *lateral* and *medial canthi.* The free margin of each lid bears the eyelashes and the mouths of sebaceous glands and, near the medial end of the upper and lower lids, on a small eminence will be found the openings of the upper and lower *lacrimal canaliculi.* Each upper lid is raised by the *levator palpebrae superioris* muscle supplied by the 3rd cranial nerve and the sympathetically innervated palpebral muscle of Müller. The lids are closed by the *orbicularis oculi* muscle supplied by the 7th cranial nerve.

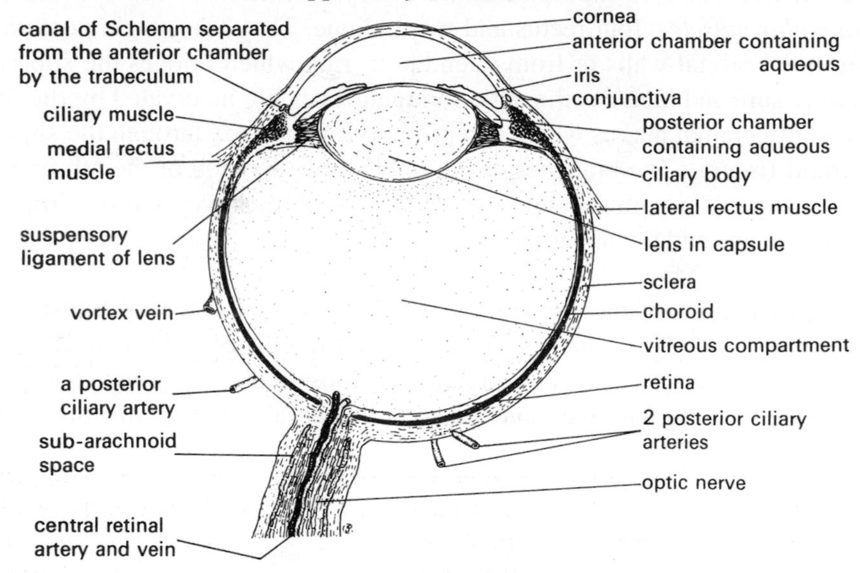


Fig. 2.2 The eye in horizontal section.

The axial length of the normally sighted eye is approximately 24mm (22-27mm). Measuring posteriorly along the surface from the limbus, the anterior termination of the sensory retina lies at 8mm, the equator at 16mm and the posterior pole at 32mm. The globe has three main layers (Fig. 2.2 p6). The outer fibrous supporting coat in front, is the clear cornea which is continuous with the white opaque sclera behind. The middle vascular coat or *uvea* consists of the *choroid,* the *ciliary body* and the *iris* which has a central opening or *pupil.* The inner sensory coat, *the retina,* has a many cell layered neural membrane and a single celled outer membrane, the *pigment epithelium.* A fenestrated opening in the *sclera* 1.5mm in diameter and 3mm medial to the posterior pole transmits the fibres of the optic nerve, mainly the axons of the ganglion cells of the retina. The *lens* is a transparent structure, suspended immediately behind the iris by fine fibres, forming the *zonule* or suspensory ligament, which runs from the surface of the ciliary body to the periphery of the lens. The anatomy of all these structures is considered in more detail in the relevant sections.

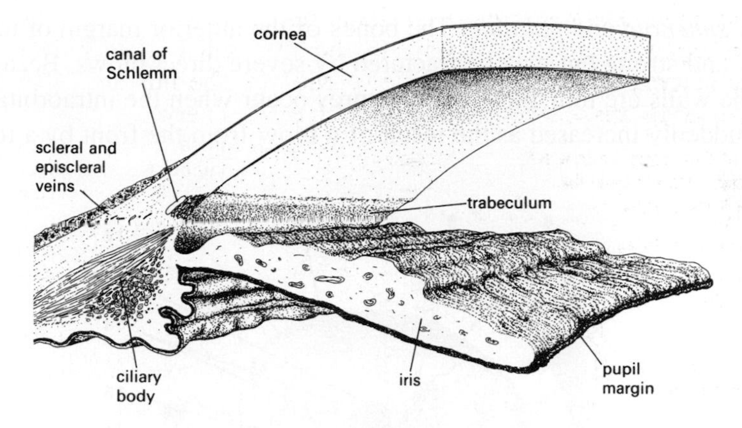


Fig. 2.3 Anatomy of the eye - view into the anterior chamber showing part of the surface of structures in the anterior and posterior chambers as well as sectional view of the ciliary body, iris diaphragm, trabecular meshwork and canal of Schlemm with connections to veins conveying aqueous fluid to the blood stream.

The eye is described as having three compartments. The *vitreous compartment* is bounded by the lens and zonule and the ciliary body anteriorly and by the retina and optic disc elsewhere. It contains the jelly-like vitreous body (or humour). The other two spaces contain aqueous fluid. The posterior chamber is small and is enclosed by the lens and zonule behind and the iris in front. The anterior chamber is continuous with the posterior chamber through the pupil and has the cornea in front and the iris behind (Fig 2.3 p7). Aqueous fluid (or humour) produced by the ciliary body epithelium into the posterior chamber passes through the pupil into the anterior chamber where it mainly drains through a filtration meshwork, the trabeculum, in the lateral wall of the anterior chamber angle, to an annular canal, Schlemm's canal. From here efferent vessels pass aqueous into the scleral and episcleral veins. With magnification some of these can be seen on the surface as aqueous veins. (*the trabecular outflow*). Some aqueous also drains through the uveal tissue and sclera to the orbital veins (*the uveo-scleral outflow*).

## The bony orbit (p533)

The eyes rest in two bony cavities, the orbits, on either side of the nose. The orbits are pyramidal in shape, roughly quadrangular in front and triangular behind. The medial walls of the orbits are parallel and the lateral walls diverge at about 45°. As indicated in Figs 2.4 p8, 2.5 p9, 2.6 p10, 2.7 p11, seven bones contribute to the orbit: the *maxilla, frontal, zygomatic, lacrimal, ethmoid, sphenoid* and *palatine*. The bones of the anterior margin of the orbit are thick and strong but may be fractured by severe direct blows. Because the rest of the walls are thin 'blow out' fractures occur when the intraorbital pressure is suddenly increased as the result of a blow from the front by a rounded

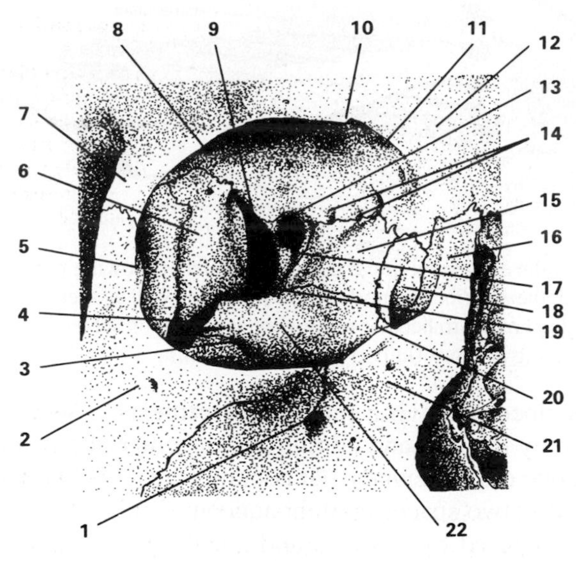


1 infra orbital foramen
2 malar or zygomatic bone
3 infra orbital groove
4 inferior orbital fissure
5 lateral orbital tubercle
6 orbital plate of great wing of sphenoid
7 lateral angular process of frontal bone
8 fossa for lacrimal gland
9 superior orbital fissure
10 supraorbital notch
11 trochlear fossa
12 frontal bone and orbital plate of frontal bone
13 optic foramen
14 anterior and posterior ethmoidal foramina
15 ethmoid
16 frontal process of maxilla
17 small wing of sphenoid
18 lacrimal bone and fossa for lacrimal sac
19 palate bone
20 lacrimal tubercle
21 maxilla
22 maxilla orbital plate

Fig. 2.4 Bones of the right orbit from the front.

object such as a small stone or a squash racquets ball. Usually the fracture occurs in the floor or medial wall with prolapse of the orbital contents (p541). On the anterior margin of the lateral wall there is a thickening, the *lateral orbital tubercle.* This gives attachment to the hammock-like suspensory fascia of the eye (the *ligament of Lockwood*), which on the medial side is attached to the bone behind the lacrimal sac. The groove for the *lacrimal sac* lies in the medial part of the front of the floor of the orbit bounded by two *lacrimal crests*, formed in front by the maxilla and behind by the lacrimal bone. The *medial canthal ligament* of the eyelids divides to be inserted into the crests and thus encloses the *lacrimal sac,* as does the orbicularis muscle. Anteromedially in the roof of the orbit is the *trochlea* which forms the pulley for the tendon of the superior oblique muscle.

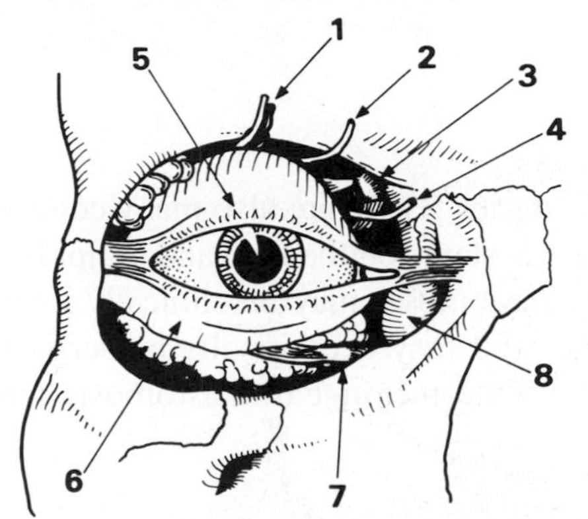


Fig. 2.5 Structures in the anterior part of the orbit viewed from the front.

At the apex of the orbit the body of the sphenoid contains the *optic foramen* which transmits not only the optic nerve but the ophthalmic artery and sympathetic nerve fibres as well. Just lateral to the optic foramen is the *superior orbital fissure or sphenoidal fissure*, which separates the greater and lesser wings of the sphenoid and is traversed by the motor nerves of the ocular muscles, the branches of the ophthalmic division of the trigeminal nerve and the ophthalmic veins. The *inferior orbital fissure* between the lateral wall and the floor of the orbit transmits the infraorbital and zygomatic branches of the maxillary division of the 5th nerve and veins linking the inferior ophthalmic vein and the pterygoid plexus. The superior and inferior orbital fissures may also transmit anastomoses of variable size between the branches of the internal and external carotid systems e.g. between the lacrimal and middle meningeal arter-

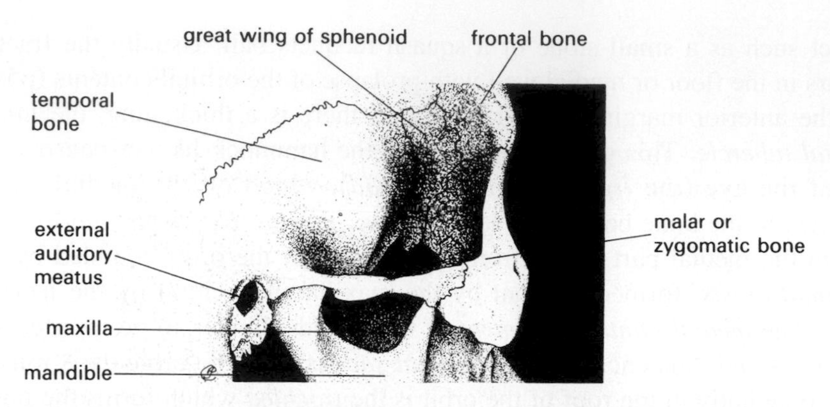


Fig. 2.6 Lateral view of the skull showing the right temporal fossa and the bones of the lateral wall of the right orbit.

ies and the lacrimal and infraorbital arteries. There are also many communications of the terminal branches of these systems on the face and scalp. In some cases variations in the size of these anastomoses may be clinically important, e.g. much of the blood supply of the orbit may occasionally be derived from the external carotid artery via the middle meningeal anastomosis with the lacrimal artery.

## The orbital fascia

The orbital contents are supported by connective tissue which is thickened in places to form definite layers and compartments which are important in surgery and in the spread of haemorrhage and inflammation:

-the *periosteum of the orbit* is continuous with the dura mater which also provides at the margin of the optic foramen the dural sheath of the optic nerve,
-the *orbital septum* stretches from the bony margins of the orbit to the tarsal plates and prevents orbital fat from herniating into the lids,
-*Tenon's capsule* (bulbar fascia) constitutes a fascial socket in which the eye moves and is continuous with the fascia around the extraocular muscles. Its lower portion forms the suspensory ligament of the eye.

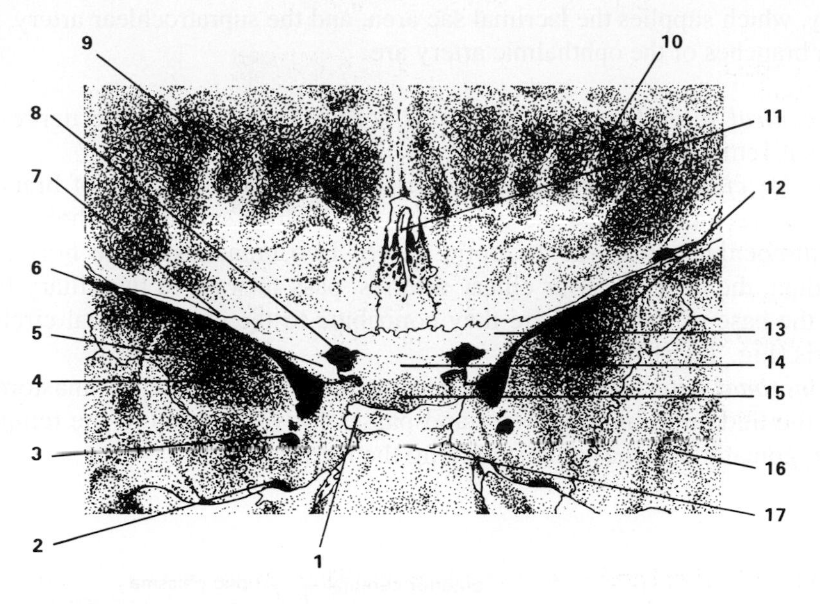


**1** posterior clinoid process
**2** foramen ovale
**3** foramen rotundum
**4** middle cranial fossa
**5** superior orbital fissure
**6** anterior clinoid process
**7** sphenoidal ridge
**8** optic foramen
**9** anterior cranial fossa
**10** orbital plate of frontal bone
**11** ethmoid and crista galli
**12** great wing of sphenoid
**13** small wing of sphenoid
**14** optic groove
**15** body of sphenoid and sella turcica
**16** temporal bone
**17** dorsum sellae

Fig. 2.7 The supero-posterior relations of the orbit .

## Blood supply of the eye and orbit

*Arteries* (Fig. 2.8 p12)

The main blood supply of the eye and orbit is from the *ophthalmic artery.* There are contributions from the external carotid artery and many variations. The ophthalmic artery arises from the convexity of the curve of the internal carotid artery as it passes out of the cavernous sinus. It enters the orbit through the optic canal below and lateral to the optic nerve and then curves above the nerve with the nasociliary branch of the 5th cranial nerve and runs along the medial wall of the orbit to terminate by dividing into the dorsonasal artery, which supplies the lacrimal sac area, and the supratrochlear artery. The main branches of the ophthalmic artery are:

-the *central retinal artery* which runs forward under the optic nerve and enters it 1cm from the globe,
*-posterior ciliary arteries* which give rise to a variable number of branches piercing the globe around the optic nerve and which supply the choroid and the intraocular portion of the optic nerve. Two branches in the horizontal meridian, the long posterior ciliary arteries, pass forward to the ciliary body near the base of the iris where they contribute to the major arterial circle of the iris (Fig. 2.9 p13)
-the *lacrimal artery* gives off a recurrent meningeal branch to anastomose with the middle meningeal artery and passes forward to end in the temporal and zygomatic branches after supplying the lacrimal gland.

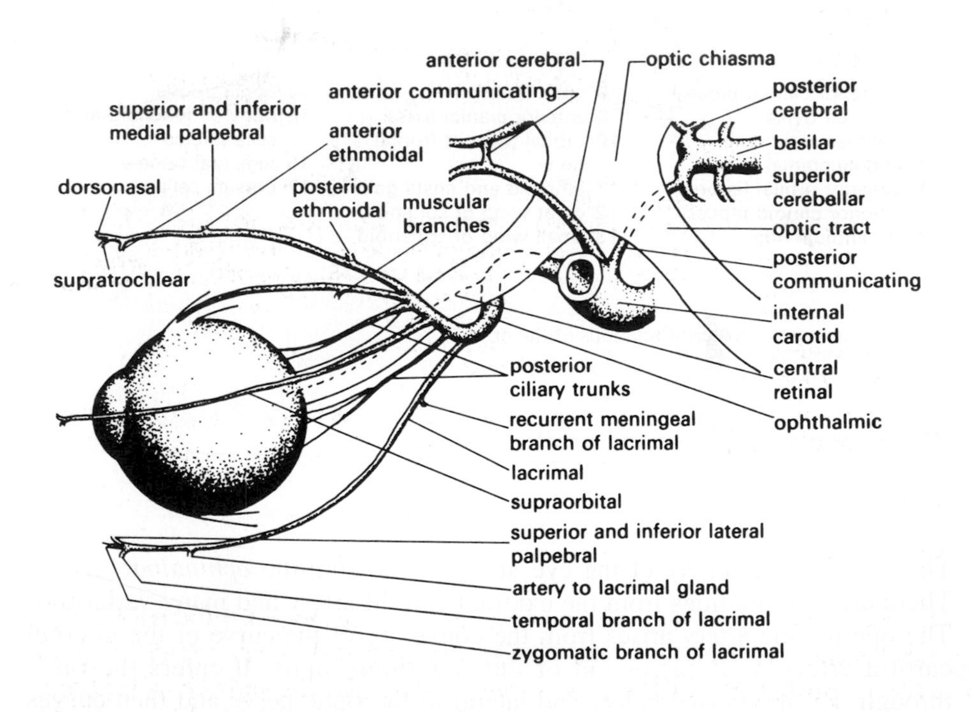


Fig. 2.8 Arterial blood supply of the left eye and orbit viewed from above .

-*muscular arteries* pass along the rectus muscles and continue forwards as the anterior ciliary arteries which end in the pericorneal arcade. They give off large branches which penetrate the sclera about 4mm from the limbus and with the long ciliary arteries form the arterial circle at the base of the iris to supply the iris and ciliary body.
-the *supraorbital artery* passes around the margin of the orbit at the supraorbital notch to anastomose with the vessels of the scalp.
-the *posterior* and *anterior ethmoidal arteries* pass through the foramina of the same name and may have meningeal branches.
-The *superior* and *inferior medial palpebral arteries* anastomose with corresponding branches of the lacrimal artery in the lids.

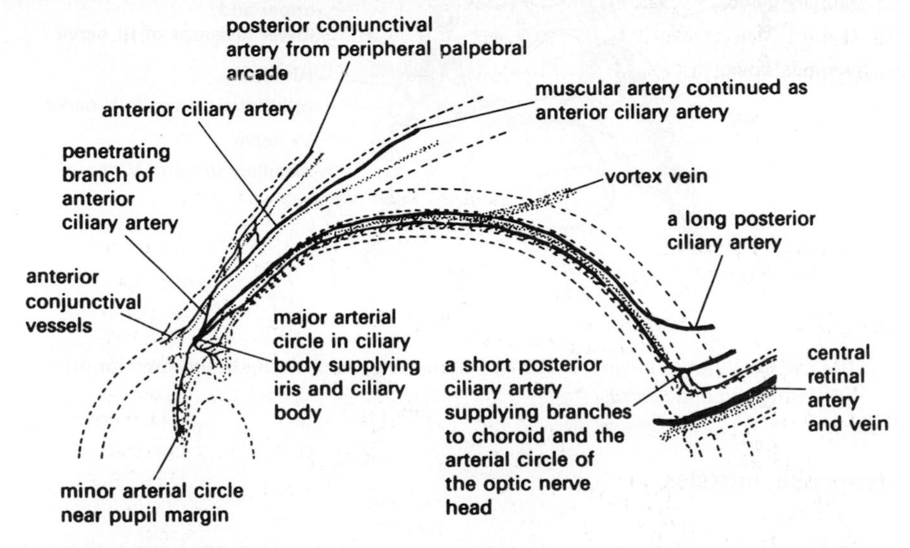


Fig. 2.9 Blood supply of the left eyeball viewed from above. Horizontal section at the level of the optic nerve.

*Veins and lymphatics*

There are three main veins within the orbit: (1) The *superior ophthalmic vein* receives the two *superior vortex veins* and drains into the *angular vein*. It connects posteriorly with the *cavernous sinus*. (2) The *inferior ophthalmic vein* drains to the *pterygoid plexus* through the *inferior orbital fissure*, having received the *inferior vortex veins*. (3) The *central retinal vein* leaves the optic nerve 1cm behind the globe to join the *ophthalmic veins*. The *cavernous sinus* lies within a splitting of the dura mater on either side of the body of the sphenoid bone. Its important relations are shown in Fig 2.10 p14. *Lymphatics* from the medial part of the lower lid drain to the submaxillary lymph nodes, while those from the lateral part and most of the upper lid drain to the preauricular nodes.

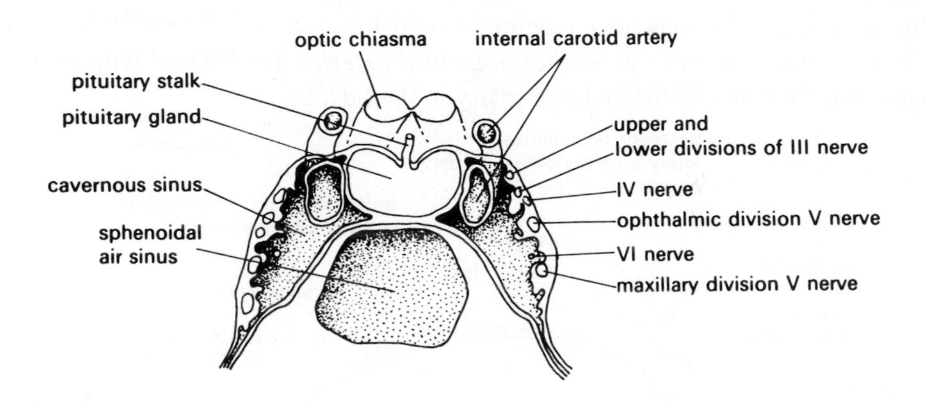


Fig. 2.10 The structures in relation to the cavernous sinuses, posterior to the superior orbital fissure and optic canal.

## Extraocular muscles

The origin of the four *rectus muscles* is a tendinous ring enclosing the optic foramen and the medial end of the sphenoidal fissure, giving attachment to the sphenoid bone (Fig 2.11 p15). They are inserted into the sclera 5-8mm from the limbus. The *superior oblique muscle* also arises posteriorly but from bone just above and medial to the optic foramen. Its tendon passes through a pulley, the *trochlea*, a U-shaped fibro-cartilage attached to the frontal bone (Figs. 11.12 p240, 2.5 p9). The tendon then passes backwards and laterally, fanning out to be attached to the upper part of the postero-lateral aspect of the globe. The *inferior oblique muscle* arises anteriorly from the maxilla just within the orbital margin lateral to the lacrimal sac. It follows a similar direction to the reflected tendon of the superior oblique, also to be inserted postero-laterally but lower down, extending medially to the posterior pole of the eye. The lateral rectus muscle is innervated by the 6th cranial nerve, the superior oblique by the 4th and the other extraocular muscles by the 3rd cranial nerve. The *levator palpebrae superioris* is described on p6.

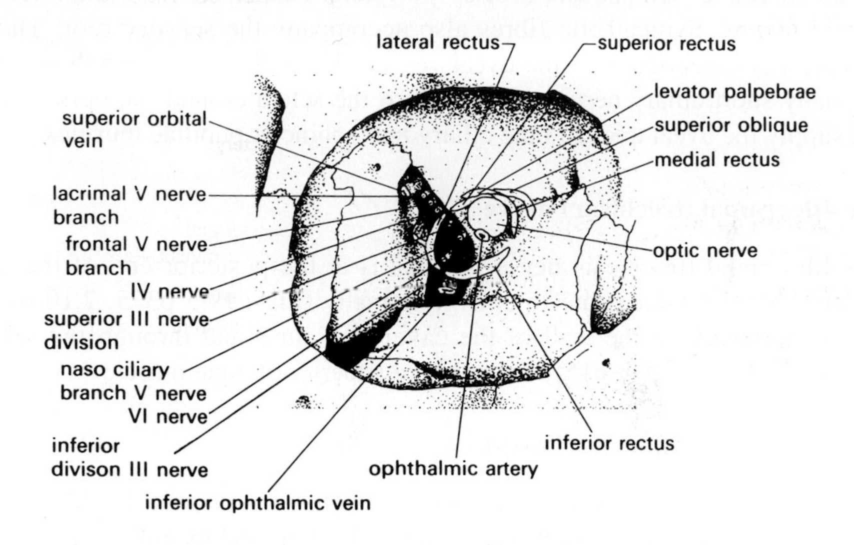


Fig. 2.11 Structures in relation to the right superior orbital (sphenoidal) fissure.

## Nerves of the eye and orbit

### The 3rd cranial (oculomotor) nerve

The nucleus of the 3rd cranial (oculomotor) nerve is situated in the midbrain beneath the aqueduct of Sylvius. Anteriorly is the nucleus of Perlia which is concerned with convergence, and antero-laterally are the nuclei of Edinger-Westphal from which parasympathetic fibres are derived. The nuclei of the 3rd, 4th and 6th cranial nerves are linked by the medial longitudinal fasciculus (Fig. 19.21 p403). The 3rd nerve leaves the midbrain between the cerebral peduncles (Fig. 19.5 p375), passes anteriorly in the lateral wall of the cavernous sinus (Fig. 2.10 p14) and enters the orbit through the lower part of the sphenoidal fissure (Fig. 2.11 p15), dividing to supply the levator palpebrae superioris, medial and inferior rectus muscles and the inferior oblique. The branch of the 3rd nerve to the inferior oblique muscle conveys the parasympathetic fibres which leave it as the motor root of the *ciliary ganglion.* The sensory root of the ganglion is contributed by the nasociliary branch of the 5th nerve and its sympathetic root by postganglionic fibres from the internal carotid plexus. Sympathetic fibres also accompany the sensory root. These mediate vasoconstriction of the uveal vessels. From the ciliary ganglion arise the many short ciliary nerves which pierce the sclera around the optic nerve and supply the uveal tract and the ciliary and sphincter pupillae muscles.

**The 4th cranial (trochlear) nerve**

The 4th cranial (trochlear) nerve nucleus is at the posterior end of the 3rd nucleus. Its fibres decussate to emerge dorsally (Figs. 19.5 p375, 2.10 p14), and run forwards in the wall of the cavernous sinus and through the sphenoidal fissure (Fig. 2.11 p15) to supply the superior oblique muscle.

**The 6th cranial (abducens) nerve**

The 6th cranial (abducens) nerve nucleus lies below the floor of the 4th ventricle and the nerve emerges ventrally between the pons and medulla to run anteriorly. It then makes a sharp turn over the petrous ridge, which renders it vulnerable to trauma and increased intracranial pressure, runs inside the cavernous sinus lateral to the internal carotid artery (Fig. 2.10 p14) through the sphenoidal fissure within the tendinous ring to supply the lateral rectus muscle. (Fig. 2.11 p15)

**The 5th cranial (trigeminal) nerve**

The 5th cranial (trigeminal) nerve (Fig. 2.12 p18) is the sensory nerve of the head and face but is also the motor supply for the muscles of mastication. The nucleus of the motor part of the nerve is near the floor of the aqueduct of Sylvius and the principal sensory nucleus concerned with tactile impulses is in the pons near its lateral surface. The nucleus of the spinal tract of the 5th cranial nerve, which mediates sensations of pain and temperature extends down to become continuous with the substantia gelatinosa of the dorsal horn of the spinal cord in its 2nd cervical segment. The sensory root of the trigeminal nerve carries fibres from the trigeminal (Gasserian, semilunar) ganglion and a few from the ciliary ganglion to the sensory nuclei and enters the lateral surface of pons. Ascending fibres enter the main sensory nucleus and descending fibres form the spinal tract of the nerve and end in its nucleus. The trigeminal ganglion lies in a depression in the petrous temporal bone (Fig. 19.5 p375) within a cleft of dura mater (Meckel's Cave) and anteriorly has three branches, ophthalmic, maxillary and mandibular. The *ophthalmic nerve* runs forwards in the wall of the cavernous sinus (Fig. 2.10 p14), divides into frontal, lacrimal and nasociliary branches which enter the orbit through the sphenoidal fissure (Fig. 2.12 p18). The *frontal nerve* passes anteriorly near the roof of the orbit and emerges as the supraorbital and the supratrochlear nerves to supply the skin of upper lid, conjunctiva and scalp (Fig. 2.5 p9). The *lacrimal nerve* accompanies the lacrimal artery along the upper border of the lateral rectus muscle to supply the lateral parts of the upper and lower lids. It receives a twig derived from the maxillary nerve which conveys secretory fibres to the lacrimal gland. The *nasociliary nerve* conveys sensory impulses from the eyeball. After passing through the tendinous ring it crosses above the optic nerve with the ophthalmic artery and runs along the medial wall of the orbit which it leaves via the anterior ethmoidal canal, enters the anterior cranial fossa, runs in the roof of the nose emerging on the face as the external nasal nerve to supply the skin of the nose. In its course it contributes the sensory root of the ciliary ganglion and the two long ciliary nerves which accompany the long posterior ciliary arteries, conveying sensory fibres from the uvea and sympathetic fibres to the dilator pupillae. The nasociliary nerve then gives off the anterior and posterior ethmoidal nerves, the infratrochlear nerve (Fig. 2.5 p9), medial and lateral nasal nerves and the external nasal nerves which convey sensory fibres from the medial part of the lids, conjunctiva and the nose.

The *maxillary* branch of the trigeminal nerve passes from the cranial cavity through the foramen rotundum to the pterygopalatine fossa (Fig. 2.13 p22) and distributes sensory nerves directly or via the sensory root of the sphenopalatine ganglion to the skin of the upper face and nose, the lower lid and its conjunctiva. It also supplies the mucous membranes of the nose, maxillary sinus and mouth; also the teeth, periosteum of the orbit and the dura mater of the middle cranial fossa. In addition it receives post-ganglionic secreto-motor fibres from the sphenopalatine ganglion which are distributed via the zygomatic nerve and its communication with the lacrimal nerve, to the lacrimal gland.

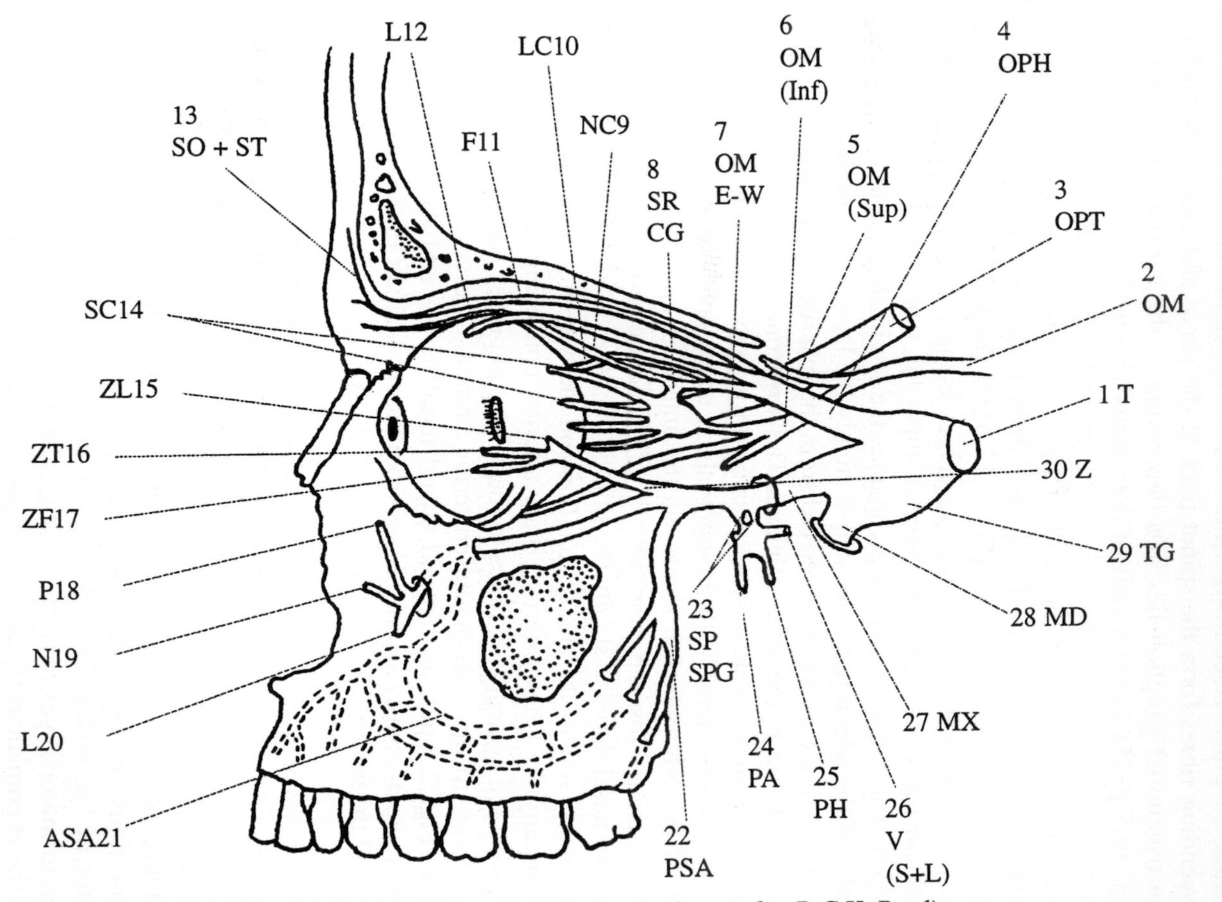


Fig.2.12 The 5th Cranial (Trigeminal) Nerve. (redrawn after B.C.K. Patel)

Legend to 5th nerve diagram (Fig 2.12 p18)

1. 5th cranial (trigeminal) nerve.
2. 3rd cranial (oculomotor) nerve.
3. optic nerve.
4. ophthalmic division of the 5th cranial nerve.
5. superior division of the 3rd cranial nerve.
6. inferior division of the 3rd cranial nerve.
7. motor root of ciliary ganglion containing autonomic fibres from the Edinger - Westphal nucleus of the 3rd cranial nerve.
8. sensory root of the ciliary ganglion from the nasociliary branch of the ophthalmic nerve.
9. nasociliary branch of the ophthalmic nerve contributing sensory root of the ciliary ganglion.
10. one of two long ciliary branches of the nasociliary nerve.
11. frontal branch of the ophthalmic nerve.
12. lacrimal branch of the ophthalmic nerve.
13. supraorbital and supratrochlear branches of the ophthalmic nerve.
14. short ciliary nerves, six to ten in number.
15. communication from zygomatic nerve conveying secretory fibres from the sphenopalatine ganglion to the lacrimal nerve.
16. zygomatico - temporal nerve.
17. zygomatico - frontal nerve.

18, 19, 20. palpebral nasal and labial branches of the infraorbital nerve.

21. Anterior superior alveolar nerves.
22. Posterior superior alveolar nerves.
23. Infraorbital nerves.
24. Spheno-palatine nerves (two in number).
25. Great palatine nerve.
26. Pharyngeal nerve
27. Vidian nerve comprised of the great deep petrosal nerve (sympathetic) and the greater superficial petrosal nerve (secretory) supplying the lacrimal gland via communication from zygomatic nerve to the lacrimal nerve
28. Maxillary division of the 5th cranial nerve, passes through the foramen rotundum
29. Mandibular division of the 5th cranial nerve, passes through the foramen ovale.
30. Trigeminal (Gasserian) ganglion.
31. Zygomatic branch of the maxillary nerve.

The *mandibular* branch of the trigeminal nerve passes through the foramen ovale to emerge into the infratemporal fossa. It has motor and sensory roots. The motor root mainly supplies the muscles of mastication. The sensory root is the sensory supply to the skin of the scalp and its auriculo-temporal branch receives secretomotor fibres from the otic ganglion for the parotid salivary gland. Its lingual branch near its origin receives from the chorda tympani both sensory taste fibres for the anterior two thirds of the tongue and secretomotor motor fibres which relay in the submandibular ganglion and supply the sublingual and submandibular salivary glands (Fig. 2.13 p22).

**The 7th cranial (facial) nerve and its connections**

The 7th cranial (facial) (Fig. 2.13 p22) nerve consists of a motor component and the pars intermedia of Wrisberg which has both a sensory and secretomotor function.

*Motor.* The motor nucleus is located in the pons lateral to the fibres of the 6th nerve and medial to the spinal nucleus of the 5th nerve. The motor fibres ascend to loop medially and dorsally around the 6th nerve nucleus in the floor of the 4th ventricle and then run ventrally to emerge at the lower border of the pons lateral to the 6th nerve. The motor part of the nerve enters the internal auditory meatus in company with the 8th cranial nerve and with the nervus intermedius which then fuses with it. The facial nerve then makes a sharp backward bend to enter the facial canal, curves over the middle ear supplying the nerve to stapedius and leaves the canal at the stylomastoid foramen, it gives branches to the digastric and stylohyoid muscles and turns forwards and divides into upper and lower branches in the parotid gland to supply the muscles of the face. The upper branch comes from the upper part of the nucleus which has cortical connections to both hemispheres while the lower part does not. As a result the muscles supplied by the upper branch including those around the eye are spared in unilateral supra-nuclear lesions.

*Facial nerve block* to immobilize the facial muscles around the eye (p613) can be achieved by injecting local anaesthetic solution against the posterior aspect of the ramus of the mandible where the nerve curves around it, with the mouth widely open to protect deeper structures (Fig. 6.6 p104).

Alternatively its branches can be anaesthetised by a more widespread injection of the upper and lower lids from the lateral side.

*Sensory.* Peripheral sensory fibres from the unipolar cells of the geniculate ganglion carry sensations of taste from the anterior two thirds of the tongue via the lingual nerve and the chorda tympani which enters the temporal bone from the pteryomaxillary region, passes in close relation to the tympanic cavity and joins the facial nerve in the facial canal a short distance above the stylomastoid foramen. Central fibres of the geniculate ganglion cells run in the pars intermedia and pass to the gustatory nucleus in the medulla where they join other taste fibres of the 9th and 10th cranial nerves.

*Secretomotor.* The pars intermedia which joins the facial nerve also conveys parasympathetic pre-ganglionic fibres from the cells of the superior salivatory part of the facial nucleus to supply secretmotor fibres to the lacrimal and salivary glands. These fibres leave the 7th nerve from the geniculate ganglion as the greater superficial petrosal nerve which runs through the petrous temporal bone to enter the cranial cavity, only to leave again through the foramen lacerum to join the deep petrosal nerve and pass through the pterygoid canal to join the sphenopalatine ganglion. Here the fibres relay and the postganglionic fibres are distributed via a branch of the zygomatic branch of the maxillary division of the 5th nerve to the lacrimal nerve to supply the lacrimal gland. Some fibres of the chorda tympani are also secretomotor to the submandibular and sublingual salivary glands after relaying in the submandibular ganglion. Other secretomotor fibres leave the geniculate ganglion and join with fibres from the 9th cranial nerve from the tympanic plexus to form the lesser superficial petrosal nerve which emerges to join the otic ganglion near the foramen ovale where they relay to supply the parotid salivary gland.

**The autonomic nervous system**

The efferent (motor) part of the autonomic nervous system has been well defined, but although clearly of great importance, e.g. the pupillary afferent supply passing to the pre-tectal nucleus, much remains to be elucidated about the autonomic afferent pathways.

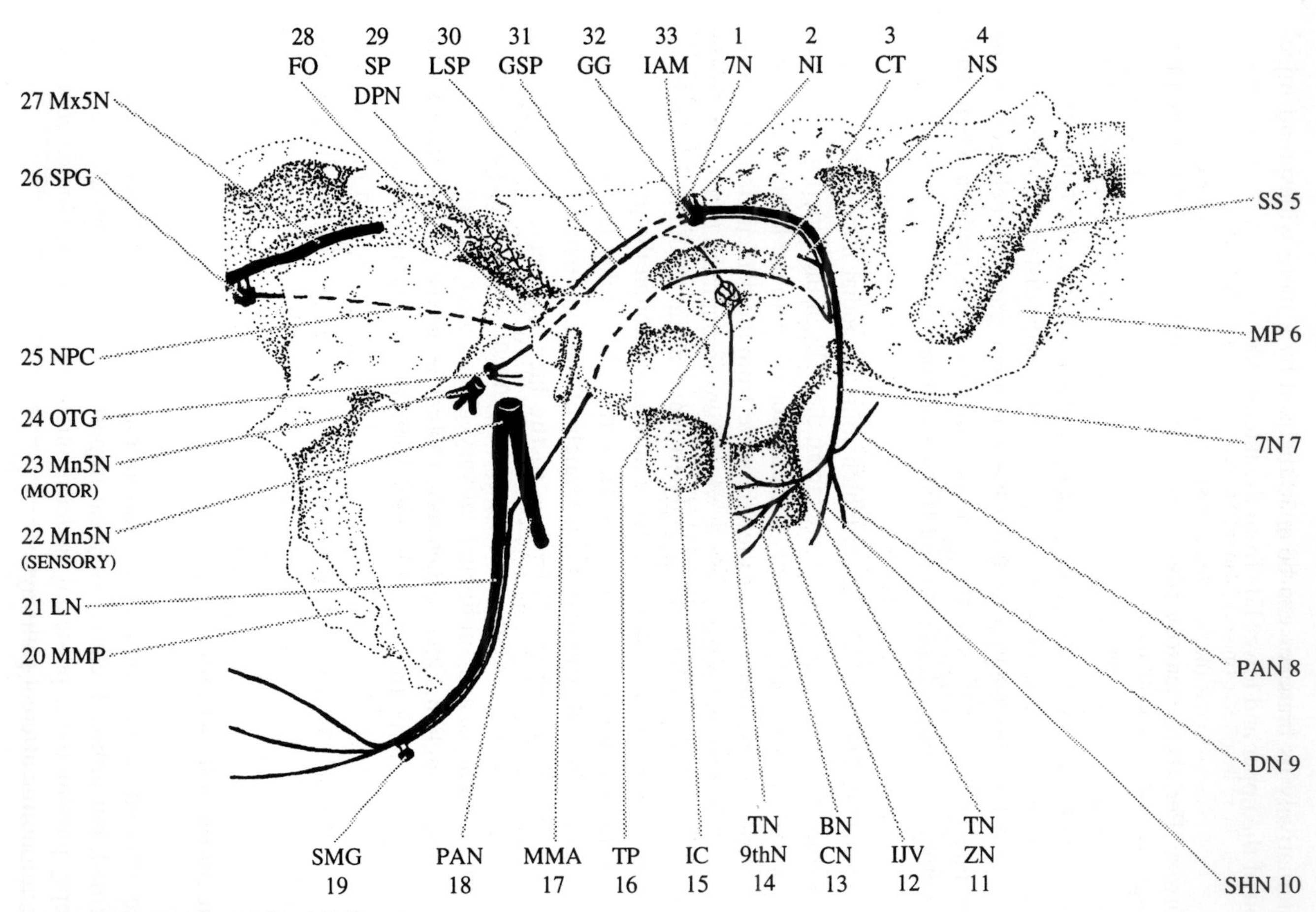


Fig. 2.13 The 7th Cranial (Facial) Nerve.

Legend to 7th nerve diagram (Fig. 2.13 p22)

1. 7th cranial nerve (motor) (also the 8th cranial nerve (not shown))
2. 7th cranial nerve (nervus intermedius) special sensory and secretomotor
3. Chorda tympani in middle ear
4. Nerve to stapedius muscle
5. Sigmoid venous sinus
6. Mastoid process
7. 7th cranial nerve emerging from stylomastoid foramen
8. Posterior auricular nerve
9. Nerve to posterior belly of digastric muscle
10. Stylohyoid nerve
11. Temporal and zygomatic nerves
12. Internal jugular vein
13. Buccal nerve and cervical nerves
14. Tympanic nerve (9th cranial) in tympanic canal joining tympanic plexus
15. Internal carotid artery
16. Tympanic plexus and branch joining the lesser superficial petrosal nerve
17. Middle meningeal artery
18. Posterior superior alveolar nerve
19. Submandibular ganglion - Cell station for secretomotor fibres from chorda tympani to submandibular salivary gland and conveying taste sensation from anterior two thirds of the tongue, via lingual nerve
20. Medial pterygoid plate
21. Lingual nerve
22. Mandibular nerve (sensory) (5th)
23. Mandibular nerve (motor) (5th)
24. Otic ganglion - Cell station for secretomotor fibres from nervus intermedius (7th) and from the 9th cranial nerve conveyed in lesser superficial petrosal nerve.
25. Nerve of pterygoid canal
26. Sphenopalatine ganglion - Cell station for secretomotor fibres from nervous intermedius to lacrimal gland via maxillary nerve
27. Maxillary nerve (5th)
28. Foramen ovale (seen in section) transmitting trunk of the mandibular nerve (5th)
29. Deep petrosal nerve from sympathetic plexus on internal carotid artery joining the greater superficial petrosal nerve to form the nerve of the pterygoid canal
30. Lesser superficial petrosal nerve
31. Greater superficial petrosal nerve
32. Geniculate ganglion
33. Internal auditory meatus

The motor part of this system supplies the stimulus for contraction of smooth muscle and cardiac muscle and the secretion of glands and, as in the case of skeletal muscle, the nervous pathway from the higher centres of the brain consists of a sequence of three neurones. For skeletal muscle all three are in the central nervous system (CNS) and the third axon is medullated and runs in the motor nerves to supply the muscle fibres, the transmitter being acetylcholine. In the autonomic system the first two neurones are in the central nervous system, but the third is situated in a peripheral ganglion and following this relay the axons (postganglionic fibres) are fine, non medullated and more numerous. At the muscle fibre or secreting cell the chemical transmitter is noradrenaline for the sympathetic postganglionic nerve endings and acetylcholine for postganglionic parasympathetic endings.

**The efferent sympathetic nervous supply to the eye and orbit (adrenergic)**

These nerve fibres originate in the hypothalamus. They pass down the brain stem and the cervical spinal cord to the level of the first and second thoracic nerves (T1 and T2) where, at the ciliospinal centre in the intermedio-lateral tract of grey matter, the first relay occurs, the axons of the second order neurones leave the spinal cord via the spinal root of T1 or T2 and enter the cervical sympathetic chain. The third relay takes place at the superior cervical ganglion where the third neurones' axons form the non medullated postganglionic fibres. These accompany the internal carotid artery into the skull and through the cavernous sinus and follow its branches into the orbit. They are also carried by the ophthalmic division of the fifth cranial nerve and its nasociliary branch and the long ciliary nerves to the dilator pupillae and ciliary muscles. Other sympathetic fibres from the carotid sympathetic plexus pass via the levator palpebrae superioris branch of the third cranial nerve to the superior palpebral muscle of Müller. This smooth muscle sheet arises from the inferior surface of the tendon of the levator to be inserted at the margin of the tarsal plate. There is a similar. less well defined muscle in the lower lid arising from the inferior rectus tendon.

**The efferent parasympathetic nervous supply to the eye and orbit (cholinergic)**

These nerve fibres originate in the hypothalamus. They then relay in the Edinger-Westphal nucleus of the third cranial nerve and are carried by the third nerve to the orbit. They accompany the branch to the inferior oblique muscle but leave it to relay to the third neurone in the ciliary ganglion. The postganglionic fibres enter the short ciliary nerves to supply the sphincter muscle of the pupil, the ciliary muscle and other smooth muscle in the eye (Fig. 19.23 p415). Lesions of the third cranial nerve, the ciliary ganglion (Holmes-Adie pupil) or the muscles themselves will result in a dilated pupil and weakness of accomodation (p417).

There is evidence that some preganglionic fibres bypass the ciliary ganglion and relay in accessory ciliary ganglia situated on the ciliary nerves and that these contribute the efferent part of the reflex arc of the pupillary 'near' reflex.

**The eyelids** (p447)

The eyelids have four tissue layers, described in two *lamellae.* (Fig 2.14 p26). Their junction is seen on the margin of the lid as the grey line and it is along this line that the lid can be split during eyelid plastic surgery.

The *anterior lamella* consists of the *skin* and the *orbicularis oculi muscle.* The eyelid skin is very thin to allow for rapid excursions of the lids during blinking and this makes the lids prone to a dramatic accumulation of tissue fluid to cause swelling or blood to form a haematoma. The eyelashes arise from modified hair follicles in the lid margin. The orbicularis muscle is the muscle of eyelid closure and is supplied by the VIIth nerve. Paralysis of the VIIth nerve leads to impairment of the blinking mechanism and exposure of the cornea. At the other extreme, dystonic spasm of the muscle may occur, leading to repeated episodes of involuntary eyelid closure (essential blepharospasm). The sensory nerves of the lids are from the ophthalmic division of the trigeminal nerve via the supraorbital nerve and to a lesser extent the supratrochlear and infratrochlear nerves and from the maxillary division via its infraorbital branch.

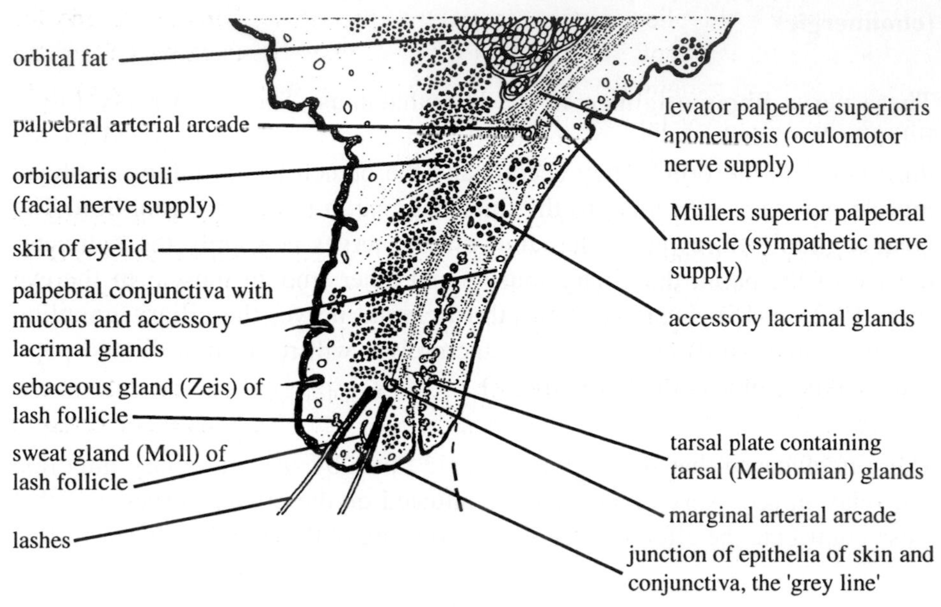


Fig. 2.14 Eyelid - sagittal section.

The *posterior lamella* consists of the *tarsal plate* and the *palpebral conjuncti-va.* The tarsal plate is the fibrous skeleton of the lid and contains a series of lipid-secreting glands which open at the lid margin (tarsal or Meibomian glands). The conjunctiva which is firmly attached to the tarsal plate provides a smooth lining to cover the cornea during lid closure

It is important to acquire the skill to evert the upper lid in order to inspect the palpebral conjunctiva as follicle formation there is of major importance in the differential diagnosis of conjunctivitis and because this is a common site for a wind-borne foreign body to lodge.

The *levator palpebrae superioris* is the muscle of eyelid opening, which aris-es from the apex of the orbit and inserts into the front of the tarsal plate by a wide tendon, or aponeurosis. A slip of muscle also inserts into the skin and provides the skin crease characteristic of non-Oriental races. The levator is a striated muscle analogous to the extra-ocular muscles but also contains smooth muscle fibres (Müller's muscle). These produce the eyelid retraction seen in the adrenergic response or in thyroid overactivity. The striated muscle of the levator is supplied by the superior division of the IIIrd nerve and the smooth muscle is supplied by sympathetic postganglionic nerves from the superior cervical ganglion. Paralysis of either leads to ptosis, or drooping of the eyelid. The lower lid retractors, analogous to the levator in the upper lid, also contain striated and smooth muscle fibres.

The *blood supply* to the lids is from the ophthalmic arterial axis via the medial and lateral palpebral branches of the ophthalmic and lacrimal arteries (Fig. 2.8 p12) and from the facial axis by the angular artery. The plentiful blood supply encourages excellent wound healing in this site.

The *lymphatics* from the lateral two-thirds of the upper lid and the lateral one third of the lower lid drain to the preauricular node. The remainder drain to the submandibular nodes.

**The lacrimal apparatus**

*The tears.* The precorneal tear film consists of an aqueous layer sandwiched between a mucus layer on the corneal surface and a lipid layer exposed to the atmosphere. The mucus layer renders the corneal epithelium hydrophilic and the lipid layer increases surface tension and reduces evaporation. The mucins are derived mainly from the conjunctival goblet cells and the lipids mainly from the Meibomian glands. The aqueous layer is the main component of the tear film and comprises 95% of its volume. *Basal tear secretion* is of the order of 1 microlitre per minute and is derived from the *accessory lacrimal glands* (of Krause and Wolfring) located in the conjunctival fornices. Protein in the tear film, averaging 1-2gm/100ml, is composed of about equal quantities of albumin, globulins and lysozyme, which is an antibacterial enzyme of limited activity against mainly non pathological organisms. The principal immune globulin is IgA with a small amount of IgG. *Additional tear secretion* is provided by the *main lacrimal gland* in response to a background of stimuli to the cornea and retina and also in response to emotions. All these influences feed into the lacrimal nucleus in the brain stem close to the facial and superior salivary nuclei in the floor of the 4th ventricle. The secretomotor nervous pathway to the lacrimal gland is described on p21, (Fig 2.13 p22). Sympathetic post ganglionic vasomotor nerve fibres arise in the superior cervical ganglion and sensory fibres are carried by the lacrimal branch of the 5th cranial nerve.

The *main lacrimal gland* lies in the lacrimal fossa in the supero-temporal quadrant of the orbit. It is divided into a larger orbital lobe and a smaller lower palpebral lobe by the aponeurosis of the levator palpebrae superioris muscle. The palpebral portion may be seen through the conjunctiva when the upper lid is everted (Plate 23.1 p465). The ducts from the gland number about 12-15 and empty into the superior conjunctival fornix, being distributed over the ocular surface by the blink. The ducts from the orbital portion pass through or close to the palpebral portion so that excision of the palpebral portion stops tear secretion from the whole gland. (Fig 2.15 p28). The lacrimal gland is supplied by the lacrimal artery, a branch of the ophthalmic artery and the lacrimal veins drain to the ophthalmic veins. Lymph vessels pass to subconjunctival lymphatics and on to the preauricular lymph nodes.

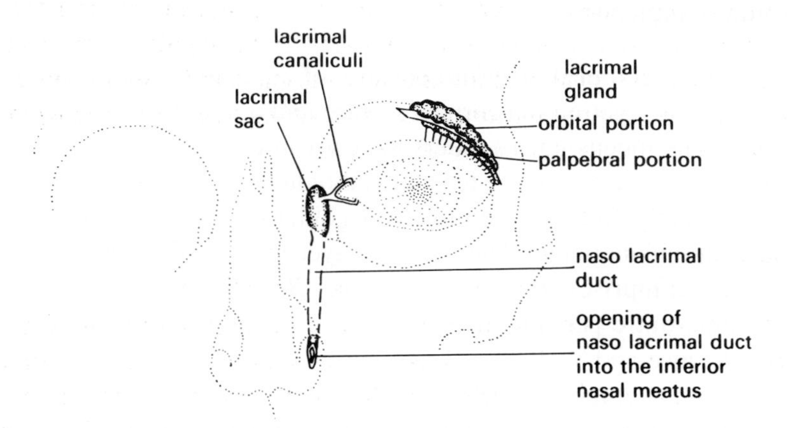


Fig. 2.15 Nasolacrimal apparatus.

The *lacrimal drainage system.* The tears drain into the puncta, which lie in papillae at the medial ends of the eyelids. The puncta open into vertically orientated ampullae about 2mm long which turn at a right angle and pass into horizontally disposed canaliculi 8mm in length. It is estimated that 70% of tears drain through the inferior and 30% through the superior canaliculus.

These unite at the medial canthus to form a common canaliculus, 4mm long which drains into the lacrimal sac. This lies in the lacrimal groove formed by the lacrimal bone and the frontal process of the maxilla. Tears from the sac pass to the nasolacrimal duct which is directed downwards, laterally and backwards to empty into the inferior meatus of the nasal cavity. To ensure the atraumatic passage of a lacrimal probe it is necesary to study the anatomy of canaliculi, sac and nasolacrimal duct very carefuly and to proceed with great gentleness (p210). False passages can give the patient endless trouble. The action of blinking provides a pumping mechanism. The lids close in a medial direction, milking the tears towards the puncta along the marginal tear strip at the posterior margins of the lids, the tears tend to enter the puncta by capillarity but pass on mainly by the action of the orbicularis muscle which occludes the puncta and compresses the canaliculi forcing the tears into the sac through the valvular opening of the common canaliculus. From the lacrimal sac the tears drain by gravity into the nose. The blood supply of the lacrimal sac region is derived from the superior and inferior palpebral branches of the ophthalmic artery, the angular artery, the infraorbital artery and the nasal branch of the sphenopalatine artery and drains to the angular, infraorbital and nasal veins. The lymphatics drain into the submandibular and deep cervical glands. The nerve supply is from the infratrochlear branch of the nasociliary nerve and the anterior superior alveolar nerve.

**The conjunctiva**

The *conjunctiva* is a thin vascularised mucous membrane consisting of a non keratinising, stratified, columno-squamous epithelium and a substantia propria. It leaves the posterior surface of the eyelids from which it is reflected forwards at the fornices, to cover the anterior sclera. The conjunctiva thus forms a potential space, the conjunctival sac, which is open at the palpebral fissure. The conjunctival epithelium is continuous with the corneal epithelium at the limbus and a muco-cutaneous junction is formed at the eyelid margins.

Anatomically the conjunctiva has five regions which are continuous with each other, viz. palpebral, forniceal, bulbar, limbal and the plica semilunaris (Fig. 2.16 p30). The *palpebral conjunctiva* lines the posterior surface of the eyelids being transitional in the *forniceal conjunctiva* with the *bulbar con-*

*junctiva* which covers the anterior sclera to which it is loosely attached. About 3mm from the cornea it is attached more strongly forming the *limbal conjunctiva.* The *corneo-scleral limbus* is the junction between sclera and cornea whereas the *conjunctival limbus lies* 1mm anteriorly. At the limbus the substantia propria ends leaving only an epithelial layer which is continuous with the corneal epithelium.

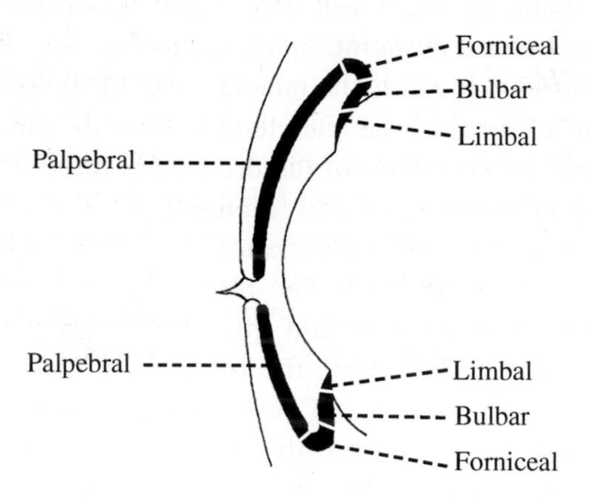


Fig. 2.16 The regions of the conjunctiva.

*The conjunctival epithelium.* Most of the conjunctiva has a non-keratinised stratified columnar epithelium. The epithelial cells at the limbus include a local reservoir of cells which can migrate to cover any acquired corneal epithelial defects. As shown in Fig.2.17 p31 the intrinsic cells of the conjunctival epithelium are:

- epithelial basal wing and superficial cells, the latter possessing microvilli
- melanocytes which are pigmented cells in the basal layer
- Langerhans antigen presenting cells also found in the basal layer
- goblet cells which are unicellular mucous glands containing mucin for discharge on the surface, which are more numerous in the fornices and plica
- the second mucus secreting system cells which supply both mucin and lipid for the precorneal tear film.

The *extrinsic cells* are lymphocytes, macrophages, granulocytes and mast cells. The conjunctival substantia propria, (Fig. 2.17 p31) is a layer of fibrovascular tissue underlying the epithelium. The fibroblast is the structural cell of the substantia propria, but the extrinsic cells present in this layer provide it with great potential in combating infection. In addition, a variety of extracellular immunoglobulins (IgG, IgA, and IgM) have been identified in normal conjunctiva. Lymphocytes may be grouped into nodules, particularly in the orbital conjunctiva, these are not true lymphatic follicles, just collections of lymphocytes. Thus the conjunctiva forms part of the *mucosal-associated-lymphoid-tissue (MALT) system* . The deeper layers of the palpebral conjunctiva are continuous with the tarsal plates . The tarsal region contains most of the conjunctival vessels and nerves, the non-striated palpebral muscle and accessory glands. The substantia propria of the limbus forms small extensions of connective tissue which radiate from the corneal periphery into the conjunctival epithelium. Small blood vessels, lymphatics and unmyelinated nerves run in these extensions. Small *accessory lacrimal glands* are found in the deep substantia propria their ducts opening on the surface.

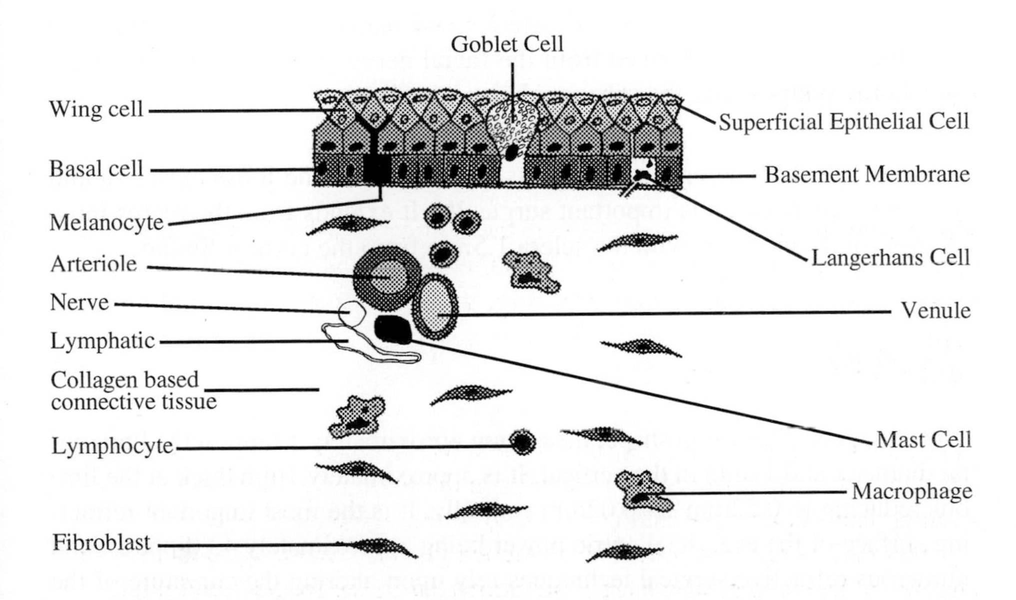


Fig. 2.17 A schematic cross sectional diagram of the conjunctival epithelium and superficial substantia propria showing a number of the cell types.

The *blood supply* is from the lacrimal and terminal ophthalmic artery branches which form arcades supplying the bulbar conjunctiva to within 4mm of the limbus where they anastomose with the anterior ciliary arteries which are the principal supply of the limbal region. Conjunctivitis mainly causes dilatation of the bulbar vessels (conjunctival injection) whereas scleritis, keratitis and uveitis cause dilatation of the deeper anterior ciliary vessels (ciliary injection). The *conjunctival veins* drain into the ophthalmic, angular and lacrimal veins. *Aqueous veins* are also present in the conjunctiva, emerging from the sclera and draining aqueous into the episcleral veins in which a laminar blood column can be seen with the slit lamp before the aqueous mixes with the blood. The *lymphatics* are arranged in a superficial and deep plexus in the conjunctival substantia propria and are important in the mediation of immune reactions. They drain medially to the submandibular glands and laterally to the preauricular nodes.

*Nerves of the conjunctiva.* Sensory nerves are derived from the 1st and 2nd divisions of the trigeminal nerve and form plexuses below and between the cells of the epithelium. Sympathetic (adrenergic) vasoconstrictor nerves from the carotid plexus accompany the arteries and parasympathetic (cholinergic) vasodilator nerves are derived from the facial nerve (nervus intermedius) via the chorda tympani and the sphenopalatine ganglion.

*Tenon's capsule* is a dense collagenous layer between the loose episclera and the conjunctiva which is important surgically. It extends from the rectus muscle insertions and fuses with the sclera 1.5mm from the corneal limbus.

### The cornea

The cornea is elliptical in shape, measuring aproximately 12mm in the horizontal diameter and 11mm in the vertical. It is approximately 1mm thick at the limbus reducing to 0.52mm +/- 0.02mm centrally. It is the most important refracting surface of the eye, the dioptric power being approximately 43 dioptres, and numerous refractive surgical techniques rely upon altering the curvature of the corneal front surface. It is comprised of five layers: epithelium, Bowman's layer, stroma, Descemet's membrane and endothelium (Fig. 2.18 p33).

The *epithelium* is a stratified non-keratinised squamous epithelium consisting of about five layers of cells which are columnar at the level of the basement membrane but become progressively flattened towards the surface. The epithelial cells are continually replaced by mitosis in the deeper layers.

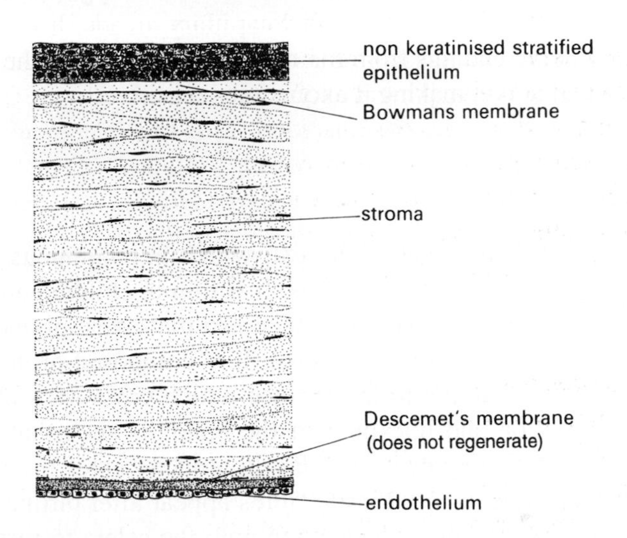


Fig. 2.18 The corneal layers.

*Bowman's membrane* is a thin structureless layer which is an irregular condensation of the superficial layers of the stroma. The *stroma* consists of layers of collagen bundles (lamellae) arranged in a precise way. Successive lamellae have collagen fibrils aligned at 90° to each other and the spacing or periodicity of the collagen fibrils is highly ordered. Between the lamellae are found cells *(keratocytes)* responsible for the production of collagen and ground substance - mucopolysaccharides and glycosaminoglycans. *Descemet's membrane* is a distinct structure and is a tough layer relatively resistant to both infection and trauma. The *endothelium* is only one cell layer thick and, unlike the epithelium, the cells do not regenerate if injured. If endothelial cell loss occurs, the remaining cells become flattened and their increased surface area allows the defect to be covered, but endothelial function is reduced.

*Corneal transparency* is due to the regular arrangement and relative dehydration of the collagen fibres within the stroma. This transparency is impaired if water passes into the collagen bundles and increases their separation. The epithelium and the endothelium prevent the passage of fluid into the stroma by acting as barriers and in addition the endothelium removes excess corneal tissue fluid by pumping it into the aqueous by an active transport process (the *endothelial pump).* The cornea derives its nutrition and oxygenation from the tears, from the aqueous and from the limbal capillary arcade. It is richly supplied with sensory nerve endings from the nasociliary branch of the first division of the trigeminal nerve, making it exquisitely sensitive.

**The sclera** (p493)

The sclera is the white outer coat of the eye which is continuous anteriorly with the cornea at the corneo-scleral limbus. It is thin in children and in some disease processes e.g. osteogenesis imperfecta, so that uveal pigment can be seen through it giving it a bluish appearance. It is composed of dense bands of fibrous tissue mainly parallel to the surface which cross each other in all directions, the orientation and thickness of the bands is such that they give maximum strength where required e.g. muscle tendons enter the sclera and fan out to give strong attachment. Elastic fibres appear after birth and are situated on the surface of the fibrous bands and help the sclera to resist permanent distension by intraocular pressure.

The sclera is about 1mm thick posteriorly and becomes thinner as traced forward. At the site of the exit of the optic nerve from the eye the sclera is fenestrated for bundles of nerve fibres to pass through. This is the weakest area and tends to stretch in the face of raised intraocular pressure being one factor in cupping of the optic disc and by this distortion, strangulating the blood vessels and nerve fibres.

The canal of Schlemm runs circumferentially at the limbus just peripheral to the trabecular meshwork which is attached behind to a projection of the inner sclera, the scleral spur. Aqueous fluid passes through the trabeculum into the canal of Schlemm and via efferent vessels into the scleral, episcleral and conjunctival veins. There is a constant flow of fluid across the sclera due to the intraocular pressure, which may be blocked by scleral inflammation or reduced by hypotony resulting in a serous choroidal detachment. Posteriorly the sclera transmits the long and short posterior ciliary vessels and nerves

(Figs. 2.2 p6, 2.9 p13). Just behind the equator the four vortex veins draining the choroid, leave the eye and form the ophthalmic veins and about 4mm from the limbus the sclera is traversed by 7-8 anterior ciliary arteries which are a continuation of the muscular arteries supplying the rectus muscles, which then anastomose with the long posterior ciliary arteries to form the major arterial circle of the iris.

## The lens

The lens is a transparent, bi-convex structure suspended from the ciliary body by the zonular fibres and situated between the iris and the vitreous (Fig. 2.2 p6) The lens is about 9mm in diameter and about 4mm thick at the centre. The thickness varies with accommodation. The anterior surface of the lens is less convex than the posterior surface. The most anterior part of the lens is called the *anterior pole,* the periphery is called the *equator* and the most posterior axial area is the *posterior pole.* The lens consists of capsule, epithelium and lens fibres. Under the *anterior capsule* lies a layer of epithelial cells. The rest of the lens consists of the lens fibres derived from other epithelial cells. There is thus no posterior epithelium. The central lens fibres are called the *nucleus* and the more peripheral fibres the *cortex.* In embryonic life the foetal lens is initially seen as a thickening of the surface ectoderm overlying the optic vesicle. This ectodermal thickening forms the lens vesicle, from which the lens is developed. New lens fibres are laid down throughout life and the nucleus of the lens becomes progressively harder and amber coloured with age *(nuclear sclerosis)*.The constant laying down of lens fibres causes the lens size to increase gradually. The anatomy of the lens can only be understood by reference to its curious development (p46, Figs. 3.2 - 3.8 pp 44 - 47 ).

The lens is an avascular structure depending for its nutrition on diffusion from the aqueous and vitreous into which the metabolic products of the lens diffuse in the opposite direction. The lens capsule acts as a semi-permeable membrane so that the lens will swell if put in a hypotonic medium and shrink in a hypertonic medium. The lens fibres and the epithelium of the lens possess active transport mechanisms which tend to keep the lens dehydrated under normal circumstances. The lens contains proportionately more potassium and less sodium than the aqueous or vitreous.

## The vitreous

The vitreous is a hydrogel and the water which makes up 99% of its volume is kept in a gel state by a collagen fibril framework and hyaluronic acid. It occupies the posterior segment of thc globe bounded by the posterior surface of the lens and zonule, the pars plana of the ciliary body, the retina and the optic disc. The outer part of the vitreous (the cortex) has the greatest concentration of collagen fibrils. The majority of these run parallel to the surface of the retina, but a few turn at 90° to be inserted into the internal limiting membrane of the retina. These collagen fibrils are responsible for the firm attachment of the vitreous gel to certain parts of the peripheral retina, the pars plana of the ciliary body and the optic disc. Less commonly, firm attachments also occur between the vitreous cortex and the retinal vessels and the macula. These firm attachments are important because traction on the retina occurs at these points if the vitreous contracts, resulting in a retinal tear. The centre of the gel contains less collagen than the cortex and the collagen fibrils are condensed into tracts. These tracts run forward from the disc. There are three major tracts. The hyaloid tract inserts into the back of the lens. The other two major tracts insert into the ciliary body. One important attribute of these tracts is their ability to confine a vitreous haemorrhage between them rather than allowing the diffusion of haemorrhage throughout the whole gel. The development of the vitreous is described on p46.

A blood/vitreous barrier is present, similar to the blood/aqueous or blood/brain barrier. This prevents free exchange of the larger molecules between the plasma and the vitreous gel. The barrier is due to the tight junctions that exist between the endothelial cells of the retinal vessels, the cells of the pigment epithelium and the inner endothelium of the ciliary processes. This barrier is impaired in disease of the retinal vessels, e.g. diabetic retinopathy. The barrier also influences the penetration of systemic antibiotics into the vitreous when the ophthalmologist is attempting to control intraocular infection.

## The uvea

The uvea is the vascular middle coat of the eye, being a continuous layer consisting of iris, ciliary body and choroid (Fig. 2.2 p6).

*The iris*

The iris is a circular diaphragm forming the posterior boundary of the anterior chamber. Its variable pigmentation determines the 'colour' of eyes. It has a central hole, the *pupil.* The iris is attached peripherally to the anterior surface of the ciliary body and the pupillary border rests upon the lens behind, the iris becoming tremulous when the support of the lens is removed. As will be seen (p49) the iris consists of a stroma and posteriorly an epithelium of two layers derived from the optic vesicle, one from the neural and one from the pigmentary layer. In this situation both layers are deeply pigmented. The anterior limiting membrane of the iris is a condensation of the stroma. The collarette (p49) marks the division of the stroma into pupillary and ciliary zones and is the site of the minor arterial circle of the iris formed by the radial vessels running inwards from the major circle. The iris muscles are situated in the stroma just anterior to the epithelium and curiously arise from neural ectoderm.The circular muscle around the pupil (sphincter pupillae) constricts the pupil. It is innervated by parasympathetic fibres of the third cranial nerve. The radial muscle (dilator pupillae) is supplied by sympathetic nerves and dilates the pupil. The pupil controls the amount of light entering the eye and is constantly varying in size depending on the action of these muscles. Tests for pupillary reactions give information regarding the nervous pathways involved (p414).

*The ciliary body*

This is the part of the uvea between the iris and choroid. It is triangular in antero-posterior section and the anterior surface into which the root of the iris is inserted forms part of the angle of the anterior chamber (Fig. 2.3 p7). The outer surface is in relation to the sclera and the anterior part of the inner surface is plicated to form the *ciliary processes,* about sixty in number (Plate 2.19 p38). Where the inner surface is smooth posteriorly it is called the *pars plana.* The junction between the retina and the pars plana is scalloped and is described as the *ora serrata* (Fig. 2.20 p39).

At the ora serrata the neural layer of the retina abruptly changes to a single layer of non-pigmented epithelium lying immediately within the continuation

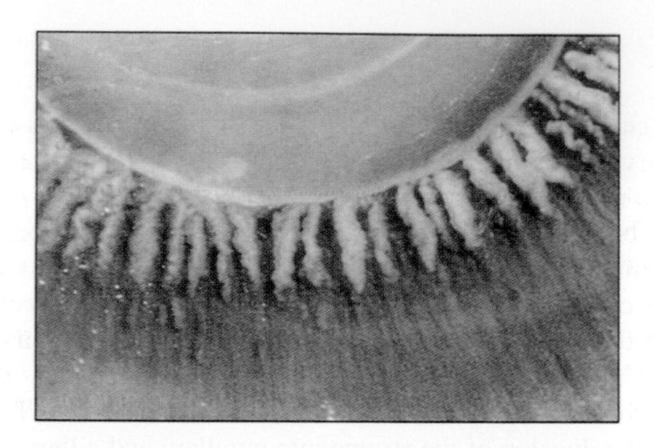

Plate 2.19 View of surface of the ciliary processes and pars plana of the ciliary body (seen from inside the eye looking forwards).

of the pigment epithelium of the retina. Together the two epithelial layers constitute the *ciliary epithelium* which covers the heavily vascularised ciliary processes and secretes aqueous. The ciliary processes enormously increase the secreting area. In the outer part of the ciliary body is the *ciliary muscle*, the fibres of which arise from an annular tendon blending with the scleral spur. The fibres are in three groups:

1. an *outer longitudinal layer* which passes backwards to the choroid
2. an *oblique layer* running towards the ciliary processes, the anterior part being so oblique as to appear circular in section, which gains insertion into the heads of the ciliary processes
3. longitudinal fibres passing forwards into the root of the iris

The muscle is mesodermal and non-striated and is supplied by the short ciliary nerves which convey postganglionic parasympathetic fibres from the ciliary ganglion. Preganglionic fibres originate in the Edinger-Westphal nucleus in the midbrain and pass via the inferior division of the third cranial nerve to the ciliary ganglion. There is also a weak reciprocal sympathetic innervation of the ciliary muscle (p24). Contraction of the ciliary muscle draws the suspensory ligament of the lens forwards and slackens it. The lens becomes more convex by contraction of its elastic capsule and focuses the eye for near distances. This is the act of *accommodation*.

### *The choroid*

The choroid is in contact on its inner surface with the pigment epithelium of the retina. It is entirely mesodermal except for the innermost layer, the cuticular part of the membrane of Bruch, which originates from the retinal pigment epithelium. The remainder consists of blood vessels which on the inner side form a fine network, the *choriocapillaris,* to nourish the outer third of the retina, while externally there are large veins leading to the vortex veins. Among the vessels are numerous elastic fibres and pigment containing cells, *chromatophores,*which give a dark brown colour to the choroid and prevent light diffusing into the eye to fog the images focused on the retina.

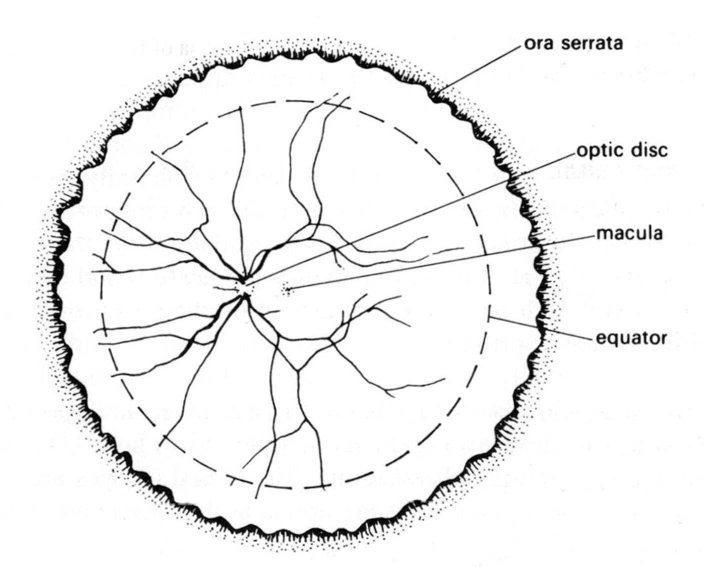


Fig. 2.20 Fundus oculi.

### *The blood supply of the uveal tract*

The blood supply of the uveal tract comes from the short and long posterior ciliary arteries and the anterior ciliary arteries (Fig. 2.8 p12). The short posterior ciliary arteries, which are variable in number, supply the choroid and important branches to the vascular circle surrounding the optic nerve head. The two long posterior ciliary arteries pass forward in the horizontal meridian within the suprachoroidal space. The anterior ciliary arteries are continuations of the arteries of the rectus muscles which continue in the episclera and perforate the sclera 4mm from the limbus to join the long posterior ciliary arteries (Fig. 2.9 p13). These form the major arterial circle of the iris which lies in the ciliary body supplying both ciliary body and iris. One anterior ciliary artery emerges from the lateral rectus and two from each of the other rectus muscles. Veins from all the uveal tract drain posteriorly to form four large vortex veins which traverse the sclera behind the equator to join the superior and inferior ophthalmic veins.

## The retina

### *Anatomy*

As viewed by the ophthalmoscope the fundus appears as in Fig. 2.20 p39, LCW 1.22, 1.23 p16. The macula is situated at the temporal side of the optic disc and the main retinal blood vessels which leave and enter it. Here the light sensitive cells, mainly cones, are closely packed. Peripherally the retina terminates at the scalloped ora serrata, although the two embryonic layers from which it developed continue forwards as the epithelium of the pars plana of the ciliary body (only the inner neural layer is pigmented) and then as the pigmented epithelium (both layers are pigmented) on the posterior surface of the iris extending to the pupillary margin. The retinal artery and vein and their divisions run superficially in the retinal nerve fibre layer, but their capillary networks run at two levels, one in the nerve fibre layer and one more deeply between the inner nuclear layer and the outer plexiform layer. The deeper network is not found in the peripheral retina. The retinal arteries are surrounded by a capillary free zone. The vessels are absent at the fovea and there are radial capillaries around the optic disc.

### *The retinal cells and layers*

The internal limiting membrane gives the sheen to the ophthalmoscopic view which becomes duller with age. The other layers are usually invisible, except for their blood vessels, down to the pigment epithelium layer. Depending on the degree of pigmentation of this layer, the choroid with its vessels and pigment is either seen as a red uniform glow when reflected light is diffused by the retinal pigment layer or, if there is little retinal pigment, as red choroidal vessels separated by pigmented stripes (the *tigroid fundus).* If there is little or no pigment in both the choroid and the retina, the retinal and choroidal circulations are both seen superimposed on the white sclera (the *albinotic fundus).* All degrees of variation occur and, except for extreme albinism, will allow normal function. There are, of course, racial differences in the degree of pigmentation. *Grouped pigmentation (likened to a cat's black footprints)* is a normal variation.

In the *macular area* the receptive elements, mainly cones, are densely packed. At the centre of the macula the nerve fibre layer is thin and almost avascular. This central pit,called the *fovea centralis,* consequently has an increased sensitivity and resolving power. Anything which makes the inner layers of the retina white or opaque, such as the combination of necrosis, oedema, and interruption of axoplasmic flow which occurs in retinal arterial occlusion or thc fatty degeneration of Tay Sachs' disease will reveal the fovea as a 'cherry red' spot where the vascular choroid shows through. (Plate 6.29 p118). The retinal cell layers are shown in Figs. 2.21 p41 and 2.22 p42.

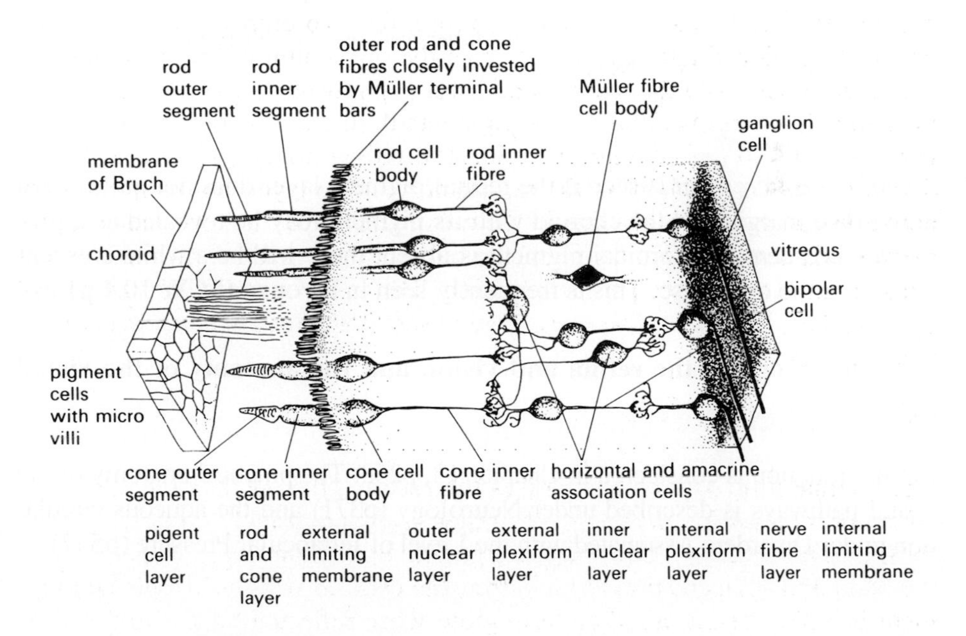


Fig. 2.21 The retinal cells and layers (redrawn after Frank Newell).

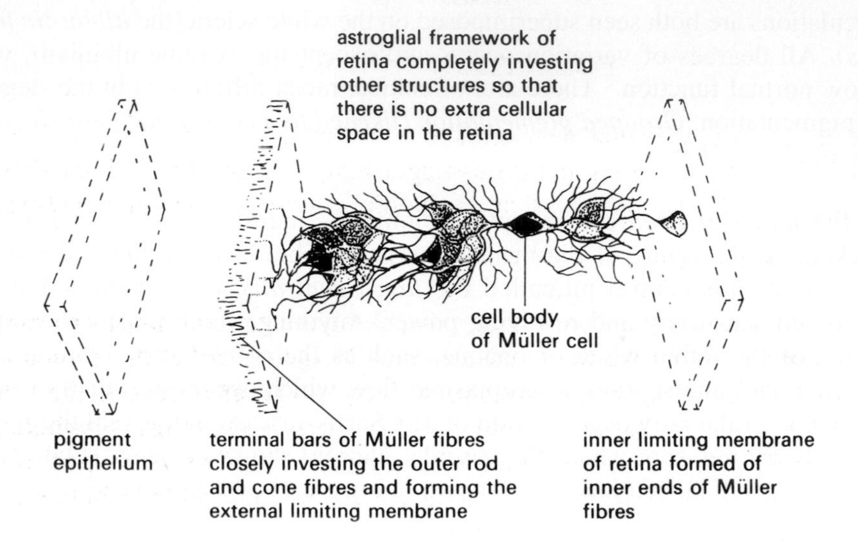


Fig. 2.22 The retinal cells and layers to show Müllers cells and their glial processes (redrawn after Frank Newell).

An *optic disc crescent* is seen if the retinal pigment layer does not quite reach to the disc margin and the choroid with its pigment may be revealed as a pigmented crescent. If choroidal pigment is absent there will be a white crescent bordering the optic disc. This is frequently seen in myopia. (LCW 10.4 p146)

## The physiology of the retina and vision and the anatomy of the visual pathways

Retinal function is considered in Chapter 13, p263. The practical anatomy of the visual pathways is described under Neurology (p371) and the aqueous circulation under Disorders Associated with the Level of Intraocular Pressure (p547).

# CHAPTER 3

# FORMATION OF THE EYE

A brief outline of ocular development is essential for understanding ocular developmental anomalies and for distinguishing them from pathological conditions. This distinction is particularly helpful when contemplating treatment.

## Development of the eye

The eye develops from both neural and surface ectoderm and from mesoderm. Surface ectoderm forms the lens and the epithelium of the cornea and conjunctiva and contributes, together with the neural ectoderm, to the vitreous body and zonule. Neural ectoderm forms the retina, ciliary epithelium, iris epithelium and its sphincter and dilator muscles, and the neural part of the optic nerve. Mesoderm gives rise to the corneal stroma and endothelium, the iris stroma, the choroid and the sclera.

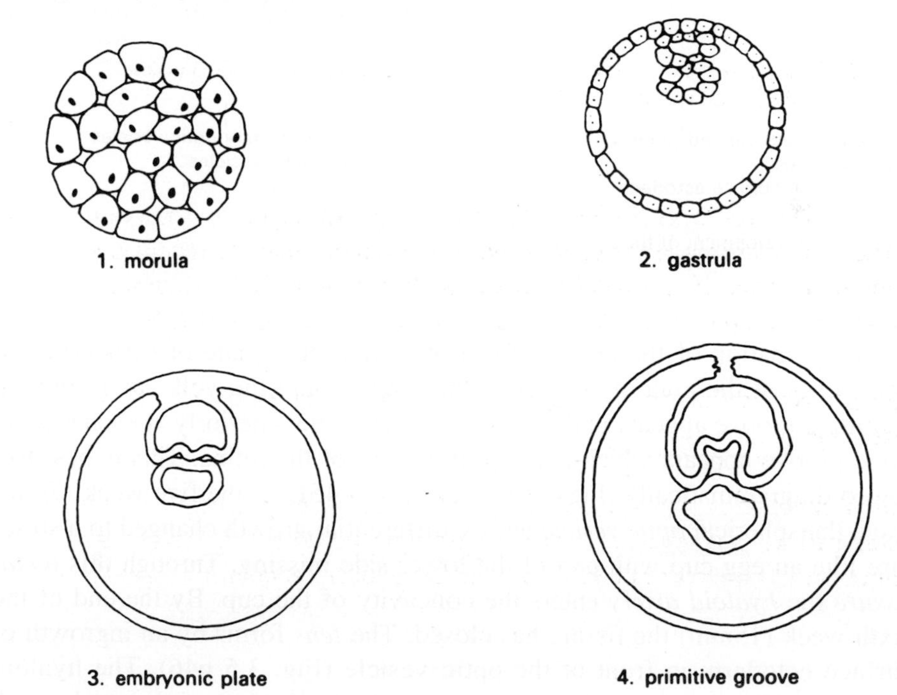


Fig. 3.1 Development of the eye (1).

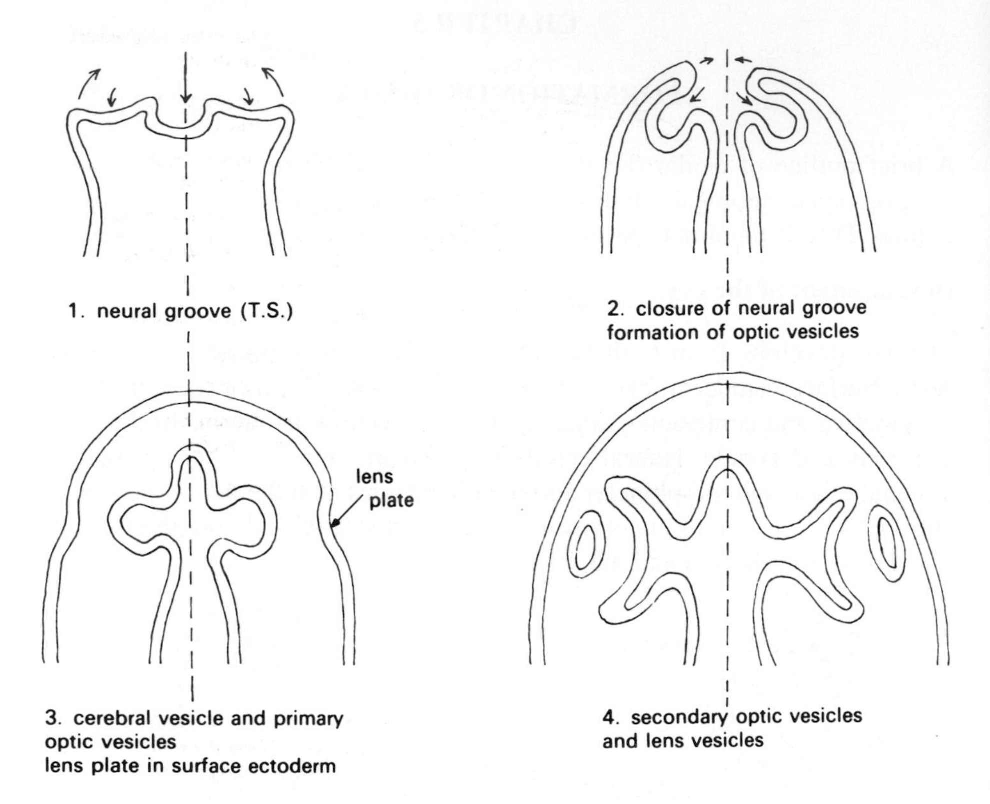


Fig. 3.2 Development of the eye (2).

In the formation of the embryo from the embryonic plate of ectoderm and endoderm, at the area of contact of the amnion and the yolk sac a midline primitive groove appears in the dorsal ectoderm and anteriorly at each side an optic groove appears which is the primordium of the retina. This is best followed diagrammatically (Figs. 3.1 - 3.4 pp43 - 45). At the five week (7mm) stage the spherical *optic vesicle* has by differential growth changed to a structure like an egg cup with part of the lower side missing. Through this *foetal fissure* the *hyaloid artery* enters the concavity of the cup. By the end of the sixth week (15mm) the fissure has closed. The *lens* forms by an ingrowth of surface ectoderm in front of the optic vesicle (Fig. 3.5 p46). The hyaloid artery ramifies on the back of the lens at the beginning of the sixth week (10mm). Meanwhile on the outer surface of the optic vesicle a network of

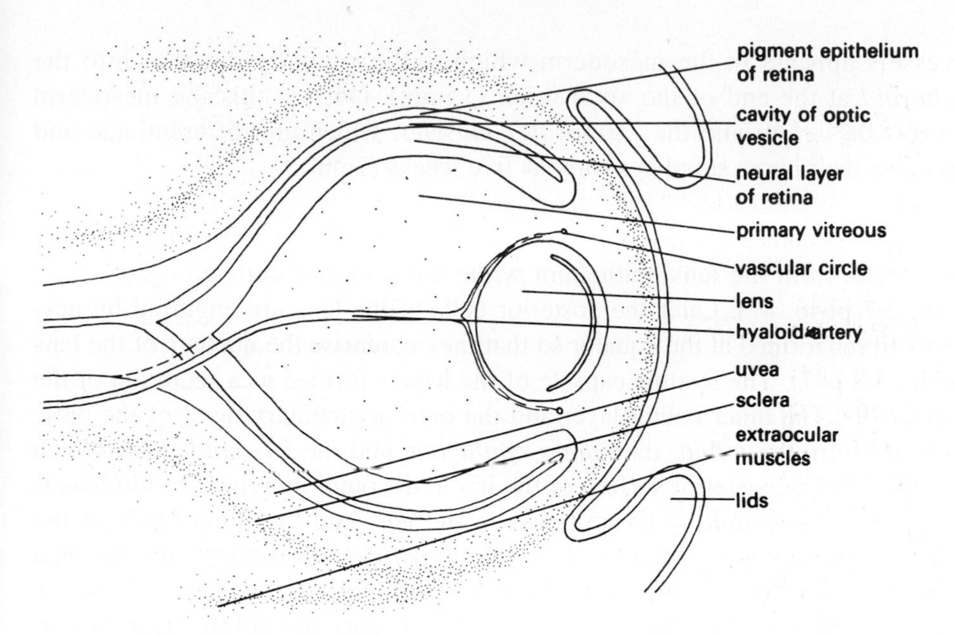


Fig. 3.3 The eye of a 10 week foetus.

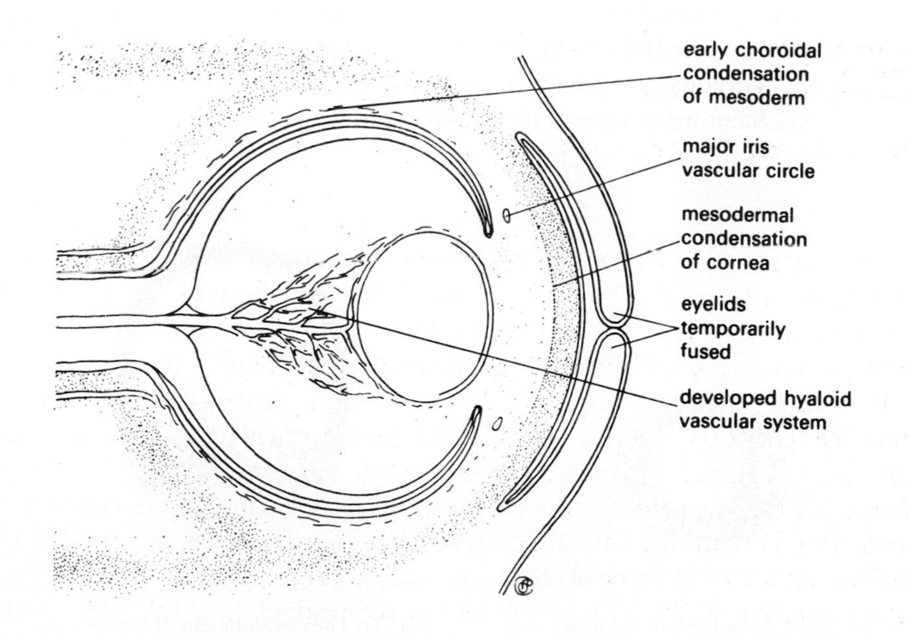


Fig. 3.4 The eye of a 20 week foetus.

vessels appears in the mesoderm which will eventually develop into the *choroid* at the end of the sixth week (15mm). Outside this the mesoderm forms the *sclera* and the *extraocular muscles,* which are differentiated and receive their nerve supply as early as five weeks (7mm).

The *lens* becomes enclosed by the edge of the developing optic cup. Its anterior cells form the lens epithelium while the posterior cells elongate (Figs. 3.6, 3.7 pp46, 47). Later the posterior cells of the lens are engulfed by new lens fibres formed at the equator so that they comprise the nucleus of the lens (Fig. 3.8 p47). The hyaline capsule of the lens is formed as a secretion of the lens cells. The inner neural layer and the outer pigmentary layer of the optic vesicle form the *retina,* the *ciliary epithelium* and the *iris epithelium* which extends on the posterior surface of the iris to the pupil margin. The *vitreous* is originally the primitive tissue between the lens and the inner layer of the optic cup. This is invaded by the hyaloid vessels. Secondary vitreous then forms from the retina which displaces the primary vitreous forwards to the ciliary region and inwards to occupy a tubular space around the hyaloid vessels. The hyaloid system regresses leaving only its retinal branches. The hyaloid artery proximal to these branches becomes the central retinal artery. Tertiary vitreous which forms the lens zonule is partly made up of tissue from the primary vitreous and partly from fibres which develop in conjunction with the basement membrane of the ciliary body (Figs. 3.3, 3.4 p45).

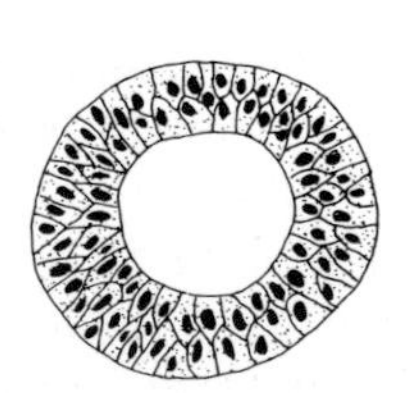

Fig. 3.5 The development of the lens - lens vesicle.

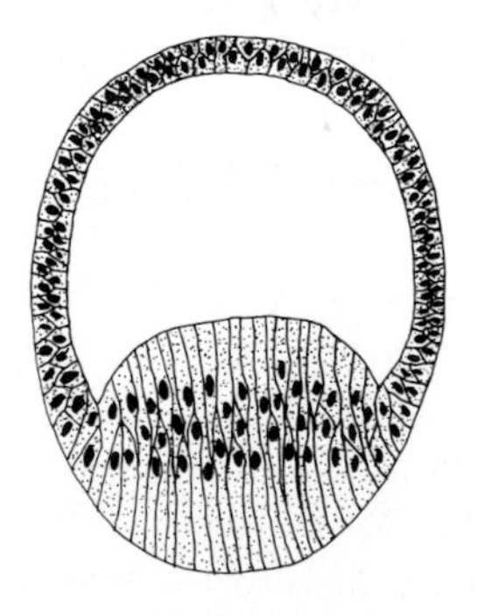

Fig. 3.6 The development of lens - elongation of the posterior cells.

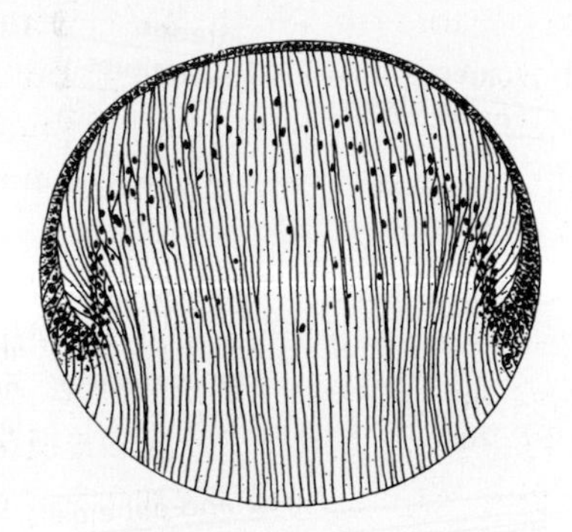

Fig 3.7 The development of the lens - nucleus.

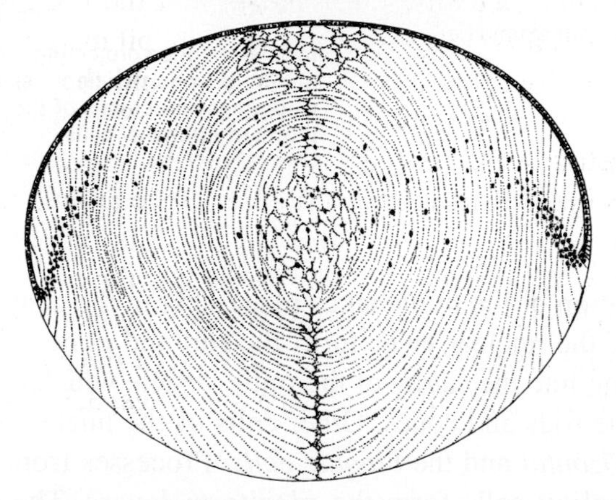

Fig. 3.8 The development of the lens - secondary lens fibres (formed at the equator and occupying a position between the nucleus and the lens epithelium).

**The development of the retina** (Fig. 3.9 p48)

The retinal neural layer of epithelial cells divides and an outer nuclear layer and an inner non-nuclear marginal layer can be distinguished. From about the sixth week (15mm) the *ganglion cells* are formed by nuclear zone cells invading the inner layer. Nerve fibres grow out from them and run to the optic stalk and thence to the brain. By 9 weeks (40mm) the chiasma has formed and the optic tracts develop. Medullation of the nerve fibres takes place from the brain distally and reaches the lamina cribrosa just before birth. The outer layer of the optic vesicle becomes pigmented at the beginning of the sixth week (10mm) and becomes flattened to form the *retinal pigment epithelium.*

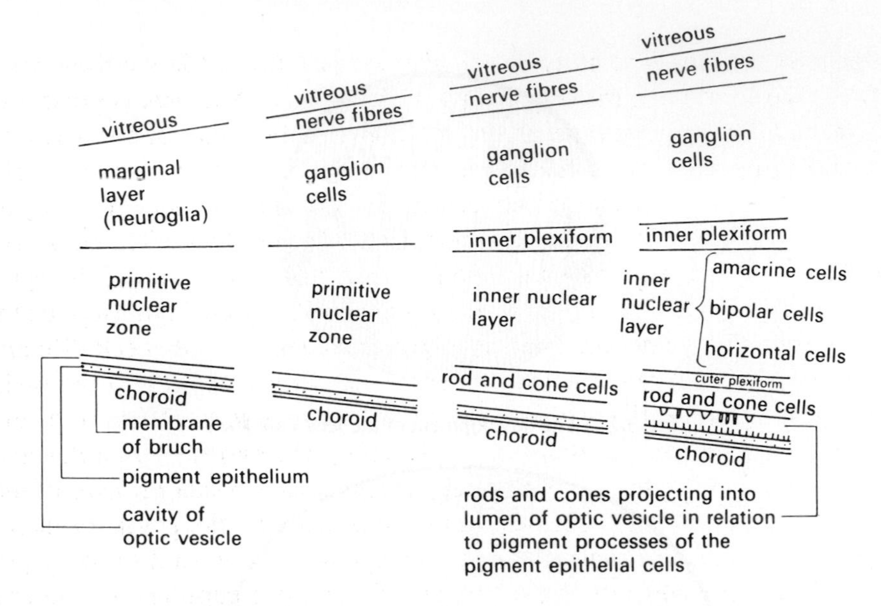


Fig. 3.9 The development of the retina.

Further differentiation of the retina takes place next in the central region. The outermost cells of the nuclear layer form the rod and cone cell bodies, they bear cilia projecting into the cavity of the optic vesicle which develop into the *photoreceptors,* the rods and cones. Other cells of the nuclear layer form the *amacrine,* the *horizontal* and the *bipolar cells.* Processes from all these cells and from the ganglion cells form the *plexiform layers.* The *macular area* forms at the 20th week (150mm) as a thickening of the ganglion cell layer and the *fovea* appears in the centre of this area in the 24th week (200mm).

## The development of other ocular structures

Mesodermal cells grow in between the lens and the surface ectoderm to form the corneal endothelium which secretes *Descemet's membrane* at about the 16th week (100mm) (Fig. 3.4 p45). Other mesodermal cells form the stroma and the surface ectoderm becomes the epithelium of the cornea. The *sclera* and *Tenon's capsule* are condensations of mesoderm around the optic cup. This process commences anteriorly where it is associated with the developing extraocular muscles. The *anterior chamber* appears as a slit in the mesoderm between the cornea and iris (Fig. 3.4 p45). The *canal of Schlemm* is present at 12 weeks (70mm). Cleavage of the structures in the angle of the anterior chamber takes place but some tissue remains to form the *trabecular meshwork* allowing communication between the anterior chamber and the canal of Schlemm. The posterior part of the mesoderm between the surface ectoderm and the lens forms the *iris stroma* where it lies anterior to the rim of the optic cup, the central part being the *pupillary membrane.* They are both supplied by the posterior ciliary arteries. There is no true iris up to the 12th week (70mm) but then there is a forward growth of neural ectoderm at the rim of the optic cup which extends to the pupillary margin by the 16th week (100mm) as the two layers of the iris. At about the 30th week (250mm) the *sphincter* and *dilator pupillae* form from the pigment epithelium. The posterior layer of the iris epithelium becomes pigmented at this time and later the pupillary membrane regresses leaving a circular ring, the *collarette,* peripheral to the pupil. Incomplete regression may leave strands of persistent pupillary membrane arising from the collarette and crossing the pupillary aperture, sometimes still adherent to the front of the lens. The iris epithelium is continuous behind with the two layers of the ciliary epithelium and mesoderm forms the stroma of the *ciliary body* and the ciliary muscle. At about the time of the formation of the iris epithelium, vascularised ciliary processes appear and between 20 and 24 weeks (150-200mm) the main arterial circle of the iris, situated in the ciliary body, arises from the two long posterior ciliary and seven anterior ciliary arteries.

Meanwhile longitudinal and then oblique (circular) portions of the ciliary muscles are forming from mesoderm and the choroidal layers and the elastic lamina of Bruch are differentiated. Choroidal pigment also forms towards the end of foetal life. The *lids* develop at the 10th week (50mm) as folds of surface ectoderm which meet loosely and separate again at 20 weeks (150mm). The *lacrimal glands* and the *sebaceous glands* of the *lids* and *conjunctiva* are formed from ingrowth of columns of ectodermal cells. The *nasolacrimal ducts* are originally a solid column of cells lying between the lateral nasal and maxillary processes. Shedding of the central cells leads to canalisation which is normally complete just before birth but may be delayed, especially at the junction of the naso-lacrimal and the nasal mucous membranes. It may occur spontaneously during the succeeding few months if infection of mucous collecting in the lacrimal sac can be prevented by frequent expression. If spontaneous canalisation does not result, the careful passage of a fine lacrimal probe frequently relieves the obstruction (pp28, 211, 467).

# CHAPTER 4

# OPTICS AND REFRACTION

## Optical definitions

*Visual Acuity*

Visual Acuity (pp77, 273) is recorded for distance and near, both with and without the appropriate correcting lenses. The refraction of the eye is dependent on the strength of the ocular lens system and the length of the eyeball.

*Refraction*

Refraction takes place when rays of light pass from one medium into another of different density. Rays striking the interface between the two media obliquely are deviated towards the normal as they enter the denser medium and away from the normal as they leave the denser medium (Fig. 4.1 p51).

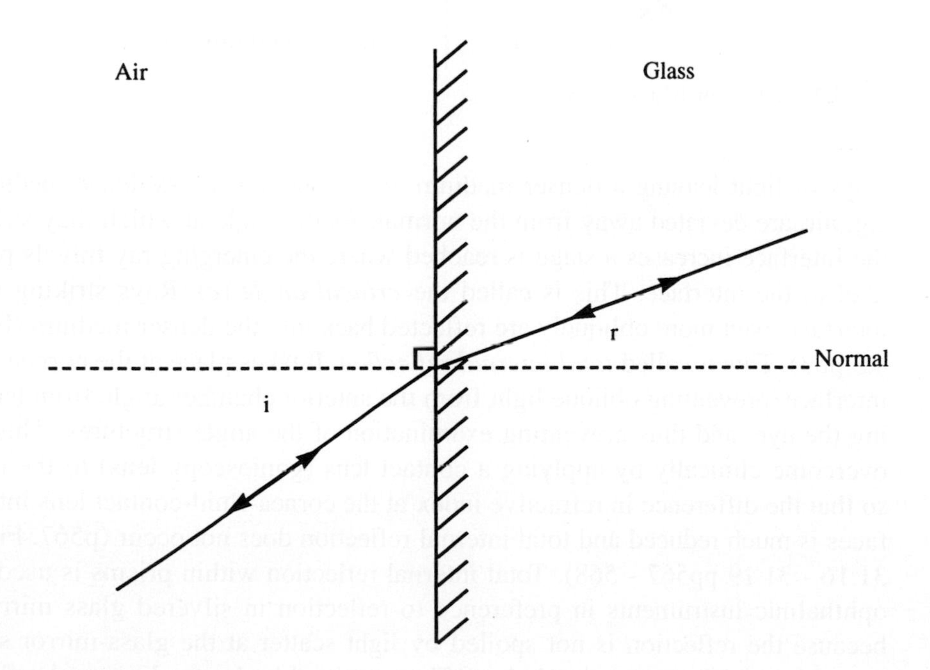


Fig. 4.1 Refraction of a ray of light.

This is because the wavefronts of light travel more slowly in the denser medium (Fig. 4.2 p52). For any two media, the angles of incidence *(i )* and refraction *(r )* are related such that sin *i* / sin *r* = a constant. If one of the media is air, the constant is termed the index of refraction of the other medium.

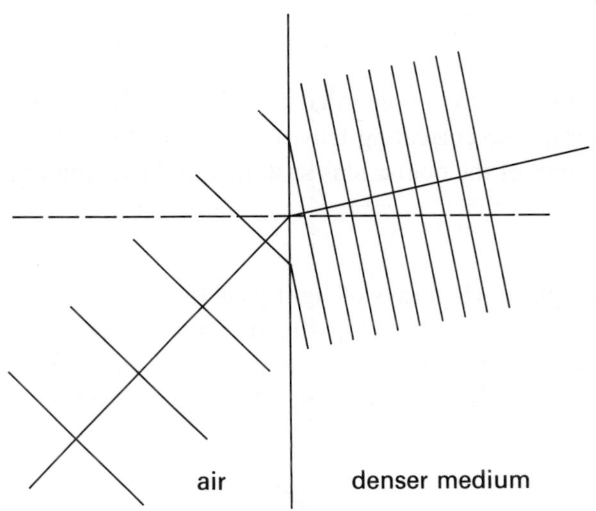


Fig. 4.2 Refraction of a wave front.

Rays of light leaving a denser medium, e.g. glass, for a less dense medium e.g. air, are deviated away from the normal. As the angle at which they strike the interface increases a stage is reached where the emerging ray travels parallel to the interface. This is called the *critical angle (c)*. Rays striking the interface even more obliquely are reflected back into the denser medium (Fig. 4.3 p53). This is called *total internal reflection.* It takes place at the cornea-air interface, preventing oblique light from the anterior chamber angle from leaving the eye, and thus preventing examination of the angle structures. This is overcome clinically by applying a contact lens (gonioscopy lens) to the eye so that the difference in refractive index at the cornea-fluid-contact lens interfaces is much reduced and total internal reflection does not occur (p567, Figs. 31.16 - 31.19 pp567 - 568). Total internal reflection within prisms is used in ophthalmic instruments in preference to reflection in silvered glass mirrors because the reflection is not spoiled by light scatter at the glass-mirror surface. Light is also transmitted along fibre-optic cables by total internal reflection.

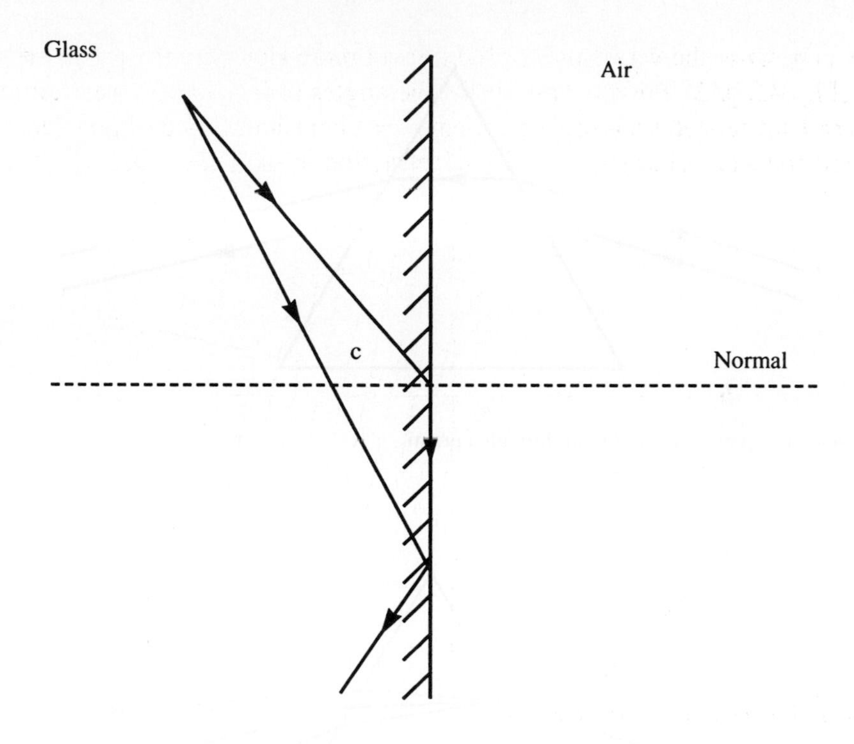


Fig. 4.3 Critical angle and total internal reflection.

## *Prisms*

A prism has two plane surfaces inclined at a finite angle (Fig. 4.4 p54). A ray of light passing through a prism is bent towards the base of the prism. In addition white light is broken up into its constituent colour spectrum *(dispersion)* because longer wavelength light is deviated less than shorter wavelength light on refraction at an optical interface (Fig. 4.5 p54). (The dispersive power of a substance is not related to its refractive index.)

The angle of deviation of a ray of light passing through a prism varies depending on the angle at which it strikes the prism (angle of incidence). However prisms used in clinical ophthalmology and refraction are thin and made with the front surface at 90° to the base. They are calibrated in the

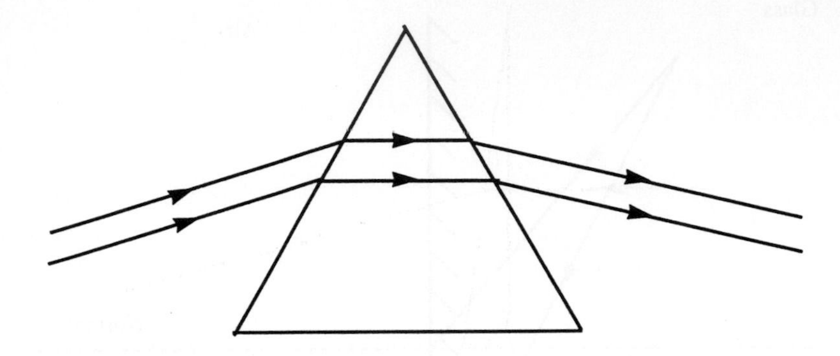

Fig. 4.4 The path of rays of light through a prism.

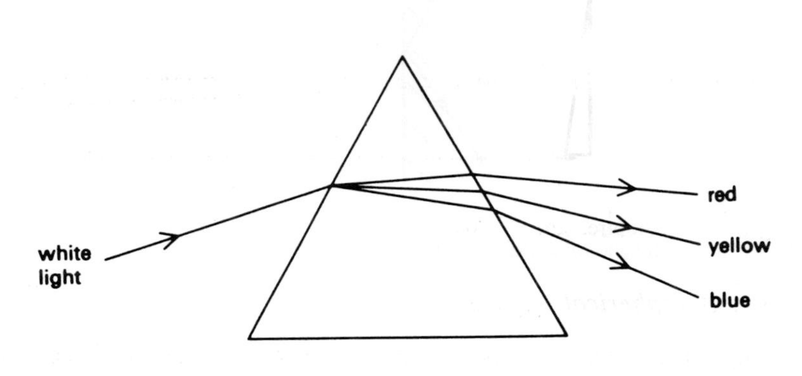


Fig. 4.5 The formation of a spectrum by a prism.

Prentice position in which light strikes the prism at 90° and all the refraction takes place at the back face of the prism (Fig. 4.6 p55).

The unit of prism power used in clinical practice is the prism dioptre. A 1 dioptre prism deviates a ray of light by 1cm along a line at 1 metre from the prism (Fig. 4.7 p55) This equals an angular deviation of ½°

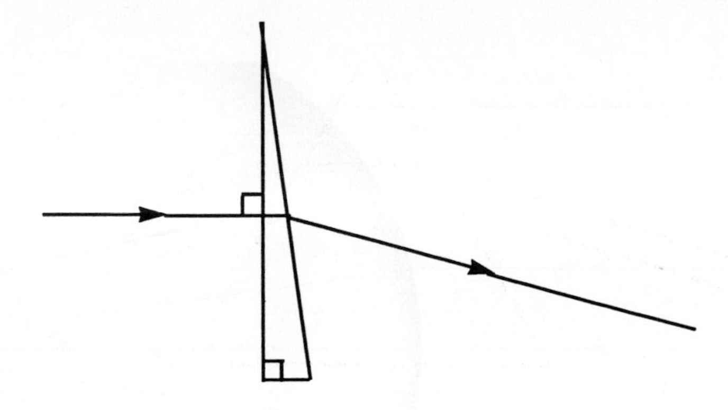

Fig. 4.6 Prentice position of an ophthalmic prism.

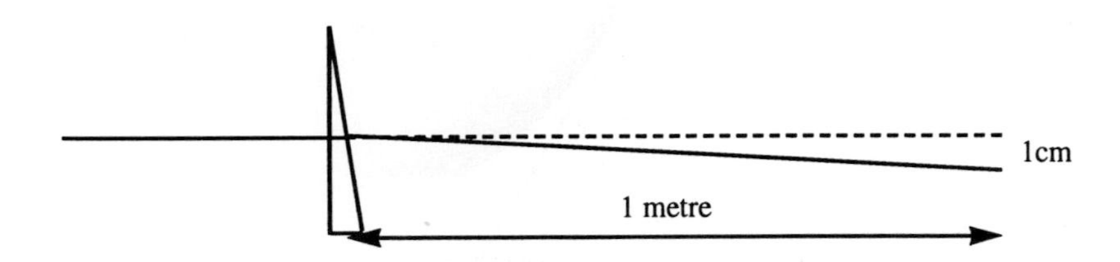


Fig. 4.7 The prism dioptre.

*Refraction at a spherical surface*

Parallel rays of light passing from air to a denser medium are deviated at a convex surface and converge to a focus within the second medium (Fig. 4.8 p56).

*Refraction by lenses*

The basic theory of thin lenses, as used in spectacles, is considered here. (For the theory of thick lenses and refracting units composed of a number of refracting surfaces separated by finite distances see standard optics texts required for ophthalmic optics examination purposes p76). A *convex* (plus) lens in section can be represented as a series of prisms, the prisms gradually becoming stronger towards the periphery (Fig. 4.9 p56). In this way rays are made to converge to a focus. The point to which parallel rays of light incident on the lens converge after passage through it is called the *principal focus* (F). The distance of the focus from the lens is the *focal length (f)*. The lens is clearly more powerful the shorter its focal length. In a similar manner a *con-*

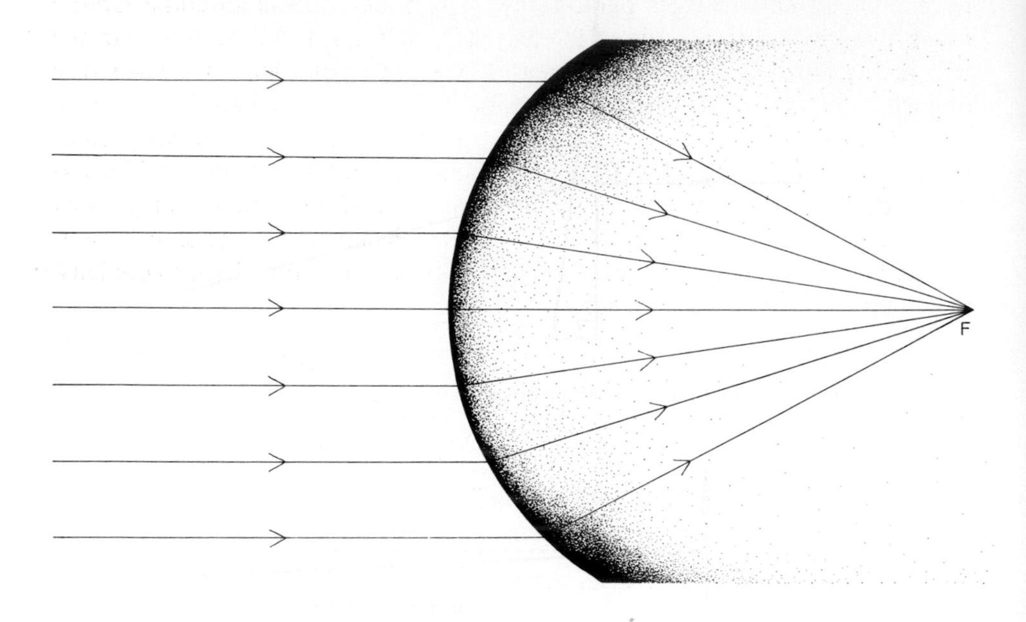


Fig. 4.8 The passage of light at a convex surface.

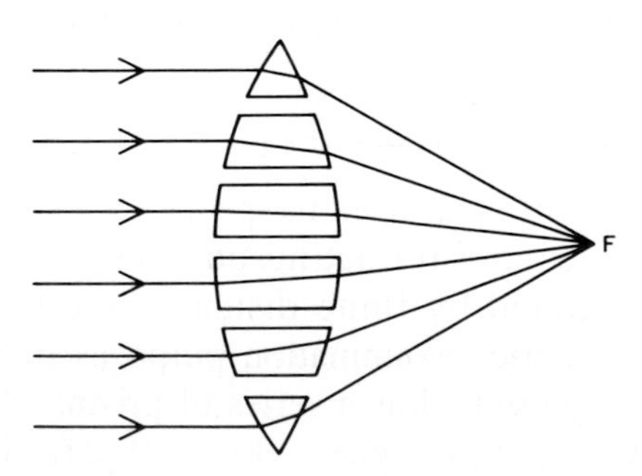


Fig. 4.9 The passage of light through a convex lens.

*cave* (minus) lens diverges parallel rays which then appear to come from a focus-F on the proximal side of the lens (Fig. 4.10 p57). As the power of the lens increases, the focal length decreases. The power of a lens is measured in dioptres (D), the dioptre being the reciprocal of the focal length in metres. Thus a 1D lens has a focal length of 1 metre, while a 2D lens has a focal length of 0.5 metre. The power of a lens depends on the curvature of its surfaces and the refractive index of the material of which it is made. Lenses may be made in several forms (Fig. 4.11 p57). The lens form having the least optical aberrations is called the *best form* lens. For most powers of spectacle lenses the best form is a *meniscus* lens.

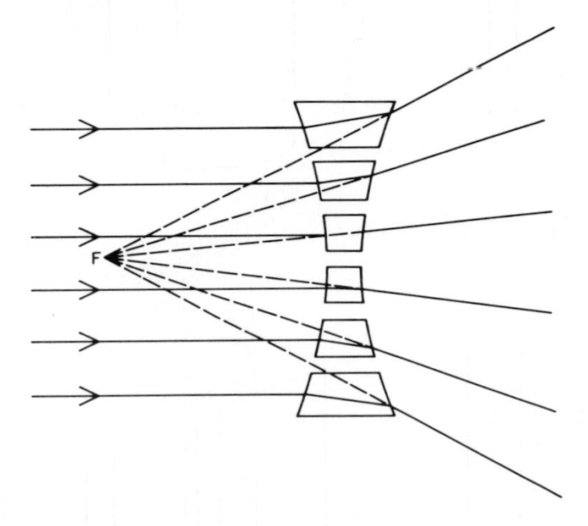


Fig. 4.10 The passage of light through a concave lens.

The surface curvature and therefore the power of a lens may not be the same in all meridia. Such lenses are called *astigmatic* lenses.

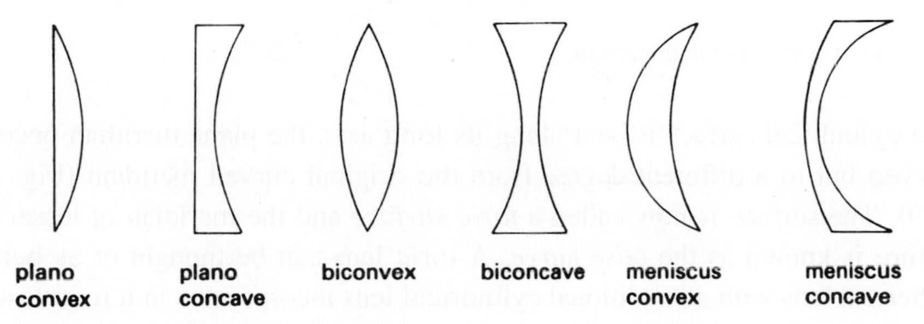


Fig. 4.11 Types of convex and concave lenses.

If one lens meridian is plane while the meridian at right angles to it is curved this is a *cylindrical* lens (Fig. 4.12, 4.13 p58). The plane meridian is called the *axis* of the cylinder and has no refractive power. The meridian of power lies at 90° to the axis.

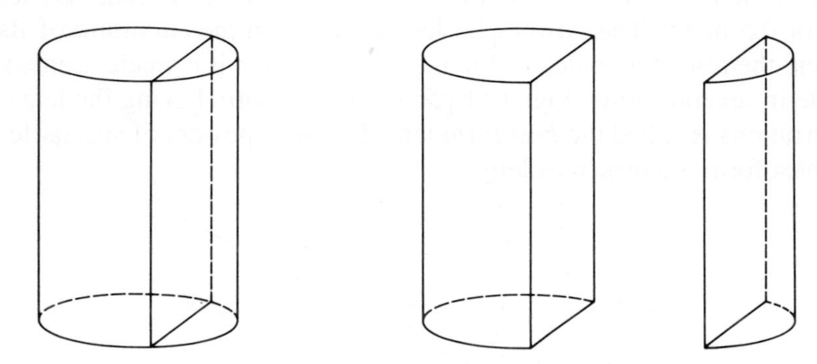

Fig. 4.12 A convex cylindrical lens.

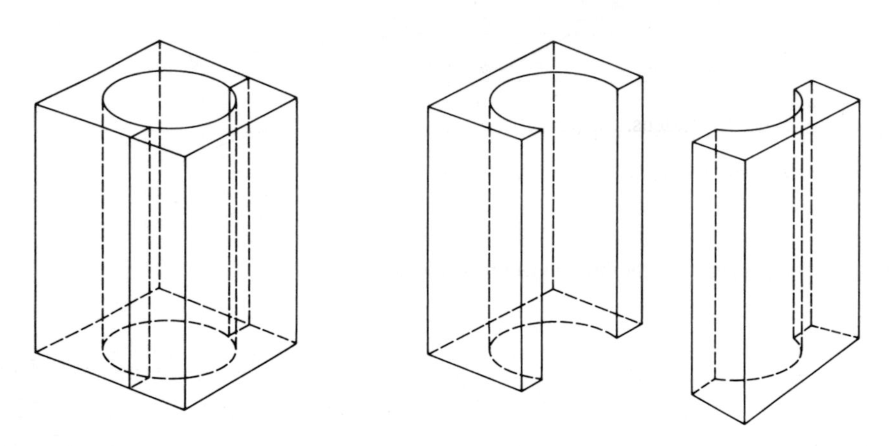

Fig. 4.13 A concave cylindrical lens.

If a cylindrical surface is bent along its long axis, the plane meridian becomes curved but to a different degree from the original curved meridian (Fig. 4.14 p59). The surface is now called a *toric surface* and the meridian of lesser curvature is known as the *base curve.* A toric lens can be thought of as being a spherical lens with an additional cylindrical lens incorporated in it to give a different power in the meridian at 90° to its axis.

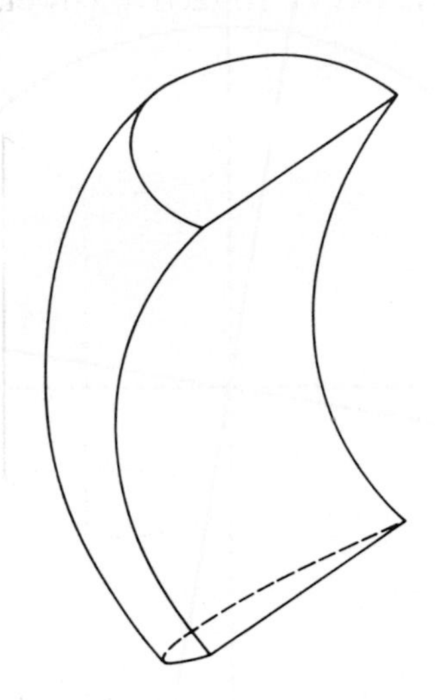

Fig. 4.14 A toric convex lens.

The orientation of the axis of a spectacle lens is described in degrees below the horizontal meridian from left to right as viewed from the front. The cylinder is expressed as its power in dioptres together with the angle of its axis. Spectacle lenses are prescribed in this way, e.g. +2.00DS/+1.00DC axis 100° specifies a toric lens with +2.00D power in the 100° meridian and a +3.00D power in the 10° meridian. Clearly such a lens could also be expressed as +3.0DS/-1.00DC axis 10°. Such a change of notation is called *transposition.* It is effected by changing the axis of the cylinder by 90° and reversing its sign, the power of the sphere becoming the algebraic sum of the original spherical and cylindrical powers (Fig. 4.15 p60).

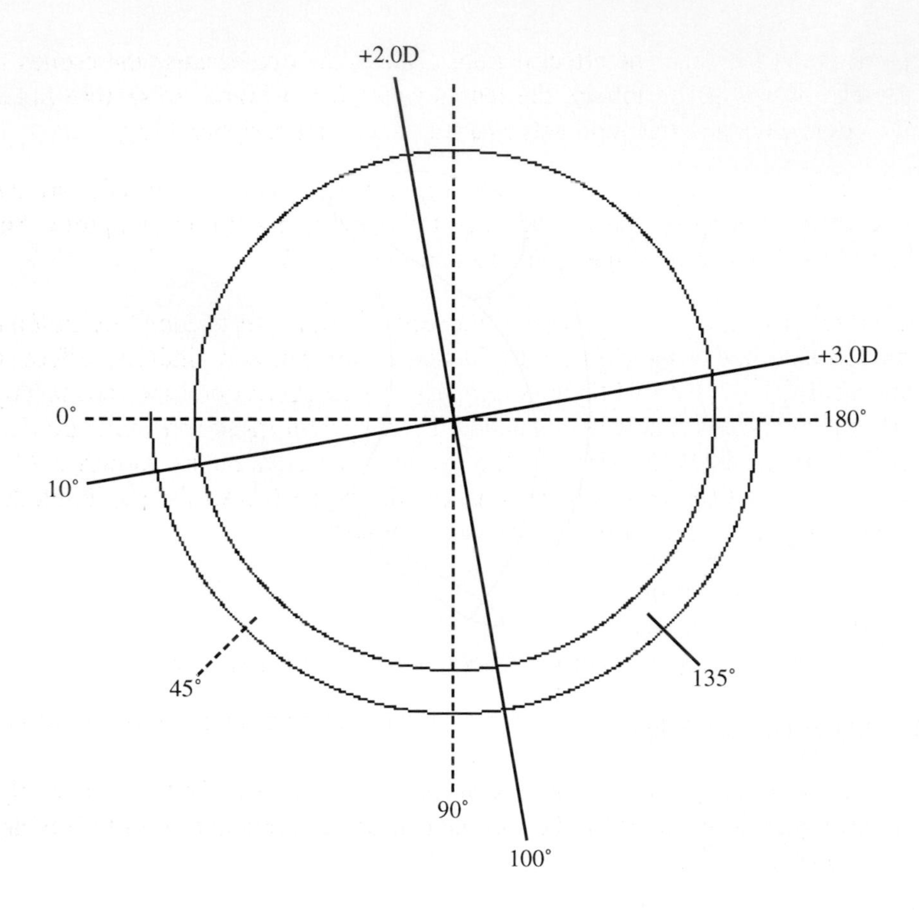


Fig. 4.15 Orientation of axis of cylinder for toric lens of power $\frac{+2.0DS}{+1.0DC \text{ axis } 100^\circ}$

By transposition the power can also be expressed as $\frac{+3.0DS}{-1.0DC \text{ axis } 10^\circ}$

*Lens aberrations*

There are many imperfections in image formation by spherical lenses and by the eye. Those which have the most practical effect are described:

-In *spherical aberration* rays of light passing through the periphery of the lens are deviated more than the central rays because of the increasing prismatic power of the lens towards its periphery, (Fig. 4.9 p56) so a perfect point focus is not formed. The effect is minimised in the eye because the cornea is flatter towards its periphery, the lens nucleus has a higher refractive index than the cortex and the pupil acts as a stop to cut out peripheral rays.

-In *chromatic aberration* the short-waved blue light is deviated more by the lens than the long-waved red light. In the eye this causes no symptoms but can be used with advantage in the duochrome test (p67).

-Aberrations occur if light passes obliquely through a lens. These are *oblique astigmatism* and *coma.* They may induce an unwanted cylindrical effect in the reading segment of bifocal spectacles. These aberrations are reduced by using meniscus spectacle lenses and by limiting light rays to the axial portion of the lens as far as possible. In the eye these aberrations are minimised by the curvature of the cornea and retina, the limiting effect of the pupil and the reduced resolving power of the peripheral retina.

*The ocular lens system*

The ocular lens system is complex but the main components are:

-the anterior corneal surface (cornea-air interface) supplies most of the refractive power (+43D).
-the anterior and posterior surfaces of the lens which supply that part of the power which is variable by the mechanism of accommodation (+15D unaccommodated).

*Accommodation*

The lens is composed of malleable fibres enclosed in an elastic *lens capsule* which tends to mould the lens so that it is effectively more convex (accommodation). This is partially prevented by the *suspensory ligament* or *zonule* which attaches the periphery of the lens to the surface of the ciliary body and tends to flatten the lens. The ciliary muscle consists of radial and circular fibres and is innervated reciprocally by the parasympathetic fibres of the oculomotor nerve (which cause contraction) and by the sympathetic. When the ciliary muscle contracts, the zonular attachment is reduced to a circle of less diameter and moves forwards. This releases the tension of the zonule and allows the elastic capsule of the lens to mould it into a more convex shape, the increased convexity mainly occurring at the anterior surface. It thus becomes more powerful and the eye is focused on objects nearer than before accommodation took place. The degree of accommodative power (or amplitude of accommodation) decreases with age, approximately from 14D as an infant to 11D at twenty, 4D at forty five and 1D at sixty five years of age.

**Emmetropia and ametropia (refractive error)**

In *emmetropia* rays of light from a distant point source which are parallel are brought to a point focus on the retina. Similarly rays of light from a point on the retina after emerging from the eye pursue a parallel course (Fig. 4.16 p62).

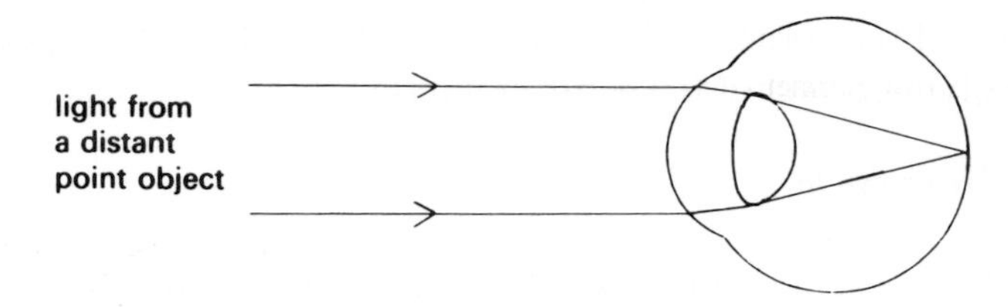


Fig. 4.16 The eye in emmetropia.

Eyes which are not emmetropic are termed *ametropic*. Types of ametropia are:

-*hypermetropia* (long sight, hyperopia) in which the eyeball is too short or the lens system too weak. Rays of light from a distance come to a focus behind the retina (Fig. 4.17 p63). The state of hypermetropia can be remedied either by an effort of accommodation, the use of a convex spectacle lens or a combination of both (Figs. 4.18 - 4.19 p63). Hypermetropia may cause aching or a feeling of strain in the eyes due to the difficulty of sustaining the degree of accommodation required.

Hypermetropic eyes are usually smaller than average and imperfect embryological development may be associated with high degrees of this condition. There is also sometimes a liability to acute glaucoma by closure of the angle of the anterior chamber because the lens and iris root may lie further forward than usual in these small eyes (pp550, 574 and Fig. 31.18 p568) showing a shallow anterior chamber at risk of angle closure). In children the accommodative effort necessary to correct hypermetropia is reflexly associated with a tendency to convergent squint, especially if one eye is handicapped in any way so that its impressions can be more easily ignored.

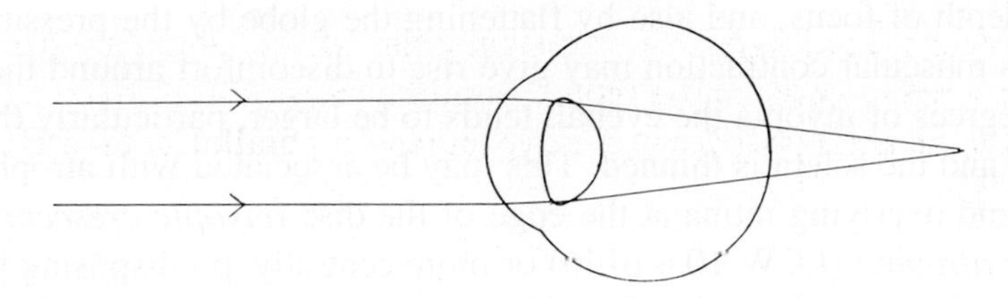

Fig. 4.17 The eye in hypermetropia.

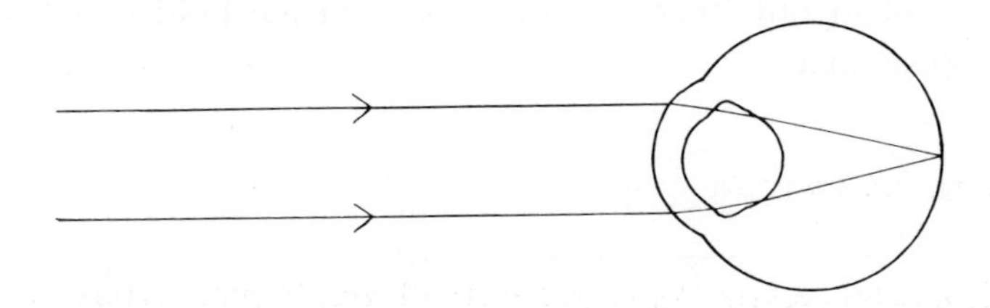

Fig. 4.18 The correction of hypermetropia by accommodation.

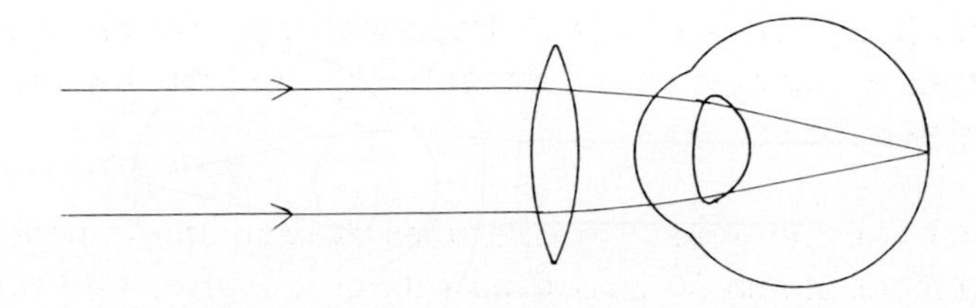

Fig. 4.19 The correction of hypermetropia by a convex lens.

-*myopia* (short sight) in which the eyeball is too long or the lens system too strong. Rays of light from a distance come to a focus in front of the retina (Fig. 4.20 p64). Myopia is corrected by concave spectacles or contact lenses (Fig. 4.21 p64). Myopia does not usually give rise to eye strain and there is merely blurred distance vision. The myope may however, screw up his eyes in an attempt to improve his acuity, both by narrowing the aperture, thus increasing the depth of focus, and also by flattening the globe by the pressure of the lids. This muscular contraction may give rise to discomfort around the eye. In higher degrees of myopia the eyeball tends to be larger, particularly the posterior half, and the sclera is thinned. This may be associated with atrophy of the choroid and overlying retina at the edge of the disc *(myopic crescent or peripapillary atrophy)* (LCW 10.4 p146) or more centrally, predisposing to macular haemorrhages (LCW 5.16 p79). The periphery of myopic eyes is also subject to degenerative changes which may give a liability to retinal holes and detachment. In extreme cases the sclera at the posterior pole may be so thinned that it bulges. This protrusion is known as a *posterior staphyloma.* The presence of myopia increases the risk of visual field loss from primary open angle glaucoma.

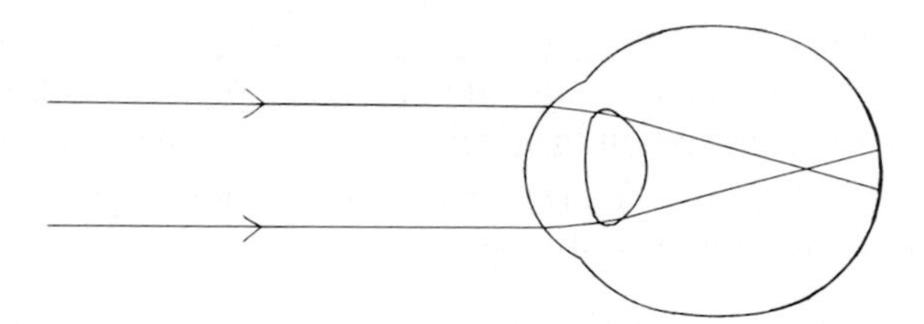

Fig. 4.20 The eye in myopia.

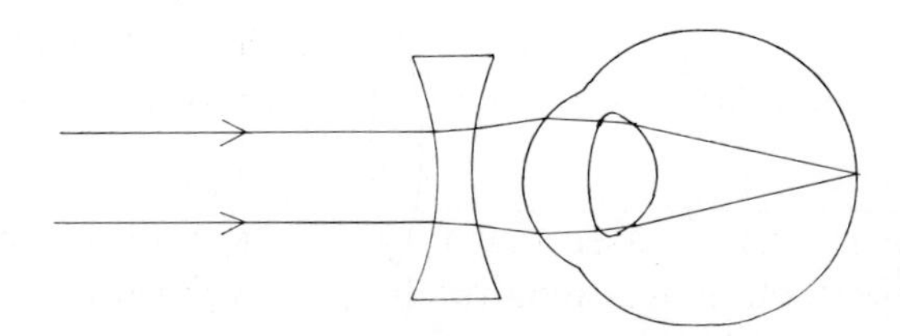

Fig. 4.21 The correction of myopia by a concave lens.

*-astigmatism* in which the refractive system of the eye has powers varying in different meridia. The maximum and minimum powers are usually at right angles. This requires correction with a cylindrical or toric lens which is correspondingly astigmatic in the opposite sense. Astigmatism may be myopic or hypermetropic in all meridia thus constituting myopic or hypermetropic astigmatism or there may be mixed astigmatism in which one meridian is hypermetropic and the other myopic. The retina of an astigmatic eye does not receive a clear image of all parts of an object. Thus when looking at the letter H, if the cross piece is in focus the uprights would be out of focus; if by the use of accommodation or a spherical lens the uprights are focused the cross piece will become out of focus.

*-presbyopia* or 'age sight' exists when the focusing of near objects becomes difficult, as in reading because of the diminishing ability to accommodate with age and patients may comment that their 'arms are now not long enough'. Symptoms of eye strain arise if more than two-thirds of the remaining amplitude of accommodation are required to focus on a near object.

*Anisometropia* exists when the refractive error of one eye is significantly different from that of the other. It is a common cause of amblyopia (lazy eye) in childhood. If both eyes are hypermetropic a difference of 1D or more will give rise to amblyopia of the more hypermetropic eye. This is because accommodation is a binocular function and only the less hypermetropic eye will have the image focused on the retina. Myopic anisometropia is better tolerated because both eyes have clear near vision unless the difference is very great, e.g. 5D or more, when amblyopia will occur.

*Aniseikonia* in which the retinal image in one eye is larger than that formed in the other. An appreciable degree, 5-10%, of aniseikonia can usually be tolerated but in unilateral aphakia corrected by spectacles there is a 30% disparity in image size. This disparity does not allow binocular single vision, although it can be achieved by the use of a contact lens which reduces this to about 7% (p74) or by an intraocular lens implant (I.O.L.) (p108),which virtually eliminates it.

The condition of *aphakia* caused much misery from aniseikonia because the strong convex spectacle lenses required to correct the high refractive hypermetropia caused considerable relative spectacle magnification (30%) and inevitably had severe optical aberrations. Many patients never adapted to their aphakic glasses after cataract extraction without lens implant insertion.

The advent of intraocular lens implants overcame these problems but because all eyes are not the same it is necessary to calculate the power of implant required to give a desired post-operative refraction for the individual eye.

*The calculation of intraocular implant lens power*

In 1980 Sanders, Ratzlaff and Kraff published a formula which has become widely used for this purpose. Known as the SRK Formula it enables the desired IOL power (P) to be calculated for a given post-operative refraction (R) if the keratometry (K) and axial length (AL) of the eye are known. A keratometer is used to measure the corneal curvature, and thus its refractive power, and the axial length is measured by ultrasound.

The SRK formula states that:

$$P = A - B(AL) - C(K) - D(R).$$

A,B,C and D are constants. The (A) constant is specific to the design of IOL in use and makes allowance for the position of the IOL within the eye. The multiplication constant (B) for the axial length is 2.5. The multiplication constant (C) for keratometry is 0.9. The multiplication constant (D) for the desired refraction is 1.25 if the IOL power for emmetropia is greater than +14 dioptres, and 1.00 if it is +14 dioptres or less. It will be seen that an inaccurate axial length measurement leads to much greater error in IOL power than an inaccurate keratometry reading because of the respective multiplication constants and for this reason an eye with a posterior staphyloma may be difficult to measure. For the principle of the keratometer see optic texts p76.

## Refractive or pathological cause for reduced visual acuity

*The pin hole test*

If the visual acuity of an eye is reduced (p79) the pin hole test is a quick method of discriminating to some extent between a refractive cause or one resulting from a pathological condition of the eye or visual pathways. The retina in ametropia receives an image which is out of focus. A diaphragm with a pin hole aperture placed in the line of sight will only allow a small pencil of light to pass through and reach the retina so that almost a point image is formed of all points of the object and renders the eye to a considerable extent independent of refractive error. Ametropia of up to 4.00D may thereby achieve an acuity of 6/6 and acuity in higher degrees of ametropia will be improved.

**Determination of refractive error**

This may be done objectively by manual retinoscopy or by the use of an autorefractor and subjectively by the use of trial lenses, either in a simple trial frame (Fig. 4.22 p68) or a refracting unit in which the lenses are changed mechanically. The basic manual methods will be described here.

*Retinoscopy*

The observer, sitting at arm's length (usually about ⅔ metre) from the patient shines a beam of light directed by a retinoscope (a mirror with a small central hole or hole in the silvering) into the patient's eye and inclines the mirror from side to side and up and down. Thus an illuminated patch of retina moves correspondingly and in turn acts as a source of light. Light from this moving area of illuminated retina will form an image at the far point of the eye and will when viewed by the observer through the hole in the mirror appear to move relative to the patient's pupil.

The myopic eye of more than -1.5D will produce a real inverted image between the patient and the observer, hence the image will move in the opposite direction both to the patch of illuminated retina and to the movement of the mirror by the observer. This is known as an *'against'* movement.
The hypermetropic eye will produce a virtual erect image behind the patient's eye, and the image seen by the observer will move in the same direction as the illuminated retina and mirror, a *'with'* movement.

The emmetropic eye forms an image at infinity, and a myopic eye of less than -1.5D forms a real inverted image falling behind the eye of the observer. In both cases a *'with'* movement is seen.

The image produced by a myopic eye of -1.5D coincides with the observer's eye at a working distance of ⅔ metre and no movement is perceived by the observer. This is the *point of reversal* of the direction of movement of the image.

Trial lenses are placed in front of the patient's eye either manually using a trial frame (Fig. 4.22 p68), or a refracting unit which mechanically changes the lenses, until the point of reversal is found. The distance refraction of the patient is calculated by correcting for the working distance (add -1.5D for ⅔ metre). The refraction may not be the same in all meridia. The difference in power between the maximum and minimum meridia and their orientation gives both a measure of the degree and the axis of the astigmatism.

In children whose accommodation is active and uncontrolled and therefore variable every few seconds, it is necessary to paralyse the accommodation before retinoscopy with a cycloplegic agent. Cyclopentolate 1% drops may be used (0.5% in infants).

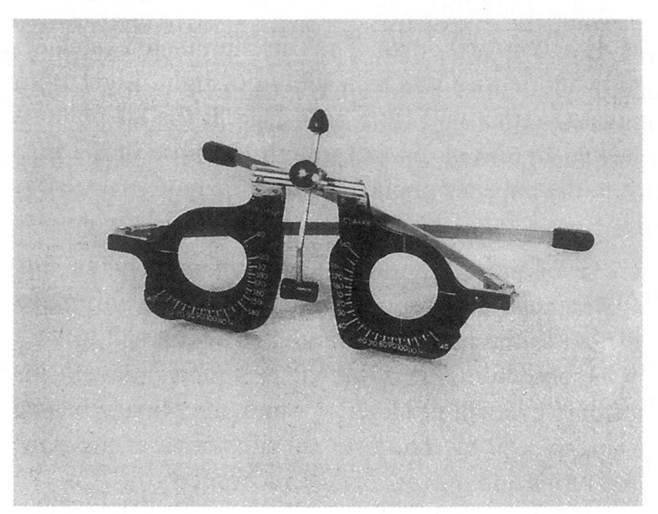

Fig. 4.22 Trial spectacle frames.

*Methods of subjective refraction*

Testing for distance vision. The patient views a Snellen distance type at 6 metres (20 feet). The trial frame is fitted ensuring that the frame is not tilted but horizontal and that the lens apertures are centred to the patient's eyes. If time allows. This is best done by measuring the *interpupillary distance*. In this way the prismatic effect of the spectacles can readily be calculated and with experience the optical centres adjusted to mitigate (but not fully correct) the strain of controlling heterophoria and relieve discomfort from this cause (p71). The refraction found by retinoscopy and corrected for working distance, is used as a starting point. One eye is occluded and the patient is offered small increments of plus and minus spherical power until the best vision is obtained. The axis of astigmatism, if any, is verified either by using a Jackson's cross-cylinder or by rotating the trial cylinder lens until best vision is obtained. The method using the cross-cylinder (Fig. 4.23 p69) is simple but requires experience in interpretation. Its theory is described in standard optics texts (p76) and its use is learned by practical demonstration. The process is repeated for the other eye. Final adjustments are made while both eyes are being used together as this usually favours relaxation of accommodation.

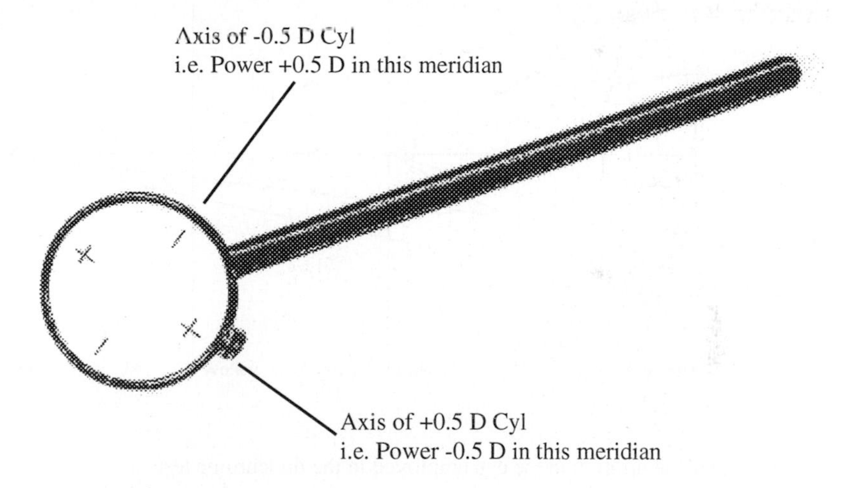


Fig. 4.23 Jackson's 1.0D cross-cylinder.

*The duochrome test*

This is a valuable 'end point' test. The eyes are subject to some degree of chromatic aberration in which light of longer wavelength is brought to a focus further away from the lens system than shorter wavelength light. This is utilised in the duochrome test. (White light passing through a prism is separated into constituent colours because they are refracted to a different extent and as we have seen a lens can be regarded as a series of prisms (Fig. 4.24 p70).

Normally the eye is focused for the rays of highest luminosity, the yellow. When letters only on a red or green background are briefly presented the red letters will be more in focus if the eye is myopic, the green will be more defined if hypermetropic and the definition of the red and green letters will be equal if the eye is emmetropic. This is the *end point.* Most patients will be accurate to ⅛ dioptre but occasionally one will be tested who seems incapable of giving reliable results. The test can be done by red/green colour defectives as it depends on the wavelength of the light and not on the appreciation of colour.

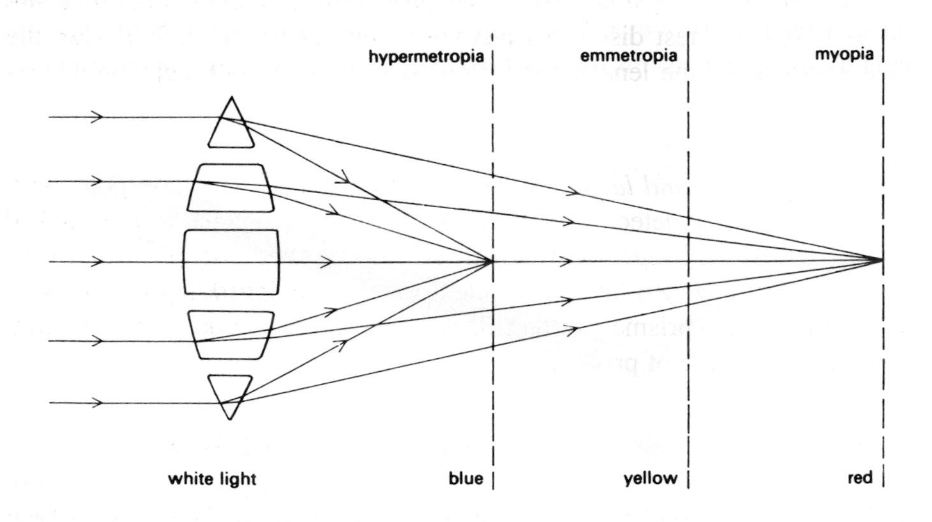


Fig. 4.24 Chromatic aberration in the eye employed in the duochrome test.

## *Near vision testing*

Using the distance correction, the ability to read near vision test type at the patient's normal reading distance is next tested and if necessary convex lenses are added equally to each eye until clear vision is achieved. It is important not to give too strong a reading addition as this is a very common cause of spectacle intolerance. A safeguard against this is to ensure that patients can read medium print (N8) (Fig. 5.1 p78) at the distance they would hold a newspaper with the proposed reading prescription. They tend otherwise to strain to read the very small print when under test and hold the reading test type closer than their habitual reading distance. The resulting stronger lenses would give rise to symptoms because of their reduced depth of focus. They also exaggerate aberrations and will tend to increase any difficulties due to exophoria. Habit or occupational considerations however, as in the case of a pianist or painter of miniatures, for example, will require lenses appropriate to the working distance.

If the prescription is of 5 dioptres power or more, the distance between the trial lens and the eye *(Back Vertex Distance)* must be recorded because any variation from the test distance when the spectacles are made will alter the effective power of the lens and the dispensing optician will make an adjustment for this.

*Detection of squint and latent squint.* An essential element in carrying out a refraction test is the detection of a manifest deviation *(squint, strabismus)* of the eyes and the presence and degree of a latent deviation *(heterophoria)* which may be alleviated by appropriate prisms or the decentring of spectacle lenses to achieve a prismatic effect (Decentration of a 1 dioptre lens by 1cm gives 1 prism dioptre of prismatic power).

*The cover test and uncover test* (p229) should always be performed.

*Other tests for heterophoria.* The principle is to dissociate the eyes without the patient being aware of this.

-In the *Maddox Rod Test* a spot of light viewed through a series of rods is seen as a thin streak at right angles to the rod. The rod or a series of rods (or of corrugations in a glass plate) usually coloured red (the Maddox Rod) is placed in front of one eye. A spot of white light visible to the other eye then appears as a red streak to the eye looking through the Maddox Rod and the latent deviation is tested both with the streak vertical and horizontal. The patient is deceived into thinking that the red streak and the white spot are two different lights and the eyes take up independent positions of rest. The strength of prism placed in front of either eye which makes the red line appear to go through the white light is a measure of the degree of the heterophoria and is usually expressed in prism dioptres. The red streak appears on the same side as the Maddox Rod in esophoria but crossed over in exophoria.

-In the *Maddox Wing Test* (Fig. 4.25 p72) one eye is presented at 30 cm with a white vertical arrow and a red horizontal arrow and the other with a white horizontal and a red vertical row of numbers. A combination of screens prevents a realisation that the display is not seen binocularly so that the eyes take up their position of rest but are accommodating to keep the arrows and numbers in focus. The patient reports to which white number the white arrow is pointing and with which red number the red arrow is level. The numbers measure in prism dioptres the degree of heterophoria for near vision. A degree of exophoria at reading distance is common and only requires correction if the patient is experiencing eye strain. Appropriately small decentration of the lenses may increase comfort however.

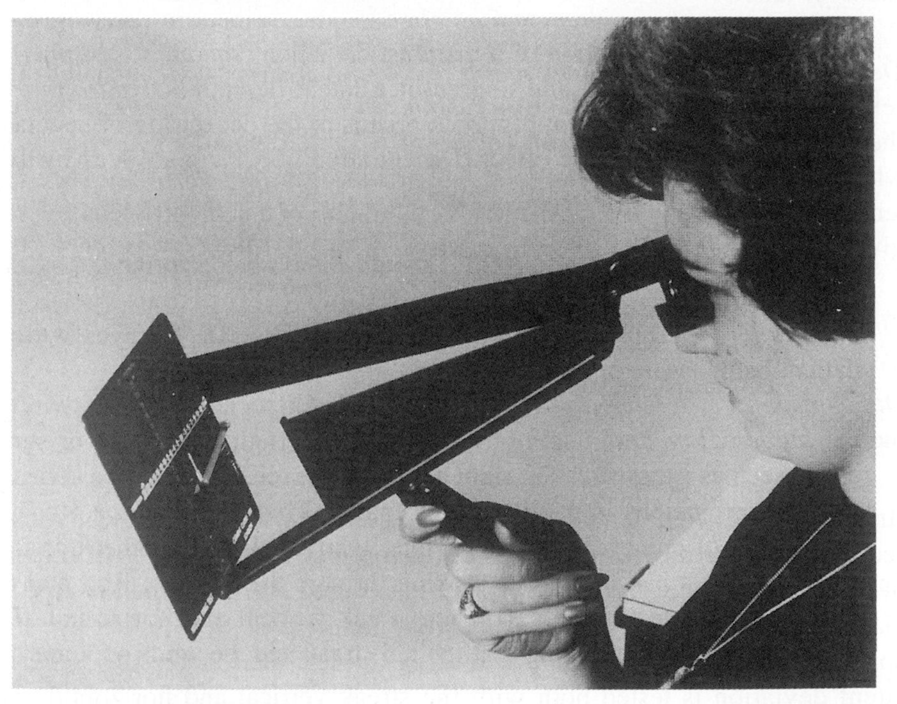

Fig. 4.25 The Maddox Wing Test.

Vertical latent deviation for distance or near may require fairly full correction but usually an undercorrection of the prismatic power is more acceptable, especially in spectacles worn intermittently and the prescription of prisms requires careful judgement.

*Spectacle Lenses* are made as 'best form' lenses. The lens form which gives least aberrations is chosen for each power of lens. For most powers the best form is a meniscus but for very high power minus lenses (-10D or more) it is plano-concave.

*Presbyopic lenses*

An additional correction for near vision can be incorporated in the lens for presbyopic patients. This may be as a *bifocal lens* with a different reading power in the lower part of the lens, or as a *progressive or variable power lens* in which the lens power gradually changes from the upper distance correction to the lower reading portion. Patients who require little or no distance correction may prefer *half-glasses* for reading.

Spectacles are sometimes tinted to reduce glare, which may help patients who have early cataract or conditions such as some types of retinal dystrophy or albinism. Polaroid lenses also reduce glare by selectively blocking the horizontally polarised light reflected from surfaces such as the sea and wet roads (see below). Protective glasses or goggles are also used where exposure to harmful light radiation is a risk, e.g. in a laser room or for welding. The lenses are designed to absorb the harmful wavelengths involved.

*Measurement of lens power*

The power of a spectacle lens may be measured either by neutralisation or on a focimeter. Neutralisation is done by viewing cross-lines through the lens and watching the movement of the image as the lens is moved in a particular meridian. The lens of opposite sign from the trial lens set which abolishes the movement gives the power of the lens in that meridian. Details of the optical principles and of the use of the focimeter is described in standard optics texts (p76).

*Polarisation of light*

The oscillations of the wave motion of the rays of light in ordinary daylight are oriented at random. A polarising filter blocks all the rays except those oscillating in one plane only, and the emerging light is said to be polarised. This principle is used in polaroid dark glasses, in the slit-lamp microscope to reduce reflections during fundus examination and to dissociate the eyes during some orthoptic tests.

*Contact Lenses*

Theoretically the ideal corrective lens would be inconspicuous, cause little change in image size and remain clear and in position under adverse circumstances. These advantages and others are possessed by contact lenses which, however, inevitably present problems of their own. They are only mentioned briefly here because details of their provision and management is a matter for specialist study.

*Types of contact lenses*

The earliest contact lenses were *scleral (haptic) lenses* which are large and rest on the conjunctiva over the sclera with an increased convexity over the cornea leaving a fluid filled space between the contact lens and the cornea (Fig. 4.26 p74). They are rarely used today except occasionally in the management of conjunctival burns or to prevent symblepharon. Their value in these conditions is controversial.

*Corneal (micro) lenses* were introduced in 1947. They are small thin lenses which float on the corneal surface in the tear film (Fig. 4.27 p74). They provide a new anterior optical surface for the eye, and as most refraction takes place at the cornea-air interface they are a potent means of correcting refractive errors, including high myopia and aphakia. Because they effectively become part of the refractive system of the eye the problems of spectacle magnification and aberrations are greatly reduced (p65).

Advances in lens technology have resulted in most hard corneal lenses being made of gas permeable materials to facilitate corneal respiration.

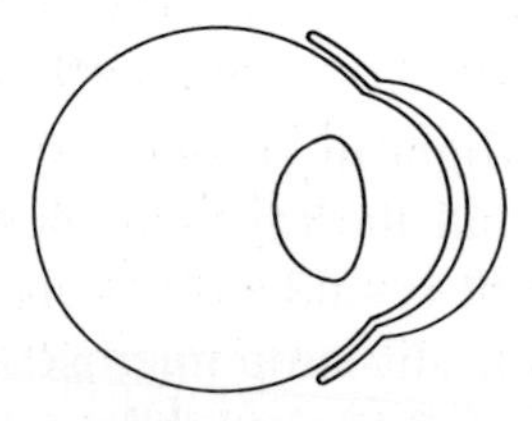

Fig. 4.26 A scleral haptic contact lens.

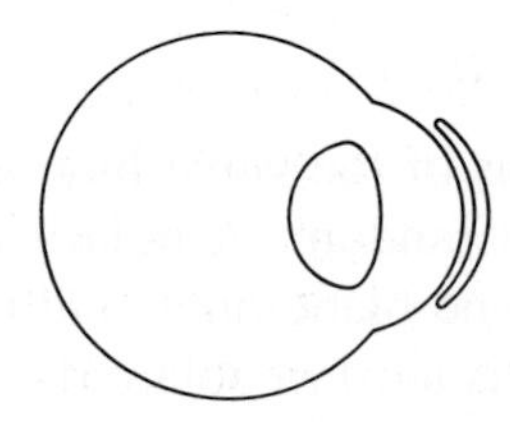

Fig. 4.27 A corneal contact lens.

*'Soft' contact lenses* are also now very popular because they are comfortable to wear. They are larger, up to 14.5mm diameter, and extend beyond the limbus. They are made of flexible hydrophilic material and conform to the shape of the eye which limits their usefulness in correcting irregularities in corneal curvature such as keratoconus and cicatricial distortion. They also require meticulous hygiene and can predispose to corneal infection or giant papillary conjunctivitis (p202) if this is lacking (LCW 10.5, 10.6 p146). Plano (of no dioptric power) soft lenses are commonly used as 'bandage' lenses to protect the cornea from abrasion by sutures or lashes and as an aid to healing and stabilisation of diseased corneal epithelium.

Specially *masked and coloured contact lenses* are useful in albinism and aniridia by restricting light to an artificial pupil and may be placed over unsightly blind eyes with a striking improvement in the patient's appearance.

**Lasers**

Laser is an acronym for *Light Amplification by Stimulated Emission of Radiation*, and this is what lasers do. Energy, usually light, is pumped into the laser substance raising the atoms to high energy levels. When an atom falls back to its lower energy level it emits a photon of light energy of a specific wavelength. The excited atom is stimulated to emit this photon if it is hit by a photon of the wavelength which it would itself emit. The atom falls to its lower energy level and the photons travel onward together. Mirrors at each end of the laser tube reflect the light back and forth and it grows progressively stronger as more and more photons are added by the above process.

What makes laser light so special is that it is monochromatic (of one wavelength specific to the lasing substance), coherent (the wave fronts travel in phase), and collimated (the rays are all parallel). The laser light may be released from the laser tube by rendering the mirror at one end partly transparent. A continuous beam of light is emitted and the laser is said to be operating in *'continuous-wave mode'*. Brief, high-power pulses of laser light can be produced by adding devices which cause the laser energy to build up within the tube and then to be suddenly released as a single pulse, *Q-switched mode* or as a train of high power pulses separated by a specific time interval, *mode locked.* (For more details see selected standard optics texts p76)

*Lasers used in ophthalmology*

Lasers are used to produce a thermal, ionising or photoablation effect on tissues.

*The thermal effect* of lasers to perform photocoagulation is used for the treatment of diabetic and other proliferative retinopathies, for maculopathies, to seal off retinal holes and to treat the trabecular meshwork in open angle glaucoma. The argon laser (blue 488nm and green 514nm), krypton laser (568nm and 647nm), dye and diode lasers are used in continuous wave mode for these purposes. Lasers emitting green light, Krypton and frequency doubled neodymium-yttrium-aluminium-garnet (Nd-YAG), are theoretically preferable for macular treatment because very little is absorbed by the macular xanthophyll pigment, unlike the argon laser blue wavelength which is strongly absorbed. Contact diode laser applications can be used for trans-scleral cyclophotocoagulation or direct application to the ciliary processes in eyes resistant to other means of lowering intraocular pressure (pp143, 581).

Lasers are also used to produce an *ionising effect.* A brief, very high energy pulse is delivered into a small focus causing ionisation of the constituent atoms of the target, the atoms of which disintegrate into a plasma of ions and electrons which rapidly expands causing a minute explosion which can be used to disrupt tissue. Capsulotomy and iridotomy may be performed by this means. The neodymium-yttrium-aluminium-garnet (Nd-YAG) laser is usually used in a Q-switched or mode-locked operation. Its wavelength (1065nm) is in the infra-red and not visible to the human eye so a red aiming beam is used, provided by a low power continuous-wave helium-neon laser.

*Photoablation* of tissue is possible using the argon-fluoride excimer laser (139nm, ultraviolet). At this short wavelength the energy of a photon can break intramolecular chemical bonds causing disintegration of molecules and tissue without producing a thermal effect. This laser can be used to sculpt layers off the cornea to correct refractive errors - photorefractive keratectomy (p492) and laser in-situ keratomileusis, LASIK (p492).

*Selected standard optics texts*

*-Coster DJ. Physics for ophthalmologists, Edinburgh. Churchill Livingstone, 1994.*

*-Elkington AR, Frank HJ, Greaney MJ. Clinical Optics 3rd Edition, Oxford. Blackwell Scientific Publications, 1999.*

# CHAPTER 5

# SIMPLE METHODS OF EYE EXAMINATION

## History and symptoms

An accurate ophthalmic and general medical history is essential as well as a knowledge of the patient's current medication. An ophthalmic history is frequently brief but the presence of a small set of specific symptoms can provide important clues to diagnosis. Patients with eye disorders commonly present with one or more of three complaints:

-pain,
-visual loss,
-a change in the appearance of the eye e.g. redness or the development of a mass.

Combinations of the above symptoms may suggest only a small list of differential diagnoses. For example: a complaint of pain, visual loss and redness of the eye may suggest acute angle closure glaucoma, keratitis or severe iritis (p151). However, painless visual loss with no change in the appearance of the eye may indicate central retinal vein occlusion or retinal detachment if of sudden onset, or development of cataract if the vision declines slowly (p91).
Features in the patient's general medical history are of importance e.g. a history of diabetes mellitus is of great relevance in a patient presenting with a vitreous haemorrhage (p111). A full history of medication is also essential e.g. a history of long term systemic steroid treatment may provide an explanation for the early development of cataract (p97) or of chronic secondary glaucoma. (p582).

## Examination of function

*1.Visual acuity*

The function of an eye is to see. A most important loss of visual function due to disease is reduction of visual acuity (p273). A record of visual acuity is therefore essential and should be the first part of the examination for both clinical and forensic reasons. At the bedside visual acuity may be assessed for practical purposes by noting if the patient is able to read small, medium or only large newspaper print held at an average distance from each eye in turn, with the aid of reading glasses if worn. Special reading test types (Jaeger,

Law) have been devised. They have a notation for different sizes of type e.g. J1, J2 etc. or N5, N6 etc. (Fig. 5.1 p78) corresponding to printers' fonts. (LCW 1.2 p12)

Distance vision is best recorded with a Snellen chart or news headlines of equivalent size at a distance of about 6 metres or 20 feet (the foot is the unit used in the U.S.A.) unless the patient's acuity is too low to read the largest print at 6 metres (Fig. 5.2 p80, LCW 1.1, 1.3 p12). The top letter is 6/60 or 20/200 and 'standard normal' visual acuity is 6/6 (20/20) or better. The figure 6 (or 20) in the 'numerator' refers to the distance in metres (or feet) of the chart from the patient and the 'denominator' to the distance at which a patient with an acuity of 6/6 (or 20/20) in a good light would be able to distinguish the let-

N.24

DISPLAY and advertisements

N.18

LARGE headlines and children's primers.
nose—one—cause—even

N.14

Book titles and paragraph headings in newspapers are often set in type like this, usually in CAPITAL LETTERS.
were—crone—our—summer

N.12

Books printed on very large pages and having many words on each line frequently use letters similar to this specimen.
name—use—means—arose

N.10

Novels, magazines, text-books and printed instructions are generally set in characters of about this size.
near—can—remove—sure

N.8

The news columns in most of the daily papers use this as the average size of print. Sometimes, the letters are larger than this, but seldom are they smaller.
crow—verse—see—renew

N.6

This is the smallest size type in general use. It is used for the classified advertisements in some papers, telephone directories, time tables, pocket diaries, and similar lists and books of reference.
assume—once—vane—sum

N.5

Printing of this size is only used for special purposes, for example, the small advertisements and financial columns in some journals, for small index lines and references, and pocket bibles and prayerbooks.
aware—eaves—sea—cream

Fig. 5.1 Law reading test type (Gilkes).

ter in question. Very poor vision will be recorded either by bringing the test type closer e.g. 3/60 (10/200) or as 'counting fingers' (e.g. CF at 2 metres), 'hand movements' (e.g. HM at 1 metre), or 'perception of light' (PL). Ultimately there may be 'no perception of light' (NPL). The distance acuity is tested with and without distance glasses. If the corrected acuity is not normal it may be checked by asking the patient to look through a pin-hole (an adequate pin hole can easily be made by passing a pin through a card) (LCW 1.4 p12). If the acuity is improved thereby it is likely that some or all of the reduced acuity can also be improved with the appropriate lenses (see p273 for the basis of visual acuity and p66 for its improvement, if necessary, by refraction).

*Visual acuity tests in children*

Assessment of visual acuity in very young children is inevitably approximate but fortunately, because most will look at a light and follow it, the accuracy of these movements can be interpreted to give a good indication of the state of vision. Slightly older children will reach for toys: the size of the stimulus required gives some indication of acuity e.g. the response to rolling white balls of different sizes. A history from the mother as to the child's visual behaviour e.g. navigation, the recognition of familiar faces and the size of toys which child can locate can be very valuable.

*Partially objective tests for children of one to three years of age*

*Cardiff Cards* (Fig. 5.3 p81) In this popular simple test which takes about fifteen minutes, each card has as a target an outline of a simple object *either at the top or bottom of the card.* This is a white line bordered on each side in black to make its average luminance equal to that of the grey background of the card. The widths of the outlines vary to correspond to visual acuities between 6/60 (20/200) and 6/6 (20/20). The cards are presented at 50cms or 1 metre depending on age and if seen by the child the eyes will be directed upwards or downwards (preferential looking). This movement is observed by the examiner, who does not know in advance the position of the target. The end point is either a wrong assessment by the examiner or the absence of a precise eye movement by the child.

The *Catford Drum* utilises the presence of optokinetic nystagmus (p408) when the child looks at a rotating broad black band on a white background 60cms from the eye. The band has steps (offsets) in its margin of different depths calibrated to Snellen acuities. The smallest step which elicits the fol-

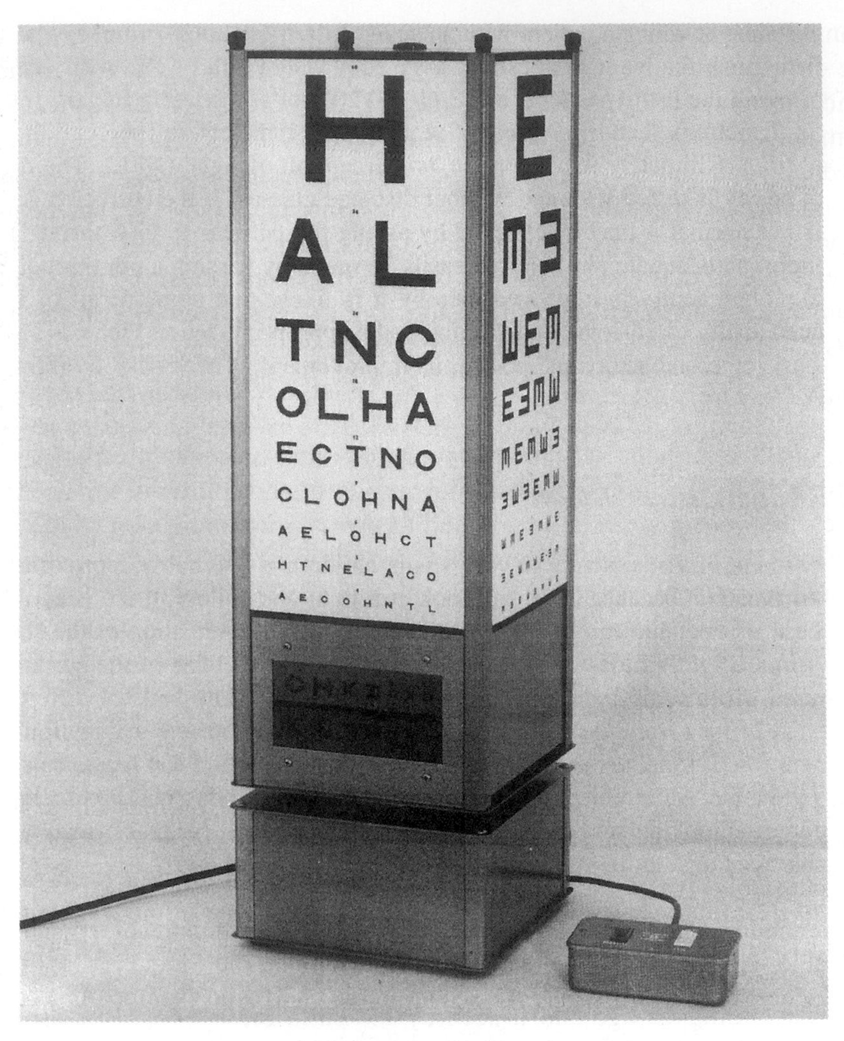


Fig. 5.2 Snellen distance test type. Also shows a duochrome test below and an 'E' test chart.

low movement of optokinetic nystagmus gives a useful measure of visual acuity of the vernier type (misalignment of edges). Being independent of volition it is also useful in patients of all ages with functional impairment of visual acuity (p441).

*Objective testing for children by electro-physiological techniques* is only available in special units.

Fig. 5.3 The Cardiff Cards Test for visual acuity estimates, especially for children aged 1 to 3 years.

Older children are able to do subjective tests more directly comparable with adult Snellen acuity results. One of the Sheridan-Gardiner Symbols are shown to the child at a suitable distance who then points to the same symbol on a card. The E test can be quite accurate for the slightly older child. The child orientates a cut-out E to match the attitude of different sizes of E displayed *singly* on a chart or conveniently on the sides of a cube held by the examiner at an appropriate distance. The Landolt C can be similarly used.

*Examination of Function*

*2. Colour vision*

Change in the appreciation of colours by one eye relative to the other is an important symptom . Disease of the optic nerve e.g. optic neuritis, commonly produces apparent desaturation of colours as an early sign before the appearance of a deep central scotoma and marked decline in visual acuity. Similarly, early optic nerve compression as in thyroid ophthalmopathy is often heralded by reduced colour appreciation. Retinal macular disease can also depress colour vision.

Colour vision can be assessed informally by asking the patient to view a richly coloured scene (colour television screen) with either eye. Alternatively, patients can be asked to comment on any differences in the appreciation of the colour intensity of a red target alternately viewed with either eye. Differences between eyes are significant and will not be due to any congenital impairment of colour vision ('colour blindness'). More formal tests of colour vision include the Ishihara colour plates and hue tests (p266).

*Examination of function*

*3. Visual field*

The visual field is the projection in space of the seeing areas of the retina.

*Types of visual field defect*

There are five principal patterns of field loss to look for and it should be the initial aim of the examiner to place any visual field defect in one of the following categories:

-*central scotoma* e.g. optic neuritis,
-*peripheral constriction* often reflecting general depression of sensitivity e.g. cataract,
-*hemianopic* or *quadrantic defect* which may be homonymous or bitemporal, where a boundary of the scotoma runs along the vertical meridian of the visual field. e.g. cerebral infarct or chiasmal compression, as by a pituitary tumor, respectively,
-*arcuate scotoma* where the defect follows the pattern of the nerve fibre bundles converging towards the optic disc e.g. chronic glaucoma,
-*altitudinal defect* in which there is an upper or lower scotoma bounded by a straight horizontal line through the centre of the field e.g. ischaemic optic neuropathy.

Classification in this way suggests specific diagnoses underlying the functional changes and helps to provide a framework for the interpretation of the sometimes confusing results of visual field testing.

*History and symptoms*

Patients may subjectively notice and describe the shape of a field defect (scotoma) especially if there is an acute or sub-acute onset. This is especially likely for scotomas which are a result of retinal disorders e.g. retinal detachment or age related macular change when patients often describe the defect as 'darker' than the surrounding view ('positive scotoma'). Patients with macular change involving irregularities of the retina may also describe corresponding distortion of central vision in association with a 'dark' central positive scotoma. Patients are often able to draw the shape of a positive scotoma: this can be assessed more formally using paper marked with a regular grid of lines which will additionally allow any distortion to be recorded (Amsler grid). In conditions affecting the visual pathways from the optic nerve to the visual cortex however, field defects are either unnoticed or perceived only as areas 'missing' from the surrounding view.

*The examination of visual fields*

There are two main methods of examining visual fields:

1. *by confrontation.*
2. *by special apparatus.*

The confrontation test is more relevant to this section. For other methods see p273.

*Confrontational visual field testing*

The test is rapid, requires the minimum of equipment and gives good control of fixation. The examiner sits at arm's length in front of the patient who has one eye covered. The examiner closes his eye opposite the patient's covered eye and asks the patient to look at the other. It is important to make sure that the general lighting does not dazzle the patient and that the background is uniform and dark enough to contrast with the target.

*-finger counting:* Each quadrant of the field is tested for the ability to count fingers by flicking one or two fingers up rapidly from a closed fist, each quadrant being tested two or three times. When patients have difficulty in

counting fingers they may not say that they cannot see them but hesitate or try to look at them.

*-finger movements:* If they cannot count fingers, finger movements are tested. When these are not seen in part of the field, the edge of the scotoma to finger movements is determined by introducing the fingers from the blind area towards the seeing part of the field. The periphery of each field of vision is tested by finger movements bringing the continually moving finger from behind the patient's head forwards to define the edge of the field.

*-red headed pin:* A hat pin with a bright red head is used although any small bright red target is acceptable e.g. red top of an eyedrop bottle. The patient looks at the target and notes its colour. It is then explained that during the test the examiner must be told if and when it loses or regains its colour, not whether he can merely see the object. Each quadrant of the field near fixation is then tested and any relative scotoma found is mapped out. The depth of any central scotoma may already be quantified by prior measurement of the visual acuity but its extent can now be defined with the red bead. Although a central scotoma may prevent really accurate fixation the patient can usually look in the direction of the examiners eye sufficiently well for testing. Very early depression of the central field in one eye may be suggested by the reduction in saturation of the red colour when the patient looks at it with each eye alternately.

*-white headed pin:* (LCW 1.24 p18) Arcuate and hemianopic scotomas are most accurately tested using a white headed pin of 3mm diameter as the stimulus. In particular they should be suspected when an upper nasal loss has been found with the other methods of testing, because an arcuate scotoma frequently widens towards the nasal side and an upper defect is more common in glaucoma. The normal blind spot (which corresponds to the photoreceptor-free nerve head) is first defined by moving the stimulus outwards from within the blind spot, the head of the pin being held equidistant from patient and observer. The size and position of the blind spots of observer and subject should then approximately coincide. The arcuate areas above and below fixation are then tested in the same way and then the periphery of the field.

Although these simple methods can be quite accurate when carried out by an experienced observer, small relative scotomas may be missed on confrontation testing and formal perimetry is always necessary to confirm the findings.

*The testing of children*

Children are often unable to maintain fixation but can do the finger counting test quite well if the fingers are only shown momentarily. The 'two toy test' can be used in infants. One toy is used to attract the child's attention and the other is brought in from the side. The point at which the child looks at the new toy is noted. Time and patience are of course required.

*Significance of field defects*

For a summary of the significance of the types of field defect see p282.

*Other visual field tests* (p266)

*Examination of function*

*4.Ocular position and movement*

The patient should be asked if they have noticed diplopia. There may be evidence of restriction of ocular movement in some directions of gaze or the presence of a squint in the primary position (Fig. 11.1 p229) which may be intermittent. The cover test is a simple and invaluable test in these cases, if properly carried out (p229).

**Examination of structure**

An account of the use of a hand magnifier, the slit lamp microscope and the direct ophthalmoscope in the examination of the eye is given on p255. It is helpful to examine the eyes and surrounding structures in a logical order from anterior to posterior as outlined below.

*The eyelids*

Local lesions or oedema of the lids may be seen. Asymmetry of the lids may be due to lid retraction, as in dysthyroid eye disease (Fig. 16.3 p307) or in proptosis (eyeball pushed forwards) by orbital space-occupying lesions (Plate 19.11 p382, Fig. 30.2 p544) or to a large globe as in myopia or buphthalmos (Plate 31.21 p573). Conversely, a narrowing of a palpebral fissure may be the result of ptosis (dropped upper lid) (Fig. 22.4 p450), blepharospasm (spasm of the lids) or enophthalmos (eyeball displaced backwards into the orbit) as may occur following 'blow -out' fractures of the orbit (p541).

*The conjunctiva*

The conjunctiva may appear red due to dilated vessels. This is usually due to conjunctivitis but dilatation of the vessels deep to the conjunctiva at the limbus (circum-corneal or ciliary injection) may indicate some deeper inflammation of the anterior segment, e.g. keratitis, corneal inflammation with perhaps corneal ulceration (Plate 7.13(b) p165), iritis (Plate 7.23 p177), or acute glaucoma (raised pressure in the eyeball) (Plate 7.27 p186).

*Presence of discharge*

The eye may be watering or there may be a discharge of mucus or pus which may even be blood stained. Watering will result from any cause of ocular irritation but discharge implies the presence of conjunctivitis (Plate 8.1 p193) or lacrimal sac inflammation.

*The cornea*

The normal cornea should be bright and clear. If the symptoms concern only one cornea it should be compared with the cornea of the other eye. A torch may reveal a loss of shine or reduced transparency. Local whitening of the cornea suggests scarring or infiltration (Plate 25.6 p486). Blood vessels in the normally avascular cornea should be looked for (Plate 17.5 p322). Lastly, the cornea should be stained with a trace of fluorescein and with Rose Bengal drops. The taking up of a fluorescein stain occurs where the cornea is denuded of epithelium and this area shines with a luminous green colour when the fluorescein is sufficiently diluted with saline (Plate 7.5 p157). Rose Bengal stains red the degenerate epithelial cells themselves (Plate 10.2 p219). Irregularity of the corneal surface as in *keratoconus* (p485) can be revealed by distortion of the corneal reflection of the rings of a target-like pattern placed in front of the eye and viewed through a convex lens mounted in an aperture at the centre of the disc - Placido's Disc (Fig 5.4 p87).

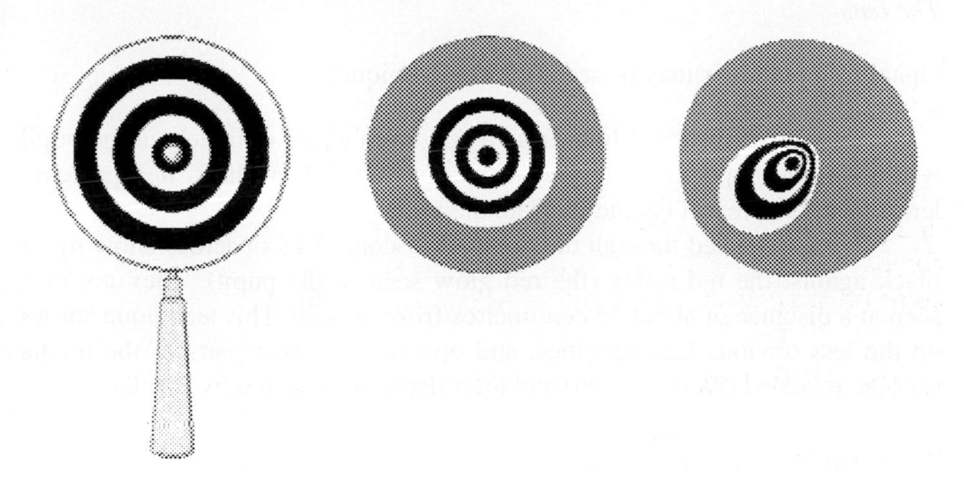

Fig 5.4 Placido's Disc. Reflection from a normal cornea. Reflection from a keratoconic cornea.

*The pupil*

-the size of the pupil on each side is compared.

-the pupil margin may be: (a) irregular if there are adhesions of the iris to the lens as in iritis (Plate 7.23 p177). The effect is more obvious if an attempt is made to dilate the pupil with a mydriatic. (b) comma shaped if the cornea is perforated with an iris prolapse (Plate 7.6 p157). (c) dilated and unreactive as in acute glaucoma and often vertically oval (Plate 7.27 p186). The figure also shows patches of iris atrophy suggestive of previous subacute attacks.

-the pupil reaction in both pupils should be equal to direct light stimulation. A bright light is required for this test. If the reaction is reduced or absent, the reaction to consensual light stimulation and to accommodation/convergence should be tested. Failure to react to direct light could be due either to inability of the pupil to react (due to iris damage or loss of motor nerve supply) or failure of that eye to receive and transmit the light stimulus to the brain (afferent defect). This may be demonstrated by the ability to react to light shone into the opposite eye (consensual reaction). For confirmation of an afferent defect see 'swinging flash light test' (p420).

*The lens*

Opacities in the lens may be seen by two techniques:

1. a focused beam of light from a pen torch shone obliquely at the pupil will show the pupil as being grey or white rather than black, if significant lens opacity is present behind it (Plate 6.2 p95).
2. when viewed through the ophthalmoscope, lens opacities show up as black against the red reflex (the red glow seen in the pupil). They are best seen at a distance of about 15 centimetres from the eye. This technique shows up the less obvious lens opacities, and opacities in other parts of the media are also revealed (Plate 6.3 p96) and their depth is indicated by parallax.

*The vitreous*

Pathological changes in the eye may cause the vitreous gel to separate into protein condensations which float in the fluid from which they have been separated, giving the appearance of opacities (dots, blobs, hairs, irregular ring) which can be seen by the patient, especially against a uniform light background. They can also be seen by an observer using an ophthalmoscope (p501). These opacities are usually seen to float on movement of the eyes and are often referred to as floaters or '*muscae volitantes*' p444. Visual appearances noticeable by the patient are known as *entoptic phenomena*. Diffuse opacities in the vitreous are difficult to identify with a direct ophthalmoscope especially if they are dense. Difficulty in viewing the fundus in the presence of a normal anterior segment may be due to a vitreous haemorrhage or marked inflammation. Indirect ophthalmoscopy is more effective in this respect.

*The retina, choroid and optic nerve*

These can only be seen with an ophthalmoscope. For methods of ophthalmoscopy see p256. Details of the findings are discussed in the relevant sections.

*Measurement of the intraocular pressure (tonometry)*

The intraocular pressure may be estimated crudely with no additional equipment by palpation, but much more accurately by instrumental tonometry. The methods are described on p555.

*Palpation* involves asking the patient to look downwards without closing his eyes. The examiner supports both hands with the middle fingers resting on the patient's forehead and gently indents the eye from above alternately with each index finger while the other steadies the eye. Even experienced observers may find difficulty unless it is markedly raised or asymmetric between the two eyes.

Simple portable indentation (Schiötz) tonometers (p556, Fig. 31.2 p556) can be useful in some circumstances, but measurement can be made much more accurately by applanation tonometry by the Goldmann technique using a portable instrument (Perkins) (Figs. 31.6 p559, 31.4, 31.5 p558). Tonometric methods are described fully in Chapter 31 p555 et seq.

# CHAPTER 6

# PAINLESS IMPAIRMENT OF VISION (IN THE WHITE EYE)

Impairment of vision, usually without pain in the eye, may be the result of:

A. - Refractive error (pp62, 91)
B. - Opacities in the media (p92)
C. - Diseases of the retina (p111) and choroid (p129)
D. - Primary open angle glaucoma (p129)
E. - Diseases of the optic nerve (p145)
F. - Diseases of the intracranial pathways (p149)

There may be pain elsewhere as, for example, in the scalp in temporal arteritis.

## A. Refractive error (p62)

If refractive error is the only cause of the patient's reduced unaided vision then this should improve to normal levels when the appropriate spectacle correction is prescribed. A useful and simple clinical test for refractive error is the *pinhole test.* If, when the patient looks through a 1.5mm or 2.0mm pinhole the vision improves, then refractive error is responsible for at least part of the reduced unaided vision. The pinhole, however, can adequately correct only approximately 4 to 5 dioptres of ametropia and therefore is only a guide to the presence of a refractive error. Other causes of impaired vision may coexist. In addition, *irregular astigmatism* arising from conditions in which there is corneal distortion as in keratoconus or changes in refractive error in the lens of the eye due to incipient cataract, may prove impossible to correct fully with either a pinhole or a spectacle prescription. Contact lenses, even as a diagnostic trial only, will indicate the best corrected visual acuity in corneal cases. Certain disease processes may be associated with changes of refractive error, e.g. lens opacities (increased myopia in nuclear sclerosis), elevation of the retina at the posterior pole (increased hypermetropia in central serous chorioretinopathy) and fluctuating blood sugar in diabetic patients (osmotic variation in lens hydration and thus in the curvature of its surfaces).

## B. Opacities in the media

*Corneal opacity*

*Diagnosis*

Discrete corneal opacities are seen usually as white areas in the cornea on direct slit lamp examination. A general loss of corneal clarity and optical brilliance or shine may occur if there is diffuse disease such as epithelial oedema or widespread corneal surface irregularity. Corneal opacities show up as dark shadows against the red reflex of the ophthalmoscope or retinoscope, or with slit lamp retroillumination, as do all opacities in the media. Any central opacity causing an irregularity of the corneal surface will be revealed by distortion of the Placido disc rings, (Fig. 5.4 p87) keratometer mires or the videokeratoscopic image.

*Congenital corneal opacities* may result from maternal rubella. Corneal oedema may be present at birth as the result of trauma to the endothelium of the cornea during a difficult delivery. Slit lamp examination may reveal splits in Descemet's membrane (Haab's striae). Infantile glaucoma (p572) may present with corneal oedema in early infancy and characteristically the affected eye is enlarged (buphthalmos).

*Corneal dystrophies* are uncommon conditions which result in reduced corneal clarity. They are usually bilateral and, as the majority are autosomal dominant conditions, enquiry may elicit a family history (p483).

*Band keratopathy* is the deposition of calcium salts at the level of Bowman's layer as a band within the interpalpebral aperture. Commonly it is idiopathic when the grey calcium layer tends to be diffuse but a roughened corneal surface can also result, especially in eyes which have been traumatised or have suffered chronic inflammation. It is seen also in hypercalcaemia from whatever cause, e.g. hyperparathyroidism, sensitivity to Vitamin D (as in sarcoidosis) or chronic renal failure (p487, Plate 6.1 p93, LCW 3.26 p49).

*Corneal scarring* may follow previous corneal inflammation (p165) or trauma - including surgical trauma e.g. refractive surgical procedures such as radial keratotomy (RK) (p491), photorefractive keratectomy (PRK) (p492) or LASIK (p492).

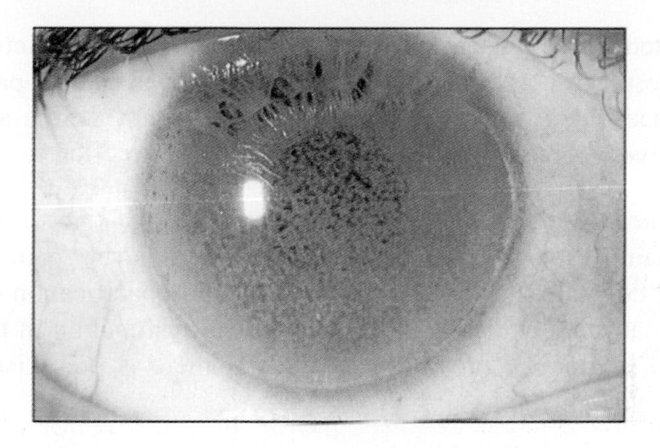

Plate 6.1 Band opacity of the cornea.

Corneal scarring may however, be an incidental finding when the patient presents with an additional cause of loss of acuity, e.g. cataract or refractive error. The original inflammatory episode or trauma may have been long-forgotten. Former interstitial keratitis due to, for example, congenital syphilis or viral infection may be revealed in this way.

*Anaesthesia of the cornea* may result from damage to the ophthalmic division of the trigeminal nerve, or specifically its nasociliary branch, by tumour, trauma or involvement by the herpes zoster virus (p171). Corneal inflammation due to the herpes simplex virus (p168) is particularly liable to cause corneal anaesthesia. The epithelium of an anaesthetic cornea is often hazy and roughened, thereby causing reduced vision, and is vulnerable to ulceration. This may be painless but is usually associated with hyperaemia (neuroparalytic keratitis) (p174).

*Perforation* (non-traumatic) is painful in most cases but gross thinning of the cornea at the limbus may occur in collagen vascular disease, especially in rheumatoid arthritis ('melting' conditions). This thinning results from loss of collagen and may occur in a white, quiet eye, especially if the patient is receiving corticosteroids which impair normal corneal repair mechanisms.

*Corneal oedema* may be the result of an endothelial dystrophy such as Fuch's dystrophy (p485). It may also follow damage to the endothelium in iridocyclitis indicated by the presence of keratic precipitates (p178). Active keratitis, e.g. as the result of herpes simplex or herpes zoster, is usually painful. If the cornea is anaesthetic from previous episodes, pain may not be a feature but usually the eye is red.

*Chronic glaucoma* (p129, p575) is usually painless and is not associated with signs of inflammation, unlike acute glaucoma which is very painful and characterised by corneal oedema. Corneal oedema and opacification occur rarely in chronic glaucoma as the result of endothelial dysfunction in the presence of increased intraocular pressure especially if there is pre-existing corneal endothelial dystrophy.

## Opacity of the lens

The practical anatomy and physiology of the lens is described on p35.

### Cataract

Cataract has been defined as 'any opacity in the crystalline lens' but the use of the term implies that the opacities are clinically significant. When less advanced, the term 'lens opacity' is preferable and less alarming for the patient.

*Symptoms.* The opacity may reduce the visual acuity directly or produce an uneven change in the refractive index of the lens causing irregular astigmatism with light scatter and sometimes *monocular diplopia* not correctable by spectacles. Patients may benefit from wearing a peaked cap in bright sunlight in order to minimize disability from glare. These symptoms may precede any obvious opacity and a progressive change in refractive error (typically an increase in myopia in the case of nuclear sclerosis which gradually advances to become a brownish nuclear cataract) may suggest incipient cataract. Colour vision sensitivity and discrimination may be reduced.

*Signs.* A blurred fundus view with the direct ophthalmoscope is a useful indicator of clinically significant cataract. Greyish white opacities can be seen in oblique illumination by a torch (Plate 6.2 p95) or by a slit lamp in optical section (Plate 20.1 p426). They are visible as dark areas in silhouette against the red reflex in the pupillary area when viewed with the ophthalmoscope at a distance of about 15 cm. These opacities are also revealed during retinoscopy (Plate 6.3 p96, LCW 4.2 p58).

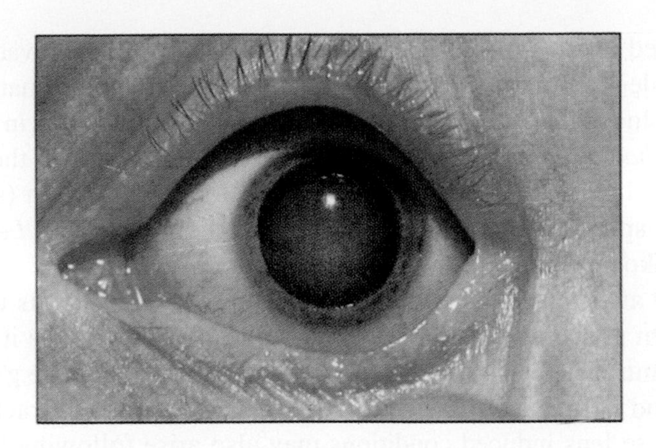

Plate 6.2 Lens opacity as seen by general illumination.

*Types of cataract.* Cataract may be classified on the basis of *stage of development* (e.g. intumescent, mature, hypermature), *anatomical position of the opacity* (e.g. cortical, nuclear, subcapsular), or *aetiology* (diabetic, traumatic)

In clinical practice all 3 classifications tend to be used when describing a cataract e.g. *marked corticosteroid-induced posterior subcapsular opacities.*

1. *Intumescent:* When lens fibres degenerate to form a cataract there is breakdown of lens proteins into smaller molecules so that water passes into the lens which then swells. Such a lens may become so large that it causes severe shallowing of the anterior chamber and may close off thc angle of the anterior chamber causing *'phakomorphic' glaucoma.*

2. *Mature:* A cataract is described as mature when the whole lens is opaque and no clear cortex is visible with a slit lamp. The mature cataract appears white and although the visual acuity is reduced to light perception, the ability to identify the direction of the light source is retained *(normal projection to light).* (LCW 4.4 p58) This important observation must always be recorded in these cases, as it gives some indication of the visual prognosis of a possible cataract extraction.

3. *Hypermature:* In this stage the lens capsule in a mature cataract becomes permeable to liquefied lens matter which then leaks out into the aqueous. The cataract shrinks leaving a small brownish nucleus surrounded

by a wrinkled capsule *(Morgagnian cataract).* Macrophages invade the aqueous and the lens from the systemic circulation, engulf the lens matter and subsequently block the trabecular meshwork resulting in a rise in intraocular pressure *('phacolytic' glaucoma).* Sometimes the cells sink to the bottom of the anterior chamber to produce the appearance of a hypopyon (sterile). The degree and speed of the pressure rise is also very variable. *Lens induced uveitis* (phakoanaphylactic uveitis), due to the development of antibodies to lens protein after lens matter leaks into the anterior chamber, is to be distinguished from phacolytic glaucoma, although it may complicate it. After control of the intraocular pressure with osmotic diuretic agents e.g. diamox or mannitol and of the uveitis with corticosteroids, a lens extraction is performed. These lens induced conditions may also arise following lens capsule rupture due to trauma or as a complication of surgery.

*Causes of the opacity (other than systemic disease and other ocular disease)*

*Congenital cataract.* Children may be born with partial or complete opacification of the lens which may be hereditary or the result of intrauterine infection, e.g. rubella. Sometimes only the nucleus of the lens is affected *(nuclear cataract)* and the opaque centre is surrounded by normal clear cortical lens fibres. If only a few lamellae of lens fibres become opaque owing to some temporary adverse influence during pregnancy, a central opacity is seen surrounded by clear lens *(lamellar cataract)* (p428, Fig. 20.3 p428, Plate 6.3 p96, LCW 8.8 p124).

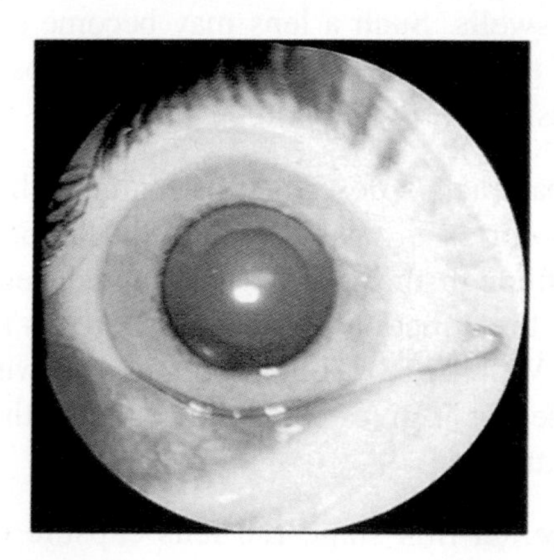

Plate 6.3 Lens opacity as seen ophthalmoscopically silhouetted against the 'red reflex' background.

Anterior and posterior polar lens opacities may be associated with persistent pupillary membrane or persistent hyaloid remnants respectively. *Posterior lenticonus* is characterized by a dense posterior polar cataract which may be fused with the posterior capsule of the lens. Removal can be difficult technically and it may be impossible to retain an intact posterior capsule. Some congenital cataracts are accompanied by other congenital ocular abnormalities, e.g. microphthalmos, and sometimes by generalised abnormalities, e.g. mongolism.

*Senile cataract.* Senile cataract is a result of the ageing process without any obvious underlying hereditary or metabolic disorder. Progressive nuclear sclerosis may result in an increased refractive power of the lens nucleus causing progressive myopia leading to a brownish nuclear cataract (LCW 4.3 p58). In the early stages vision can be improved with new spectacles. However, this progression may be very asymmetric leading to non-tolerance of any spectacles prescribed because of anisometropia (causing unequal right and left image sizes). Eventually the opacity causes progressive deterioration of visual acuity. Cortical opacities affect the visual acuity mainly when they involve the visual axis but the less common axially placed subcapsular opacities often affect visual acuity at an early stage. (LCW 4.1 p58) Cortical opacities and posterior subcapsular opacities can cause significant visual disability because of the associated glare even in the presence of good visual acuity as tested in clinic conditions. For this reason, patients with significant cataract will often benefit by wearing dark glasses or by shading their eyes from direct light with a peaked cap.

*Traumatic cataract* (pp153, 158)

*Radiation cataract.* The lens may be damaged by radiation by either the longer wavelengths of light and heat (red and infrared) or by ionising radiation (X- and gamma rays). Glass blowers' and furnace workers' cataracts result from working in front of furnaces. True exfoliation of the lens capsule may occur. Cataracts due to X-rays are usually the result of radiotherapy to malignancies of the eye or orbit. This applies also to the more recent proton beam treatment used for posteriorly placed ocular malignancies. Despite every effort to shield the lens from any extraneous radiation, the site of the tumour sometimes makes this impossible.

*Cataract due to corticosteroids.* Increased levels of corticosteroids in the eye also increase the leak of potassium ions from the lens and are associated with cataract formation. This is typically posterior subcapsular in location. The common cause is the treatment of a wide variety of diseases with systemic steroids, although raised levels of corticosteroids are found also in tumours of the adrenal cortex. In addition, prolonged use of topical corticosteroid eye-drops can be associated with lens opacity, although this may be due also to the underlying condition being treated.

**Cataracts associated with systemic disease**

*Diabetic cataract*

Diabetes may affect the lens in three ways (p312):

1. *osmotic effects:* fluctuations in the blood sugar level may cause variation of lens thickness and curvature from osmotic influences. This alters the state of refraction and for this reason spectacles are only prescribed when diabetes is stabilised. Frequent changes of refractive error are sometimes a useful clue to hitherto undiagnosed diabetes. The lens absorbs water and the eye tends to become myopic in the presence of a high blood sugar because the normal hexokinase controlled pathway for glucose metabolism in the lens becomes saturated and the excess glucose is converted to sorbitol and fructose. These molecules build up in the lens causing water to enter the lens from the aqueous by osmosis. Once blood sugar is normalised the refraction will return almost to normal in most cases.

2. *acute juvenile diabetic cataract:* some juvenile diabetics develop an extremely high blood sugar and this may be associated with the formation of vacuoles and snowflake opacities. Rapid control of the blood sugar may sometimes be rewarded by the reversal of the lens changes.

3. *early onset of senile cataract:* diabetics tend to develop senile cataract at an earlier age than others so that diabetes must be excluded in all patients with lens opacities.

*Galactosaemia* is a recessively inherited deficiency of either the enzyme galactose-1-phosphatase uridyl transferase or galactokinase. The resulting build up of galactitol in the lens leads to cataract formation. Prompt diagnosis is essential followed by the withdrawal from the diet of milk and any foods containing lactose which forms galactose on digestion.

*Hypoparathyroidism* is a relatively uncommon condition, one cause of which is inadvertent and unavoidable removal of the parathyroid glands during thyroidectomy. The resultant low serum calcium results in increased excitability of peripheral nerves and, in severe cases, tetany. Deficiency of calcium ions in the lens results in a leak of potassium ions and progressive lens opacification.

*Myotonia dystrophica* is an uncommon condition characterised by weakness and wasting and slow relaxation of muscles. It is classically demonstrated by the patient's inability to relax his grip after a handshake. Other findings include cataract, loss of facial expression due to atrophy of the facial muscles, premature balding and in male patients testicular atrophy.

*Eczema.* Severe atopic eczema is associated in a few patients with the development of shield-shaped anterior subcapsular cataract.

**Cataract formation associated with other ocular disease**

*Retinitis pigmentosa* (p520) may be associated with posterior subcapsular cataract formation. In the presence of retinal disease it can be difficult to determine the contribution of the cataract to the visual disability.

In *acute angle closure glaucoma* the rapid rise in intraocular pressure can result in the formation of superficial greyish lens opacities which may remain after the intra-ocular pressure has been brought under control (Glaucomflecken). The presence of Glaucomflecken indicates that the pressure rise was rapid and severe with significant ischaemia.

Any *prolonged uveitis* will result eventually in cataract formation presumably due to disturbance of lens metabolism, e.g. cataract changes complicating the chronic uveitis of *Still's Disease* (p359)

A *long-standing retinal detachment* will be associated usually with a low grade uveitis and cataract formation. Functionally *disorganised blind eyes,* often with low intraocular pressure due to reduced aqueous secretion, almost invariably develop cataract.

*Heterochromic cyclitis* (p580)

**The management of cataract and indications for surgery** (p495)

A brief outline of the principles of cataract management is necessary, even for those who are not going to perform cataract surgery themselves, in order that informed advice can be given to patients. The advent of the operating microscope has revolutionised cataract surgery and indeed most ophthalmic operations. (LCW 4.18 p64)

*Preventive measures*

Prevention of cataract is assisted by good control of any associated medical condition such as diabetes or uveitis. It is also prudent to avoid radiation lens damage when practicable and when this is not possible, to wear goggles to prevent the absorption of infrared and some ultraviolet wavelengths, especially when exposure is repetitive as in the case of certain occupations (glass blowers, laser workers and welders) p163. Much research on lens metabolism is in progress and medical methods of inhibiting lens opacification may result.

*Management of cataract in children*

*Congenital cataract*

*Bilateral congenital cataract.* The management of bilateral congenital cataract depends on the degree to which the lens opacity is reducing retinal image formation and the presence of other ocular malformations or systemic disease. General hypoplasia or microphthalmos may be a contraindication to any treatment because of the limitation of retinal function. Electrodiagnostic tests are of great importance therefore in assessing infants with congenital cataract.

*Minor opacity:* any other associated ocular condition, e.g. strabismic amblyopia or significant refractive error should be treated if possible.

*Moderate opacity:* in the case of denser central opacity the visual handicap is more difficult to assess. Some children benefit from dilatation of the pupil with a mydriatic or even a sector iridectomy *(optical iridectomy)* was sometimes used in the past as this allows the light rays to pass through the unaffected clear lens cortex.

*Severe opacity:* if bilateral cataracts are causing significant visual loss, the longer the child is deprived of clear foveal vision the less chance there is of the visual system developing normally *(deprivation amblyopia)* (p224), and vision will remain poor even if the cataract is removed successfully at a later date. The development of pendular nystagmus is an indication of appreciable visual impairment. As neonate and infant lens matter is usually soft, the treatment is *anterior capsulectomy (discission) followed by irrigation and aspiration* of the lens matter with great care to clean the posterior capsule thoroughly while avoiding any damage to it. Both eyes are operated on during the first few weeks of life. The resulting aphakic refractive error must be corrected as soon as possible, either with spectacles or preferably constant wear contact lenses (carefully monitored), as uncorrected aphakia will also cause deprivation amblyopia. The lower age limit for intraocular lens implantation has been reduced as longer term experience with these lenses has been gained and some surgeons implant lenses in infants in the hope of avoiding the numerous problems that spectacles and contact lenses engender in this age group. Sometimes these children are predisposed to glaucoma developmentally and this may increase the risk of secondary glaucoma following surgery. In any event, these children require careful follow-up, preferably in a special centre, with repeated refraction (if necessary under anaesthetic). The treatment of any strabismic amblyopia, a frequent complication of congenital cataract, by occlusion of the better eye for prescribed periods is essential.

*Uniocular congenital cataract.* The presence of a moderately dense congenital cataract even in an otherwise normal eye carries a very poor prognosis for vision in that eye when the fellow eye is normal. The child will always prefer to use his normal eye and the cataractous eye will become amblyopic despite early surgery and correction of the subsequent refractive error with a contact or intraocular lens. Operation may nevertheless be indicated in some patients at the discretion of the ophthalmologist. It could, for example, be argued that any improvement in the cataractous eye is worthwhile since there is no guarantee that the better eye will remain so throughout life (accident, infection etc).

*Acquired cataract*

*Bilateral acquired cataract in children.* The development of cataract in children requires investigation for underlying metabolic disorders, e.g. galactosaemia or underlying ocular disorder, e.g. uveitis. The principles of management are as for bilateral congenital cataract.

*Uniocular acquired cataract in children.* The uniocular acquired cataract is frequently due to trauma. Provided that the rest of the eye has not been extensively damaged, the early removal of the cataract and correction of the refractive error with a contact lens or intraocular lens implant (if feasible technically) gives the best chance of retaining vision in that eye. In general the older the child at the time of injury, and therefore the more mature the visual system, the greater the chance of attaining binocular vision and the smaller the chance of amblyopia supervening. All children under the age of seven will need close ophthalmic supervision and patching of the better eye to reverse any amblyopia.

*Management of cataract in adults*

The development of cataract in the adult patient is an indication for screening for any general predisposing condition, especially diabetes. In the management of the cataract it is important to determine the degree to which the cataract itself is the cause of visual loss, rather than corneal, retinal or optic nerve disease, as these may co-exist.

*Bilateral cataract.* Cataract extraction is usually delayed until the visual loss is such that the patient's way of life is affected. This is a relative indication and will vary considerably from patient to patient. The type of cataract is important because, as stated above, a posterior subcapsular cataract may be associated with marked disability from glare and light scatter even though distance visual acuity under test conditions is relatively good. It is important to refract the patient carefully and to note both the distance and the near vision, because in central lens sclerosis reading vision is frequently disproportionately better than distance vision would indicate. In making a recommendation for cataract extraction it is essential to know the patient's way of life and visual requirements. Testing the visual acuity under glare conditions will be helpful. For example, a professional driver will need relatively good distance acuity whereas an elderly patient who goes out very little but can read, perhaps even unaided because of nuclear sclerosis-induced myopia, may feel deprived following operation even though it is technically successful for distance.

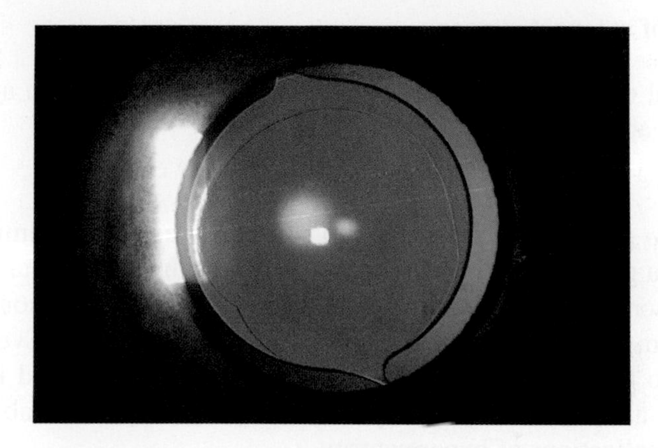

Plate 6.4 Posterior chamber lens implant in the capsular 'bag'.

*Unilateral cataract* may require extraction if the patient has a desire for or occupational requirement for, binocular vision, or if the cataract is becoming hypermature. In such cases contact lenses or a plastic lens implant will allow reasonable equalisation of image size and the possibility of binocular vision. Intraocular lens implants are, ideally, placed within the capsular bag in the posterior chamber (Plate 6.4 p103). Should circumstances prevent this an anterior chamber implant supported in the angle of the anterior chamber is an alternative (Plate 6.5 p103)

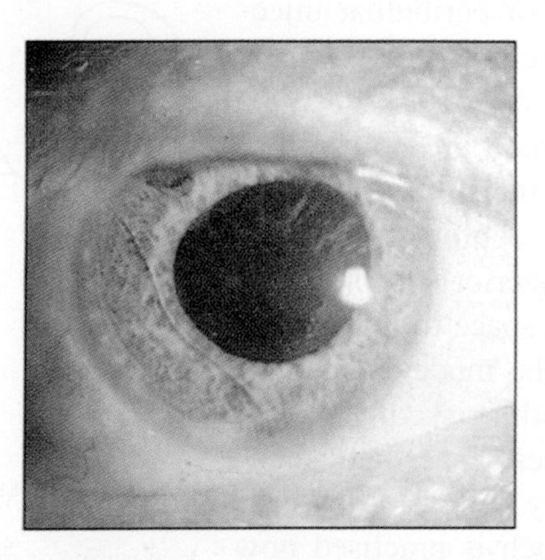

Plate 6.5 Anterior chamber lens implant (Choyce).

## Principles of cataract surgery

The removal of the opaque lens is termed a cataract extraction and this renders the eye aphakic.

*Pre-operative assessment*

*Projection to light.* In the presence of a dense cataract with a limited view of the fundus, a guide to the integrity of the visual pathways is obtained by testing 'projection to light'. The patient is asked to point to the source of light when a beam is shone on the eye from different directions. Even when the projection to light is poor, cataract extraction may be indicated in some circumstances, and very exceptionally even if light perception is absent, to prevent the complications of hypermaturity.

Prior to dilating the pupils it is important to assess pupil reflexes. A relative afferent pupillary defect is indicative of optic nerve dysfunction or extensive retinal disease.

*Anaesthesia for cataract operations*

Most young patients will require general anaesthesia. However, after achieving akinesia of the orbicularis oculi (p20), local anaesthesia using topical anaesthetic eyedrops with or without sub-Tenon's capsule, retrobulbar or peribulbar injections have advantages in many cases (Figs. 6.6 p104, 6.7, 6.8 p105). For sub-Tenon injections which are now sometimes preferred, an incision is made in the conjunctiva and a blunt cannula is positioned in the posterior subconjunctival and sub-Tenon's space to inject fluid around the globe. The moderately blunt cannula minimises the risk of perforation of the globe. These anaesthetic techniques are especially suited to day-case cataract surgery which is practised now in many cases.

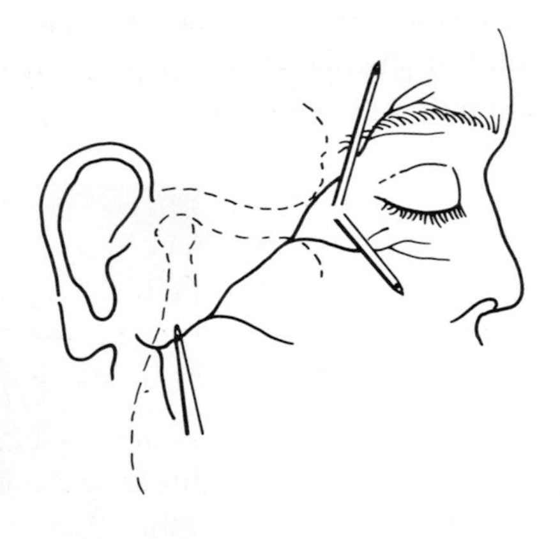

Fig. 6.6 Akinesia of orbicularis oculi. (see also p20)
1. at ramus of mandible
2. at lateral part of orbicularis oculi

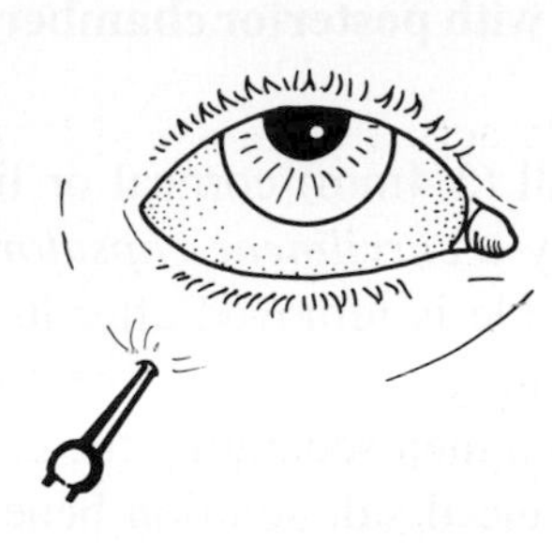

Fig. 6.7 Retrobulbar injection (1).

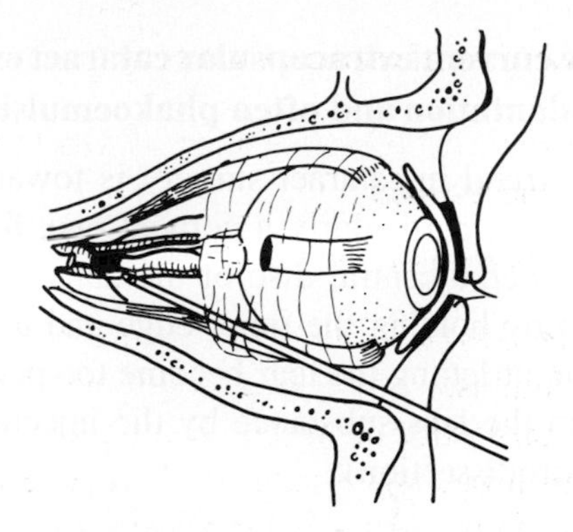

Fig. 6.8 Retrobulbar injection (2).

*Surgical techniques*

Cataract extraction is carried out using an operating microscope (LCW 4.18 p64, 4.6 p59) either by the *extracapsular* method in which all the lens capsule remains intact except a central anterior capsular disc, or by the *intracapsular* method by removal of the entire lens including its capsule. Certain complicated cataracts may require *lensectomy* (p504).

The great advantage of the extracapsular technique is that the posterior capsule and the zonule not only support the vitreous but the capsular bag will usually contain the posterior chamber lens implant.(LCW 4.5 p59) While the main disadvantage of the extracapsular method was originally the opacification of the posterior part of the capsule which then required surgical discission or "needling" (incision with a needle type of knife), this is now treated with a brief high energy pulse of the Nd YAG laser (p76) with the patient sitting at the slit lamp microscope, although in the case of particularly thick capsules, it may still be necessary to carry out a surgical posterior capsulectomy. However, the risk of posterior capsule opacification is now rarer, particularly when using certain types of lens implant.

In the intracapsular operation (p495) there is always a risk of a capsular tear or vitreous loss. In addition, although anterior chamber implants were sometimes used, they were subject to complications and do not compare with posterior chamber implants in the capsular bag. The most highly developed technique of extracapsular extraction will here be first described but the earlier extracapsular methods and the intracapsular technique are subsequently outlined because they may have practical advantages where surgical facilities are less developed as they do not require such refined equipment (p498).

## The current extracapsular cataract extraction with posterior chamber lens implantation and often phakoemulsification

The trend in cataract surgery is towards a small (3-4mm) corneal or limbal incision, particularly a corneal one, followed by a *curvilinear capsulorhexis* in which a central disc of the anterior lens capsule is removed after its incision by holding the inner edge and tearing it with great care in a neat circle without letting the tear become too peripheral and then separating the capsule from the lens substance by the injection of balanced salt solution beneath it (hydrodissection).

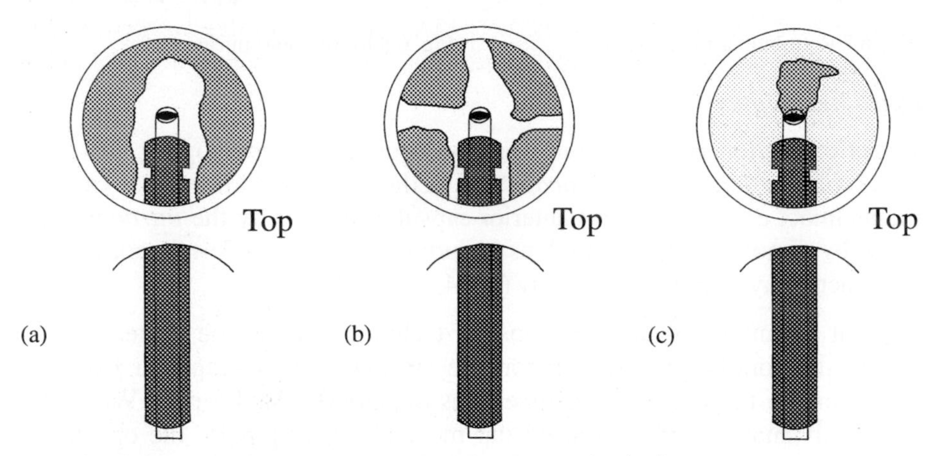


Fig. 6.9 Phakoemulsification cataract extraction (surgeon's view).

(a) A scleral pocket/tunnel is illustrated. Having removed the anterior capsule using the technique of continuous curvilinear capsulorhexis, the phakoemulsification probe is used to produce deep grooves within the lens nucleus. A corneal incision is also frequently employed.

(b) The nucleus has been split into 4 segments, which can then be removed piecemeal by continued phakoemulsification.

(c) The final piece of lens nucleus is phakoemulsified. The residual soft lens cortical material is aspirated manually or with a customised probe and the capsular bag is then ready to receive the lens implant.

In younger patients up to about 20 years of age who do not usually have a hard lens nucleus, irrigation and aspiration is effective in removing the lens matter from within the capsule.

If, however, the lens nucleus is hard it may be removed by manual expression requiring an enlarged limbal or corneal incision, but in the technique of *phakoemulsification*, which is increasingly used but more demanding, the hard nucleus is emulsified by a probe vibrating at ultrasonic frequencies incorporating both an irrigation and aspiration facility by which the emulsified nucleus is removed. (Fig 6.9 p106) The remaining soft lens matter is then aspirated using a special probe designed to avoid damage to the posterior capsule. A silicon or acrylic foldable lens implant is inserted into the capsular bag and supported there by flexible curved extensions (Figs. 6.10, 6.11 p107, Plate 6.4 p103, LCW 4.21 p64). As the very small limbal or corneal incision is designed to be self sealing, post operative astigmatism due to corneal distortion during healing is virtually eliminated. Before capsulectomy and again

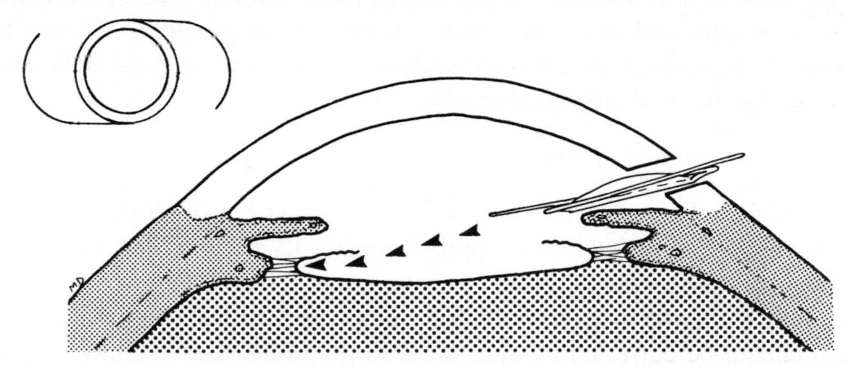

Fig. 6.10 Extracapsular cataract extraction with lens implant (saggital section - top to right) is showing the insertion of a non-folding posterior chamber lens implant into the capsular bag. A folding implant would be introduced in a folded state and allowed to unfold into the capsular bag. Plan view of implant is shown above.

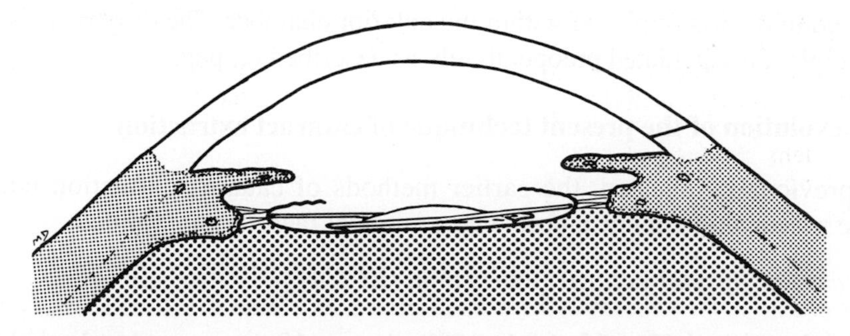

Fig. 6.11 Extracapsular cataract extraction with implant - Posterior chamber lens implant held centrally within the lens capsular bag by its flexible curved extensions.

before implant insertion, most surgeons inject visco-elastic fluid into the anterior chamber to protect the corneal endothelium and iris and make manipulations easier. This is subsequently aspirated.

*Complications of cataract extraction* (see p495 for more detail)

Cataract extraction, athough exacting, is now one of the most successful forms of surgical intervention. Possible complications however include vitreous loss due to per-operative posterior capsule rupture, haemorrhage, infection, uveitis, raised intraocular pressure, flat anterior chamber, macular oedema, retinal detachment and corneal oedema. Modern, small incision phakoemulsification techniques assisted in many cases by the introduction of visco-elastic into the anterior chamber at certain stages, provide a closed anterior chamber throughout the procedure and as a consequence reduce these problems. As can be appreciated, scrupulous attention to equipment and good surgical technique are essential for consistently good results.

*Refractive correction of aphakia*

In patients for whom an intraocular lens is not possible the high refractive hypermetropia of the aphakic eye (approximately +12 diopters) can be corrected by the use of spectacles, but the strong convex lenses required result in severe optical aberrations some of which affect the field of vision. Unilateral aphakia cannot be corrected by spectacles to give binocular vision owing to the large disparity in image size. This aniseikonia can be reduced to tolerable limits by contact lenses (pp65, 74) and virtually eliminated by an intraocular implant ideally supported within the capsular bag. If the bag has been damaged, the lens implant can be inserted between the capsule and the iris or sutured to the iris or placed within the anterior chamber. The dioptric power of the implant is calculated preoperatively as described on page .

## The evolution of the present technique of cataract extraction

As previously indicated, the earlier methods of cataract extraction may be more practical in circumstances where surgical facilities are limited.

*Basic extracapsular cataract extraction. (ECCE)*

Formerly (Figs. 6.12, 6.13, 6.14 p109) a longer 10mm corneal or limbal incision was necessary to remove the hard lens nucleus. This required closure by

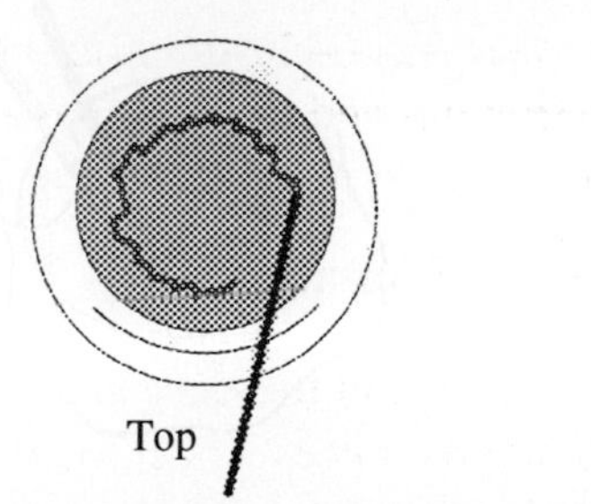


Fig. 6.12 (ECCE) A reverse slope corneal incision has been made centred on the '12 o'clock position' and the anterior capsule of the lens is incised with a bent needle (or cystotome). This has now advanced to curvilinear capsulorhexis when possible (p106)

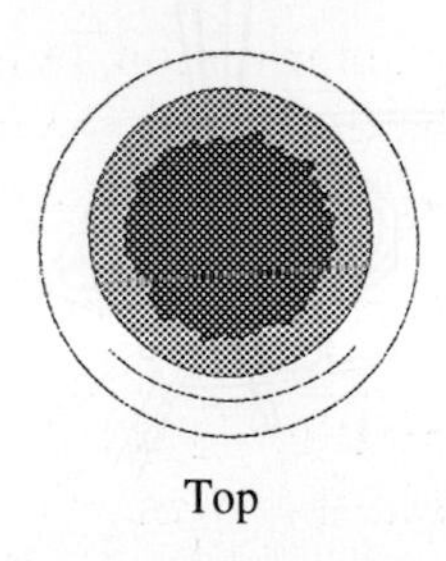


Fig. 6.13 (ECCE) The circular flap of anterior capsule has been removed and the corneal wound opened to its full extent (approximately 10mm).

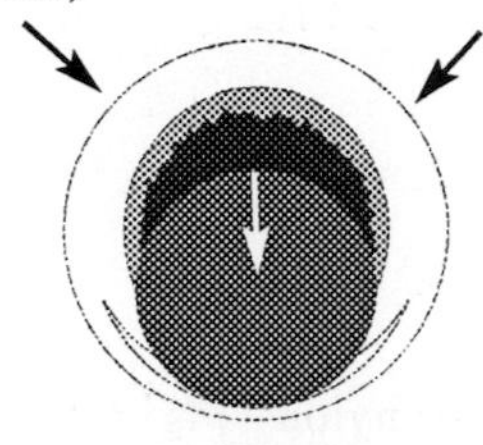


Fig. 6.14 (ECCE) The lens nucleus is expressed through the corneal section by external pressure from the '6 o'clock' position. Irrigation of the anterior chamber is helpful at this point and obviates collapse of the cornea at the final point of expression.

sutures and was accompanied by the risk of leakage of aqueous or the entry of infection subsequently. It also increased the tendency to corneal distortion during healing and thus post operative astigmatism. Strong convex spectacle lenses (10-12D) were used which, due to optical aberrations and different image sizes in unilateral aphakia (aniseikonia), were difficult for patients to wear. Contact lenses were a great improvement but for various reasons patients often found them unacceptable.

*The intracapsular cataract extraction (ICCE)*

After a wide (10-12mm) limbal or corneal incision and peripheral iridectomy the lens was held by forceps, suction probe or cryoprobe (LCW 4.20 p64) and its suspensory ligament was ruptured, sometimes after it had been weakened by the instillation of alpha-chymotrypsin. The lens in its capsule was then

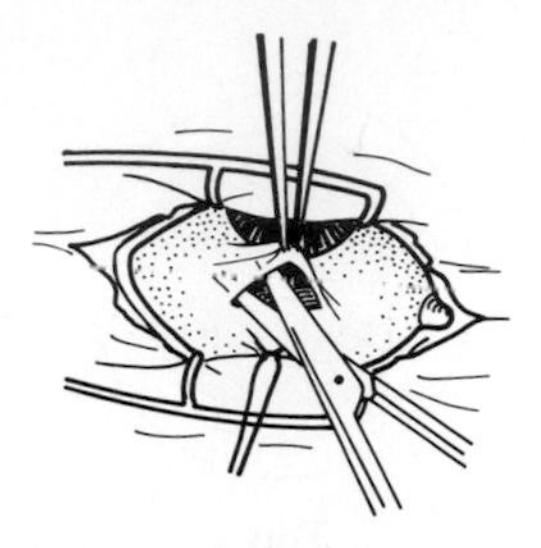

Fig. 6.15 ICCE - conjunctival and Tenon's capsule incision and separation by scissors.

Fig. 6.16 ICCE - groove made in cornea at the limbus by diamond knife or metal blade.

Fig. 6.17 ICCE - preplaced silk or nylon sutures across groove and full thickness section completed with scissors.

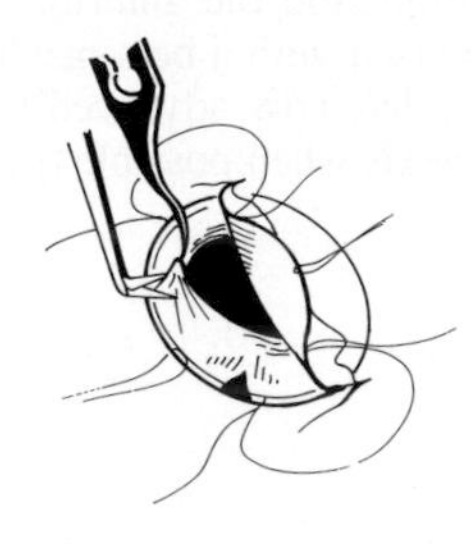

Fig. 6.18 ICCE - peripheral iridectomies. (Alpha chymotrypsin may be injected through iridectomies to weaken the lens suspensory ligament.)

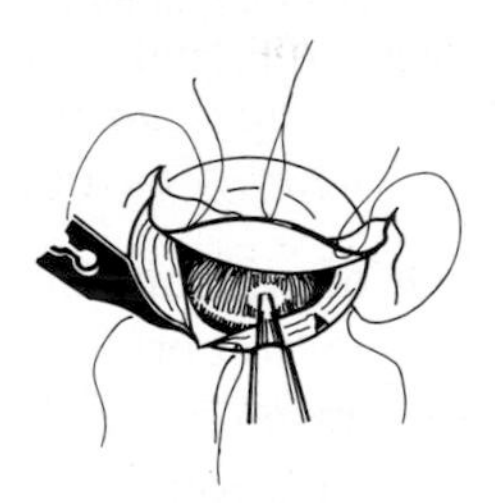

Fig. 6.19 ICCE - a cryoprobe freezes on to the lens capsule and subjacent lens fibres which is then extracted by sliding out.

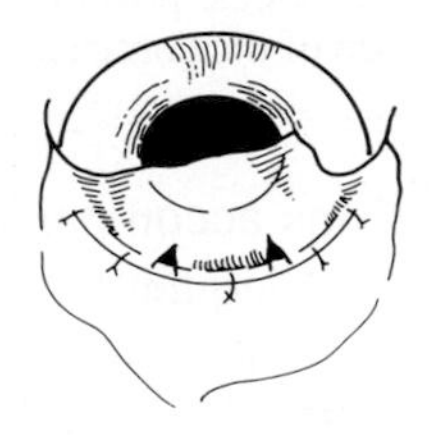

Fig. 6.20 ICCE - Closure. A monofilament 10/0 nylon continuous suture may also be used and the preplaced sutures removed.

extracted very carefully by sliding to avoid capsule rupture or vitreous loss (Figs. 6.15 - 6.20 p110, LCW 4.20 p64). Capsule forceps may be used but there is greater risk of capsule rupture. Some surgeons prefer to 'tumble' the lens by grasping the capsule with forceps nearer the inferior pole. If successful, the intracapsular method usually gave a visual result superior to the original extracapsular operation but required a higher degree of expertise.

## C. Diseases of the retina and choroid

### C (a) Retina

**Vitreous haemorrhage i.e. haemorrhage into the vitreous** (pp291, 316)

*Clinical picture*

The symptoms are sudden, complete or partial loss of vision in one eye which may be preceded by the appearance of showers of black spots in front of this eye. The fundus view is usually poor and there may be complete loss of the red reflex. If the cornea and lens appear clear on examination, then it is reasonable to attribute the poor view of the fundus to a vitreous haemorrhage. The diagnosis may be supported by the knowledge of a predisposing cause, e.g. diabetic retinopathy.

*Causes of vitreous haemorrhage*

Various ischaemic conditions of the retina can cause *new vessel formations* which ramify on the posterior surface of the detached vitreous, being delicate, these vessels are liable to recurrent haemorrhage:

-*Diabetic retinopathy* of the proliferative type (p313).
-*Retinal vasculitis* (p518) which is an inflammatory condition of the retinal vessels, usually of obscure aetology. Perivascular exudates can be seen and the condition leads to vascular stasis and occlusion, ischaemia and neovascular formation (Plate 6.21 p112). Recurrent vitreous haemorrhage due to this cause used to be called Eales' disease.
-*Sickle cell disease* (p297), in which red cells with abnormal haemoglobin aggregate to occlude vessels particularly under conditions of anoxia is practically confined to patients of African extraction.
-*Central retinal vein occlusion* (p121) in which ischaemia may cause new vessels to form. These may subsequently rupture (Plate 6.32 p121).

*Retinal detachment:* a retinal tear, as the result of vitreous traction, will sometimes proceed to a detachment. Should the tear involve a blood vessel, the resulting vitreous haemorrhage makes it difficult or impossible to see the detachment.

*Trauma*, either blunt or penetrating, to the eye can obviously result in vitreous haemorrhage (p153).

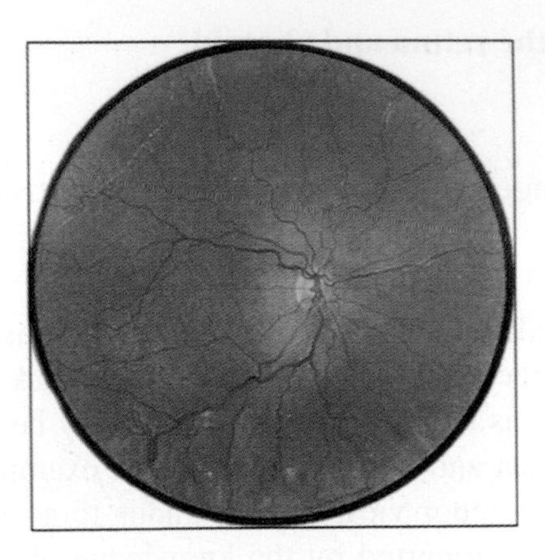

Plate 6.21 Retinal vasculitis.

*Hypertension* predisposes to retinal vascular occlusions which may lead to proliferative retinopathy and vitreous haemorrhage.

*Bleeding diatheses* cause retinal haemorrhages, which may extend into the vitreous (p297).

A retinal or choroidal tumour may occasionally present as a vitreous haemorrhage.

### *Management*

The cause is sought and treated where possible. If the retinal view is poor an ultrasound scan should be performed to determine whether a retinal detachment or other lesion is present requiring treatment. The haemorrhage may absorb to allow a fundus view in a few days but resolution of a larger haemorrhage may be slow and may take up to a year in some cases. In a non diabetic, repeated ultrasound scans are performed while waiting for the haemorrhage to clear. As the cause becomes visible its treatment can be undertaken, e.g. laser treatment for proliferative retinopathy or a retinal tear.

If the haemorrhage fails to absorb, the vitreous can be removed by the operation of *vitrectomy* using a suction cutter but the functional results of this operation depend on the nature of the underlying cause of haemorrhage. Vitrectomy is usually considered if spontaneous resolution has not occurred after six months.

**Retinal detachment with a retinal hole or tear** (syn. rhegmatogenous retinal detachment)

Other types of retinal detachment are less common (p525)

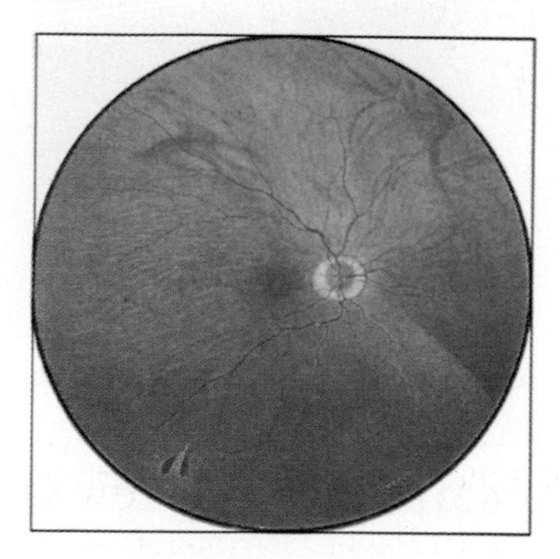

Plate 6.22 Rhegmatogenous retinal detachment which appears greyish with darkened blood vessels. Retinal tear below.

*Symptoms*

The classical symptoms of impending retinal detachment are sensations of flashing lights accompanied by black floating dots and streaks, followed by the appearance of a blur in the field of vision encroaching from the periphery, described as a 'shadow' or 'curtain'. When the detachment crosses the macular region, the field loss progresses to the sudden loss of central visual acuity. The time interval is usually a few days or weeks but it may be longer when the detachment is below. The flashing lights are due to traction of the vitreous on the peripheral retina because the retina has no pain fibres and responds to abnormal stimulation by giving a sensation of light. The black spots are due to haemorrhage into the vitreous from the retinal tear. In the case of a small retinal hole or holes the onset may be insidious with gradual loss of peripheral visual field; very gradual onset lower retinal detachments may even be discovered accidentally when unsuspected.

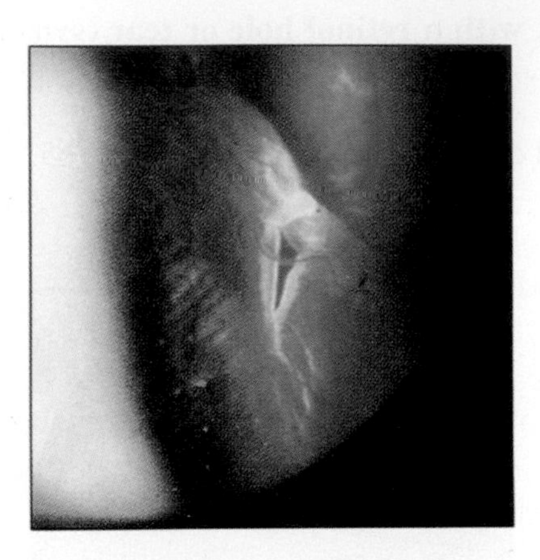

Plate 6.23 Retinal tear and detachment with retinal arteriole bridging across tear.

*Signs* (Plates 6.22 p113, 6.23 p114, 6.24 p115, LCW 5.19, 5.20 p80)

The detached area of retina appears darker and greyer than the surrounding fundus and the vessels on its surface are darker and more tortuous. The detached retina is elevated and often mobile. This can be detected with practice even when using the monocular ophthalmoscope. The elevation can be shown by the different lens required in the ophthalmoscope to focus on the detached retina as compared with the normal retina. The retinal tear appears as a red area where the choroid contrasts sharply with the greyish detached retina. NB. The detached area of retina is in the opposite direction to the field loss, e.g. Loss of upper field of vision indicates a *lower* detachment.

*Causes*

Some retinal tears follow a contusion or perforation of the globe but most tears are due to vitreous traction. In either case the tear may proceed to a retinal detachment as the fluid in the vitreous passes into the potential space between the neural part of the retina and the pigment epithelium allowing the inner neural layer of the retina to fall away from the retinal pigment epithelium which remains attached to the choroid. In myopic eyes with lattice degeneration of the retina (p520, Plate 6.24 p115), there are multiple points of adhesion between the retina and vitreous. As the vitreous fluid detaches the neural layers of the retina from the pigment epithelium these adhesions may cause multiple small round retinal holes on the edge of the lattice. Retinal tears or holes may however seal themselves spontaneously so that 'flat holes' are sometimes found on routine examination. Fibrosis and contraction of the posterior surface of the vitreous may cause a traction detachment, commonly seen in diabetic retinopathy. The detached retinal layers are separated from the choroid from which they receive much of their oxygen and nutrition, but if the retina is replaced it will function again. The sooner the retina is replaced the better the chance of good function. *A retinal detachment is an 'emergency' when the macular area has not yet detached but is threatened.* because although greatly improved vision may result from the treatment of even total detachments of the retina, the chances of good visual acuity are much reduced if the macula has become separated.

*Differential diagnosis of rhegmatogenous retinal detachment*

The most important conditions to be distinguished from rhegmatogenous retinal detachment are malignant melanoma of the choroid (p508), disciform degeneration of the retina (p529) and retinoschisis (p526). The basis of distinction from these three conditions is the presence of a retinal hole supported by certain other signs. The appearance of a red glow in the pupil when the eye is transilluminated in a darkroom by a light in contact with the lids or sclera (which is seen in rhegmatogenous retinal detachment) may be absent from the 'solid detachment' of malignant melanoma. (See Plates 28.1 p508, 28.2, 28.3 p509, 28.4 p512). Some subretinal fluid may also be found adjacent to a malignant melanoma and new vessels and haemorrhages may be

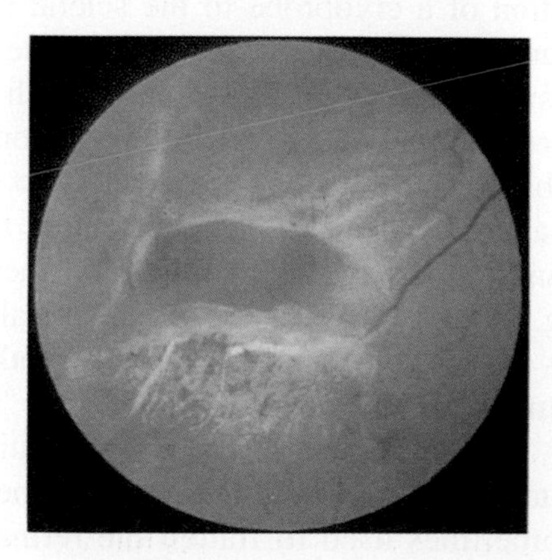

Plate 6.24 Retinal tear in area of detached retina showing lattice degeneration.

seen on its surface, which may vary from slaty grey to chocolate brown. In the early stages a melanoma may appear as a circumscribed dark 'mound' rising from a normal fundus background but it is frequently not suspected until it is well advanced when it affects central vision. Disciform degeneration of the retina may give rise to diagnostic problems but occurs typically in the macular region, the central areas of both retinae showing degenerative changes and the darkish raised central area is typically surrounded by haemorrhages. Ultrasonography and fluorescein angiography show suggestive appearances in malignant melanoma. Retinoschisis tends to be symmetrical and peripheral and, while often shallow, may form a prominent cyst with a very thin inner wall which may occasionally develop holes (p526).

*Treatment of rhegmatogenous retinal detachment*

Every effort must be made to prevent subretinal fluid from involving the macular area, because if it does, the visual acuity even after successful replacement rarely exceeds 6/18. As gravity plays a part in the separation of the retinal layers the patient's head should be appropriately placed so that the retinal hole is at the most dependent part of the eye. This will help prevent extension of the detachment and in some cases the retina will settle back. A careful fundus drawing is made, followed by the planning of surgical treatment which is carried out as expeditiously as possible. The operation primarily aims to seal the retinal hole, usually by a combination of scleral buckling to indent the choroid, with the retinal pigment epithelium adherent to it, towards the detached portion of the retina at the site of the tear, then drainage of subretinal fluid and the application of a cryoprobe to the scleral surface over the tear. This causes a 'cold burn' which results in adhesion of the layers when brought into apposition (Figs. 6.25, 6.26 p117). In cases with multiple peripheral holes, a ridge is formed internally by an encircling silicone strap, under which plastic sponge ‘plombs’ may be placed appropriately to increase the indentation locally, to help apposition of the detached retina to the choroid in the region of any holes and to increase the likelihood of their adhesion with closure of the holes. (Fig. 6.27 p117) When there is practically no subretinal fluid the area around a 'flat' tear or hole can be sealed by cryotherapy or laser alone. (Fig. 6.28 p117). In more complicated detachments an "internal" approach is used with vitrectomy, internal drainage of subretinal fluid through the retinal hole and the internal tamponade of the retina with air, inert gas or silicone oil. "Heavy liquid" is sometimes used to flatten the retina temporarily before inserting the longer term tamponade.

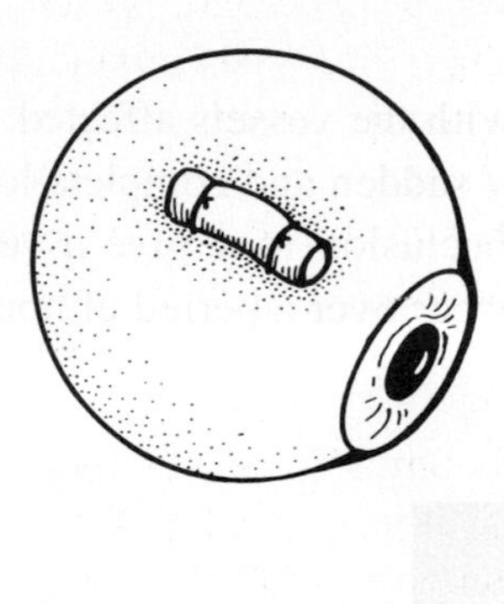

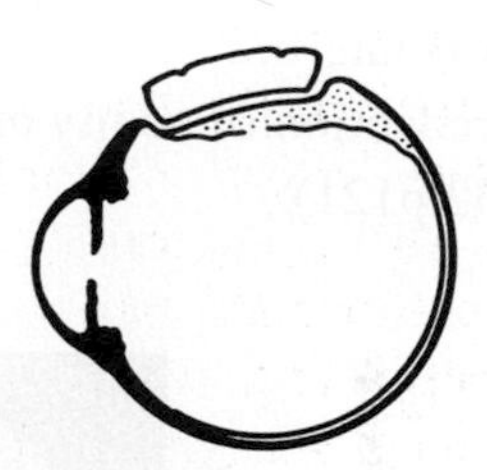

Fig. 6.25 Replacement of detached retina by radial plomb. Margins of retinal tear will be in apposition to the pigment epithelium and choroid when the stitches around the plomb are tightened.

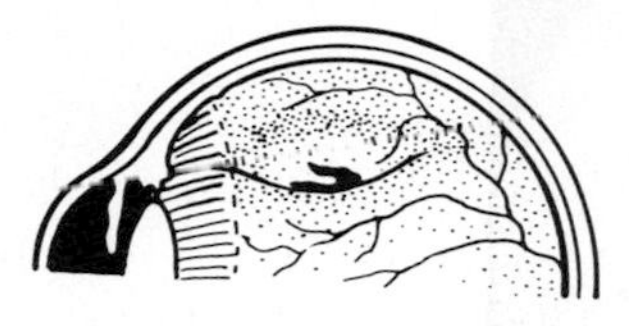

Fig. 6.26 Replacement of detached retina by local indentation with a radial plomb sutured to the sclera directly over the position of the retinal tear. Depending on the nature of the tear it may be placed radially or circumferentially.

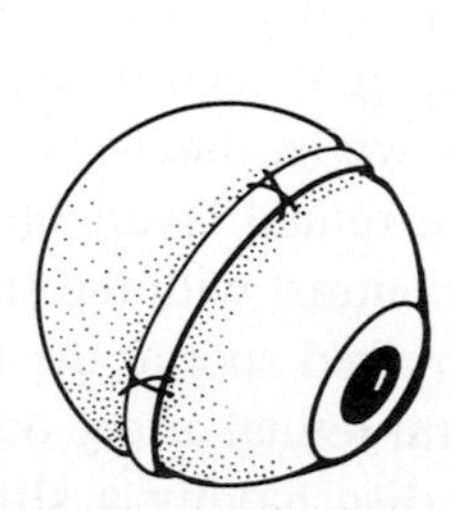

Fig. 6.27 Replacement of detached retina, indentation by encirclement.

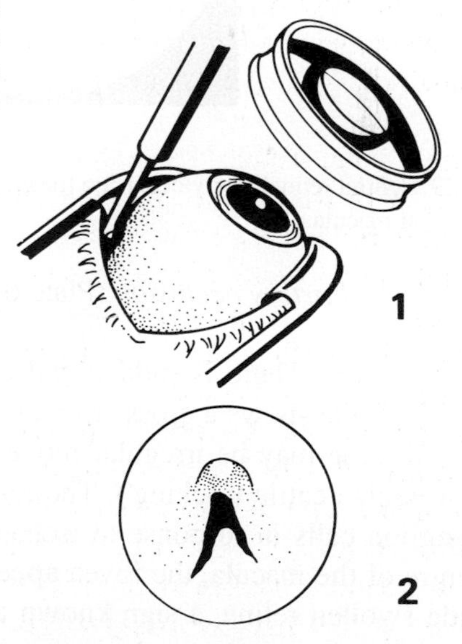


Fig. 6.28 Replacement of detached retina 1.prophylactic cryotherapy 2. fundus view of cryoprobe after drainage of subretinal fluid, or initially in the case of a 'flat tear'.

## Vascular disease of the retina

The nature and degree of visual impairment vary with the vessels affected. If the central retinal artery is occluded there is usually sudden and complete loss of vision (Plate 6.29 p118). Central retinal vein occlusion may give a very variable reduction of acuity and its onset tends to occur over a period of hours (Plate 6.32 p121).

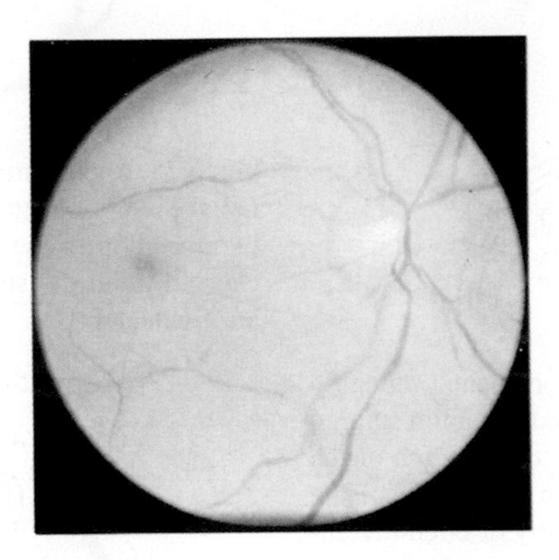

Plate 6.29 Central retinal artery occlusion (recent) oedema and necrosis and 'cherry red spot' at macula.

*Central retinal artery occlusion* (Plate 6.29 p118, LCW 5.4 p75)

*Clinical picture.* There is sudden and complete loss of vision. The pupil on the affected side shows a gross afferent defect. The retinal arterioles are attenuated and there may be irregular movement of the segmented blood column in the vessels ('cattle trucking'). The retina appears white, due to swelling of the ganglion cells in response to ischaemia. As the retinal layers are thin at the centre of the macula, the fovea appears red in contrast with the surrounding pale swollen retina, a sign known as the 'cherry red spot at the macula'. Finally, after about two months in a complete central retinal artery occlusion, the changes of optic atrophy occur with a pale disc having a slightly ill defined edge, the walls of the empty vessels appearing as white strands (Plate 6.30 p119).

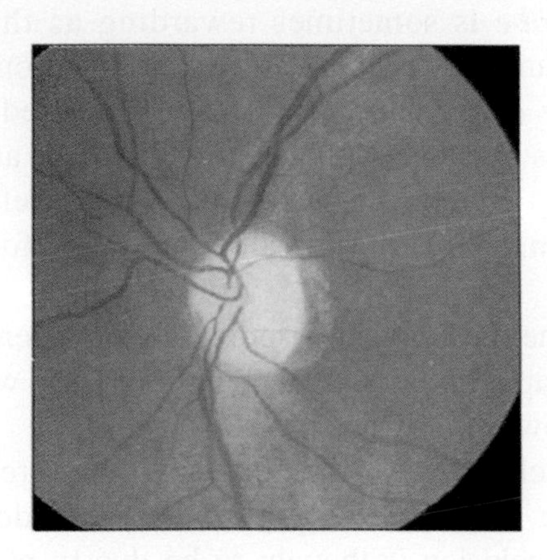

Plate 6.30 Left optic atrophy following central retinal artery occlusion.

*Causes.* 1. *Thrombosis:* usually complicating pre-existing atheroma, perhaps associated with hypertension or diabetes. 2. *Embolic disease:* any cause of vascular emboli can lead to occlusion of the central retinal artery. The emboli may come from a diseased heart valve or from atheromatous plaques in the carotid. Emboli may cause transient obscuration of vision as they temporarily block the central retinal artery and then either fragment, or are forced on to the periphery of the retina. Platelet emboli are known to do this. Episodes of *amaurosis fugax* (fleeting loss of vision) give rise to anxiety as they may sometimes precede permanent occlusion or stroke (Plate 6.31 p120).

*Management.* If retinal infarction has occurred the prognosis for vision is poor. The condition should be treated as an emergency but even after 12 hours, treatment should be tried. The time factor is critical. Partial occlusions may improve to some extent with treatment even after twelve hours. When there is a suspicion of temporal (giant cell) arteritis, treatment with steroids should be commenced without waiting for the results of ESR or biopsy tests. In general, the aims of treatment are to enlarge the lumen of the vessel and if there is an embolus, to move it on to the periphery and also to improve the retinal circulation by decreasing the resistance of the intraocular pressure.

-The *intraocular pressure is reduced* by giving intravenous acetazolamide (Diamox) e.g. 500 mg.

-*Massage of the globe* is sometimes rewarding as this both reduces the intraocular pressure and may cause some vessel dilatation.
-*Paracentesis of the anterior chamber* may be carried out if facilities are available. A needle is passed into the anterior chamber and some fluid drawn off. The sudden drop of intraocular pressure causes reflex dilatation of the retinal arterial system. (NB this can really only be performed by an ophthalmologist).
-Carbon dioxide is the most potent vasodilator of the cerebral circulation and re-breathing into a paper bag is an emergency measure which may be helpful in restoring blood flow.
The causes of the occlusion should be diagnosed and treated where possible. It is important that: 1. The blood pressure is recorded. 2. Glycosuria is excluded. 3. If the occlusion is thought to be due to platelet emboli carotid bruits are listened for and if present, carotid stenosis may require investigation and possible surgical treatment. Aspirin may be used to *reduce platelet stickiness*. 4. Evidence of giant cell arteritis (p124) is sought and an ESR is carried out, systemic steroid therapy being commenced even if there is only slight suspicion of this condition. However giant cell arteritis usually affects the posterior ciliary arteries causing optic disc ischaemia rather than a central retinal artery occlusion.

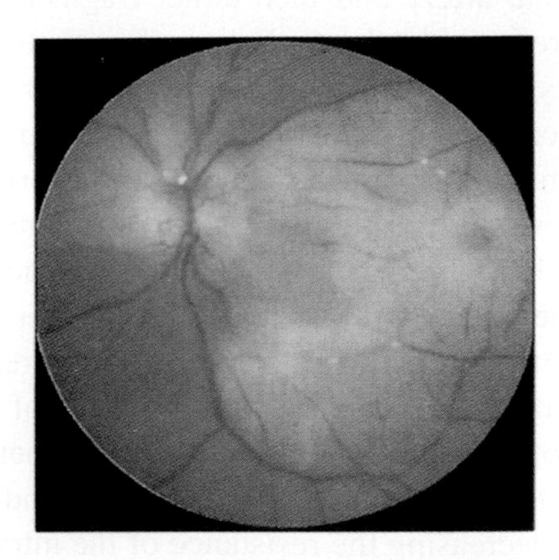

Plate 6.31 Left retinal arteriolar occlusion by multiple cholesterol emboli. Retinal necrosis and oedema with a dull red spot at the macula.

*Branch central retinal arterial occlusion*

The symptoms depend on which branch is occluded. A field defect is present corresponding to the area of infarcted retina. If the macular region is affected, central vision is also lost. A sector of whitish retinal oedema can be seen corresponding to the area supplied by the occluded artery and an embolus may be seen. Cholesterol emboli appear white and shining but do not usually occlude the vessel. (Plate 6.31 p120).

**Central retinal vein occlusion** (p290, Plate 6.32 p121, LCW 5.5 p75)

*Clinical picture.* There is usually blurring of vision coming on over a period of some hours. The loss of acuity is usually severe (less than 6/60) but it depends on the degree of occlusion of the vein and whether the resulting haemorrhages and oedema affect the macula. On ophthalmoscopy it can be seen that the retinal veins are engorged and there is oedema in the affected area with scattered haemorrhages and sometimes retinal infarcts (cotton wool spots). The disc is usually swollen.

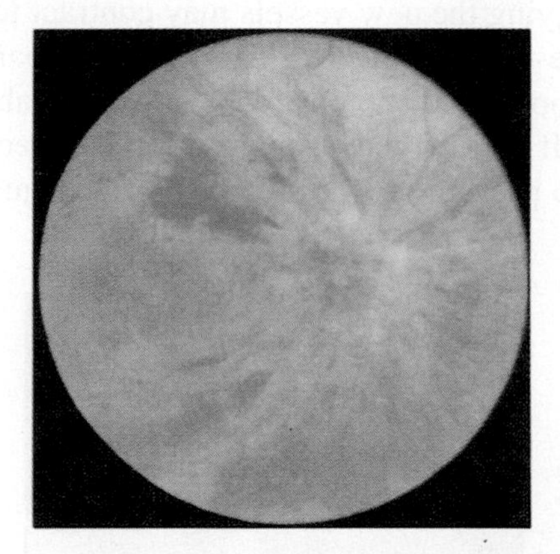

Plate 6.32 Central retinal vein occlusion, widespread haemorrhages and generalised retinal oedema.

*Predisposing factors.* A raised intraocular pressure predisposes to venous stasis and the pressure should always be measured. Treatment to reduce a raised intraocular pressure may save the fellow eye from the same fate or from eventual glaucomatous visual field loss. Atheroma is common and diabetes and hypertension predispose to this condition. In many patients the occlusion may be due basically to diminished arterial flow and consequent venous stasis. Hyperviscosity states, e.g. macroglobulinaemia and polycythaemia, may also be factors. It usually occurs in the elderly, although a similar picture in younger patients is considered to result from retinal phlebitis. The contraceptive pill may also predispose to thrombo-embolic disorders in this age group.

*Management.* It is probably reasonable to limit the investigations to taking the blood pressure, testing the urine for glycosuria, examining a blood film and making a plasma protein estimation. Some believe that retinal perfusion may be improved by using acetazolamide to lower the intraocular pressure. Treatment by anticoagulants or other direct therapy for the acute stage has not yet been proved to be effective.
*Complications:* 1. *Rubeotic glaucoma* (syn. '100 day' glaucoma, neovascular glaucoma): a total central retinal vein occlusion causes ischaemia of the affected tissues. This leads to new vessel formation by a presumed chemical mediator. In some cases new vessels form in the iris and the angle of the anterior chamber (Plate 6.33 p122, 16.11 p316, LCW 4.15 p63). The fine connective tissue accompanying the new vessels may contract to close the angle and prevent the drainage of the aqueous. This causes a particularly intractable glaucoma. It takes about three months for this to come about, hence the term '100 day' glaucoma. If the iris neovascularisation is detected at an early stage it may be possible to prevent progression by laser coagulation of anoxic retina in the periphery.

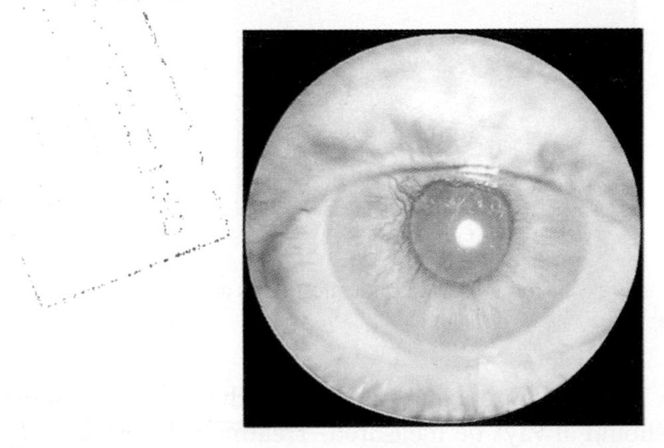

Plate 6.33 Rubeotic glaucoma - neovascularisation of iris following central retinal vein occlusion.

2. *Proliferative retinopathy:* new vessels may form as a response to ischaemia and are an indication for laser treatment before they lead to vitreous haemorrhage. 3. *Chronic macular oedema:* the anoxic vessels may leak for months after the original occlusion. In some cases photocoagulation of the leaking areas may reduce the macular oedema and improve vision.

*Branch retinal vein occlusion* (Plate 6.34 p123, LCW 5.6 p75)

This produces a similar picture to the central vein occlusion affecting one or more tributaries. If the occlusion does not affect the macular area, little immediate damage to central vision occurs. However, new vessels may still form requiring laser treatment to avoid vitreous haemorrhage. Rubeotic glaucoma is rare in uncomplicated branch vein occlusion.

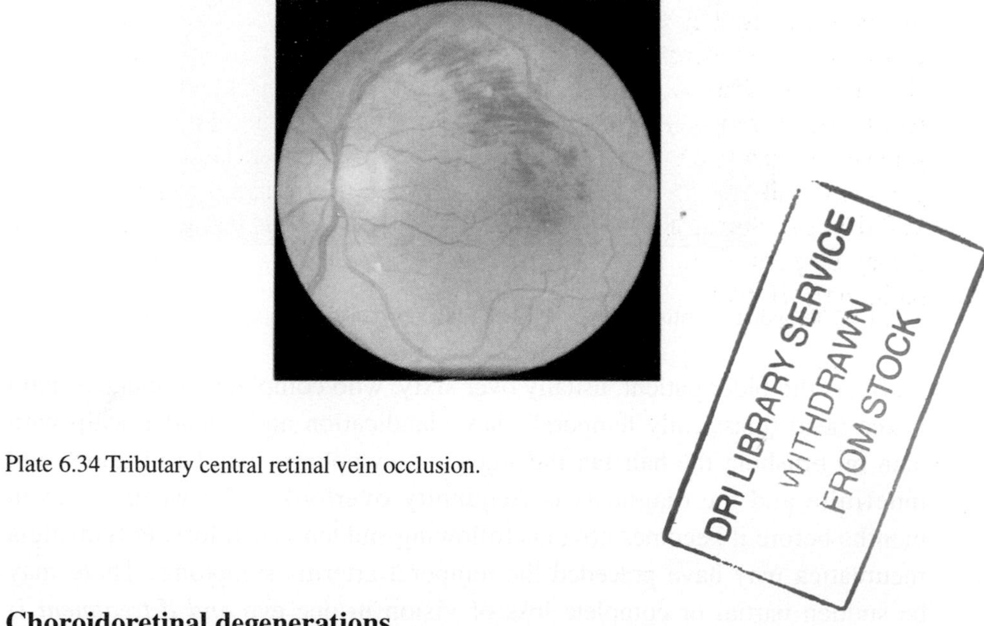



Plate 6.34 Tributary central retinal vein occlusion.

## Choroidoretinal degenerations

These mainly affect central vision and are not uncommon in the elderly, especially disciform degeneration of the retina (p529). Central serous retinopathy (p528) and pigment epithelial detachment may occur at younger ages (p529).

## Retinal changes associated with systemic disease

*Giant cell arteritis and polymyalgia rheumatica - Polymyalgia arteritica* (syn. Cranial or Temporal or Horton's Arteritis) is one cause of central retinal artery occlusion although it more usually causes similar symptoms by posterior ciliary artery occlusion with signs of an infarction of the anterior end of the optic nerve *(anterior ischaemic optic neuropathy)* (Stereo plates 19.13, 19.14 p385). It is an occlusive arteritis of medium and large vessels which is self-limiting and of obscure origin.

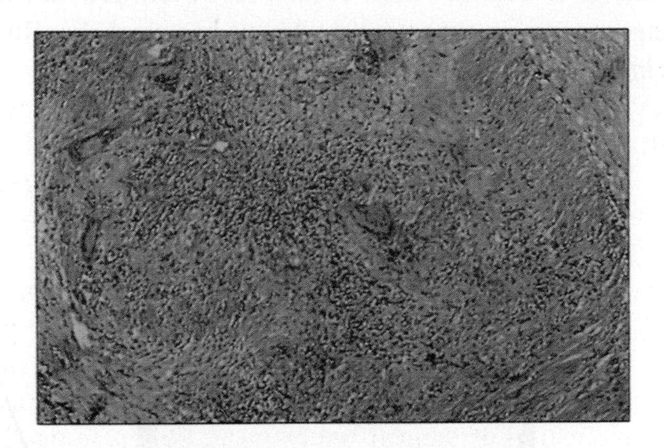

Plate 6.35 Temporal arteritis - biopsy specimen showing giant cells.

It affects the older patient, usually over sixty, who complains of malaise, pain in the head (classically temporal), jaw claudication and a tender scalp with pain on brushing the hair but the signs and symptoms may be insidious and indefinite and the diagnosis is frequently overlooked for weeks or even months before it becomes obvious following sudden visual loss. Polymyalgia rheumatica may have preceded the temporal arteritis symptoms. There may be sudden partial or complete loss of vision in one eye *and if treatment is delayed this can be followed by the same disaster in the other eye*. In the cases of ischaemic optic neuropathy at the disc due to involvement of the posterior ciliary arteries, the loss of visual field may be less extensive and the optic disc may be oedematous, the retinal vessels appearing normal for this age group. In addition to tenderness of the scalp over branches of the temporal and occipital arteries, these arteries may be thickened and pulseless, and the skin over them may be swollen and reddened in places. There may also be neurological signs such as a paralytic squint or hemiparesis due to similar lesions of other vessels. The ESR is the only helpful investigation and it is usually considerably raised in this condition due probably to an increase of certain plasma proteins. Plasma immunoglobulins, however, do not show any constant change. Confirmation of the diagnosis can only be made by a biopsy of a branch of the superficial temporal artery preferably in a segment which is tender and swollen. If temporal arteritis affects the segment of the vessel examined, biopsy will show infiltration by lymphocytes, plasma cells, macrophages and the occasional giant cell and eosinophil. Medial and intimal damage may be seen which may have caused thrombosis. Biopsy should be done as soon as is practicable *but it is important not to wait for this before commencing treatment for what is an ophthalmic medical emergency* (Plate 6.35 p124).

*Treatment.* Steroids systemically are given *without delay*, e.g. an intravenous injection of 10 mg dexamethasone followed by 60 - 80 mg of prednisolone daily. This may be reduced to a variable maintenance level (about 10 mg daily) when the disease is under control. Continuation of therapy has to be based on a combination of clinical findings and ESR readings however, an increased ESR has been shown to be an unreliable sign of impending relapse and often is unchanged during a relapse. Long continued steroid therapy should be accompanied by treatment to try to prevent osteoporosis in these patients who are usually elderly.

*Polymyalgia rheumatica* typically presents as painful stiffness in and around the shoulder joints, worse on wakening in the morning, in a patient feeling somewhat debilitated. It is often very disabling but the symptoms are dramatically relieved following cortico-steroid treatment. A biopsy from a variety of sites has, in some cases, revealed the histological changes of giant cell arteritis. Polymyalgia rheumatica must always be considered in relation to giant cell artertis and they are now often regarded as part of a multisystem pathological process *polymyalgia arteritica* and the early recognition of this may prevent blindness from retinal or optic nerve arterial occlusion.

*The significance of the early symptoms of polymyalgia rheumatica is often overlooked, as indeed those of temporal arteritis may be. It is essential for the clinician to be alert to the risks of arterial occlusion and blindness which can be prevented by the timely use of cortico-steroid treatment.*

*Diabetic retinopathy* (p313)

*Hypertensive retinopathy* The general effects of vascular disorders are considered in Chapter 14 p289. A prognostic clinical grading of hypertensive retinal changes into four grades, which requires experience for their interpretation, is that of Keith, Wagener and Barker (p295).

*Benign hypertensive retinopathy.* Grade I has the hypertensive vessel changes of arteriolar sclerosis alone. In Grade II these are more marked when the hypertension is prolonged and especially if associated with atheroma. Ischaemia may cause deep round haemorrhages and oedema is associated with scattered 'hard exudates' *(arteriosclerotic retinopathy).* The vessel changes include thickening of the vessel wall and intima with narrowing and irregularity of the blood column, brightness of the reflection streak from the vessel wall and 'nipping' of the veins at arteriovenous crossings by the thickened artery obscuring the vein or actually compressing it. Arteriolar tone increases in response to hypertension which will cause narrowing except in segments which cannot contract due to pre-existing atheroma. Vascular accidents are liable to complicate the picture with signs of retinal venous or arterial occlusion, partial or complete.

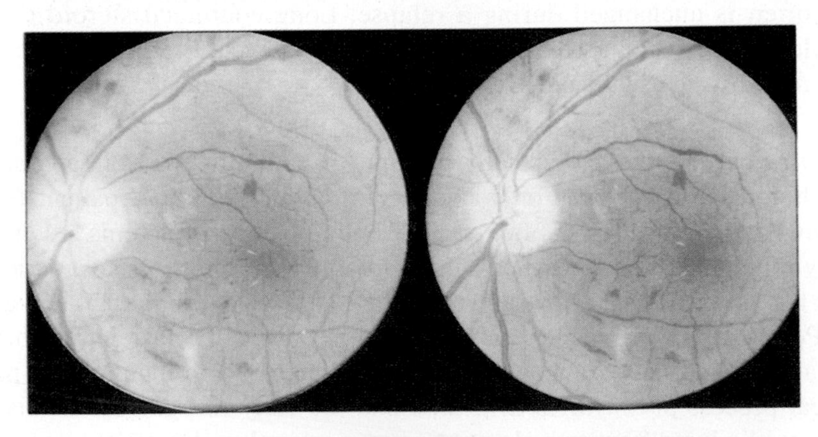

Stereo Plate 6.36 Malignant hypertensive retinopathy KWB Grade III (see p4).

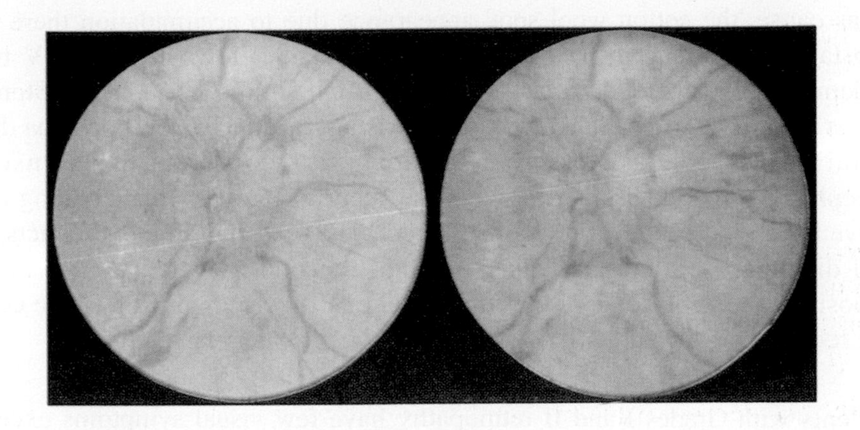

Stereo Plate 6.37 Malignant hypertensive retinopathy KWB Grade IV (see p4).

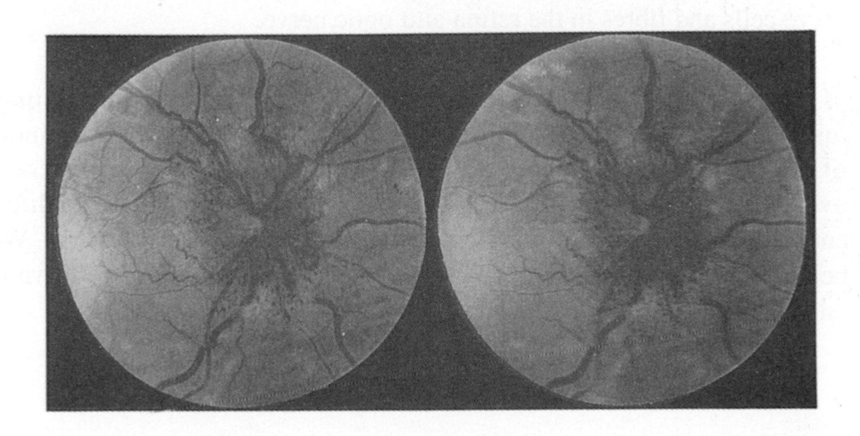

Stereo Plate 6.38 Malignant hypertensive retinopathy KWB Grade IV. Fluorescein angiography (see p4).

*Malignant hypertensive retinopathy.* Grade III is characterised by retinal oedema, superficial haemorrhages and cotton wool spots (retinal infarcts). These result from the vessel changes of arteriolar necrosis *(fibrinoid necrosis)* in which the vessel walls show hyaline thickening in association with spastic constriction and leak plasma and blood (Stereo plate 6.36 p126). Ischaemic areas cause 'the cotton wool spot' appearance due to accumulation there of substances involved in the interrupted axoplasmic flow. In Grade IV the oedema is so marked as to cause swelling of the optic nerve head (Stereo Plates 6.37, 6.38 p127). In a few cases this is aggravated by papilloedema due to raised intracranial pressure associated with coincident hypertensive encephalopathy. The haemorrhages in the superficial layers line up along the nerve fibres and are linear or 'flame shaped'. The lipid and protein products of cell disintegration and the leakage of plasma condense gradually to form deposits *(exudates),* some of the superficial exudates ,may line up in the central retina along the nerve fibres to produce the 'macular star' picture.

Patients with Grades I and II retinopathy have few visual symptoms except for those which may accompany a complicating local vascular occlusion. In Grades III and IV the haemorrhage, exudates and infarcts cause appreciable blurring of vision and ultimately the loss of sight depends on the damage to the nerve cells and fibres in the retina and optic nerve.

*Renal retinopathy.* This is a severe hypertensive retinopathy with a similar fundus picture. Exudative features may be marked in this as in the retinopathy of pregnancy toxaemia where an exudative retinal detachment may occur in severe cases. Formerly the five year prognosis for sight and for life in patients with Grades III and IV hypertensive retinopathy was very poor. With the better methods of blood pressure control now employed, the preservation of vision and life expectancy have both improved.

**C (b) Choroid**

***Posterior uveitis*** (p179)

***Neoplasms of the retina*** (p531) ***and choroid*** (p508)

**D. Primary open angle glaucoma** (p551)

This includes: 1. Chronic simple glaucoma (CSG) with intraocular pressure more than 21 mm Hg and 2. Low tension glaucoma (LTG) (Syn. Normal pressure glaucoma (NPG)) with intraocular pressure 21 mm Hg or below.

*The nature of primary open angle glaucoma.* It is a form of *anterior optic neuropathy* with characteristic changes in the optic nerve in the region of the optic disc and lamina cribrosa, which leads to visual field impairment of a nerve fibre bundle type (usually depression or loss of sensitivity in the arcuate region). The angle of the anterior chamber is open but although the intraocular pressure (IOP) is frequently raised above statistically normal levels, in the majority of cases it is only mildly raised, although this is not a diagnostic criteria as it is sometimes within the 'normal' range. *The IOP is nevertheless a most important risk factor*, but the vulnerability of the nerve to IOP is of great importance. Increased vulnerability may result from a pre-existing poor blood supply or adverse blood quality, weak supporting tissue in the laminar region or inadequacy of the nerve tissue itself. Many factors may also interact to aggravate the situation. (see p552, p553)

*Presentation*

An important and dangerous aspect of primary open angle glaucoma is its insidious nature and the condition is usually discovered during routine eye examinations with a view to the provision of spectacles. It may be asymptomatic even when well advanced. Other occasional presentations include:

-*Loss of central vision* in one eye at a late stage of the disorder. However, because glaucoma is usually bilateral, the loss of vision in one eye may give an opportunity to diagnose and treat the less affected eye.
-*Loss of visual field:* patients do not usually notice the slow constriction of their peripheral field, or the presence of an arcuate scotoma, but sometimes this may be the initial symptom.

*-Headaches and chronic ocular pain* may occur although only occasionally.
*-Haloes and blurred vision* are characteristic of acute angle closure glaucoma (p186) but can occasionally occur in chronic glaucoma if there is an unusually steep rise in intraocular pressure. A halo may be seen around light sources as a misty ring with spectral colours and is due to diffraction by the subepithelial droplets in corneal oedema.
*-Retinal vein occlusion:* a raised intraocular pressure (IOP) predisposes to retinal vein occlusion and should always be measured in such patients. The rubeotic glaucoma that may result from central retinal vein occlusion is quite different (p122).
*-Glaucoma detection* (p555) glaucoma of all types affects about 2% of a white population over the age of 40 and primary open angle glaucoma is found in about 1% of the same population. In those of African extraction the prevalence is usually higher, 4% or more. Because of its insidious onset and progress, screening or case finding programmes are practicable in certain circumstances if applied to higher risk groups. Diagnosis will depend on tonometry - raised intraocular pressure, ophthalmoscopy - cupping and pallor of the optic disc, and perimetry - arcuate scotomas with some general depression of sensitivity of the visual field. In addition to these three main signs, small haemorrhages may be seen on the disc margin and advanced cases may show reduced visual acuity or corneal oedema. Although they are separate disorders, chronic closed angle glaucoma and certain forms of secondary glaucoma may in some cases give somewhat similar signs. Contrast sensitivity and motion detection tests may show impairment of visual function and acquired colour vision anomaly affecting the blue-yellow mechanism is commonly found and may precede field loss detectable by conventional perimetry, even when automated. The practical value of these tests and laser scanning ophthalmoscopy in the detection of glaucoma is being assessed.

*Measurement of the intraocular pressure* (p555)

## Mechanisms

### *Mechanisms of the rise in intraocular pressure*

Aqueous is produced by the epithelium of the of the ciliary body. It is responsible for maintaining the intraocular pressure which usually ranges between 10 and 20 mm Hg and results from a balance between the production and drainage of aqueous. The average intraocular pressure (IOP) is 15.5.mmHg with a standard deviation of 2.5, in persons over 40 years if age. It is considered statistically normal in this age group if less than 2 standard deviations from the mean i.e. there is a chance of glaucoma of 1 in 20 (5%) if the IOP is equal to or less than 21 mm Hg. The intraocular pressure is measured by using a tonometer (p555, Fig. 6.40 p132).

- *trabecular outflow of aqueous.* Aqueous mostly passes through the pupil into the anterior chamber and then via the trabecular meshwork, just anterior to the angle between the cornea and the iris, into the canal of Schlemm, leaving the eye via the aqueous veins which join the plexus of veins in the sclera and conjunctiva (Fig. 6.39 p131).

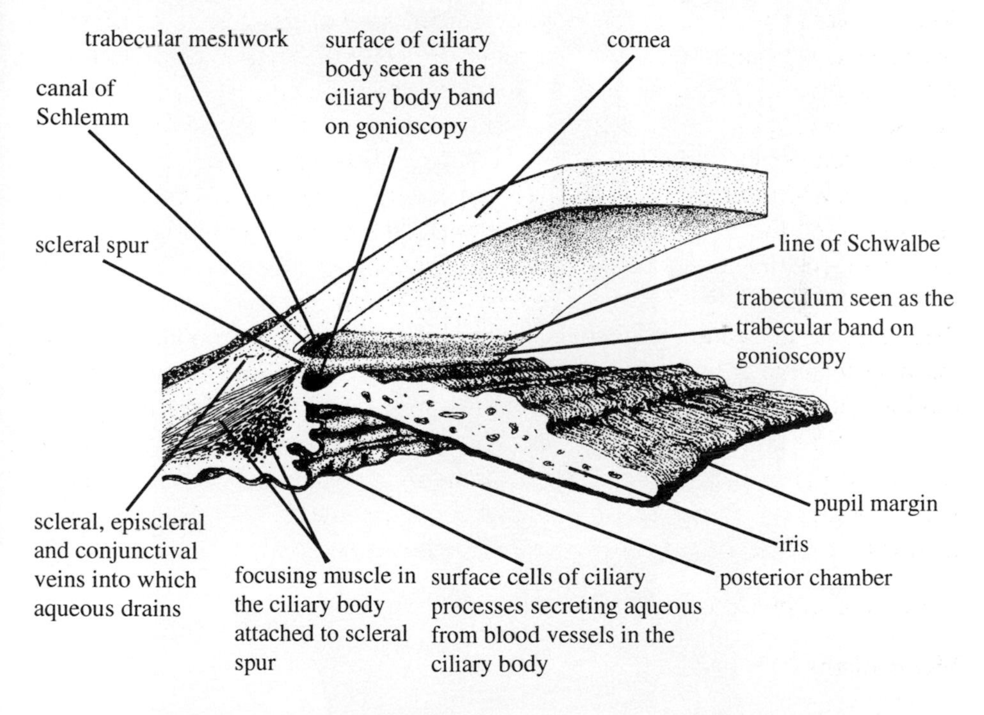


Fig. 6.39 The angle of the anterior chamber showing a normal open angle as seen through the gonioscope (pp52, 567) in relation to the underlying structures seen in section. The position of the canal of Schlemm is sometimes indicated by a line of pigment on the trabecular band or occasionally by a reflux of blood into it from the scleral veins due to pressure by the goniocope. The scleral spur shows as a whitish ridge on gonioscopy.

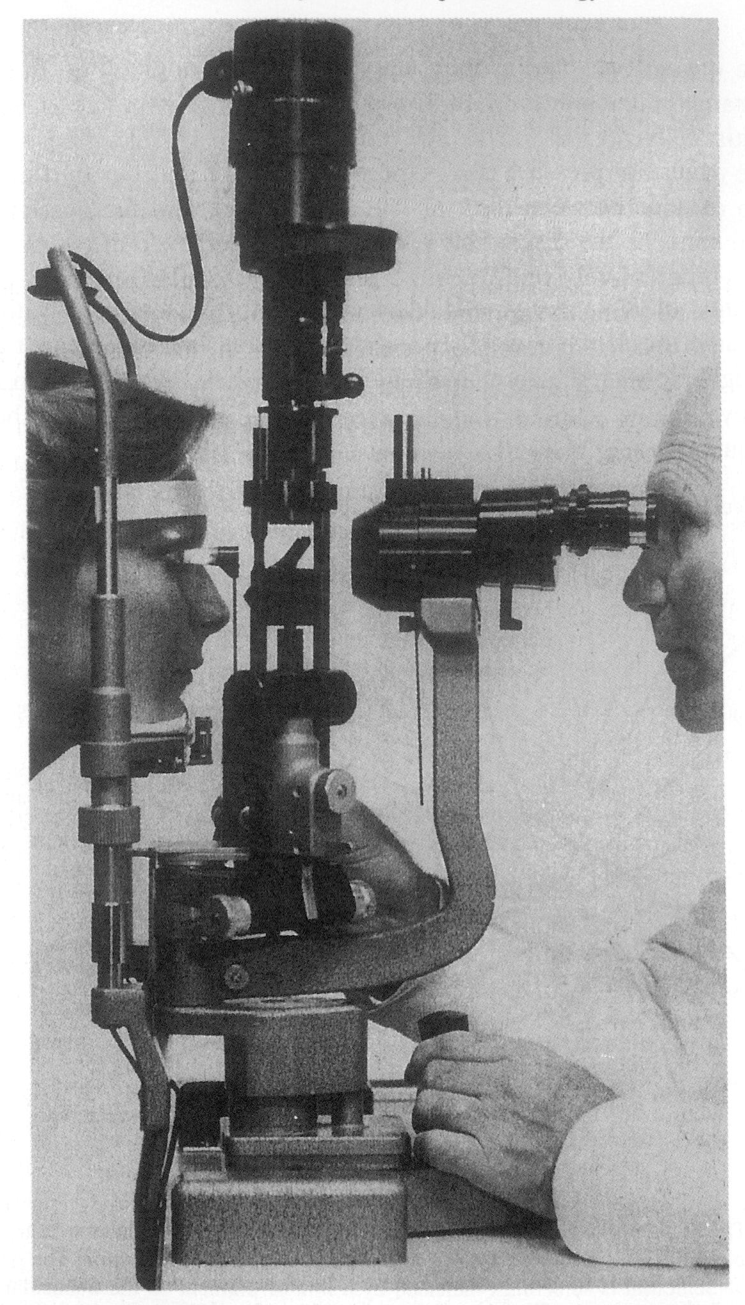

Fig. 6.40 Measurement of intraocular pressure using the Goldmann applanation tonometer. (slit-lamp mounted)

- *uveo-scleral outflow of aqueous.* A minor proportion of aqueous (normally about 15%) passes through the uveal tissue and permeates the sclera to drain eventually into the orbital veins.

Gonioscopy will reveal an open anterior chamber angle (Fig. 6.39 p131, p567)

The main cause of pressure rise is the obstruction of the trabecular outflow of aqueous due to tissue resistance between the trabecular meshwork and the outflow to the aqueous veins. As the susceptible patient grows older, the outflow mechanism becomes progressively inefficient so that the intraocular pressure rises, but only slowly, allowing the corneal endothelium time to adapt to the pressure so that corneal oedema is rarely a presenting sign in this condition. Reduced efficiency in drainage may sometimes be revealed by a water drinking test. Normally drinking a litre of water in five minutes produces negligible rise in intraocular pressure. If the rise is more than 8 mm Hg after 30 to 60 minutes, this may indicate a reduced reserve drainage. The test is now considered an unreliable indicator. The uveo-scleral outflow can only partially compensate for obstruction of the trabecular channels (but see latanoprost p141).

*Mechanisms of the changes in the optic disc.* The circulation of blood to the optic nerve head comes mainly from the posterior ciliary arteries with a possible minor contribution from the central retinal artery. The flow of blood through the anastomoses around the optic disc is mainly at the level of the lamina cribrosa and choroid (Fig. 2.9 p13) and therefore the intraocular pressure exerts a force on the vessels tending to close them. Blood from the posterior ciliary arteries can be shunted to branches less exposed to pressure, and thus the blood supply to the optic disc is particularly vulnerable to raised intraocular pressure. However, while the intravascular pressure remains higher than the intraocular pressure, blood flow continues. This delicate balance may be disturbed when the intraocular pressure rises or the blood pressure falls. It may also mechanically cause sliding of weak scleral laminae, both pinching off the small capillary vessels supplying the disc region and compressing the optic nerve fibres themselves. In chronic simple glaucoma the intraocular pressure is higher than the statistical upper limit of normal (usually considered to be 21 mm Hg) and is accompanied by a glaucomatous type of optic nerve atrophy. There is a great variation in the tolerated intraocular pressure level between patients. *In addition to pallor due to loss of optic nerve fibres and their blood supply, there is cupping of the optic disc.* This is partly due to loss of fibres, but there is also collapse of the glial framework supporting the nerve fibres due either to posterior ciliary ischaemia or the effect of raised intraocular pressure on a weak sclera and lamina cribrosa, so that the normal physiological cup is greatly increased in size and, in advanced cases, may extend to the disc margin and excavate further beneath it (Stereo Plates 31.9, 31.10 p562, LCW 4.10, 4.11 p61).

In the presence of a poor optic disc circulation or other factors reducing the viability of the nerve tissues, a relatively low intraocular pressure may further contribute to optic nerve damage of a glaucomatous type. When a glaucomatous optic disc and visual field defects occur even in the presence of a 'normal' intraocular pressure, the condition is called *'low tension glaucoma'* in which the vulnerability of the optic nerve is high and the influence of the intraocular pressure less, although of course some degree of intraocular pressure is necessary to maintain the eye as an optical instrument. Conversely an intraocular pressure of 25 mm Hg or even 30 mm Hg may be tolerated for long periods, indeed decades, without optic disc impairment, but there is nevertheless an appreciable increased liability to this. Such patients are described as having *'ocular hypertension'*. This is by definition not glaucoma, because there is no detectable optic nerve damage, but it is a potent risk factor especially at over 30 mm Hg. A non-progressive field loss associated with an atrophic optic disc may also follow a temporary severe drop in blood pressure, e.g. during a gastrointestinal haemorrhage.

*Explanation of field defects in chronic glaucoma.* The most commonly described type of field defect in chronic glaucoma is the arcuate scotoma, proceeding if untreated, to constriction and finally loss of the field of vision. The field changes are due to damage of the optic nerve fibres at the optic disc, causing a *nerve fibre bundle* type of defect usually an arcuate scotoma which is a patch of visual depression or loss that extends in an arc either above or below the fixation point. The reason for the shape and position can be understood with the aid of a diagram of the path of retinal nerve fibres and the knowledge that the central point of a visual field chart represents the macula (Figs. 6.41 p135, 6.42, 6.43 p136, LCW 4.7 p60). In the early stages of the development of an arcuate scotoma there is often patchy loss of vision along the arc of any affected nerve fibre bundle due to partial loss of nerve fibre function. If the intraocular pressure can be controlled then those patchy areas of field loss may recover. If it remains high these areas will coalesce to form an arcuate scotoma. An arcuate scotoma may be found in both upper and lower hemispheres of the visual field, but not usually to the same degree. A characteristic feature as shown in the figures mentioned above and in Figs. 13.17 p284, 13.13 p281 is a straight boundary on the nasal side on the horizontal meridian. This is called a *nasal step*. In the earliest stages of development of the arcuate scotoma, it is only possible to detect the changes by using a small target or stimulus, but as the arcuate defect progresses it will increase in density and area so that it can be found using larger targets. If the disease remains untreated the arcuate scotomas will increase in number and become confluent. Less frequently fibres running into the nasal side of the optic disc are affected giving rise to a 'temporal wedge' field defect. While arcuate depression is a characteristic feature, a generalised loss of retinal sensitivity is also important.

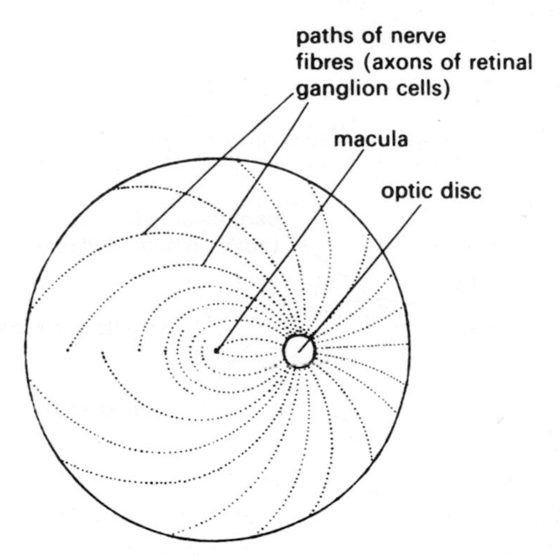


Fig. 6.41 Anatomical basis for upper arcuate scotoma R eye (N.B. This is R fundus view).

*Loss of peripheral field.* Even an early arcuate scotoma may be accompanied by a peripheral field defect, usually in the upper nasal quadrant, due to the damage to nerve fibres of ganglion cells in the peripheral retina which takes place at the disc where they occupy the peripheral zone. The macular fibres are usually the last to be affected so that the untreated patient is eventually left with only a small island of central visual field causing 'tunnel vision'. Without treatment all sight will eventually be lost. (LCW 4.8 p60)

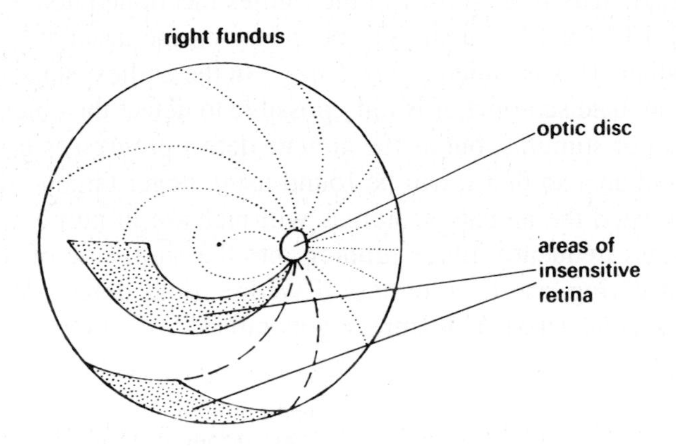


Fig. 6.42 Anatomical basis for upper arcuate scotoma R eye. (N.B. This is R fundus view).

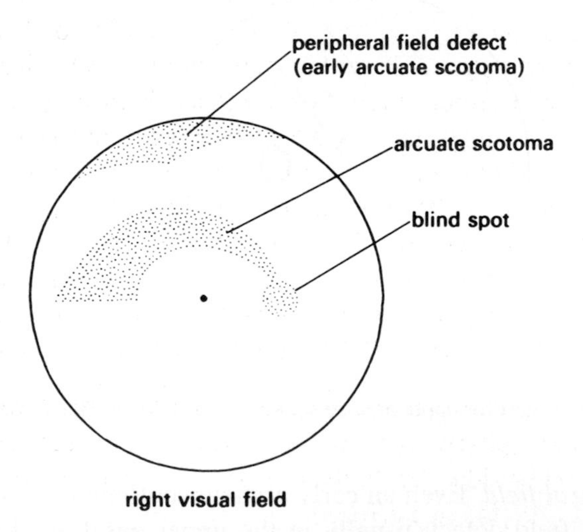


Fig. 6.43 Anatomical basis for upper arcuate scotoma R eye.
(N.B. This is a R visual field chart corresponding to the loss of optic nerve and retinal nerve fibres shown in Fig. 6.42 above).

*Management*

Successful management involves a provisional estimate, based on experience of the factors involved, of a *target level of intraocular pressure* at which to aim at first and, subsequently, the control of the intraocular pressure at a level which either allows the maintenance of the field or sometimes the maximum retardation of field loss. The disease is chronic so the patient must remain under regular observation for the rest of his or her life. An operation may make further medication unnecessary but *regular* review is important to confirm that all is well. It is important that when using eye drops patients are advised to administer them in a manner which will minimise their drainage via the tear ducts to the nose, thus reducing any side effects due to the ready absorption of their active principle into the blood stream through the nasal mucous membrane. (p617)

*Treatment to lower intraocular pressure*

*1. Medical treatment. A review of available medications*

*Pilocarpine* (p592). The longest established treatment is by pilocarpine drops 1 to 4% used two to four times daily. These are thought to act by causing a contraction of the ciliary muscle which is attached to the scleral spur, into the other side of which the layers of the trabecular meshwork are inserted. This opens the pores of the trabeculum allowing aqueous to flow more freely through them. There may well be other more subtle mechanisms. Pilocarpine has undesirable side effects including miosis which may reduce the light entering the eye with delay in dark adaptation when entering dim surroundings. It also causes discomfort and appreciable temporary myopia by contraction of the ciliary muscle. Patients who find it difficult to remember to use the drops regularly may use a slow release lamella *(Ocusert)* which provides continuous medication and is renewed weekly. A viscous base for pilocarpine drops has also been shown to prolong its effect. (*Pilogel*). This is used at night and reduces the side effects experienced the following day.

*Anticholinesterase agents* can also be used as an alternative to pilocarpine in the form of eserine (physostigmine) eye drops 0.25% but they cause more discomfort and other side effects than pilocarpine and are very infrequently used. (see pp594, 595 for side effects).

*Carbachol drops* 3% can also be used as an alternative to pilocarpine if allergy prevents its use.

*Adrenaline* (p591, p618), is used less often now but if other drops are contraindicated and if pilocarpine fails to reduce the intraocular pressure to acceptable levels or if its side effects are troublesome it can be combined with or replaced by adrenaline drops 0.1% to 1% once or twice daily. Adrenaline increases outflow and reduces production of aqueous by a combination of vasoconstriction and other unknown effects. It is very useful but sometimes frontal headaches may make its continued use impractical. Continued use may tend to make the eye constantly red and watery and can cause fibrous closure of the lacrimal punctum. Adrenaline is useful in patients with lens opacities who may be severely handicapped by the small pupil if pilocarpine or other miotic drops are used. Adrenaline is contraindicated in eyes with narrow angles because it may occasionally cause mydriasis and angle closure. In aphakic patients adrenaline may lead to diminished visual acuity from cystoid macular oedema. This is fortunately reversible in early cases on withdrawal of the drug which is nevertheless best avoided in aphakia. Adrenaline drops may oxidise to pink adrenochrome or brown melanin and they should then be replaced. *Dipivalyepinephrine* which penetrates the eye readily and is converted into adrenaline in the eye can also be effective in lower concentrations than adrenaline itself. Some of the superficial side effects of adrenaline such as conjunctival hyperaemia are thus minimised. Combined pilocarpine and adrenaline drops are found to be effective when used twice daily.

*Carbonic anhydrase inhibitors* (p596). Acetazolamide (Diamox) has been given systemically for chronic glaucoma by mouth, in tablets of 250 mg qds or in slow release capsules (250 mg bd) or by injection and materially reduce the production of aqueous, but it is now mainly used to lower high eye pressures as an urgent measure as in acute angle closure glaucoma or secondary glaucoma by mouth or by intravenous injection (500 mgm) and is now only rarely used for the regular treatment of primary open angle glaucoma mainly because of its side effects (p596).

*Dorzolamide (Trusopt)* (p596) is a carbonic anhydrase inhibitor suitable for topical use as a 2% eye drop. It is used twice daily as an adjuvant to beta blocking drops or three times daily as a single treatment. The duration of effect is about 8 hours and the reduction of aqueous production when used as monotherapy is comparable to but a little less than in the case of timolol and their combined effect is greater than either when used alone. Dorzolamide drops are in general well tolerated. They may sting on application or cause lid and conjunctival irritation and a local allergic reaction in patients sensitive to sulphonamides. There is some systemic absorption but due to the relatively small amounts absorbed, they only rarely cause in a mild form the problems encountered with acetazolamide.

*Brinzolamide (Azopt)* (p597) is an alternative topical carbonic anhydrase inhibitor eye drop in the form of a 1% suspension.

*Timolol and dorzolamide combination drops (Cosopt).* This single drop of 0.5% timolol maleate with 2% dorzolamide hydrochlor used twice daily is claimed to have approximately the same eye pressure lowering effect as the sum of each used separately and also would favour compliance (p597).

*Sympathetic beta blocking agents.* Paradoxically, agents which stimulate and those which block the alpha and/or the beta sympathetic receptors in the eye both lead to a reduction of intraocular pressure, and it is clear that their method of action is not fully understood. Propanolol used locally or systemically reduces intraocular pressure but it has a local anaesthetic effect which precludes its topical use. It will often lower the intraocular pressure even in a dosage which has little or no effect on the systemic blood pressure so that in the treatment of arterial hypertension in patients who also have glaucoma this effect may make beta blocking agents the medication of choice for hypertension,especially the beta-1 blocker *atenolol.* This is effective topically but it has poor penetration and its duration of action is only five hours. *Timolol* which is a non-selective beta blocking agent used topically as 0.25% or 0.5% drops of timolol maleate was the first to be licenced to lower the intraocular pressure and does not affect pupil size, visual acuity or blood pressure, which is much appreciated by the younger patients especially those with lens opacities. Only twice daily administration is necessary because the effect lasts nearly 24 hours and is additive to that of other therapy. Pressure fall averages about 12 mm Hg. initially, reducing gradually to about 6 mm Hg. and this is then maintained. It has been shown to act by reducing aqueous production.

*Long acting timolol maleate in a gel-forming solution has been introduced as Timoptol-LA* in order to achieve once daily administration (and hence increased compliance), with augmented intraocular drug concentrations and reduction of side effects. When Timoptol-LA comes into contact with tears it becomes a gel, giving it the convenience in use of a solution and the improved ocular contact of a gel. The intraocular pressure reduction resulting from the use of 0.25% Timoptol-LA drops once daily was found to be equivalent to that following 0.25% Timoptol ophthalmic solution twice daily The effects of 0.5% concentrations were similarly equivalent. Nyogel, another gel preparation, is used in a concentration of 0.1% timolol maleate. Other non-selective beta blocking drops are *levobunolol and carteolol.* Timolol, levobunolol and carteolol applied to the eye may be absorbed systemically to some extent and aggravate occlusive lung disease, asthma, heart block and congestive cardiac failure and in insulin dependant diabetics they may tend to mask the symptoms of incipient hypoglycaemia and reduce hepatic glucose mobilisation. They are contraindicated until this has been discussed with the physician. Insomnia, vivid dreaming and impotence have also been reported. One side effect of Timoptol-LA is a temporary blurring of vision which may last from about 30 seconds to rarely more than 5 minutes which is found to be acceptable to most patients. Peak plasma concentrations of timolol were less, following eye drops of Timoptol-LA than after drops of Timoptol ophthalmic solution. This led to a corresponding reduction in systemic side effects, but nevertheless the same contraindications and precautions must be observed in both cases. A beta-1 sympathetic blocking drop *betaxolol* in solution or suspension is less likely to have effects on breathing and the heart but also must be used with care (pp144, 618). It has a neuroprotective effect in experimental nerve injury.

*Sympathetic alpha 2 agonists*

*Apraclonidine (Iopidine)* is a selective agonist of alpha-2 sympathetic nerve receptors. One drop of 0.5% apraclonidine produces a rapid fall of intraocular pressure of short duration. It is useful in controlling the spike of intraocular pressure rise following anterior segment laser therapy.

*Brimonidine (Alphagan)* is similarly an alpha-2 agonist and is even more selective than apraclonidine. It is dispensed in the form of brimonidine tartrate 0.2% eye drops for use twice daily. It lowers intraocular pressure by both inhibiting aqueous production and increasing its uveoscleral outflow and the degree of reduction is similar to that of timolol drops. Research suggests that it may have some neuroprotective influence on the optic nerve. The drops may cause hyperaemia, stinging and corneal epithelial disturbance, in addition ocular allergic reactions occur in about 10% of patients after 3-9 months of use. It usually has little effect on blood pressure, pulse rate or pulmonary function but upper respiratory and gastrointestinal symptoms have been reported. Both brimonidine and apraclonidine are however, contraindicated in patients with severe cardiovascular disease and in those taking certain antidepressants and antihypertensive medications. Compared with the earlier types of alpha 2 agonists, clonidine and apraclonidine, brimonidine is very much more selective for alpha 2 receptors. It is also much less likely to cause dizziness and depression because it is less able to pass the blood-brain barrier.

*Prostaglandins*

*Latanoprost (Xalatan)* Uveo-scleral outflow can be appreciably increased by latanoprost. It was the first of a different class of eyedrops to be approved for reducing intraocular pressure in glaucoma and ocular hypertension. It is a prostaglandin ester. Prostaglandins (PG's) are a series of naturally occurring fatty acids which have diverse biological effects. They are formed when mechanical or chemical stimuli cause a release of arachidonic acid in tissues, which is then broken down by enzyme action. They then bind to appropriate prostanoid receptors. PGF 2 alpha was found to reduce the intraocular pressure in monkeys and then in human volunteers.

The next development was an ester of phenyl substituted PGF 2 alpha subsequently called latanoprost. This functioned as a pro-drug which, hydrolysed by the enzymes in the cornea, produced the acid of latanoprost, an epimer of which is the active principle causing a fall in intraocular pressure. It acts by facilitating the passage of aqueous through the iris, ciliary body and scleral tissues to the orbit where it drains eventually into the orbital veins. By increasing this uveo-scleral outflow, its effect is additional to that of other medications acting primarily to open the trabecular pathway or to decrease the production of aqueous. There is no demonstrable effect on the blood-aqueous or the blood-retinal barriers. The latanoprost acid remains in the eye for a long time, but any absorbed systemically has a short plasma half-life and is completely cleared by the liver to inactive metabolites.

Having a 24 hour duration of action, latanoprost need only be used once daily and trials indicate no tachyphylaxis. The drops are used in a concentration of 0.005% and studies reveal that an intraocular pressure of 25mm Hg is, on average, reduced by about 25% to about 18mm Hg and one of 13mm Hg by about 15% to 11mm Hg. It can be used in combination with other drops and useful additional reduction in intraocular pressure has been documented. It is effective in primary open angle glaucoma including normal pressure glaucoma, in ocular hypertension and in pseudocapsular and pigmentary glaucoma patients.

There are no reported side effects on the blood circulation or respiration.

Latanoprost drops do however sting on application and may cause superficial punctate impairment of the corneal epithelium and while careful testing suggests that there is no interference with the blood-aqueous barrier, caution is necessary as cystoid macular oedema can occur in phakic or pseudophakic patients. Uveitis has also been reported rarely as a complication of this therapy. A gradual change in iris colour due to the accumulation of brown pigment in its periphery, mainly noticeable in blue or hazel eyes, was originally worrying but this appears to result from increased synthesis of melanin in melanocytes. While this may be permanent, it is not known so far to have other than a cosmetic effect. The use of these drops tends to stimulate growth of the eyelashes and other hairs nearby. Rarely a skin rash, an aggravation of asthmatic tendency and dyspnoea have been reported. A combination eye drop of latanoprost 0.005% with timolol maleate 0.5% (*Xalacom*) was introduced to combine the mode of action of both medications to enhance the hypotensive effect and achieve once daily application. It is, of course, necessary to be aware of the side effects of both components.

*Travoprost (Travatan) 0.004% and bimatoprost (Lumigan) 0.03%* eye drops have been introduced subsequently with actions and side effects similar to latanoprost (Xalatan). Another prostaglandin metabolite, *unoprostone (Rescula)* has been used. It is tolerated well but less potent.

### 2. *Laser treatment - laser trabeculoplasty* (p76)

Laser applications to the trabeculum have been found to increase drainage of aqueous possibly by causing multiple focal scars which by contraction draw open adjoining trabecular spaces or the canal of Schlemm.

### 3. *Surgical treatment - trabeculectomy*

If the field of vision continues to deteriorate despite all reasonable medical treatment which can be tolerated and sometimes after laser therapy if found to be inadequate, then surgery in the form of fistulising procedures becomes necessary, usually a trabeculectomy (Fig. 6.44 p143, LCW 4.19 p64). Following a trabeculectomy or other drainage operation aqueous escapes into the subconjunctival tissue from which it is absorbed back into the blood stream thus by-passing the trabeculum and canal of Schlemm and acting as a

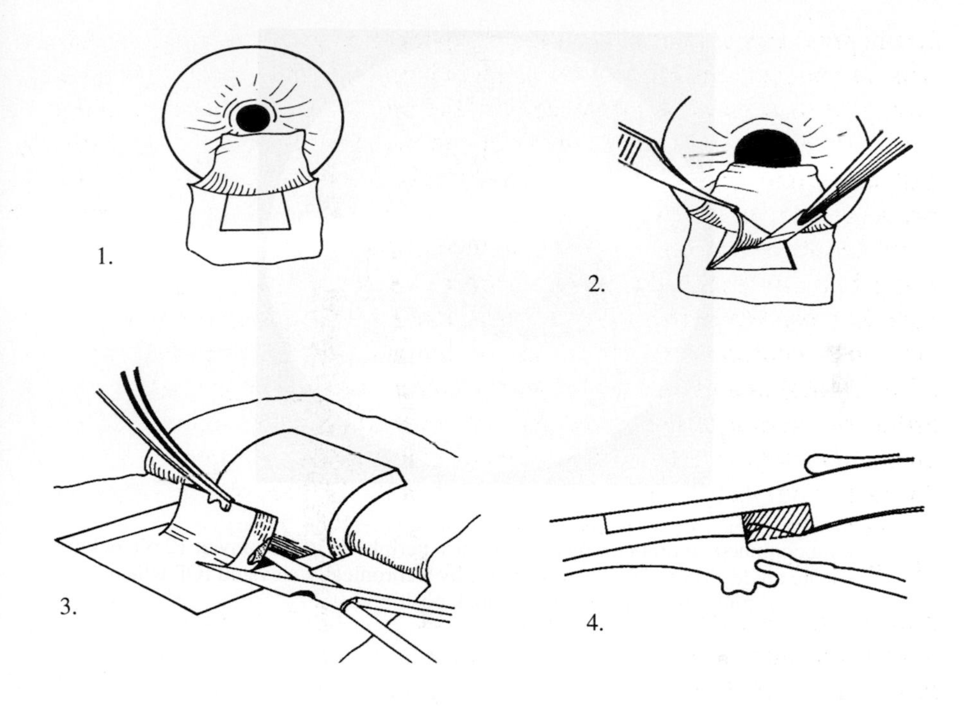


Fig. 6.44 Drainage operation for glaucoma - Trabeculectomy. 1. Conjunctival flap reflected. 2. Partial thickness scleral flap reflected. 3.Full thickness scleral flap including trabeculum lifted and excised, peripheral iridectomy followed by 4. swift flap replacement secured by two pre-placed sutures. The conjunctiva is then closed by a continuous suture. There are many minor variations in the technique but the principle is the same.

safety valvc The site of the drainage fluid can be seen as a raised diffuse bleb of conjunctiva near the limbus under the upper lid (Plate 6.45 p144). While fibrosis of the trabeculectomy bleb may occasionally occur in any patient the probability is increased in younger patients and those of African extraction. In these, the anti mitotic agents 5 - fluorouracil or mitomycin C, if applied briefly to the relevant area of sclera followed by irrigation, have improved the results of glaucoma surgery. However, there are side effects associated with these drugs, including hypotony and cystic blebs which are susceptible to infection. Newer agents such as antibodies to growth factors, such as transforming growth factor beta 2 offer a more physiological inhibition of scarring. If facilities are not available for a trabeculectomy operation, or if for

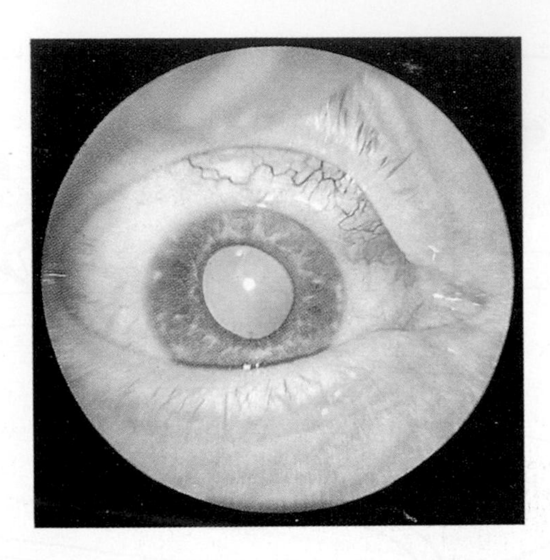

Plate 6.45 Subconjunctival drainage bleb. (There is a peripheral iridectomy at 12 o'clock for acute angle closure. This was followed by a chronically elevated IOP which required trabeculectomy in the 1 o'clock position.)

various reasons it is judged unlikely to be successful, including an eye in which previous operations of this type for primary open angle glaucoma have failed due to conjunctival scarring, significant pressure reduction and less need for medical treatment have been achieved by *trans-scleral or intraocular diode laser cyclophotocoagulation (CPC)* (p581). This appears to have a more lasting effect and less risk of hypotony and corneal or retinal complications than cryo-applications which it has replaced. It may also be repeated.

Another approach to an eye with a scarred conjunctival bleb is a drainage tube (Molteno tube), sometimes with a valve, through which aqueous drains from the anterior chamber subconjunctivally to empty into the loose connective tissue of the orbit and thus returns to the bloodstream.

In aphakic eyes a *cyclodialysis* operation can be performed which aims to drain aqueous into the suprachoroidal space, while reducing the activity of the ciliary body by separating a segment of it from the sclera by a spatula. In a phakic eye such a procedure would have a considerable risk of inducing lens opacities.

## A practical regime for the treatment of primary open angle glaucoma

It is well known that a somewhat raised IOP can be tolerated for a long time, but the risk of damage is on average noticeably increased from levels about 21 mm Hg. Especially above 30 mm Hg, most ophthalmologists now regard treatment to be necessary despite the presence of an apparently normal optic nerve. The ophthalmologist chooses a *provisional target level* of IOP for the individual eye, which requires experience in assessing the influence of the presenting optic nerve damage as seen by ophthalmoscopy and measured by perimetry and considering these in relation to of the level of the presenting IOP, its initial response to treatment and its subsequent fluctuations and the central corneal thickness which affects the interpretation of the recorded pressure. The patient's age, general attitude, the ability to put in drops and to remember when to use them, will all affect compliance and may mean that arrangements have to be made for the administration of treatment to be supervised closely by a relative or carer.

The range of possible medications for POAG raises the question of what régime should be advised after diagnosis. Pilocarpine and/or adrenaline drops are now much less prescribed because although often effective, their side effects can be troublesome and make compliance difficult. Usually one of the beta blocking drops are advised for use twice daily. If there is evidence of respiratory or cardiac problems, a beta-1 selective eye drop is more likely to be tolerated (p140). This should still be used with caution in susceptible patients. To achieve the target level of IOP it may be necessary to add or substitute drops of a carbonic anhydrase inhibitor, a prostaglandin or sympathetic alpha 2 agonist depending on the ophthalmologist's opinion of what is most suitable for the particular patient. Should the medical treatment prove to be inadequate or cause unacceptable side effects, laser trabeculoplasty may be considered. It may be necessary to advise drainage surgery and in certain circumstances diode laser photocoagulation but the patients age, rate of visual field loss, general health and all the factors considered above in setting the target pressure are again reviewed in addition to specific surgical considerations, to ensure that the risk to sight during the patient's anticipated lifetime is minimised.

*The prevention of blindness from glaucoma* depends on early detection by performing glaucoma tests at intervals on people from the age of 40 years and especially as an extension of refraction examinations assisted by the recognition of risk factors such as age, race, family history, high myopia and diabetes (p571). If enquiry reveals these factors then testing before the age of 40 would be indicated. Detection needs to be followed by regular monitoring and medical or, if necessary, surgical treatment to adjust the intraocular pressure to a level at which the progression of visual field loss is arrested or maximally retarded. Research into methods of decreasing the vulnerability of the optic nerve head to pressure are being actively sought and trials of oral aspirin and calcium channel blockers and the treatment of Raynaud's disease to effect improvement in the optic nerve circulation are in progress, especially for low tension glaucoma. In addition, intensive work is in progress on neuroprotection.

## E. **Diseases of the optic nerve** (p375)

Disease of the optic nerve is revealed by visual impairment with characteristic visual field loss, accompanied by pallor and sometimes swelling of the optic nerve head. In addition to the clinical signs and normal X-rays, computerised tomography scanning (CT), ultrasound and magnetic resonance imaging (MRI) may be valuable investigations for possible causes of optic nerve disease which include:

-*pressure effects (including intraocular pressure in glaucoma)*
-*inflammation*
-*toxic substances*
-*deficiency states*
-*vascular disturbances* (including the ischaemic aspects of primary open angle glaucoma)
-*trauma*

Optic nerve changes are often painless, except in active optic neuritis, when it may occur on eye movement, although the cause of the optic nerve damage may itself cause pain, e.g. acute glaucoma or tumour. If the optic nerve fibres are damaged, the optic disc will become pale due to loss of fibres, leaving only white glial tissue which both obliterates and covers the blood vessels. This is called *optic atrophy*.

*Pressure effects* (p376)

Intracranial lesions may involve the optic nerve indirectly by causing raised intracranial pressure leading to papilloedema which may not affect vision until a late stage. Raised pressure in the orbit from Graves' disease or orbital masses may also cause papilloedema. Visual loss in Graves' disease is an indication for immediate measures to reduce the intraorbital pressure (p311). Tumours or inflammation, either intracranial or in the orbit, and supraclinoid aneurysms may compress the optic nerve directly with early visual disturbance in the form of a central scotoma which enlarges to join any increasing peripheral field defect. This eventually leads to loss of sight unless treatment is effective.

*Inflammation (optic neuritis)* (p383)

Inflammatory disease of any part of the optic nerve is described as optic neuritis. The most common cause of optic neuritis is demyelinating disease (other inflammatory and infective causes are discussed on p383). In the active phase when the optic disc *(optic papilla)* is involved it will appear swollen and hyperaemic. This is sometimes described as 'papillitis' (LCW 7.1 p107). Swelling of the optic nerve head may be due to oedema caused by raised intracranial pressure, or local inflammation or vascular disturbance. The term 'papilloedema' is best reserved for disc oedema due to raised intracranial pressure *(plerocephalic oedema)* described on p377, which does not itself give visual symptoms until advanced (though its basic cause may do so), whilst early visual loss is usual in disc swelling due to inflammation, demyelination, toxins and vascular disturbance.

If the inflammatory focus is further up the nerve, these ophthalmoscopic signs will not be seen although in either case optic disc pallor especially affecting its temporal parts will eventually appear if the demyelination is appreciable. This is frequently called '*retrobulbar neuritis*' although its nature is the same.

*Symptoms.* There is rapid and marked loss of vision, occasionally so severe that the patient may not even be able to see light, associated with tenderness of the eye and pain on eye movement which is thought to be due to traction on the inflamed meninges of the optic nerve.

*Signs*

-*Loss of visual acuity* which is frequently about 6/60 but all degrees of defect through to loss of light perception may occur.
-A *central scotoma* which may be very large (Fig. 13.14 p282).
-*Afferent pupil defect* which indicates optic nerve or retinal disease (the swinging flashlight test) (p420).
-*Swelling of the optic disc* if demyelination is near the optic nerve head.
-*Optic atrophy* of any degree from negligible to severe will eventually develop.
-Evidence of *retinal vasculitis* in a quarter of the cases as shown by sheathing of vessels and leakage of fluorescein.
-*Delay in conduction of visually evoked responses* in the electroencephalogram (p286). Eventually the VERs return to normal in about half the cases.
-Plaques in the brain stem can be demonstrated by computerised tomography and even better by MRI where they are revealed as abnormally bright areas. MRI can also be adapted to show plaques in the optic nerves which can be differentiated from the appearance in granulomatous optic neuropathy and Leber's disease.

*Prognosis.* The majority of cases will improve spontaneously. There is usually some permanent defect of vision, but it may only be a slight relative central scotoma, especially to red stimuli. The degree of recovery in each case is unpredictable and a few unfortunate patients may suffer a severe permanent loss of vision. Many patients only have an isolated episode of retrobulbar neuritis, especially children and those with a bilateral presentation and they may never develop marked signs of multiple sclerosis. Other patients may have recurrent attacks with further impairment of the field of vision and visual acuity. The frequency of attacks is extremely variable. Unfortunately, for some patients the episode of retrobulbar neuritis is the first sign of progressive general demyelination, although this may be delayed for many years.

*Treatment.* The variable spontaneous recovery which occurs without treatment makes it very difficult to assess the value of therapy. Systemic steroids and orbital injection of steroids have been used. If demyelinating disease is basically an immune vasculitic disorder, steroids might be expected to be beneficial and reduce the subsequent damage. Vitamin B12 injections have also been used. During the course of multiple sclerosis recurrent small lesions may occur in the optic nerve which do not give florid signs. This results in a variable degree of optic atrophy particularly affecting the macular fibres and leading to a relative central scotoma. A unilateral central depression of sensitivity to red targets is characteristic and this may also be demonstrated by pseudo-ischromatic plates as used in Ishihara's colour vision test. *It is most important nevertheless to regard the picture of 'chronic retrobulbar neuritis' as optic nerve compression until proved otherwise.*

*Toxic substances (toxic optic neuropathy or syn. toxic amblyopia)* (p383)

Toxic effects on the optic nerve can occur from tobacco or tobacco combined with ethyl alcohol, from methyl alcohol or lead or as a side effect of various drugs used therapeutically. Leber's hereditary neuropathy is included here because it may indirectly have a toxic basis.

*Deficiency states* (p355)

*Vascular causes of optic neuropathy* (p384)

*Trauma*

Blunt or penetrating injury to the eye and orbit can result in compression, laceration or even avulsion of the optic nerve with local pain. The signs and symptoms depend on the type and degree of trauma, but after head injury (p374) not directly involving the eye there may be sudden loss of vision which persists in about half the cases. The remainder may show sectorial defects or general loss of sensitivity in the visual field. When the optic nerve lesion is in the optic canal associated with a fracture which may be difficult to demonstrate, the optic disc, which may at first appear normal, develops pallor after about two weeks. Surgical decompression of the nerve is rarely rewarding. High doses of steroids systemically are usually prescribed.

## F. Disease of the intracranial visual pathways involving the optic chiasma, optic tracts, optic radiations and visual cortex (p390 et seq)

Patients with intracranial disease may present to the ophthalmologist with visual disturbance as the first sign of an intracranial lesion. However, it is probably more common for these patients to be referred by the physician when they complain of visual difficulties in addition to other relevant symptoms.

The visual defect is usually that of field loss with or without impairment of visual acuity. It is essential to chart the visual fields accurately, as the type of impairment can give a clue to the site of the lesion which may be precisely revealed by radiological methods and MRI. Careful standardised recording of the fields and subsequent follow-up examination are invaluable in assessing the progress of the lesion.

Diseases of the intracranial part of the optic nerve are described on p380.
Lesions of the optic chiasma, the optic tracts, the radiations and the visual cortex are considered on p390 et seq.

The loss of field due to a particular lesion is predictable if a simple plan of the anatomy of the visual pathways is understood. A simplified diagram of the visual pathways is given in Figs. 19.18 p391, 19.19 p396.

*Pituitary lesions* affect the chiasma and classically cause a bitemporal hemianopia. However, depending upon the position of the lesion a single optic nerve may be involved causing a central scotoma. Also, a homonymous hemianopia can occur due to optic tract compression. In these homonymous cases field loss which crosses the vertical midline is strongly suggestive of a chiasmal lesion.

A *right cerebral lesion* may produce a left hemiplegia or a left homonymous hemianopia, or both.

*Lesions of the lateral geniculate body* and the visual pathway anterior to it will cause optic atrophy. However, more posterior lesions in the optic radiations or occipital cortex, do not, although when they produce papilloedema, this may itself be followed by optic atrophy.

# CHAPTER 7

# PAINFUL IMPAIRMENT OF VISION (IN THE RED EYE)

Impairment of vision, in association with a red eye, which is usually painful, is most frequently due to:

A. - ocular trauma (p151)
B. - corneal inflammation - keratitis, sometimes with conjunctivitis (p165)
C. - anterior uveitis - often with keratitis or posterior uveitis (p176)
D. - acute primary angle closure glaucoma and secondary glaucoma (p186)

## A. Ocular trauma

### Blunt injury (contusion)

The bony structure of the orbit frequently protects the eye from the potential damage of blunt injury, the brunt of which may be taken by the eyelids alone. However, trauma associated with small objects, such as golf or squash raquets balls can result in severe damage to the globe and/or orbit. In all cases the lids, globe and orbit should be examined adequately.

*Eyelids.* Usually the bruising is relatively minor ("black-eye"), requiring conservative management alone. Associated abrasions should be cleaned to avoid infection and traumatic tattooing. Following a severe blow, however, periorbital haematoma and oedema may obscure more posterior trauma. When necessary, Desmarres retractors should be used very gently to part the lids. Fluid may spread subcutaneously to produce bilateral lid oedema - but this assumption should not be made automatically and both eyes should be examined.

*Anterior segment trauma.* Conjunctival oedema (*chemosis*) and subconjunctival haemorrhage (Plate 7.1(a) p152) can be managed conservatively. Subconjunctival haemorrhage without a posterior limit should alert the examiner to the possibility of an orbital fracture. Conjunctival and corneal abrasions are treated with topical antibiotics and usually heal quickly. Corneal abrasions are very painful.

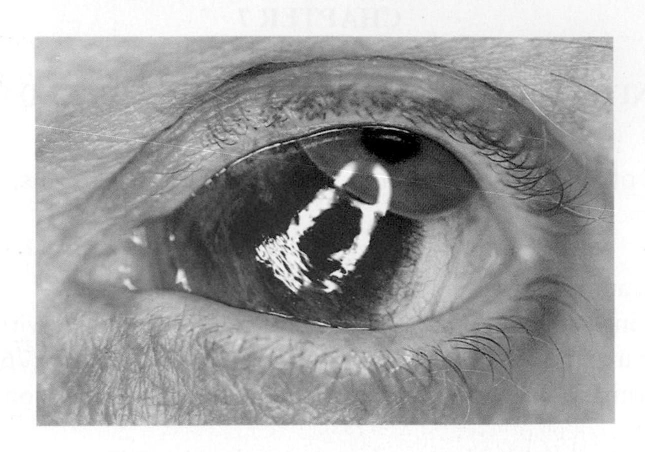

Plate 7.1(a) Subconjunctival haemorrhage.

Iris irritation causes photophobia and pain but is eased with cycloplegia. Moderate trauma may result in iris pigment liberation with deposition on corneal endothelium, anterior lens capsule and trabecular meshwork. More severe iris trauma (Fig 7.1(b) p153) may result in *pupil sphincter rupture* (seen as a notched pupil margin) or *iridodialysis* (iris root disinsertion seen as a peripheral red reflex together with a D-shaped pupil) (LCW 9.18 p138). Sometimes surgical repair is required (pupilloplasty), but the eye remains at risk of late-onset glaucoma, irrespective of management. Haemorrhage from an iris vessel into the anterior chamber (hyphaema) (LCW 9.15, 9.16, 9.17 p137) may range from red blood cells visible only with the slit lamp microscope to a total filling of the chamber. A moderate hyphaema is seen as settled with a flat upper margin (Fig 7.1(b) p153) and most clear spontaneously. Recurrent haemorrhage, usually 2 - 5 days later, may occur and is associated with a worse outcome. A persistent hyphaema, especially one complicated by raised intraocular pressure requires surgical evacuation to prevent corneal blood staining.

*Traumatic glaucoma*. Secondary traumatic glaucoma may be acute due to trabecular meshwork blockage, direct injury of the drainage angle or as a complication of lens subluxation or dislocation (LCW 9.19 p138). Alternatively, angle recession (separation of the longitudinal and circular ciliary muscle fibres) diagnosed at gonioscopy is associated with late-onset glaucoma and is an indication for long-term follow-up. Other causes of secondary traumatic glaucoma include (a) posterior synechiae formation, iris bombé and angle closure, (b) peripheral anterior synechiae formation and angle closure, and (c) ghost cell glaucoma. (p583)

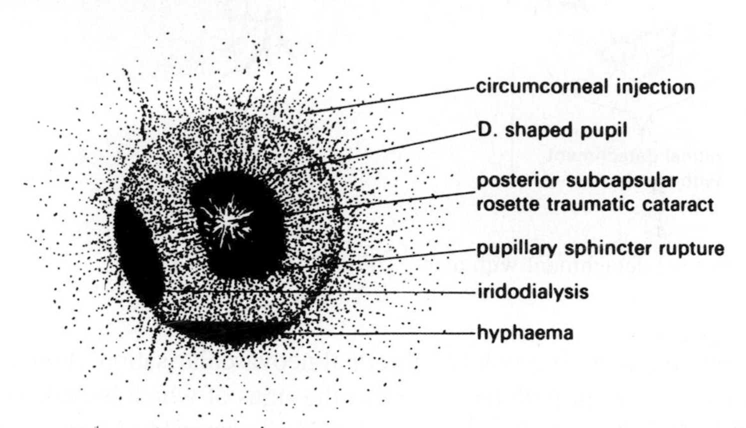


Fig. 7.1(b) Ocular effects of contusion injury.

*Traumatic (concussional) cataract.* Lens opacification following blunt trauma is usually axial and of the posterior subcapsular type. In most cases the cataract is initially a localised rosette of opacification (Fig 7.1(b) p153) which may or may not progress slowly to involve the whole lens. Cataract development may be complicated by lens subluxation which can make extraction more hazardous.

*Posterior segment trauma.* Careful inspection of the fundus is necessary in all cases. Retinal haemorrhages and whitish patches of retinal oedema *(commotio retinae)* (p527, LCW 9.20 p138) may be seen. Resolution usually occurs without sequelae, but fundoscopy should be performed for a month following the trauma to exclude treatable retinal breaks and/or detachments (Fig. 7.2 p154, Plates 6.22 - 6.24 pp113 - 115). Posterior vitreous detachment (PVD) may be induced and give rise to the symptom of floaters. There may be an associated vitreous haemorrhage (p111). Retinal breaks frequently seen following trauma include retinal dialyses, retinal tears and macular holes (p527). Occasionally rupture of the choroid occurs, frequently concentric with the disc (Fig 7.3(a) p154, Plate 7.3(b) p154). The rupture usually heals with scarring and visual

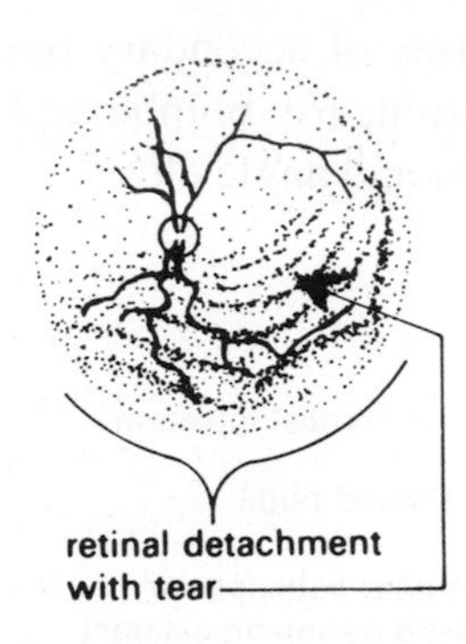


Fig. 7.2 Retinal detachment with tear.

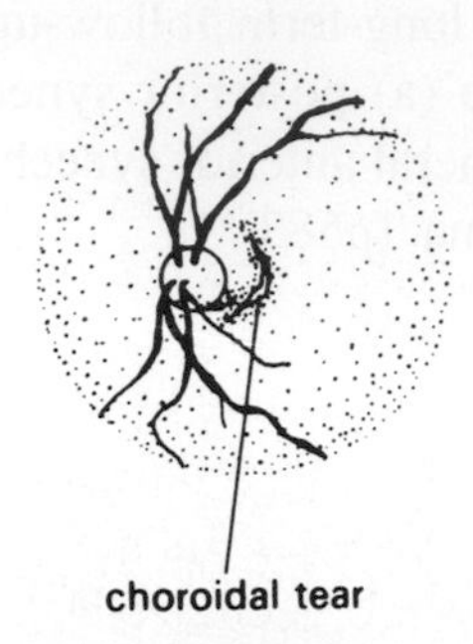


Fig. 7.3(a) Choroidal tear.

impairment, but there is a risk of subretinal neovascularisation, *Avulsion of the optic nerve* may occur with blunt trauma, the signs of which include whitening and depression in the peripapillary region. Very severe blows may cause rupture of the globe with prolapse of contents requiring early surgery.

*Orbit.* Orbital fractures, including "*blow-out fracture*" (pp540, 542, Fig 30.1 p542, LCW 9.23 p139) may occur with blunt trauma and it is important to assess ocular motility in all cases of significant peri-ocular blunt trauma.

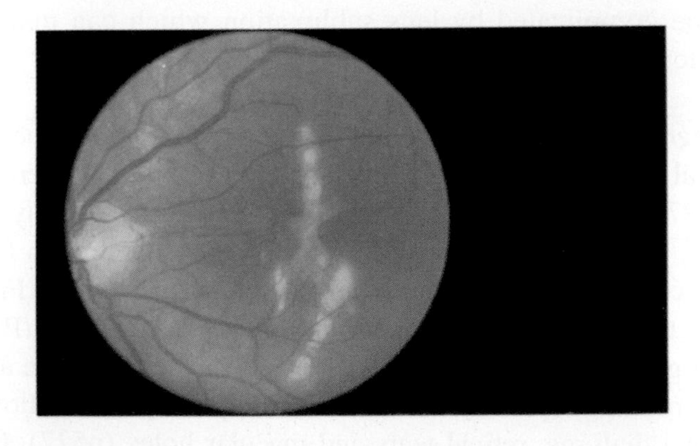

Plate 7.3(b). Choroidal tear.

**Non perforating corneal injury**

*Corneal abrasion* (Plate 7.4 p155). Traumatic loss of corneal epithelium is a common and very painful condition. Diagnosis is aided by demonstrating the epithelial defect as bright green staining of the underlying Bowman's layer or stroma when a trace of fluorescein from a sterile impregnated paper strip is introduced into the tear film. In most cases the epithelial loss is only partial and providing the stroma of the cornea is not involved, healing occurs quickly without scarring. Involvement of Bowman's layer or the corneal stroma results in a scar which may affect vision permanently. In all cases the presence of a subtarsal foreign body should be excluded. Treatment is simple. Topical antibiotic (e.g. Oc.Chloramphenicol 1%) is applied to prevent infection and a pad can be employed to aid relief of the discomfort. Local anaesthetic drops, except to allow examination, are harmful because they reduce the rate of healing and may allow further damage to occur. Mydriasis (e.g. with G. cyclopentolate 1% or G. homatropine 1-2%) will relieve the pain of iris spasm. Healing usually occurs between 12 and 48 hours depending mainly on the size of the abrasion. Failure of healing may be due to infection, a dry eye, a persistent foreign body or a corneal dystrophy (p483). Following a large corneal abrasion, particularly those caused by sharp edges (e.g. paper), the cornea is at increased risk of the *recurrent corneal erosion* syndrome in which there is imperfect healing, revealed by the presence of small epithelial vacuoles. This allows a recurrence of epithelial loss with a sudden return of symptoms, which characteristically occurs on waking and which is ascribed

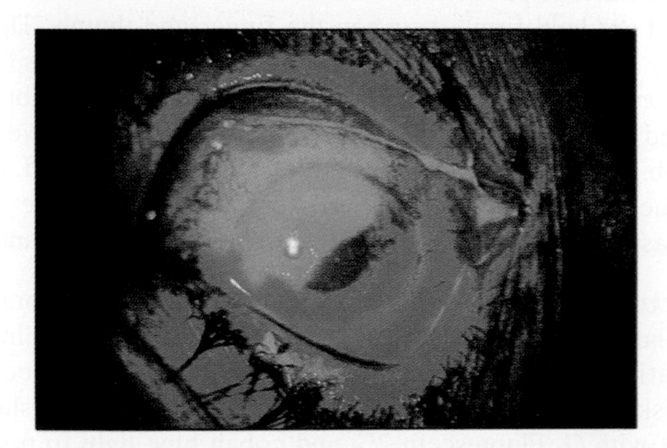

Plate 7.4 Corneal abrasion staining with fluorescein.

to adherence of the upper palpebral conjunctiva by mucus to the surface of the imperfect epithelium which is disturbed at this time. It is treated in the same way as the original abrasion and may resolve. Some cases may require removal of the affected epithelium if there is repeated recurrence.

*Corneal Foreign Bodies.*(LCW 9.5 p134) A superficial corneal foreign body can usually be seen. If recent and superficial, it can be lifted off the cornea with a sterile hypodermic needle after local anaesthetic drops have been instilled into the conjunctival sac. Subsequent management is as for a corneal abrasion (see above). Iron containing foreign bodies rapidly rust and the *rust ring* which forms in the cornea around them produces a great deal of local irritation and can be very difficult to remove. A rotary drill brush can be useful for this purpose. Careless removal of a foreign body can produce more scarring than the original injury and if incomplete, the risk of infection is increased. *In general, all corneal foreign bodies in the visual axis and certainly all deep rust rings should be referred to an ophthalmologist for removal at the slit-lamp with a sterile needle or drill brush.*

*Conjunctival foreign bodies.* (LCW 9.4 p134) Foreign bodies may lodge under the eyelids causing persistent irritation and corneal abrasions and examination and removal may make a drop of local anaesthetic desirable eg. Ophthaine (p611) although this should not be done if there is suspicion of a perforating injury. Subtarsal foreign bodies under the upper lid are easily missed unless the upper lid is everted. Eversion is a simple procedure in a co-operative patient. The patient should be asked to look down and the lashes of the upper lid are held firmly between the finger and thumb. The lid is then pulled gently downwards and away from the globe and a glass or plastic rod or the thumbnail of the other hand is held against the lid just above the tarsal plate. The lid may then be pulled upwards by the lashes and everted over the rod or thumb. Any subtarsal foreign body may then be removed. Conjunctival foreign bodies can usually be removed with a twist of sterile cotton wool. Significant associated abrasions should be treated with topical antibiotic.

*Partial thickness corneal abrasion.* Any suspected non-perforating corneal laceration should be carefully examined to exclude a self-sealing full thickness laceration, the management of which has to take account of the increased risk of endophthalmitis (pp158, 166). Examination should include Seidel's test with G. Fluorescein 2% and cobalt blue light to reveal any tiny leaks (Plate 7.5 p157). Management of partial thickness lacerations is similar to that of corneal abrasions, although unstable wounds may require a bandage contact lens or sutures. Complications include scarring and astigmatism.

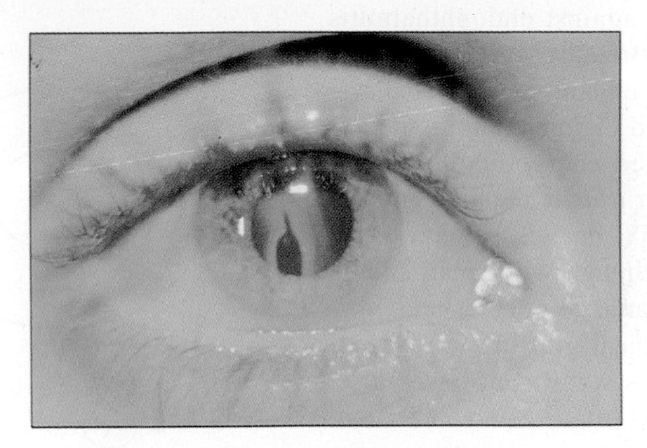

Plate 7.5 Seidel's test for leaking corneal abrasion.

## Perforating injury

*Full thickness corneal laceration.* Useful signs which aid the diagnosis of a perforating injury include; distortion of the pupil with iris prolapse into the wound (Plate 7.6 p157, LCW 9.12, 9.13 p136), a shallow anterior chamber, blood in the anterior chamber or vitreous, lens opacity and iris perforation (best seen with retro-illumination). A patient with a suspected penetrating injury should be referred to an ophthalmologist immediately. Management

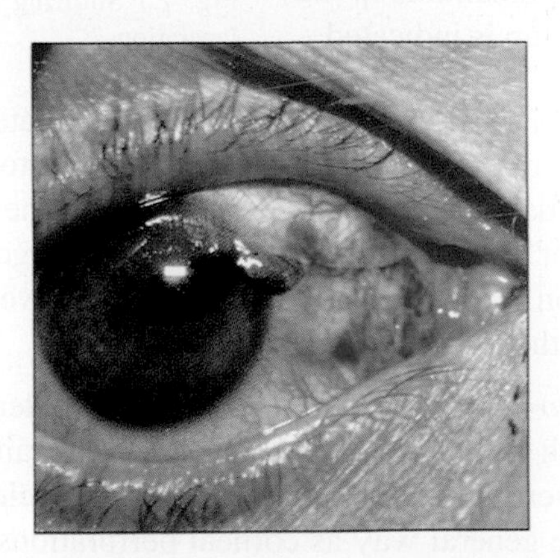

Plate 7.6 Corneal perforation and iris prolapse.

depends on the stability of the wound and extent of the lesion and includes prophylaxis against endophthalmitis and tetanus. Broad spectrum antibiotics should be given both topically and systemically and endophthalmitis if present is treated energetically (p166). Wound repair usually requires suturing (Fig 7.7 p158) or closure with tissue adhesive, although with small simple wounds a bandage contact lens may suffice. Corneal grafting is required when there is significant tissue loss. Prolapsed iris often requires excision, although attempts are made to preserve tissue when this appears viable. Vitreous incarceration should be relieved by anterior vitrectomy to avoid the risk of endophthalmitis, chronic inflammation, cystoid macular oedema and retinal detachment. When perforating trauma has involved any part of the uveal tract the patient is at risk of sympathetic ophthalmitis (p182) and long-term follow-up is indicated.

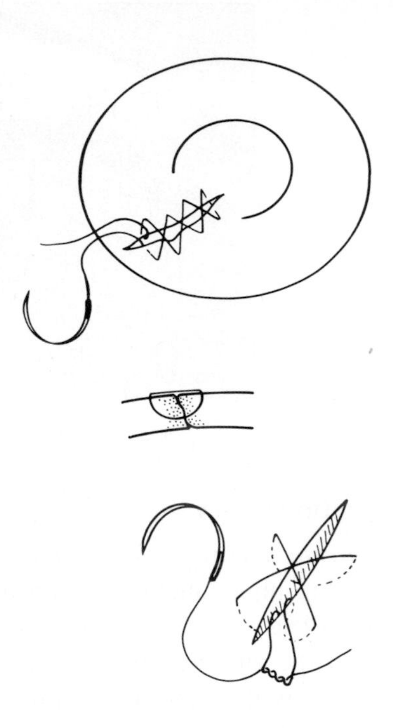

Fig. 7.7 Suturing technique for a corneal laceration.

*Traumatic cataract following capsular perforation.* (Plate 7.8 p159, LCW 9.6 p134*)* If the lens capsule is ruptured, aqueous passes into the lens resulting in a rapid swelling and destruction of the lens fibres and the whole lens becomes opaque in a matter of hours or days, the rate being largely dependent on the size of the perforation. Rarely, small tears seal themselves so that the cataract remains localised in the region of the original injury.

*Full thickness corneo-scleral or scleral laceration.* Scleral involvement may be concealed beneath a subconjunctival haemorrhage and should be considered in cases of severe trauma. Corneo-scleral and scleral perforations are managed in the same general way as corneal perforations and retinal involvement managed by a specialist vitreo-retinal surgeon.

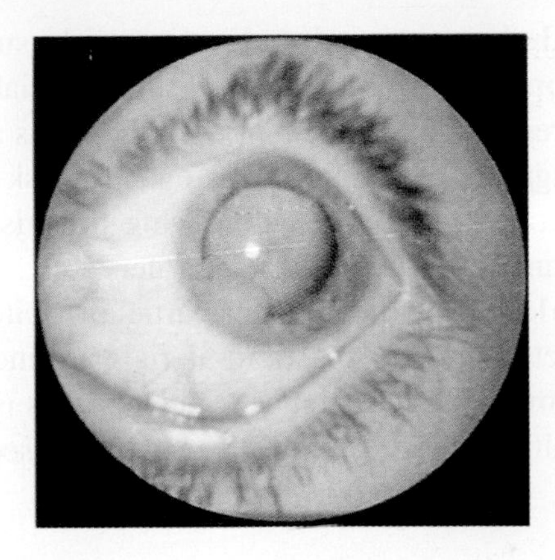

Plate 7.8 Lens matter in the anterior chamber following a perforating injury.

*Intraocular foreign body (I.O.F.B.).* The diagnosis of this condition can be difficult but if missed can result in the loss of an eye and a claim for negligence. Small foreign bodies may pass through the conjunctiva and sclera and occasionally even through the cornea with minimal signs. On suspicion of an intraocular foreign body, the whole eye should be examined very carefully after pupil dilatation (Plate 7.9 p159, LCW 9.8 p135) and the eye should undergo ultrasound scanning and/or X-ray examination. If an intraocular for-

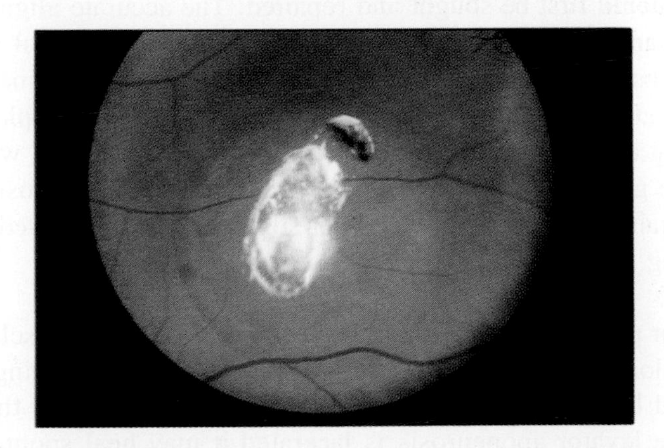

Plate 7.9 Metallic intraocular foreign body lying in the vitreous in front of impact area of commotio retinae and circle of sub-retinal fluid or retinal oedema.

eign body is revealed, accurate localisation aids subsequent removal. Management is as for any perforating injury and invariably requires removal of the intraocular foreign body to prevent complications and toxicity. If a ferrous intraocular foreign body is retained, the eye is at risk of *siderosis* because iron salts diffuse throughout the eye, colouring the iris brown and causing widespread toxic changes, particularly in the nerve cells of the retina (p582, LCW 9.7, 9.9, 9.10 p135). Retained copper particles excite a purulent reaction. Occasionally, uninfected glass or plastic materials are inert and in theory can be left in the eye. Removal of intraocular foreign bodies is achieved using fine intraocular forceps, although occasionally a magnet is used for ferrous intraocular foreign bodies.

**Eyelid lacerations** (LCW 9.11, 9.14 p136)

Detailed examination to assess the extent of the injury and careful records for medico-legal purposes are important in this type of case. Removal of foreign materials is followed by minimal debridement because the blood supply is good and infection rare but antibiotic and antitetanus prophylaxis is advisable in the case of dirty wounds.

Orbicular and septal lacerations in the line of the muscle fibres close spontaneously but those at right angles will gape and are closed by absorbable sutures. Prolapsing orbital fat is clamped, excised and cauterised. Any globe laceration should first be sought and repaired. The accurate alignment of the tarsal plate and the lid margin is necessary to prevent corneal or watering problems subsequently. This is achieved by aligning the grey line with a fine silk suture (6.0) leaving the suture ends long, closing the tarsal plate with two or three long acting absorbable sutures and the rest of the skin with fine silk sutures. The grey line suture is secured away from the corner using the long ends and retained for 10 days. The other silk sutures are removed after about 3-5 days (Fig. 7.10 p161).

If the levator palpebrae is lacerated the severed ends of the muscle are joined with buried long acting absorbable sutures, the proximal end being more easily identified by upward gaze under local anaesthetic. If less than half the width of the levator aponeurosis is lacerated it may heal spontaneously. If there has been loss of lid tissue primary closure is still applicable if not more than a quarter of the lid length (a third in the elderly) has been lost, but other-

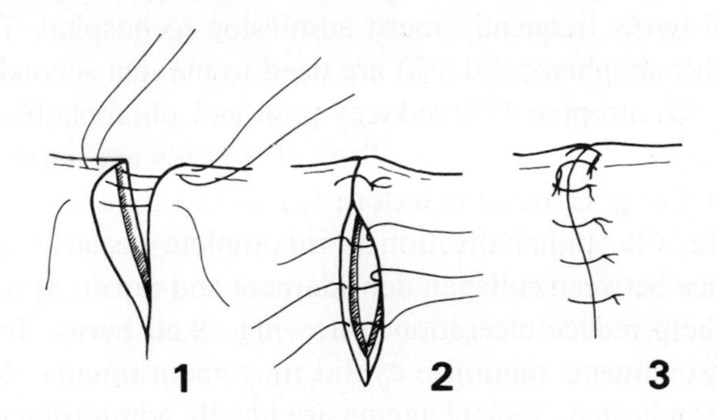


Fig. 7.10 Repair of lid wound. 1 Lid Margin. 2 Tarsal plate . 3 Skin.

wise a lateral canthotomy and mobilisation of a skin flap is necessary. Medial canthus lacerations may occur, especially from dog bites. If the medial end of the tarsus is avulsed it is sutured to the periosteum behind the lacrimal crest and the anterior part of the orbicularis tendon is sutured directly, marsupialising the proximal portion of the canaliculus to the conjunctival sac.

**Orbital trauma is considered on** p540

**Burns**

*Chemical.* Poorly managed chemical burns have the potential to cause permanent and severe visual impairment. Immediate irrigation using the nearest source of water should be performed in ALL cases. As soon as feasible the patient should be transferred to hospital for instillation of topical anaesthetic (e.g. G. amethocaine 1%) and a further 20 minutes of irrigation with a buffered solution. Both superior and inferior fornix should be adequately irrigated. Particles (e.g. lime) should be lifted or scraped from the conjunctiva because they can become firmly embedded and may cause continuous damage to the ocular surface. Litmus paper can be used to assess pH (normal tear pH = 7.4) and monitor the effect of irrigation. Alkali burns have the potential to be more devastating than acid burns because alkalis readily penetrate the eye, whereas acids coagulate superficial proteins producing a barrier to further penetration. Ischaemia of more than a half of the limbus, severe uveitis and early onset cataract are signs of poor prognosis.

Management following irrigation depends on the nature and severity of the damage. Alkali burns frequently merit admission to hospital. Topical antibiotics (e.g. G. chloramphenicol 0.5%) are used to prevent secondary infection. Mydriasis (e.g. G. atropine 1%) reduces pain and photophobia and helps to prevent the complications of uveitis. Topical steroids are important in reducing inflammation (e.g. G. dexamethasone 0.1%) but should be used with caution after 10 days if epithelialisation is incomplete because they adversely affect the balance between collagen debridement and repair. Systemic and topical ascorbate help reduce ulceration following alkali burns. Tear substitutes and lubricating ointments minimise eyelid movement trauma. Anti-glaucoma therapy may be indicated. Topical agents are ideally administered in an unpreserved preparation to avoid toxicity problems. Prevention of forniceal conjunctival adhesions (*symblepharon*) is achieved with sweeps of a lubricated probe or the use of a ring conformer. Long-term complications include corneal scarring, non-epithelialisation, cicatrisation, dry eye, secondary glaucoma and cataract. Management may include corneal grafting, conjunctival auto-grafting or limbal auto-grafting. (Plate 7.11 p162, LCW 9.1, 9.2, 9.3 p133).

*Thermal.* Fortunately in conscious patients the blink reflex tends to prevent burns of the globe itself and the damage occurs to the eyelids. Immediate treatment consists of sterile dressings, local antibiotics and relief of pain. Later the lids are carefully cleaned and fully examined. Management depends on the extent and degree of burn and may require the skill of a specialist ocu-

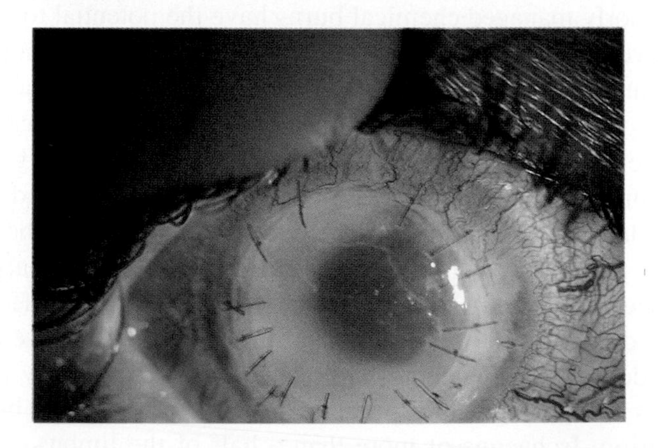

Plate 7.11 Graft rejection following corneal graft for opacification from an alkali burn.

loplastic surgeon. Contraction of scar tissue can result in ectropion, entropion or corneal exposure. A tarsorraphy (Fig 16.6 p311) may be required and frequently severe burns causing contracture require early skin grafting.

*Radiational.* While the visible parts of the electromagnetic spectrum can be designated in photometric units the rest can only be expressed in radiometric units. The irradiance of a source is measured in watts/cm$^2$ and radiant exposure in joules/cm$^2$ and when these are known, safe exposure times can be calculated.

*Ultra-violet.* Ultra-violet radiation is customarily specified as UVA at wavelengths ($\lambda$) of 380-320 nm, UVB 320-290 nm and UVC 290-200 nm. UVA is less damaging to living tissues than UVB or UVC.

*Corneal effects* . The cornea is the most affected by $\lambda$s of 210 – 315 nm (i.e. UVC and UVB) with peak absorption at 270 nm. Acute photokeratitis follows, after a latent period of some hours, the use of a sunlamp or arc welding torch (arc eye) or to reflection from snow field (snow blindness). It is characterised by punctate erosions of the corneal epithelium and decrease in corneal

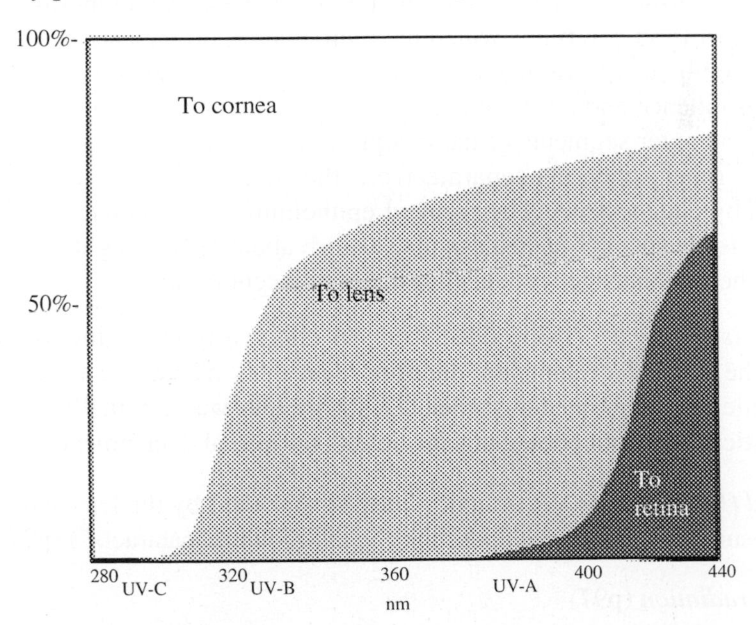


Fig. 7.12 The transmittance of UV-A, UV-B and UV-C to cornea, lens and retina. (After Charman)

sensitivity. If the exposure is severe it can also affect the endothelium with a permanent loss of endothelial cells, this is caused maximally at λ 300 nm (i.e. UVB). The eyelids and conjunctiva are reddened with a severe foreign body sensation, photophobia, watering and blepharospasm. A single application of a local anaesthetic drop will relieve the pain temporarily and with an antibiotic ointment, (e.g chloramphenicol 1%) and a firm pad for 24 hours there is abatement of symptoms after another day, usually with early and complete resolution. Heavy exposure may also result in transient signs of uveitis.

*Lens and retinal effects.* (Fig. 7.12 p163) These are most affected at λs of about 320 nm with a range of 375 – 290 nm. The cornea acts as a filter for the lens and retina. The lens absorbs much of the UV spectrum up to of λ 375 nm but it is particularly likely to induce cataract change at λs of 325-300 nm. Epidemiologically there is evidence that the incidence of cataract is related both to exposure to sunlight and to UV exposure for persons living at different latitudes but whether this is cause and effect is unproven. The retina only receives 1% of radiations of less than λ 340 nm and 2% between λs 340 and 360 nm, nevertheless at high radiant levels at these wavelengths and for repeated low level exposures which have little effect on the cornea, the retina is considerably at risk. Retinal exposures are difficult to quantify due to variable transmittance and retinal image size. The retinal changes involve vacuolation of the outer segments of the receptors with loss of their lamellar structure. The inner segments separate from the outer segments and both are removed by phagocytosis. The pigment epithelium is also damaged by UV at close to visible spectral λs and can also absorb about 60% of light at λ 500nm causing thermal damage with destruction of adjacent tissues.

*Intense visible light.* Visible light from an effectively small intense source such as the sun, solar retinopathy *(eclipse blindness)*, may cause a focal retinal burn while not affecting the media. This principle was originally employed therapeutically in light photocoagulation e.g. for retinal detachment.

*Infra-red (IR) radiation.* IR radiation may be absorbed by the lens and continual exposure may cause lens opacities ("glass-blowers' cataracts") (p97).

*Ionising radiation* (p97)

*Protection* - If exposure to radiation is essential, it is important to wear protective lenses designed *adequately* to absorb the appropriate wavelengths.

## B. Corneal inflammation and infection (keratitis)

*Phlyctenulosis* (p324)

*Bacterial infection*

The symptoms of a bacterial corneal infection (bacterial keratitis) are pain, photophobia, variable loss of vision (depending on the site of the lesion and its severity), watering and a purulent discharge from the eye. Bacterial keratitis usually starts with loss of epithelium and the area of ulceration can be demonstrated by instillation of fluorescein. White cells then surround the area of infection which is seen as a white haze around the ulcer (Fig. 7.13 (a) p165, LCW 3.18 p47). The infection and host reaction cause local corneal swelling (oedema) with increased corneal opacification and, if untreated, there is progressive necrosis of corneal tissue resulting finally in perforation of the globe and loss of the eye. The whole process is associated with secondary iritis, sometimes with a hypopyon (Plate 7.13 (b) p165, LCW 3.19 p47).

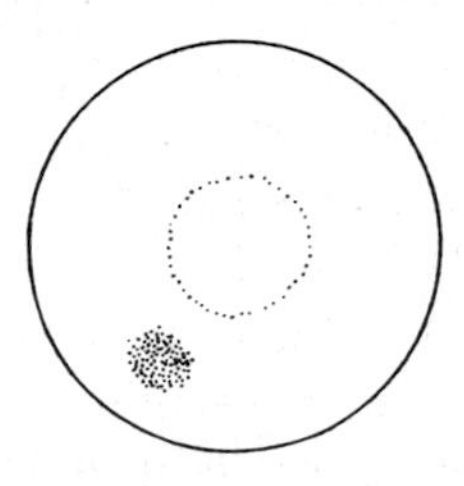

Fig. 7.13 (a) Corneal ulcer.

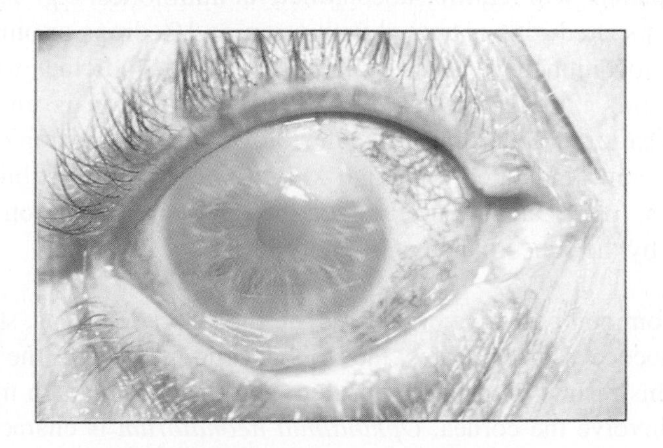

Plate 7.13 (b) Ulcer in upper nasal part of cornea with hypopyon.

Perforation is followed usually by prolapse of the iris which then 'plugs' the defect (Fig. 7.13(c) p166). This may seal the wound and the increased local vascularity may allow resolution of the infection leaving a dense white scar to which the iris is adherent - an *adherent leukoma.* However, once perforation occurs, infection will spread rapidly into the eye *(endophthalmitis)* (p185) which if not given urgent treatment will cause total loss of vision, shrinkage (*phthisis bulbi*) and often loss of the eye.

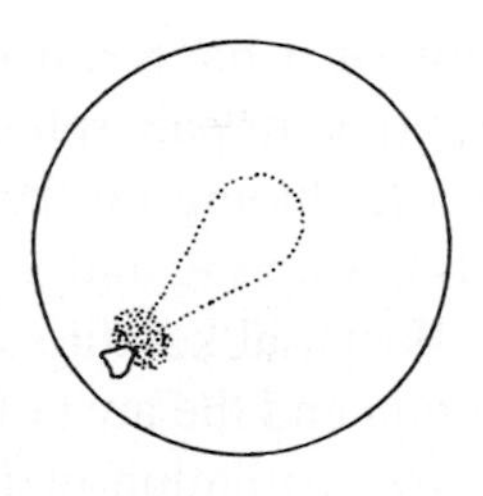

Fig. 7.13(c) Perforated corneal ulcer with iris prolapse.

*Treatment.* There should be no delay in the treatment of an established purulent corneal ulcer because it requires emergency hospitalisation. Swabs are taken for Gram's staining and culture. Intensive broad spectrum antibiotics are given at least hourly (eg fortified ofloxacin (1%-5%), ceftazidime (1%-5%), cefuroxime (1%-5%) or gentamicin (0.3%-1.5%)). If the infection is severe the patient will require subconjunctival antibiotics, e.g. 40 mg of gentamicin. Associated iritis is treated with atropine 1% drops or ointment and if necessary subconjunctival Mydricaine No 2 (p591). Judicious use of topical corticosteroids is deferred usually until the infection is under control. Therapy is tailored according to bacteriology results and topical corticosteroids are contraindicated in fungal keratitis. Systemic antibiotics should also be given (p602) and are useful even though their penetration into the eye is less than by the subconjunctival route (p586)

The most common causative organisms are the staphylococcus, streptococcus and pneumococcus. In neonates, the gonococcus may infect the eyes during delivery. This results in a severe purulent conjunctivitis which may progress rapidly to involve the cornea. *Ophthalmia neonatorum* is characterised by a conjunctival discharge within the first 21 days of life (from whatever infective aetiology). It is a Notifiable Disease (p198).

*Bacterial toxins*

Marginal keratitis is a relatively benign condition believed to be due to an immune reaction to staphylococcal toxins from the lashes (Plates 7.14 p167, 17.5 p322, LCW 3.17 p47). It is characterised by a focal, peripheral infiltration of the cornea by white cells and some loss of epithelium, detected with fluorescein. It rarely causes loss of visual acuity except in susceptible people with other eye problems (e.g. Sjögren's syndrome).

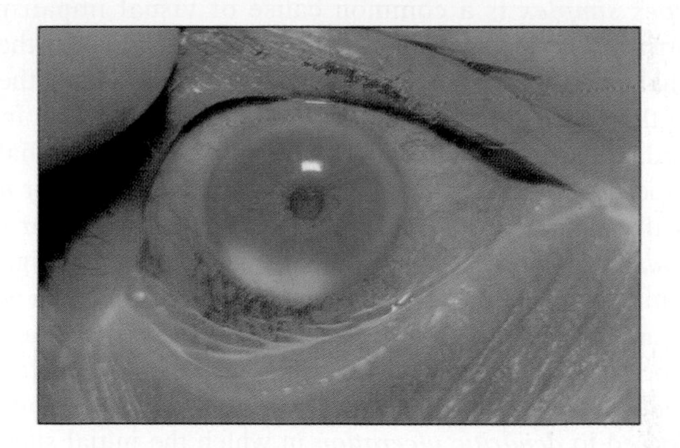

Plate 7.14 Marginal corneal ulcer (Fluorescein).

*Treatment.* This involves instruction in lid hygiene (eg by cleaning the lid margins twice a day with a weak solution of sodium bicarbonate) and the application of an antibiotic ointment to the lid margins perhaps twice per day (e.g. Chloramphenicol 1%). If the diagnosis is certain, local steroid drops may be instilled (e.g. prednisolone 0.1 %) for a few days to suppress the local immune response. It must be remembered however that *topical preparations of corticosteroids should not be used if the nature of the condition is uncertain* because they both reduce the resistance to herpes simplex infection and in addition can cause glaucoma in susceptible patients (Steroid "responders" - pp582, 606, 623). Eyes have been lost because of the injudicious use of topical corticosteroids. As a good general rule they should not be prescribed unless under the direction of an ophthalmologist who will wish to monitor the patient's progress frequently. If local treatment proves ineffective a course of tetracycline systemically may be rewarding.

*Interstitial keratitis - tuberculous p325, syphilitic p328*

## Viral infections of the cornea

Three viruses cause significant corneal inflammation: herpes simplex, adenovirus and herpes zoster. In addition chlamydial organisms (pp195, 338) cause trachoma and trachoma-resembling inclusion body conjunctivitis (TRIC). See also herpes viruses p330 and adenovirus p331. AIDS notably increases the vulnerability to herpes viruses p336.

1. *herpes simplex* is a common cause of visual impairment due to corneal scarring. The mechanism of infection is identical to that of herpes zoster ophthalmicus (see below) because latent virus within the trigeminal ganglion (in this case from a primary herpes simplex infection in childhood) is reactivated, travels down the first division of the trigeminal nerve and along its nasociliary branch to reach the cornea. If incorrectly or inadequately treated, it will cause severe visual loss. *Inadvertent treatment with topical corticosteroids can be disastrous.* The early symptoms and signs of herpes simplex keratitis are pain, photophobia, blurring of vision and a non-purulent, watery discharge which is usually scarcely more than excessive lacrimation. There may be foci of epithelial disturbance *(superficial punctate keratitis)* which respond to treatment so that the condition resolves. Some infections however proceed to *dendritic ulceration* in which the initial stage of corneal involvement is epithelial. The epithelial cells infected by the virus can be detected because they stain red with Rose Bengal which stains degenerating (virus laden) cells. When these cells die, the areas of denuded epithelium are revealed by fluorescein. The pattern of staining is usually characteristic and is seen as a branching (dendritic) lesion or occasionally as multiple stellate lesions (Fig. 7.15 p168, Plate 7.16 p169, LCW 3.9 p44). The epithelial dendritic ulcer may regress spontaneously without treatment, but it progresses usually to involve the superficial corneal stroma. Scarring follows even on resolution of the infection. Later the ulcer may become indolent with more widespread epithelial loss and lead to progressive loss of corneal stroma preceded by some swelling.

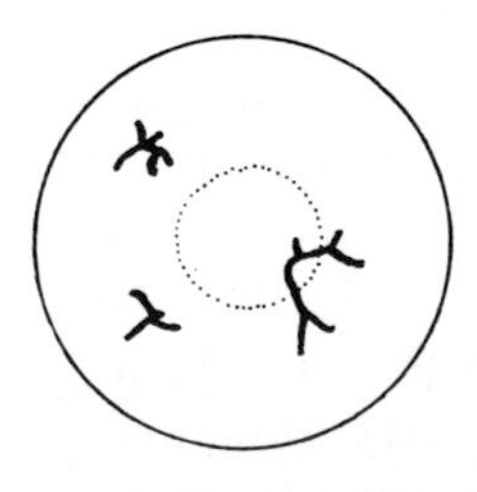

Fig. 7.15 Herpes simplex - early dendritic ulcers.

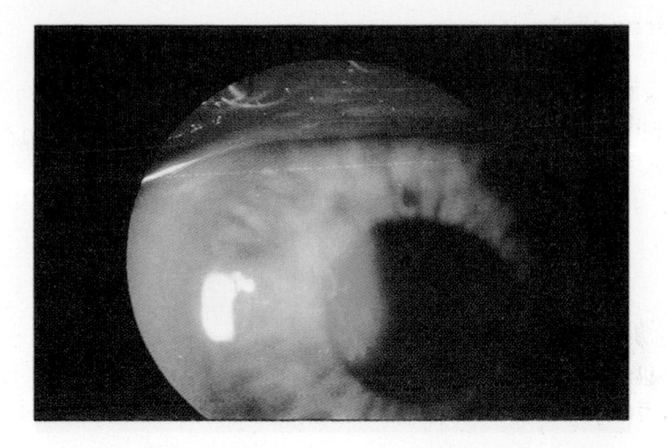

Plate 7.16 Very early dendritic ulcer (Fluorescein stain).

Treatment with topical corticosteroids will encourage this progression to the development of a geographic or amoeboid ulcer (Fig. 7.17 p169, Plate 7.18 p170, LCW 3.10 p44). A chronic dendritic ulcer promotes vascularisation of the cornea. This vascularisation not only makes the scar more dense but in addition increases the likelihood of graft rejection should the patient require penetrating keratoplasty. Eventually there may be progressive loss of corneal stroma with a risk of corneal perforation.

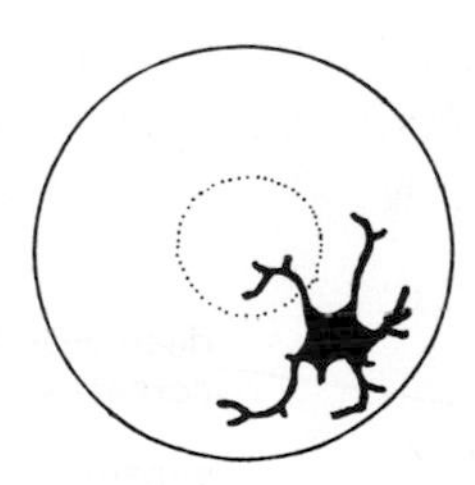

Fig. 7.17 Herpes simplex - amoeboid dendritic ulcer.

*Treatment.* Intensive local antiviral therapy (p603) is effective especially if used in the early stages. Aciclovir eye ointment 3% is used five times daily. Alternative antiviral agents are trifluorothymidine 1% drops (F3T) or idoxuridine (IDU) 0.1% drops or 0.5% ointment also five times a day. The associated iris should be treated with atropine 1% drops.

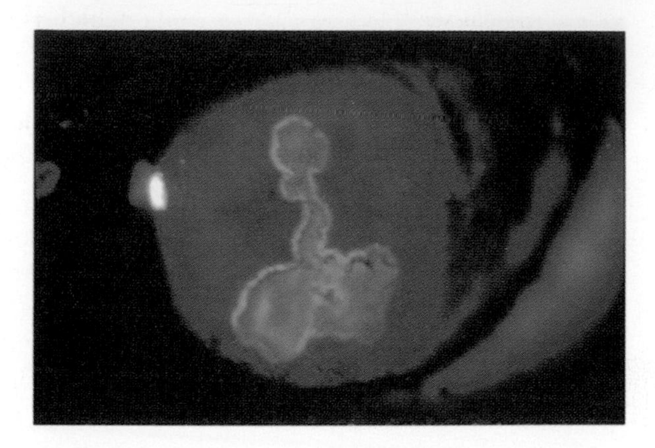

Plate 7.18 Amoeboid dendritic ulcer (Fluorescein stain in blue light).

Following the initial infection the patient is prone to recurrence of dendritic ulceration. However, another manifestation of later infections is disciform keratitis (Fig. 7.19 p170, LCW 3.24 p49). The corneal surface may not always be ulcerated but massive swelling of the stroma of the central cornea occurs, usually as a localised disc (hence "disciform"). This lesion can be associated with a moderate anterior uveitis and is thought to be a hypersensitivity reaction to the virus within the corneal stroma. It is treated therefore with topical corticosteroids under the supervision of an ophthalmologist and is one of the few justifications for the use of steroids in this disease.

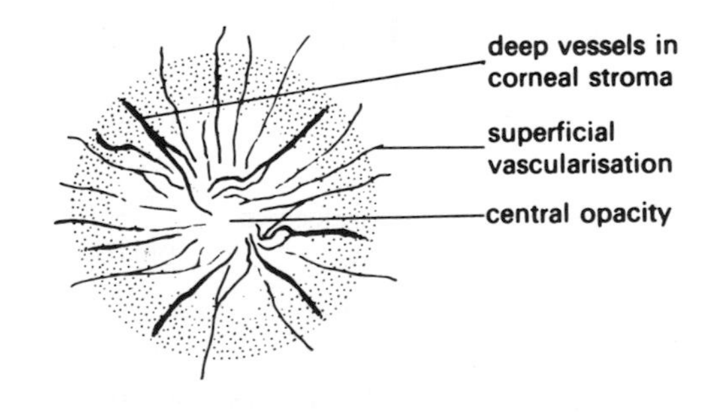


Fig. 7.19 Herpes simplex - disciform keratitis.

2. The *adenovirus* causes a kerato-conjunctivitis which is highly contagious and which may occur as an epidemic usually due to adenovirus serotype 8 or 19 (epidemic keratoconjunctivitis or EKC) (p200). There may be an associated upper respiratory tract infection (pharyngo-conjunctival fever). Initially there is a mild hyperaemia of the conjunctiva with a serous discharge which may later become purulent and in severe cases blood stained. The palpebral conjunctiva becomes roughened due to the aggregation of lymphocytes into follicles. Vision may be reduced after a week to ten days because of characteristic opacities in the superficial corneal stroma. The preauricular lymph nodes may be enlarged and tender on palpation. Photophobia and pain are common complaints as with all infective or inflammatory diseases of the anterior segment. The active conjunctivitis settles within three to four weeks but the corneal opacities may be present for much longer.

*Treatment.* No treatment has yet proved to be effective for this infection. Chloramphenicol or other broad spectrum antibiotic drops are given to prevent secondary bacterial infection. During the painful phase, due primarily to active inflammation within the superficial cornea, topical corticosteroids and non-steroidal anti-inflammatory drops (p610) can sometimes be helpful in relieving pain. It is important that the patient is informed of the highly contagious nature of the infection so that they can take appropriate precautions (by using separate towels and avoiding close contact with family members for example).

3. *herpes zoster ophthalmicus* (p330) is due to latent virus reactivation within the trigeminal (Gasserian) ganglion and concomitant infection of the first division of the trigeminal nerve. Initially, there is neuralgic pain over the distribution of the affected nerve (which extends from the periorbitum up to the vertex of the scalp) followed by a vesicular eruption in the skin. The lesions become encrusted and then resolve with a variable amount of scarring (Plate 7.20 p172, LCW 2.3 p28). The eye involvement is very variable and there is a greater likelihood of ocular involvement when the side of the nose is affected because this indicates nasociliary branch involvement (Hutchinson's sign). The commonest complication is corneal inflammation (keratitis). Initially a few small vesicles develop in the corneal epithelium and these may be followed by swelling of the corneal stroma. It is essential to try to prevent this by treatment with topical corticosteroids because the corneal swelling will result in opacification upon resolution. An additional feature is

the presence of an anaesthetic cornea (p174). Two other complications which are common in herpes zoster ophthalmicus are anterior uveitis and glaucoma. Both may occur either at the time of infection or later into the episode. Further complications include paralytic squint, optic atrophy and other neurological signs if an encephalitis supervenes. In AIDS patients a rapidly progressive outer retinal necrosis can be caused by herpes zoster (pp301, 337).

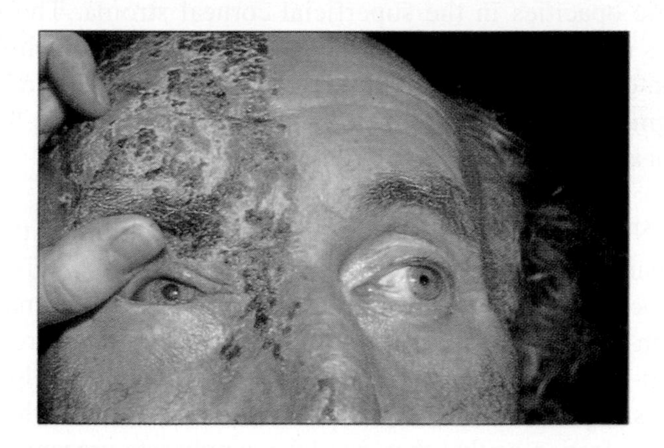

Plate 7.20 Herpes zoster ophthalmicus.

*Treatment.* The scalp lesions are best treated with a steroid and antibiotic ointment combination (e.g. betamethasone and neomycin). The ocular management involves the application of antibiotics to combat secondary infection (e.g. Oc. chloramphenicol 1%) and local steroids and mydriatics to treat the corneal complications and uveitis. For aciclovir therapy see p604. Ocular hypotensive agents such as G. timolol 0.25% or G. dorzolamide 2% or if necessary acetazolamide (Diamox) by mouth or injection are given to treat the secondary glaucoma and ample analgesics to relieve the herpetic neuralgia. In addition, antidepressants may be helpful in this very demoralising condition. All patients with herpes zoster should be investigated for the presence of an underlying malignancy e.g. a blood dyscrasia or lymphoma.

4. *chlamydial infection* (pp195, 338). Trachoma is contagious and a major cause of blindness worldwide but more so in the Middle East and tropical countries. It is due to chlamydial organisms which are characterised by basophilic inclusion bodies in affected cells. An allied condition, not uncommon in the UK, is *Trachoma resembling inclusion conjunctivitis (TRIC)* which is due to parachlamydial infection of the genitourinary tract which can be distinguished from trachoma by immunofluorescent assay. This condition does not lead to permanent corneal opacity. Trachoma is a progressive infection which initially affects the conjunctiva. There is continual irritation making the eyes chronically red, watery and photophobic. Large follicles form in the palpebral conjunctiva which necrose and heal by fibrosis (Plates 7.21 p173, 7.22 p174). The lacrimal canaliculi may become scarred increasing the

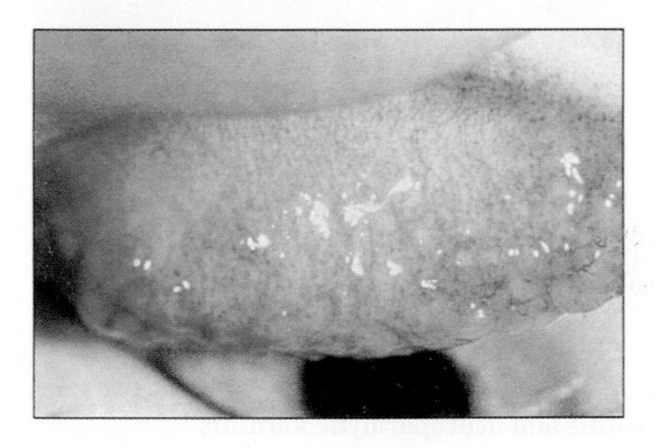

Plate 7.21 Trachoma follicles.

epiphora. The upper part of the cornea becomes oedematous and ultimately the whole cornea may be involved. Vessels grow into the cornea in the path of the oedema forming a vascularised opacity which progresses from above downwards (pannus) (Plate 8.2 p196, LCW 3.12 p45). In the absence of treatment, and largely due to secondary bacterial infection and ulceration, the cornea becomes irregular with a varying amount of thinning, scarring and vascularisation. Fibrosis of the palpebral conjunctiva causes a cicatrical inturning of the lower lid (entropion) and misdirection of the lashes (trichiasis) (Plate 22.1 p448). The eyelashes then abrade the cornea and further aggravate the corneal lesions. The diagnosis is confirmed by demonstrating inclusion bodies in conjunctival scrapings and by serological tests for antichlamydial antibodies.

*Treatment.* The condition is treated either locally with tetracycline ointment or systemically by doxycycline 100 mg daily for 10 days or by sulphonamides.

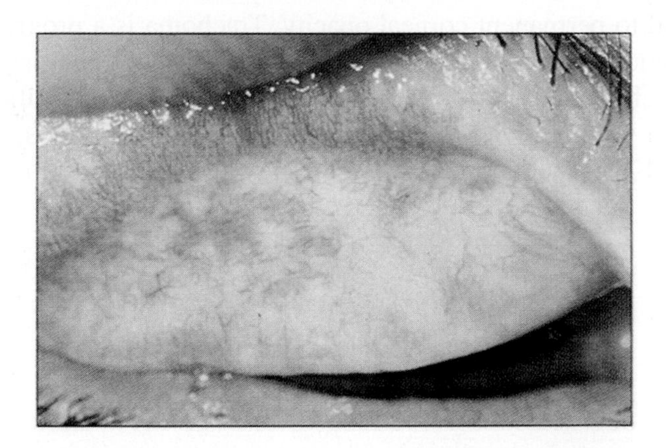

Plate 7.22 Trachoma scarring of the upper lid palpebral conjunctiva.

**Exposure keratitis and neuroparalytic keratitis**

Healthy eyelids, an intact blink reflex and an adequate tear film are all essential for the continued well-being of the cornea. Corneal exposure may result from a seventh nerve (Bell's) palsy, from eyelid malformations, malposition or scarring or from severe proptosis or exophthalmos as in thyroid eye disease (LCW 6.18 p98). The unconscious patient may fail to close the eyes fully (lagophthalmos) with resultant exposure. The initial changes of corneal exposure are those of individual epithelial cell loss seen clinically as punctate epithelial staining on instillation of fluorescein. Loss of layers of epithelial cells, or even sloughing of the whole epithelium, and a progressive drying with secondary loss of the stroma follow. If the condition remains untreated, infection and even corneal perforation may result. This condition may be managed conservatively in the first instance with tear film supplements in the form of 'artificial tears', as required throughout the day and lubricant eye ointment before sleeping. These measures and the careful closure of the eyes of an unconscious patient with adhesive tape serve to minimize the problem in this group of patients. An eye pad alone is insufficient and, if loosely applied, may rub the cornea and exacerbate the problem. Those patients with prolonged corneal exposure may require a tarsorrhaphy in order to save the cornea from perforation.

Whenever the nerve supply of the cornea is compromised, as occurs following surgical section of the fifth nerve for recalcitrant and severe trigeminal neuralgia, trauma to the fifth nerve as in the removal of an acoustic neuroma, or herpes zoster ophthalmicus, the cornea develops a keratitis with punctate epithelial erosions (neuroparalytic keratitis). Because of the importance of a nerve supply in maintaining a normal epithelium there can be loss of a large area of epithelial cells without associated trauma. In addition, serious epithelial abrasion and ulceration can result if a foreign body enters the conjunctival sac since the patient will be unaware of its presence. Management of this condition may be conservative in the first instance by protective drops or ointment. However, a tarsorrhaphy may be required ultimately. This is especially important when there is also corneal exposure as a result of a combined trigeminal and facial nerve lesion, which may occur, for example, in acoustic neuromas and in these cases should NOT be delayed. The tarsorrhaphy (Fig. 16.6 p311) should be adequate and carried out by an experienced ophthalmic surgeon.

## C. Uveitis

The iris, ciliary body and choroid are a continuous layer collectively known as the uveal tract. Uveitis is an inflammation of the uveal tract. Many classifications of uveitis are used:

*Anatomical classification*

-*Anterior uveitis.* Anterior uveitis involves the iris (iritis), the ciliary body (cyclitis) or both (iridocyclitis).
-*Intermediate uveitis.* Intermediate uveitis predominantly involves the pars plana; a subset of this group is called pars planitis .
-*Posterior uveitis.* In posterior uveitis the inflammation is mainly behind the base of the vitreous. This may be subdivided into retinochoroiditis and chorioretinitis depending on the degree of retinal involvement.
-*Panuveitis.* In diffuse or panuveitis the entire uveal tract is involved.

*Clinical classification*

Uveitis may be acute, sub acute, chronic, recurrent or non recurrent depending on the speed of onset and the subsequent course.

*Aetiological classification*

Exogenous uveitis is caused by trauma or by the invasion of the eye with micro organisms from the outside.

Endogenous uveitis may be caused by direct invasion or autoimmune reaction to micro organisms from within the patient. This group is sub divided:

-underlying systemic disease including arthritis (ankylosing spondylitis) (p359) infection (tuberculosis) (p324) and granuloma (sarcoid) (p365),
-parasitic infestation e.g. toxoplasmosis (p182),
-viral infection e.g. cytomegalovirus (p331),
-fungal infection e.g. candidiasis (p340),
-idiopathic specific uveitis cases who have unique features of their own e.g. Fuchs heterochromic cyclitis(pp179, 580),
-idiopathic non specific uveitis. These cases have no identified underlying cause and are the majority of all cases.

*Pathological classification*

Uveitis may be classified as granulomatous or non granulomatous depending on the appearance of the keratic precipitates. However there is considerable overlap in the clinical behaviour and the aetiological characteristics between the two groups.

*Clinical features*

*Anterior uveitis (iritis, iridocyclitis)* (Plate 7.23 p177, LCW 3.6 p42)

In the *acute exudative type* the onset is rapid, the sight is blurred and the eye becomes very red and acutely painful with watering and photophobia. The redness of the eye is particularly marked in the circum-corneal zone where the perforating anterior ciliary arteries are in communication with dilated vessels in the ciliary body, termed *ciliary injection.* A dense leakage of protein occurs from dilated permeable uveal capillaries into the anterior chamber giving opalescence to the aqueous which is described as an *aqueous flare* (LCW 5.1 p74), when revealed by a beam of light. The protein may be so dense that it coagulates. With adequate magnification minute shining dots can be seen in the flare. These are cells, at first polymorphs, then giving place to lymphocytes and plasma cells, which are deposited on the walls of the anterior cham-

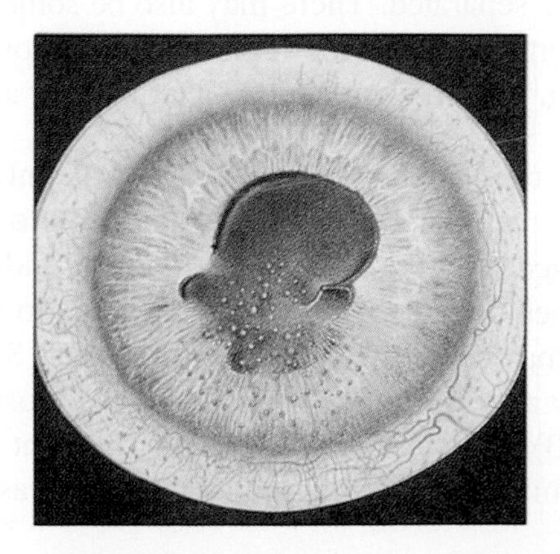

Plate 7.23 Anterior uveitis - active showing ciliary injection, KP and posterior synechiae. Pupil dilated by atropine drops.

ber and are just visible as fine keratic precipitates (or KP) diffusely scattered on the back of the cornea. The somewhat constricted pupil is due to iris oedema and irritation of the sphincter (LCW 3.7 p43). Adhesions called *posterior synechiae* are liable to form and bind the iris to the lens near the pupil margin. Involvement of the posterior uvea is usually mild in this type of uveitis.

*Subsequent course and complications*

*Synechiae:* Anterior uveitis may result in organisation of the fibrinous adhesions between iris and lens (posterior synechiae) (LCW 5.2 p74).

*Secondary glaucoma:* The inflamed ciliary body tends to secrete less aqueous and the pressure is usually low or normal. Sometimes however, the intraocular pressure becomes raised due to obstruction of the trabeculum in the angle of the anterior chamber by protein and cells, sometimes so severe that a hypopyon forms (LCW 4.16 p63, 5.3 p74). The angle may also be narrowed when the iris bows forward (iris bombé) due to the pressure of aqueous trapped behind it, which is likely if synechiae around the pupil are extensive (p580). *Peripheral anterior synechiae* may then form across the angle.

*Vitreous opacities:* Exudation into the vitreous is accompanied by a breakdown of its gel structure. Solid particles and filaments are seen floating in the fluid from which they have separated. There may also be some organisation of exudates which then exert traction in the region of choroidoretinal scars to which they may be adherent. This may precipitate a retinal tear and detachment.

*Corneal changes:* Keratitis may itself be a part of the inflammatory condition as in herpes zoster or the interstitial keratitis of tuberculosis or syphilis pp325, 328 but any prolonged uveitis damages the corneal endothelium which normally keeps the cornea dehydrated. The stroma tends to swell and the epithelium becomes oedematous followed by opacification. Subepithelial deposits of calcium salts may also form as a band across the most exposed part of the cornea (Plate 6.1 p93). This *band opacity* is a prominent feature of the uveitis associated with juvenile rheumatoid arthritis (Stills disease) (p359).

*Cataract:* Lens opacities complicating uveitis typically occur posteriorly just in front of the capsule often giving a play of iridescent colours in the beam of the slit lamp microscope *(polychromatic lustre)*. They may also spread from the sites of iris adhesions. Cataract is also a common complication of a type of chronic uveitis called *heterochromic cyclitis* associated with a depigmentation of the iris, the absence of synechiae, and secondary glaucoma (p580).

*Intermediate uveitis (cyclitis)*

In intermediate uveitis the symptoms are of blurred vision and floaters. The signs are of vitritis (cells in the vitreous) and 'snowballs' which is the term given to clumps of inflammatory cells in the inferior vitreous. An exudate on the inferior pars plana is called a 'snowbank' and if this is present the disease is called 'pars planitis'.

*Posterior uveitis (choroiditis)*

As in intermediate uveitis, the main symptoms are blurred vision and floaters. The signs are of vitritis, choroiditis and vasculitis. The inflammation may be diffuse but usually one or more yellowish white inflammatory patches are seen in the fundus and in the active phase these will tend to be obscured by vitreous haze (Plate 7.24 p179). Although the seat of the inflammation is the choroid the overlying retina is also involved. If the lesion is near the optic disc, papilloedema may be present associated with surrounding retinal oedema *(juxtapapillary choroiditis)* and a variable degree of optic atrophy may result. Healing occurs by fibrosis leaving a white scar with defined edges

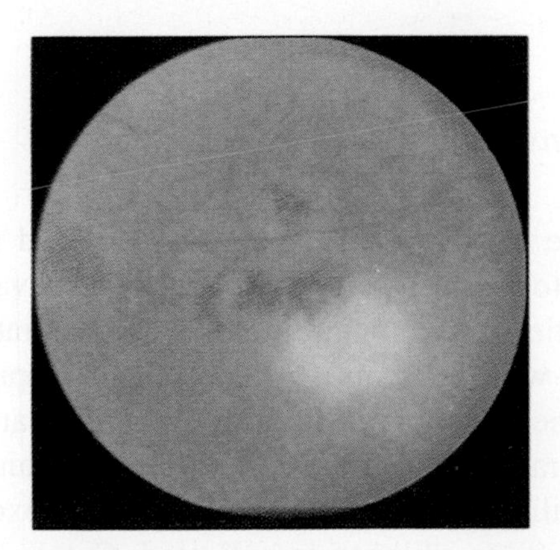

Plate 7.24 Active choroiditis - showing foci of inflammation seen through haze of inflammatory exudates in the vitreous.

(Plate 7.25 p180, LCW 5.7 p76). Frequently irregular patches of choroidal and retinal pigment lie in and around the scar. Vasculitis usually involves the veins *(periphlebitis)* rather than the arterioles *(periarteritis)*. The retinal veins may be of irregular calibre and surrounded at places by whitish exudates. Leakage of fluid from inflamed blood vessels in the macular area can cause cystoid macular oedema.

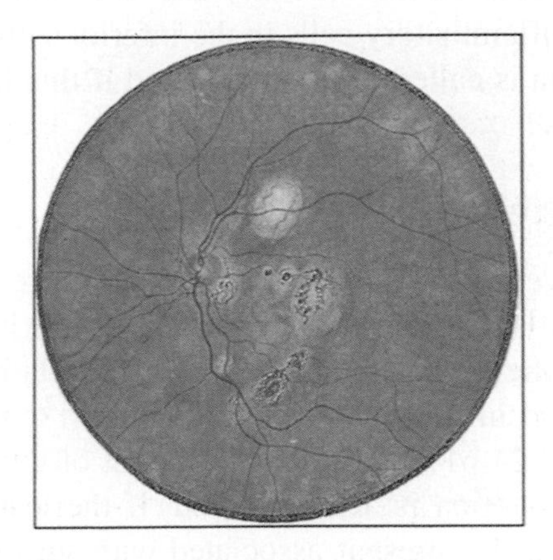

Plate 7.25 Scarring following choroiditis.

*Investigation of uveitis*

An underlying cause for the uveitis may be suspected from the ocular and general medical history and examination. Systemic evaluation is essential, with particular attention to the skin, joints and the respiratory system. The relevant investigations will then confirm the clinical suspicion. The following tests are frequently helpful: 1. Chest X ray for sarcoid and tuberculosis 2. X ray of the lumbosacral spine and sacroiliac joints for ankylosing spondylitis 3. Serology for syphilis 4. Full blood count 5. ESR 6.Toxoplasma antibodies In selected cases, a vitreous biopsy is indicated. Cytology, culture and sensitivity of this biopsy may establish the diagnosis. Frequently no cause will be found.

**Treatment of uveitis**

The aims of treatment are to relieve the patient's discomfort, prevent complications and to treat the underlying cause. Mydriatics and corticosteroids are the cornerstones in the treatment of uveitis. Mydriatics dilate the pupil. This relieves the pain caused by ciliary spasm. It also tends to prevent the formation of posterior synechiae and may break synechiae that have already formed. Mydriatics used include tropicamide 1%, cyclopentolate 1%, atropine 1% and phenylephrine 10%. They are usually given topically but some cases require subconjunctival injection (p591). Corticosteroids are used to reduce the inflammation. They may be given topically, subconjunctivally, by orbital floor injection or systemically. Topical and subconjunctival steroids are effective in the treatment of anterior uveitis but are not used for posterior uveitis as they do not penetrate well behind the lens. The frequency of administration and the strength of steroid used is tailored to the severity of the inflammation. Localised side effects of topical steroids are raised intraocular pressure, predisposition to bacterial infection and reactivation of latent viral keratitis. Continued use may predispose to lens opacities. Orbital floor injections of steroids are used to treat sight threatening vitritis and cystoid macular oedema in intermediate and posterior uveitis. They are of particular value in unilateral or asymmetrical cases or when systemic steroids are contraindicated. An injection of betamethasone 4 mg (short acting) and methylprednisolone acetate 40 mg (long acting) is commonly used as this achieves an early onset and sustained effect. The injection may be repeated at 4 -6 week intervals, depending on the response. Orbital floor injections are contraindicated when there is raised intraocular pressure as the local steroid may cause further rise in pressure. Any such rise in intraocular pressure is treated with acetazolamide by mouth or injection.

Systemic steroids are used for posterior uveitis especially the more severe inflammations with vitritis and macular oedema, and for disease which does not respond to orbital floor injections. The initial dose is 1 - 1.5 mg/kg daily and this is gradually tapered according to the response. The visual benefits must be balanced with the potential side effects of adrenal suppression and Cushingoid features.

When uveitis does not respond to systemic steroid treatment, or when side effects of treatment are unacceptable, alternative therapy must be considered. Cyclosporin A is an immunosuppressive agent that has been effective in these cases. The dose is 5 mg/kg daily and treatment is commenced with high dose steroids. The steroid dose is gradually reduced to approximately 10 mgs per day and then the cyclosporin dose may also be gradually tapered. The main side effects of cyclosporin treatment are hypertension and nephrotoxicity and all patients must be carefully monitored.

## Sympathetic ophthalmia

Sympathetic ophthalmia is a rare panuveitis which occurs after penetrating ocular trauma or rarely after intraocular eye surgery. The traumatised eye is referred to as the 'exciting' eye and the fellow eye is called the 'sympathising' eye.

The symptoms in the sympathising eye are blurred vision and photophobia. The signs in this eye are of granulomatous panuveitis. A characteristic finding are Dalen Fuchs nodules in the fundus. These are small, deep, yellowish white nodules and consist of retinal pigment cells, epithelioid cells and sparse lymphocytes.

The inflammation must be treated vigorously with topical, periocular and systemic steroids. Enucleation of the exciting eye within 2 weeks of the injury may prevent sympathetic uveitis. However, since it is not possible to predict which cases will go on to develop sympathetic uveitis this prophylactic treatment is rarely practicable unless a potentially exciting eye is both blind and irritable.

## Toxoplasma retinochoroiditis

*Toxoplasma gondii* is an obligate intracellular protozoan parasite. The definitive host is the cat, both domestic and wild, and it is only in this animal that the parasite can complete its life cycle. Man is an intermediate host and it is estimated that 500 million people throughout the world have been exposed to the disease and have antibodies to the infection.

There are three stages in the life cycle of toxoplasma. The tachyzoite is the actively dividing form responsible for the acute manifestation of the disease, the bradyzoite is the slowly dividing form contained within tissue cysts, and the oocyst is the spore form which is excreted in the cat faeces.

The disease in man is commonly congenital but may be acquired. If a non immune mother becomes infected during pregnancy the parasites may pass through the placenta to infect the foetus. Acquired disease is usually the result of eating undercooked meat which contains the tissue cysts. A less common way of acquiring the disease is by accidental contamination of food with cat faeces which contain the oocysts.

The disease has 3 main clinical forms in the human host:

The *acute* infection is usually asymptomatic or may cause an influenzal type of illness. The tachyzoite is responsible for this stage of the infection. It travels in the bloodstream and becomes disseminated throughout the body with frequent involvement of the eyes and the brain. The immunocompetent host will mount an immune response and produce specific anti-toxoplasma antibodies. This response curtails the acute infection and will result in the production of tissue cysts. Metabolically inactive organisms, the bradyzoites, are contained within these tissue cysts and the disease is said to be inactive or *chronic. Recurrent* disease is thought to be due to rupture of the tissue cysts and release of the organisms. Hypersensitivity reaction to the tissue cysts has also been suggested as a mechanism of disease reactivation.

In the brain the chronic condition mainly affects the tissues around the cerebral ventricles and may cause widespread signs and symptoms. The lesions tend to calcify and may be revealed by brain scans. Signs of hydrocephalus and encephalitis may result including squint, hemiplegia, fits and mental retardation.

The hallmark of chronic ocular toxoplasmosis is a pigmented retinochoroidal scar (Stereo Plate 7.26 p184, LCW 5.8 p76). This is sight threatening if it occurs at the macula or in the papillo macular bundle. Recurrent disease is characterised by an area of necrotising retinochoroiditis at the edge of the old scar - the 'satellite' lesion. This active lesion is white or yellowish white and has fluffy indistinct edges. There is frequently an associated vitritis which will obscure the view of the active focus. The inflammation may spill over into the anterior chamber to produce a granulomatous or a non granulomatous iridocyclitis.

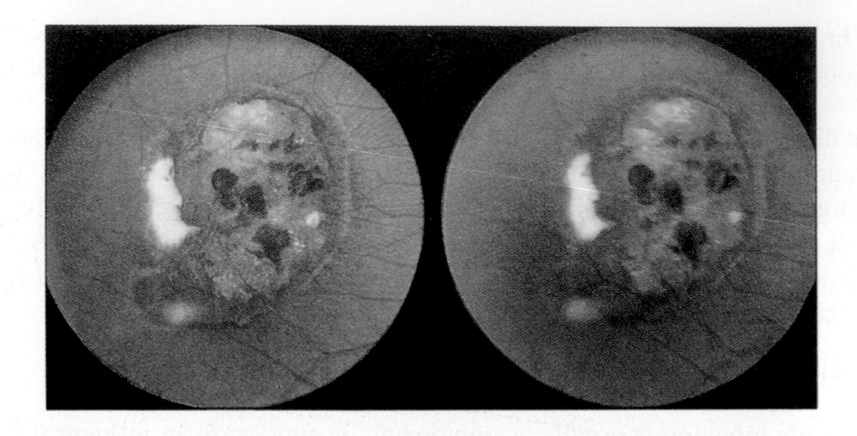

Stereo Plate 7.26 Toxoplasmosis choroido - retinal lesion (inactive) (see p4).

The diagnosis of ocular toxoplasmosis is clinical. This is supported by serological tests if a rise in titre of toxoplasma specific IgA antibody is demonstrated. The presence of IgG antibody is of limited value in the diagnosis of toxoplasmosis as a significant percentage of the population will have these antibodies as a marker of previous exposure. In selected cases serology of the aqueous humour may be required to establish the diagnosis.

Not all cases of ocular toxoplasmosis require treatment. The chronic disease is never treated as there is as yet no drug available to eradicate the tissue cysts. Recurrent disease is treated if there is visually significant vitritis or if the active focus is in a sight threatening area i.e. the macula, papillomacular bundle or the optic disc area.

The drug treatment of ocular toxoplasmosis is either sulphadiazine and pyrimethamine, or clindamycin. Both of these treatment regimens are used in conjunction with systemic steroids. Pyrimethamine may cause bone marrow suppression leading to thrombocytopenia and leucopenia so regular blood counts must be performed. The risk of this complication is reduced by giving folinic acid supplements throughout treatment. The main side effect of sulphadiazine is allergic skin rash and renal stones.

The doses used are pyrimethamine 25mgs three times daily and sulphadiazine 1gm four times daily with prednisolone 60-80 mgs daily. Treatment duration is usually 1 month during which time the steroid dose is gradually reduced. There is recent evidence that treatment with pyrimethamine and sulphadiazine will reduce the size of the residual retinal scar. This regimen should therefore be considered if the reactivated disease is close to a visually important structure.

The main, but rare, side effect of clindamycin is pseudomembranous colitis. Patients should be advised to discontinue treatment if they develop diarrhoea. The dose of clindamycin is 300 mgs four times daily used in conjunction with systemic steroids. Clindamycin is the preferred treatment if reactivation is from a peripheral focus and is associated with a significant vitritis.

**Endophthalmitis (panophthalmitis)**

Endophthalmitis is intraocular infection. This is usually a complication of perforating eye injury, but may occur very rarely after intraocular surgery, or by infection of a drainage bleb even after several years, or by metastasis from an infective lesion elsewhere in the body. In acute endophthalmitis the patient complains of pain and loss of vision. The eyelids become swollen, the eye is injected and the conjunctiva becomes oedematous (chemosis). The cornea is hazy and inflammatory cells in the anterior chamber may aggregate to form a hypopyon. Dense vitritis is also a prominent sign. Urgent treatment is required to prevent loss of the eye.

*Treatment.* An attempt must be made to identify the causative organism as success of treatment depends on eliminating this organism with the appropriate antibiotics. A vitreous sample is taken and inoculated on to culture media. At the time of sampling, broad spectrum antibiotics are injected into the vitreous cavity. A common regimen is intravitreal vancomycin and amikacin. Antibiotics and mydriatics are also given by subconjunctival injection, topical antibiotics are repeated every half hour initially and then tapered according to the response. Once the results of the vitreous cultures and sensitivities are available the antibiotics are altered accordingly. Topical steroids are given to suppress the inflammatory response; in severe cases systemic steroids may be necessary. If the intraocular pressure rises acetazolamide is given orally or by injection. An early vitrectomy is indicated if the vision is significantly reduced on presentation.

The most common causative organisms in bacterial endophthalmitis are *Staphylococcus epidermidis, Staphylococcus aureus,* pseudomonas species and proteus species. Fungal endophthalmitis may have a sub acute or chronic presentation and the prognosis is very poor. (p340)

## D. **Acute primary angle closure glaucoma** (pp550, 574)

*Symptoms*

Characteristically patients complain of severe pain in and around the affected eye often accompanied by blurring of vision with rapidly progressive visual loss. Reflex nausea and vomiting may occur. There may be a history of seeing coloured rings around lights (haloes) and of similar subacute episodes which have resolved spontaneously after a night's sleep, the miosis of sleep combined with nervous relaxation allowing normal drainage of aqueous to be resumed.

*Signs*

The conjunctival vessels are dilated causing a red eye with noticeable circumcorneal injection. The cornea is cloudy due to oedema, the semi-dilated pupil does not react to light and the anterior chamber is shallow (Plate 7.27 p186, LCW 3.8 p43, 4.12 p62). On palpation the eye feels harder than normal. Tonometry, if available, may show a raised intraocular pressure reading even up to 70 mm Hg. Early 'pressure' cataract (glaucomflecken) may be seen at the anterior surface of the lens, as irregular whitish dots.

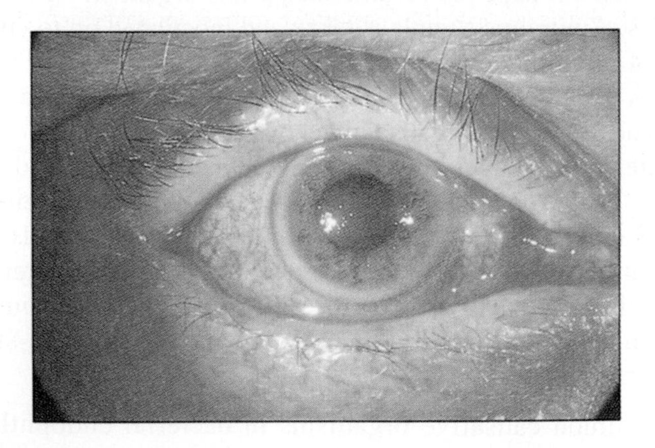

Plate 7.27 Acute angle closure glaucoma. Iris atrophy suggests previous subacute episodes (Corneal arcus coincidental).

*Detection of patients at risk of primary angle closure glaucoma (screening)*

Signs and provocative tests which may indicate liability to primary angle closure glaucoma, both the acute and chronic types, are described on p575.

*Mechanism*

Most of the aqueous normally drains back to the blood stream through the trabeculum just anterior to the angle between the iris and the cornea, the angle of the anterior chamber. In persons predisposed to acute angle closure the width of the entrance to this angle is very narrow, usually with a shallow anterior chamber. With age the lens increases in size and pushes the iris forward thus further reducing the width of the entrance to the angle. The trigger mechanism which induces the acute attack is partial dilatation of the pupil and iris congestion may also be a factor. The iris becomes less taut in mid-dilatation and the pressure of the aqueous behind the iris bows it forward. A naturally rigid iris renders the patient less liable to angle closure. As the pupil dilates the iris thickens as well as being bowed forward and both these factors contribute to complete the angle closure. When the angle is closed the outflow of aqueous stops and there is a rapid rise in pressure even up to 70 mm Hg. The raised pressure opposes the normal pumping of fluid by the corneal epithelium from the cornea into the anterior chamber so fluid is retained causing corneal oedema. The pressure also damages the anterior lens capsule causing superficial lens opacity. The most important effect of the pressure is interference with blood supply to the optic disc, as the ocular pressure becomes higher than the venous and the capillary and eventually more than the arteriolar pressure. The vortex veins as they leave the eye have the lowest pressure and collapse first causing venous engorgement and optic disc swelling. If the pressure remains high, then optic atrophy results. The pupil dilatation which precipitates the attack may be caused by anxiety and stress and by darkness or drugs, by either local mydriatic drops, e.g. atropine, or systemically by some bronchodilators and certain antidepressants containing parasympatholytics.

*Explanation of the physical signs*

*Circumcorneal injection:* about 4 mm from the cornea the anterior ciliary vessels perforate the sclera to supply the ciliary body. They therefore dilate as the result of any congestion of the anterior segment.

*Corneal oedema:* the corneal endothelium has a metabolic pump mechanism which pumps water out of the corneal stroma into the aqueous. It is usually working against an intraocular pressure of 15 mm Hg. and if the pressure suddenly rises to 40 mm Hg. or more, then the pump cannot cope and water fails to pass out of the cornea. It appears as droplets under the corneal epithelium which cause the coloured rings as a diffraction spectrum. The hazy cornea may be cleared temporarily to allow gonioscopy and a fundus view by osmotic dehydration using 50% glycerol drops.

*Dilated pupil:* closure of the angle of the anterior chamber is the cause of the acute attack, this may be precipitated by dilatation of the pupil. In turn the acute pressure rise causes paralysis of the iris muscles so that the pupil remains fixed in semi-dilatation. Previous episodes causing infarction of segments of iris may have resulted in atrophy causing distortion of iris pattern which acquires a whorled appearance.

*Optic atrophy and permanent visual loss:* this is an ischaemic change in the optic nerve head as the intraocular pressure rises above capillary and arteriolar pressure, causing at first congestion and later a failure of perfusion of vessels supplying the nerve.

*Treatment*

The intraocular pressure is lowered with acetazolamide (Diamox) which also aids the penetration of eye drops into the eye (p596). Diamox reduces the secretion of aqueous and this effect may be enhanced by timolol 0.5% drops twice daily. Pilocarpine 4% drops are used to contract the pupil as this may help the angle to open and drainage of aqueous to recommence.

As acute glaucoma is an emergency the acetazolamide should be given by intramuscular or preferably intravenous injection (500 mg). Oral osmotic agents such as glycerol are also employed if nausea does not prevent their use (p597). Glycerol is used as 1 ml/kilo body weight of glycerol made up to 50% solution with iced lime juice to reduce nausea, aided by metaclopramide (Maxolon) or perphenazine (Fentazin) if necessary. Mannitol intravenously as an osmotic agent is probably best reserved for patients who have not responded to medical treatment and then used immediately prior to surgery. Once an attack has been reversed by the above methods, a more permanent treatment is required. Laser iridotomy or peripheral iridectomy keep the pressure equal between the anterior and posterior chamber and so will stop the forward bowing of the iris. In the absence of pre-existing iris adhesions across the angle the operation will result in cure. Both eyes are predisposed to closed angle glaucoma and the unaffected eye will remain liable to an attack unless a prophylactic laser iridotomy (LCW 4.13 p62) or surgical peripheral iridectomy (Figs. 7.28, 7.29 p189, LCW 4.14 p62) is carried out on this eye too.

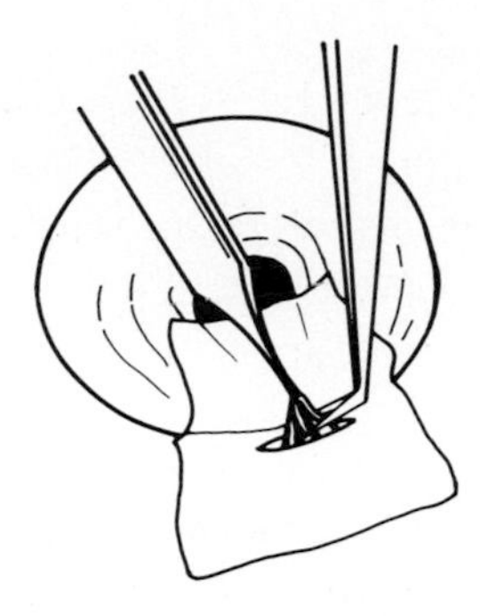

Fig. 7.28 Peripheral iridectomy - performed with angled scissors while forceps hold peripheral iris.

Fig. 7.29 Peripheral iridectomy - closure.

The differential diagnosis of acute closed angle glaucoma and uveitis with secondary rise in intraocular pressure is sometimes difficult. If delay is inevitable before an ophthalmologists' opinion is available, in either case carbonic anhydrase inhibitors can be given with advantage until ophthalmic expertise and special equipment is available, General practitioners are advised to carry acetazolamide (Diamox) for this purpose because, if there are no known contraindications to its use, if given early it may abort an acute attack of angle closure and allow the necessary laser or surgical treatment to be carried out on a less damaged and less congested eye.

*Plateau iris* p551

**Secondary glaucoma**

This is defined as an intraocular raised pressure above 21 mm Hg as a result of local ocular or general pathological conditions. It is considered in Chapter 31 p578.

# CHAPTER 8

# RED STICKY EYES (CONJUNCTIVITIS)

The most common ophthalmic symptom is one of "red sticky eyes" secondary to conjunctivitis.

*Conjunctivitis* is inflammation of the conjunctiva and is characterised by injection of blood vessels *(hyperaemia),* oedema *(chemosis),* cellular infiltration and an exudation usually containing mucus which may or may not be infected. There is frequently associated discomfort, foreign body sensation, photophobia, watering *(lacrimation)* and in allergic conjunctivitis the eyes/eyelids are very itchy.

*Classification*

Conjunctivitis can be *primary* (with no associated disease) or *secondary* to a toxic effect or local spread of inflammation from diseased adnexal tissues (e.g. an inflamed lacrimal sac - *dacryocystitis*). Keratoconjunctivitis refers to the condition when associated with corneal inflammation *(keratitis).* Blepharoconjunctivitis refers to the condition when associated with eyelid inflammation *(blepharitis).*

In addition conjunctivitis can be either *acute* or *chronic.*

*Aetiology*

The aetiology of conjunctivitis is determined from the history, the clinical findings and examination of conjunctival discharge and scrapings (including microbiology and virology). The commonest causes of conjunctivitis include bacteria, viruses and allergy, but fungal, chemical or toxic inflammation and mechanical irritation should not be forgotten.

*Conjunctival discharge.* Most bacterial inflammations are accompanied by a purulent or muco-purulent discharge which may be profuse. Mucopurulent discharge is also a feature of chlamydial infection. If the conjunctivitis has an allergic, toxic or viral aetiology the discharge is usually watery or serous with little exudate which may consist of very tenacious mucus in vernal conjunctivitis.

*Associated signs.* Viral conjunctivitis is often accompanied by preauricular lymphadenopathy. The presence or absence of conjunctival follicles, papillae, inflammatory membranes and pseudo-membranes and the distribution of conjunctival injection should be noted, as well as the presence of any obstruction to the lacrimal passages, dacryocystitis, blepharitis or keratitis. Papillae are a non-specific response to inflammation and consist of tightly packed elevations each with a vascular core. Follicles are dome-shaped elevations containing lymphocyte aggregations and no vascular core. Adherent (true) inflammatory membranes form in severe infections as exudate permeates the epithelium. Pseudomembranes are non-adherent and consist of coagulated surface exudate which can be peeled off with ease.

*Investigation.* With suspected bacterial conjunctivitis, conjunctival swabs should be obtained for microbiological investigation if an initial course of broad spectrum topical antibiotic has failed.

*Microscopy.* The material should be placed on a glass slide and allowed to dry before being stained with Gram's stain for microscopy. Giemsa stain can be used to identify the cells within the discharge which tend to be neutrophils in bacterial infection, lymphocytes in viral disease, a mixture in chlamydial disorders and eosinophils in allergic cases. Ziehl-Neelsen stain is used to identify *Mycobacterium tuberculosis.*

*Cultures.* These are usually made on blood (aerobic bacteria, fungi) and chocolate agar (neisseria, haemophilus), thioglycolate and meat broths (anaerobes) or on other special media when necessary. Sabouraud's agar and brain-heart broth may be used to grow fungi.

*Virus detection.* Viral conjunctivitis is usually diagnosed on the basis of clinical examination. However, in certain instances conjunctival scrapings are necessary, obtained by the edge of a scalpel held vertically, the resultant material being placed on a slide, allowed to dry and examined for inclusion bodies (Papanicoulou stain). The detection of circulating antibodies (serology) and virus culture may be helpful.

*Immunology.* Serum and tear IgE levels are raised in certain allergic conditions. Specific tests using fluorescein-labelled monoclonal antibodies or enzyme-linked immunosorbant assay (ELISA) are highly sensitive and specific for certain infections (chlamydial and viral).

*Conjunctival biopsy.* This can be diagnostic in ocular cicatricial pemphigoid and sarcoidosis.

## Types of conjunctivitis

-Bacterial
-Chlamydial
-Viral
-Allergic/Hypersensitivity
-Toxic
-Other

## Bacterial conjunctivitis

*Acute bacterial conjunctivitis.* The commonest causative bacteria are *Staphylococcus aureus* (in children and adults), *Streptococcus pneumoniae* and *Haemophilus influenzae* (especially in children) and others include *Streptococcus viridans* and *pyogenes.* Usually the onset is relatively acute and the eyelids are frequently stuck together with a mucopurulent exudate (Plate 8.1 p193, LCW 3.1 p41) upon waking. By the time of presentation the infection is usually bilateral. Subconjunctival haemorrhages may be seen particularly with pneumococcal or haemophilus infections.

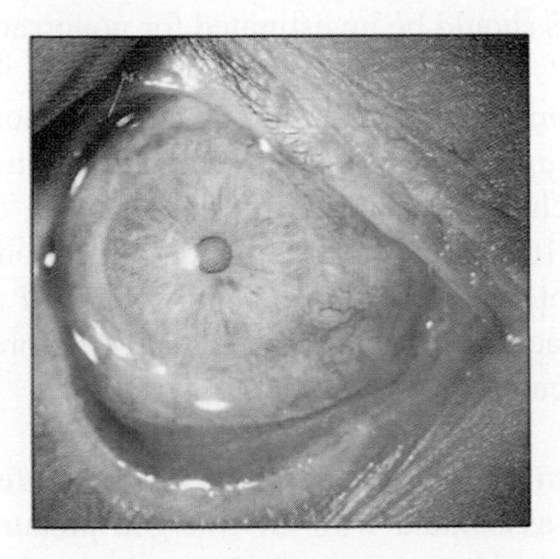

Plate 8.1 Bacterial conjunctivitis (mucopurulent discharge has been washed away).

The conjunctivitis usually responds well to a 7-10 day course of topical antibiotic drops or ointment (e.g. chloramphenicol). Combined preparations of corticosteroids with antibiotics are contraindicated.

*Hyperacute (gonococcal) conjunctivitis.* Gonococcal conjunctivitis (p322) is severe, rapidly progressive, particularly purulent and liable to lead to rapid corneal ulceration because *Neisseria gonorrhoeae* are able to penetrate an intact epithelium (unlike the majority of bacteria). Other signs include swollen eyelids, chemosis and preauricular lymphadenopathy. The conjunctivitis has an incubation period of up to five days and tends to affect adults usually as a result of infection from an acute urethritis. The infection can also infect the newborn infant of a mother with an infection of the birth canal (ophthalmia neonatorum p198). The organism can be demonstrated as Gram-negative intracellular diplococci in conjunctival scrapings at the onset and in the exudate a little later. Culture is best achieved using chocolate agar or Thayer-Martin media.

Treatment is with systemic and local antibiotics (penicillin with probenecid, spectinomycin or tetracycline) together with frequent ocular irrigation. Any accompanying keratitis and uveitis should be managed with appropriate, intensive treatment. In some cases gonococcal urethritis may cause a bacteraemia with signs of polyarthritis and tenosynovitis. Sterile conjunctivitis may also occur, which must be differentiated from Reiter's syndrome (p360). In all cases, sexual partners should be investigated for gonorrhoea.

*Chronic bacterial conjunctivitis:* Chronic infection is most often due to toxins of *Staphylococcus aureus or S. epidermidis* resident in the lid margin. Associated signs include blepharitis, inferior punctate corneal erosions and marginal keratitis (p167). Other causes include infection with gram-negative bacteria such as proteus, klebsiella, *Escherichia coli* and moraxella. Any associated chronic dacryocystitis has to be treated before an improvement in the conjunctivitis occurs.

*Tuberculous conjunctivitis* (p323) may be a primary infection appearing as a lesion in the fornix associated with an enlarged preauricular gland or may result from the spread of lupus or another cutaneous tuberculous condition of the eyelids (see also phlyctenulosis pp205, 324).

## Chlamydial conjunctivitis and keratitis

Chlamydial micro-organisms resemble bacteria in many ways but like viruses are obligatory intracellular parasites. In the eye *Chlamydia trachomatis* can cause trachoma (types A, B, Ba and C), adult inclusion conjunctivitis (AIC) (types D-K) and neonatal inclusion conjunctivitis (NIC) (types D-K), which is a form of ophthalmia neonatorum (p198).

*Trachoma*

*"Trachoma"* literally means *"rough"* and this refers to the appearance of the upper tarsal conjunctiva of patients with trachoma. Trachoma affects about 400 million people, mainly in developing countries, about 6 million of whom are blind. It is one of the commonest causes of preventable blindness worldwide. It is believed that multiple serial exposures to *C.trachomatis* are required for development of the blinding condition, whereas a single acute infection is usually self-limiting.

*Classification.* The latest World Health Organisation classification scheme is as follows:

A. Normal tarsal conjunctiva, no corneal disease

B. Trachomatous follicular inflammation (TI): Immature follicles (>5) are seen in the superior tarsal conjunctiva. (LCW 3.11 p45)

C. Trachomatous intense inflammation (TI): Mature follicles (subtarsal and limbal, papillary hypertrophy and inflammatory thickening obscuring more than 50% of the deep tarsal vessels is evident (Plate 7.21 p173, LCW 3.12 p45). There is usually a vascularised superficial corneal opacity which spreads from above *(pannus)* (Plate 8.2 p196) and often corneal punctate epithelial erosions and/or subepithelial infiltrates (usually affecting the upper third of the cornea).

D. Trachomatous conjunctival scarring (TS): Fine, branching linear scars develop in the superior tarsal conjunctiva (Von Arlt's lines). Eventually a large white subepithelial scar forms subtarsally (Plate 7.22 p174). Small depressions (Herbert's pits) mark the sites of previous limbal follicles (Plate 8.2 p196, LCW 3.12 p45).

E. Trachomatous trichiasis (TT):Subtarsal scarring results in cicatrisation with consequent entropion, trichiasis, lagophthalmos and dry eye.

F. Trachomatous corneal opacification (CO):Dry eye, exposure, secondary bacterial infection and the continual development of pannus results in corneal opacification and blindness (Plate 22.1 p448).

The WHO Expert Committee on Trachoma considers that the clinical diagnosis of trachoma requires the presence of at least two of the following signs: 1. *Superior tarsal follicles* 2. *Limbal follicles / Herbert's pits* 3. *Corneal pannus* 4. *Characteristic linear conjunctival scarring.*

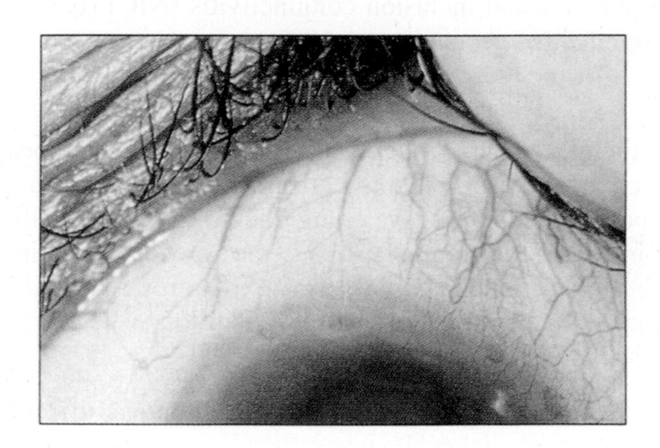

Plate 8.2 Trachoma pannus and Herbert's pits.

*Investigations.* Laboratory investigations may help to confirm clinical suspicion. Perinuclear intracytoplasmic chlamydial inclusion bodies in conjunctival cells can be demonstrated with Giemsa stain or by specific immunofluorescence (fluorescent monoclonal antibody staining) which is a more specific and sensitive test. Furthermore, type-specific antibodies may be identified in peripheral blood or eye secretions by immunofluorescence.

*Treatment.* Trachoma is an important cause of preventable blindness in the world. Community measures to improve hygiene and water supplies are essential. In addition, control programmes are important, their aim being to reduce the risks of the disease in individuals and within the community. One method of treatment is the intermittent topical application of tetracycline 1% eye ointment four hourly for six consecutive days each month, for three months. Systemic therapy with doxycycline, tetracycline or erythromycin are satisfactory for the treatment of individuals but are not recommended for large scale operations in view of adverse side effects. Vaccines have not been successful so far, but treatment with azithromycin tablets as infrequently as three times a year may prove to be an ideal preventive therapy. Surgical treatment is necessary to correct scarring of the eyelids and trichiasis which so commonly cause corneal ulceration, scarring and blindness.

*Adult Inclusion Conjunctivitis (AIC)*

'AIC' is associated with a non-specific urethritis in men and chronic cervicitis in women which may be symptomless. In most cases, adult infection is transmitted sexually but it can survive in swimming pools and cause a form of 'swimming-pool conjunctivitis'. The condition can be difficult to distinguish from adenoviral infection (p200). Within 3 weeks of exposure the conjunctivitis commences as an acute, unilateral follicular inflammation but it becomes bilateral shortly afterwards (Plate 8.3 p197). The discharge is mucopurulent and preauricular lymphadenopathy may occur. After several weeks the inflammation subsides and a persistent chronic conjunctivitis develops which lasts several months. The cornea is rarely involved (occasionally peripheral subepithelial infiltrates are seen) and eventually, in most cases, recovery is complete without conjunctival scarring. The best diagnostic investigation is fluorescent monoclonal antibody staining. The condition responds well to a ten day course of oral tetracycline, doxycycline or erythromycin. Local treatment with tetracycline is useful but less effective and must be continued for at least three weeks.

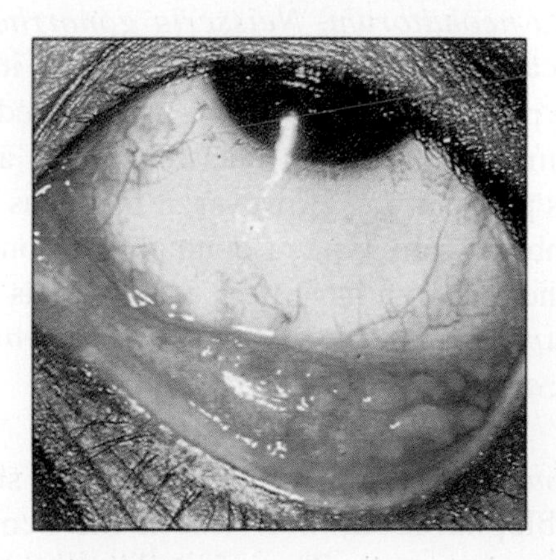

Plate 8.3 Adult inclusion conjunctivitis (chlamydial).

*Neonatal Inclusion Conjunctivitis (NIC): (see below)*

**Ophthalmia neonatorum** (Neonatal conjunctivitis) (LCW 8.10 p125)

Ophthalmia neonatorum is conjunctivitis occuring within the first three weeks of life. In the United Kingdom it is a notifiable disease and in many cases energetic treatment is necessary if permanent corneal damage is to be prevented. The most serious cause is infection with Neisseria gonorrhoeae, but in the UK it is caused more commonly by *Chlamydia trachomatis* (neonatal inclusion conjunctivitis - NIC) and by *staphylococci* or *pneumococci. Herpes simplex type 2 virus* is a relatively rare cause. In all cases a papillary reaction predominates. Follicles and lymphadenopathy do not occur since the neonate has an immature immune system.

*Neonatal Inclusion Conjunctivitis (NIC).* Chlamydia trachomatis (types D-K) can infect the newborn infant of a mother with an infection of the birth canal and is the commonest cause of ophthalmia neonatorum in the UK. Signs of infection usually occur between 5 and 14 days of age. The conjunctivitis may be uni- or bi-lateral, mucopurulent or purulent and may be complicated by an inflammatory membrane or pseudomembrane. If treated adequately, sequelae are rare.

*Bacterial ophthalmia neonatorum. Neisseria gonorrhoeae* accounts for a small proportion of cases but is important in view of its severity. The conjunctivitis is often hyperacute and there is a pronounced purulent discharge. Gonococcal ophthalmia neonatorum classically has an early onset from between 1 and 3 days post partum. However, if there has been premature rupture of maternal membranes any type of ophthalmia neonatorum may have an early onset. Many other bacteria have been identified as the causative organisms including *Staphylococcus aureus, Streptococcus pneumoniae, Haemophilus influenzae* and *Pseudomonas aeruginosa.*

*Viral ophthalmia neonatorum.* Active maternal herpes simplex type 2 infection may result in blepharoconjunctivitis in the neonate with an onset between 5 and 7 days post partum. The infected eye is at risk of herpetic keratitis.

*Management of ophthalmia neonatorum*

Early treatment is essential and diagnostic investigations must be carried out. Specimens should be obtained from the child, the mother and her sexual partner(s). Conjunctival swabs and epithelial scrapings should be obtained from the infant. Gram staining should be performed immediately and cultures set up. If possible, fluorescent monoclonal antibody staining provides the most rapid diagnosis.

Treatment of the infant consists of frequent irrigation. In some cases where the lids are tightly closed, they may have to be opened by retractors although great care is required in their use to prevent corneal injury and one should be aware that pus may spurt out under pressure. Treatment with systemic and local antibiotics is appropriate for bacterial conditions and aciclovir is used in herpetic infection. Inclusion conjunctivitis is treated with systemic and topical erythromycin. In all cases, the mother and her sexual partner(s) require appropriate treatment.

Credé's prophylaxis consisted of instillation of one drop of 1% silver nitrate into the lower conjunctival sac of each eye immediately after birth. The regimen did not prevent NIC but was able to eliminate gonococcal infection as effectively as any antibiotic. However, because NIC is the more common cause, because silver nitrate produces a chemical conjunctivitis and because topical and systemic penicillin is so effective this prophylactic therapy is rarely used today.

**Viral conjunctivitis** (p329, LCW 3.2 p41)

Infection of the conjunctiva by viruses may be part of a systemic infection or it may be a local condition limited to the epithelium of the cornea and conjunctiva. The ocular inflammation may be a very minor part of the general infection as in measles or mumps or it may be a main feature as in herpes zoster and herpes simplex or epidemic kerato-conjunctivitis. Herpetic infections are considered on pp168, 330.

*Adenoviral conjunctivitis.* Adenoviral conjunctivitis can be classified into 3 types:

-pharyngoconjunctival fever (PCF),
-epidemic keratoconjunctivitis (EKC),
-non-specific follicular conjunctivitis (NFC).

Diagnosis is clinical, but can be aided by serology, viral cultures and adenovirus detection from conjunctival specimens by electron microscopy, immunofluorescence antibody or enzyme-linked immunosorbant assay (ELISA) testing.

*Pharyngoconjunctival fever (PCF)* is usually caused by adenovirus type 3, 4 or 7, and is characterised by pharyngitis, acute follicular conjunctivitis, fever and general malaise. The painful condition is highly infectious and tends to occur in epidemics within schools or organisations, with onset about a week after initial exposure. The condition can occur as a form of "swimming-pool conjunctivitis". Features of the conjunctivitis, which is usually bilateral, include swollen eyelids, follicular conjunctivitis (affecting inferior tarsal conjunctiva in particular), marked hyperaemia, chemosis, minor subepithelial haemorrhages and a serous discharge. Preauricular lymphadenopathy is commonly evident and a keratitis characterised by scattered subepithelial (immune complex) infiltrates may develop (LCW 3.3 p41). An associated anterior uveitis is rare. Management should include scrupulous personal hygiene to avoid infecting close contacts, topical antibiotics to prevent secondary bacterial infection and paracetamol to ease pain and reduce the fever. Topical steroid therapy can reduce the discomfort of the keratitis but should only be prescribed by an ophthalmologist and in severe cases. Antiviral therapy appears to be of no benefit. Resolution usually occurs within 3 weeks, although the keratitis may persist for several months.

*Epidemic keratoconjunctivitis (EKC)* is usually caused by adenovirus type 8 or 19 and is characterised by conjunctivitis in the absence of major systemic symptoms (cf PCF p200). Type 19 adenoviral infection can be transmitted between sexual partners and is a known cause of cervicitis. The highly infectious condition can occur as a form of "swimming-pool conjunctivitis". Features of the conjunctivitis, which is usually unilateral, include swollen eyelids, follicular and papillary conjunctivitis, marked hyperaemia with or without subepithelial haemorrhages, chemosis and a serous discharge. In severe cases subconjunctival haemorrhage may be significant and an inflammatory membrane, pseudomembrane or symblepharon may form. Preauricular lymphadenopathy is commonly evident. In many cases a very painful keratitis characterised by scattered epithelial punctate lesions develops together with an associated anterior uveitis. Often the keratitis progresses to one characterised by small subepithelial infiltrates or occasionally, multiple disc shaped opacities (pp171, 331). Management is as for PCF (see above). Resolution usually occurs within 3 weeks, although the keratitis may persist for several years.

*Non-specific follicular conjunctivitis (NFC)* is a mild adenoviral conjunctivitis which is usually self-limiting and requires no treatment.

*Acute haemorrhagic viral conjunctivitis* is characterised by acute follicular conjunctivitis, often gross subconjunctival haemorrhage and a fine punctate epithelial keratitis, this rare picornavirus (Enterovirus 70) infection is highly infectious but self-limiting.

*Molluscum contagiosum.* This viral skin infection which can infect the eyelids (pp329, 456 and Plate 17.7 p329) is often associated with a secondary chronic follicular conjunctivitis.

## Allergic/hypersensitivity conjunctivitis

*Seasonal (allergic) conjunctivitis.* Conjunctival inflammation can occur in atopic individuals following exposure to pollens (hayfever) and other allergens as a type 1 (immediate) hypersensitivity reaction. Features include redness, watering, itchiness and swelling of the lids. Chemosis is often prominent and there is a diffuse papillary reaction. The cornea is not affected. The secretion contains many eosinophils but the inflammation results from mast cell degranulation and thus responds to treatment with antihistamine preparations or to G.Sodium cromoglycate 2% which prevents histamine release

*Vernal (kerato) conjunctivitis (spring catarrh)* is a florid, recurrent, bilateral conjunctival condition usually occurring in the spring or summer. It typically affects children and young adults, is more common in males and atopic individuals and is thought to be mediated by mast cells and eosinophils. The underlying mechanism is thought to be a combined type I and type IV hypersensitivity response. The major symptom is severe itchiness and this is accompanied by a tenacious mucus discharge, lacrimation, pinkish hyperaemia, foreign body sensation and photophobia.

In the *palpebral* form the tarsal conjunctiva, especially of the upper lid, is hyperplastic and there are large, flat-topped pale papillae with a cobblestone appearance (LCW 3.13 p45). Sometimes there is an associated ptosis. Subconjunctival scarring does not occur, regardless of disease duration. The *limbal* type has a 3-4mm wide elevated area of perilimbal chemosis. The limbitis has a gelatinous appearance which may develop into a concentric ring of mucoid nodules. Raised superficial, chalky-white infiltrates *(Trantas' dots)* may be seen straddling the limbus.

The major complication is keratitis which can be divided into 5 types: 1. Punctate epithelial micro-erosions 2. Geographic epithelial macro-erosions 3. Vernal plaque formation (due to the coating of macro-erosions with layers of altered mucus discharge) 4. Subepithelial scarring (usually as a ring-scar) 5. Pseudogerontoxon (resembles an arcus senilis)

The condition responds slowly to drops of sodium cromoglycate, antihistamines and cortico-steroids, although prolonged use of steroid should be avoided if possible. Vasoconstrictors, such as weak adrenaline, can reduce hyperaemia and chemosis but rebound vasodilatation and tachyphylaxis occur with overuse. Prostaglandin release has been implicated in vernal conjunctivitis and systemic aspirin has been used successfully in severe cases. Remission of the condition tends to occur before the age of 20 years.

*Giant papillary conjunctivitis (GPC)* (LCW 10.6 p146) is due to direct irritation of the superior palpebral conjunctiva. The condition is associated with the presence of contact lenses, ocular prostheses or prominent suture ends. Symptoms are highly variable but, in general, are similar to those of vernal disease (see above). Early in disease progression small (0.3mm diameter or less) non-specific papillae develop. Continual irritation results in inter-papillary fibrous septa rupture, collagen proliferation and the formation of giant (more than 0.3mm diameter) papillae, often associated with conjunctival ulceration.

The pathogenesis of GPC is complex and involves patient predisposition since not every contact lens wearer develops the condition. Management includes removal of the "foreign body" (e.g. suture or contact lens), strict lens or prosthesis hygiene (preferably with a non-preserved and hydrogen peroxide based system) and improved contact lens design, fitting and material. A change from soft contact lenses to rigid gas-permeable lenses may be all that is required. Regular irrigation with unpreserved saline relieves symptoms but the use of topical corticosteroids is not effective unless the inflammation is acute and severe.

*Atopic keratoconjunctivitis (AKC).* Atopy, characterised by eczema, asthma, rhinitis and a predispostion to other allergic disorders including allergic conjunctivitis, is common. True AKC, however, is rare, non-seasonal and is characterised by severe bilateral chronic papillary conjunctivitis, keratitis and atopic dermatitis involving the eyelids. The underlying mechanism is thought to be a combined type I and type IV hypersensitivity response. Onset is usually in adults who complain of itch, a burning sensation, lacrimation, mucus discharge, photophobia and blurred vision. Conjunctival hyperaemia is often only mild but the eyelid margins are red, indurated and scaly. The associated keratitis is usually a punctate epitheliopathy and there is an increased risk of suppurative keratitis, corneal neovascularisation, conjunctival and corneal scarring. In addition, AKC is associated with keratoconus and cataract formation. Diagnosis is clinical but can be aided by raised serum and tear IgE levels. Treatment is with systemic and topical antihistamine preparations and corticosteroids with topical sodium cromoglycate.

*Ocular cicatricial pemphigoid (OCP)* (previously called *benign mucous membrane pemphigoid (*p481) is an autoimmune disorder which may prove to result from a type II hypersensitivity reaction with deposition of conjunctival basement membrane autoantibodies. OCP fits into a spectrum of a variety of disorders encompassed by the term *"chronic progressive cicatrising conjunctivitis"* including Stevens-Johnson Syndrome (p481 and Plate 8.4 p204, LCW 3.5 p42) pemphigus, dermatitis herpetiformis, and drug-induced or pseudopemphigoid. It is most commonly seen in patients over seventy years of age but sometimes earlier. Early ocular features of OCP include those of non-specific chronic conjunctivitis (hyperaemia, chemosis, papillary inflammation and discharge). The discharge is mucoid and has a tendency to adhere to the epithelial surface. A useful early sign, to distinguish the disorder from other causes, is loss of inner canthal architecture. Acute exacerbations are associated with conjunctival ulceration which may be complicated by secondary infection. The main feature of the progressive disorder, however, is subepithelial fibrosis, forniceal shrinkage, symblepharon, secondary disorders of eyelid position, dry eye and subsequent corneal involvement which can result in blindness. Definitive diagnosis is made by conjunctival biopsy and the demonstration of basement membrane immunoglobulin. The disease may be confined to the eyes but other mucous membranes maybe involved including the mouth, oesophagus, larynx, nose, urethra and rectum.

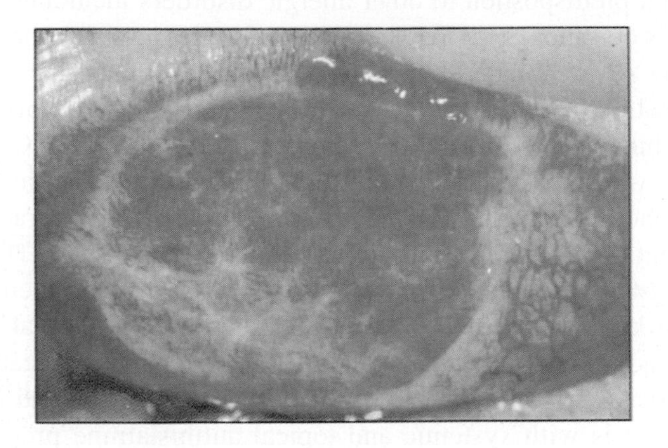

Plate 8.4 Stevens - Johnson syndrome - scarred conjunctiva of upper lid.

Treatment depends on disease stage and activity. Topical corticosteroids are used when there is evidence of acute inflammation, but are ineffective on a chronic basis. Systemic immunosuppression, however, is a useful tool in the management of chronic progressive cicatrisation. Drugs used include corticosteroids, cyclophosphamide, methotrexate, azathioprine and diaminodiphenylsulphone (dapsone). Secondary ocular surface disorders and complications are managed in a standard manner, but oculoplastic procedures should be carried out only when the disease process has been controlled, so as to avoid inflammatory flare-ups.

*Phlyctenulosis.* Phlyctenular conjunctivitis (p324) takes the form of a small vascularised nodule on the bulbar conjunctiva frequently situated near the limbus. The granulomatous condition is believed to be a type IV hypersensitivity reaction to bacterial proteins and is more common in children than adults. It was relatively common when tuberculosis was more prevalent, but in addition it can occur in association with staphylococci and less commonly with other organisms (e.g. candida, coccidioidomycosis, lymphogranuloma venereum). The lesions respond rapidly to local treatment with corticosteroids and when necessary appropriate antimicrobials. In cases presumed to be secondary to staphylococcal blepharitis, frequent eyelid hygiene is recommended.

## Toxic conjunctivitis

Adverse reactions to many topical medications can occur, either after chronic use (e.g. antiglaucomatous therapy), or in a more acute manner. Acute allergic or toxic reactions have been reported with almost all topical agents, although the preservative (benzalkonium chloride, thiomersal) or vehicle may have been the cause in many of these cases. In some instances the reaction can be hyperacute with copious mucopurulent discharge. In most cases, however, the reaction is less dramatic, being characterised by non-specific hyperaemia, papillary or follicular conjunctivitis (particularly of the inferior palpebral conjunctiva), a punctate epithelial keratopathy and a watery or serous discharge. In true allergy there has usually been repeated previous exposure to the allergen with resultant sensitisation of the immune system. In practice, without patch testing it can be difficult to differentiate between allergy and toxicity. In cases where there is no history of topical medication usage, other causes should be considered such as chemical irritation from swimming pools or exposure to other irritants. When all other causes have been eliminated one can consider self-mutilation.

Management consists of stopping the potentially responsible agent(s) when possible. If continued topical therapy is required the use of unpreserved agents is advised and may be curative. Rarely, a short course of weak topical cortico-steroids may help.

## Other causes of red sticky eyes

*Blepharo-conjunctivitis.* Persistent conjunctival inflammation secondary to blepharitis is common and a cause of chronic discomfort and misery. The most common cause of associated acute red sticky eye(s) is an acute or chronic staphylococcal infection which should be managed as for any bacterial conjunctivitis. However, until the blepharitis is controlled (p453) staphylococcus infected meibomian secretions keep the conjunctiva inflamed and at risk of further acute exacerbations of conjunctivitis.

*Rosacea kerato-conjunctivitis* (Plate 8.5 p206). Acne rosacea is a troublesome condition, more frequent in middle aged women and is characterised by vasomotor instability of the cutaneous blood vessels, especially of the face, causing chronic hyperaemia of cheeks, forehead and nose. Nasal skin involvement may result in rhinophyma. The conjunctival vessels share the instability and dilate, either as an emotional response or a reflex effect from the digestion of hot or spiced food, tea, coffee or chocolate, giving rise to episodes of facial flushing and red eyes. The eyelid meibomian glands tend to dilate, produce an altered secretion and then become the seat of a chronic staphylococcal inflammation. A chronic follicular conjunctivitis develops in the majority of cases and a keratitis in about 10%. The keratitis (Plate 8.6 p207) may be a mild

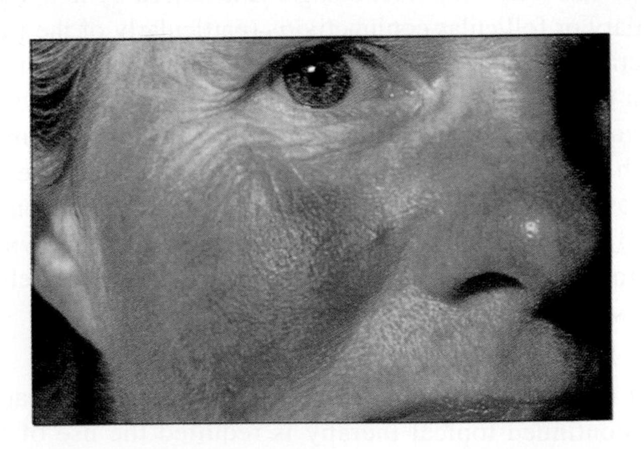

Plate 8.5 Rosacea- facial appearance.

punctate epithelial keratopathy but can be severe with subepithelial infiltrates which tend to ulcerate, peripheral scarring with vascularisation and peripheral or central corneal melting/thinning.

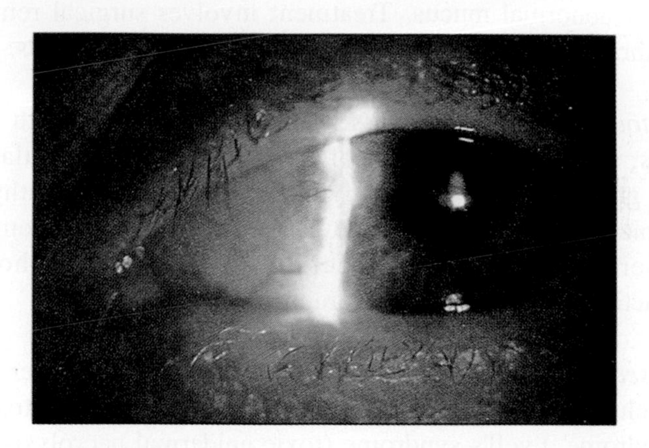

Plate 8.6 Rosacea- keratitis, peripheral infiltration advances centrally with vascularised scar liable to ulcerate.

Fortunately the acute manifestations respond well to topical steroid therapy and the chronic features improve following use of oral tetracycline 250 mg taken four times a day for 1 month followed by once daily for a period of about 6 months. The therapy has a direct antibacterial effect but, in addition appears to correct the abnormal nature the meibomian secretion. Stimuli which promote facial flushing, including dietary factors, should be avoided.

*Dry eye syndromes.* Kerato-conjunctivitis sicca (KCS) occurs due to a reduction in tear production and is frequently associated with recurrent conjunctival infections (p217) as are any causes of dry eye.

*Fungal conjunctivitis.* Fungal conjunctivitis is usually secondary to adnexal disease (p339)

*Parasitic conjunctivitis.* Certain parasitic conditions may produce conjunctivitis (p341)

*Ligneous conjunctivitis.* This is a rare disorder of unknown aetiology which affects children. The palpebral conjunctivitis involving all four lids is characterised by a massive fibrinous exudation and pseudomembrane formation with large amounts of granulation tissue. There is often an associated copious production of abnormal mucus. Treatment involves surgical removal of the pseudomembranes, intensive topical heparin and cortico-steroids.

*Granulomatous conjunctivitis.* This is rare but can occur with sarcoidosis, tuberculosis, syphilis or as a part of Parinaud's oculoglandular syndrome (monocular granulomatous conjunctivitis, local lymphadenopathy, fever and malaise) which may be secondary to cat-scratch fever, tularaemia, spirotrichosis, tuberculosis, syphilis, coccidioidomycosis, lymphogranuloma venereum, actinomycosis and other rare infections.

*Systemic disease.* Conjunctivitis may be a feature of a number of systemic disorders including rheumatoid arthritis, Wegener's granulomatosis, Stevens-Johnson syndrome, Lyell's syndrome (toxic epidermal necrolysis), scleroderma and linear IgA disease.

# CHAPTER 9

# WATERING EYES

The anatomy and physiology of the lacrimal apparatus is described on p27 and its development on p49.

*Lacrimation* is watering due to over-production of tears by the lacrimal gland, whereas *epiphora* is watering due to impairment of the lacrimal drainage system. The action of blinking provides a pumping mechanism. The lids close in a medial direction, milking the tears toward the puncta. The action of the orbicularis occludes the puncta and compresses the canaliculi, forcing the tears into the sac through the valvular opening of the common canaliculus. From the sac the tears drain by gravity to the nose.

## Causes of lacrimation

Reflex stimulation of the lacrimal nucleus, due to 1. Environmental factors e.g. dust, smoke, wind. These will always aggravate weeping due to the other causes. 2. Corneal irritation e.g. ulcer, foreign body. 3. Conjunctival irritation e.g. conjunctivitis. 4. Lid margin disease (blepharitis). 5. Dental or sinus disease.

Less commonly there are central causes such as migrainous neuralgia (cluster headache) and tabetic crisis. Aberrant re-innervation may rarely lead to anomalous lacrimation or 'crocodile tears', caused by facial nerve lesions proximal to the geniculate ganglion, because the salivary fibres regenerate into the greater superficial petrosal nerve and the eye waters in response to gustatory stimuli (Fig. 2.13 p22).

## Causes of epiphora

This may result from: 1. Pump failure e.g. facial palsy. 2. Punctal malposition e.g. ectropion. 3. Punctal stenosis 4. Canalicular obstruction. 5. Nasolacrimal duct obstruction, which is the most usual cause. 6. Nasal or sinus disease.

## Investigation of watering eyes

*Clues from the history.* Enquiry is made regarding ocular, dental or sinus disease, Bell's palsy, or previous facial injury but watering which is gradually becoming worse is the usual symptom. This is first evident in windy conditions becoming constant as the nasolacrimal duct is progressively stenosed.

*External eye disease.* Evidence of blepharitis, conjunctivitis or keratitis is sought. The upper lid must be everted to exclude the presence of a foreign body.

*Examination of the lid position and the punctal orifices*

*Fluorescein dye test.* A drop of fluorescein is placed in the eye in order to help visualise the tear meniscus and to estimate the clearance of tears. Normally very little dye is left after a few blinks. The appearance of dye at the opening of the nasolacrimal duct using a cotton bud placed in the nose is an indication of patency.

*Syringing.* The puncta are gently, slowly and progressively dilated if necessary with a Nettleship's dilator, taking great care not to rupture the dense ring of tissue around the orifice which keeps it open and aids drainage, otherwise it becomes a slit opening which is much less efficient. A Hay's punctum seeker is invaluable for stenosed puncta. A fine lacrimal cannula is then introduced into either canaliculus. The cannula is first held vertically and then turned horizontally. Lateral traction is applied to the lid to avoid trauma due to kinking of the canaliculus. Three things are noted: the quality of the stop provided by the tissues, the passage of saline to the nose and the site of any regurgitation through the puncta. In obstruction of one canaliculus, a soft stop will occur in less than 10mm, no saline will pass to the nose and saline will regurgitate through the punctum being irrigated. In *common canalicular obstruction,* a soft stop will occur, no saline will pass to the nose and regurgitation will occur through the other punctum. In *nasolacrimal obstruction,* a hard stop will occur (the tip of the cannula meeting the bony wall of the lacrimal fossa), no saline will pass to the nose and regurgitation will occur through the other punctum. In cases of *partial nasolacrimal obstruction (stenosis),* a hard stop is felt, some saline passes to the nose and some regurgitates through the other punctum. In these circumstances, fluorescein dye may be retrieved from the nose after syringing when it was absent before.

*Dacryocystography.* In certain circumstances it is useful to delineate the anatomy of the lacrimal passages radiographically by injecting the canaliculi with contrast medium. This test is performed in common canalicular obstructions, in craniofacial malformations, in post-traumatic obstructions, after failed lacrimal surgery or when the results of syringing are equivocal.

**Delayed canalisation of the lacrimal apparatus in babies**

This is usually due to non-perforation of a membrane at the lower end of the nasolacrimal duct (pp28, 49). The child presents with epiphora (LCW 8.9 p125), recurrent conjunctivitis, dacrocystitis or a mucocele. If spontaneous resolution is to occur it is generally within six months and rarely after a year. Controversy exists as to the timing of surgical intervention. The school of early intervention argue that the nasolacrimal duct should be probed soon after presentation in order to prevent chronic inflammation of the lacrimal passages and thus improve the ultimate prognosis and also to prevent the possible complications of dacryocystitis and fistula formation. The school of late intervention counter that resolution is likely in any case, and that probing carries a risk to the canaliculi. They advocate treatment with prophylactic topical antibiotics and regular massage in the sac region to empty any mucocele and to encourage distal patency by hydrostatic pressure.

A reasonable middle course seems to be to manage the child conservatively until the age of six months and then carry out probing of the lacrimal duct. If the child is not symptom free in one month then a dacryocystogram should be carried out. If this confirms a distal membranous obstruction then probing may be repeated or, in some centres, the system may be stented with soft silicone tubes for six months. Failure of two probings, or of a probing and an episode of intubation, is an indication for dacryocystorhinostomy (DCR). If the dacryocystogram obtained after the first probing shows atresia of the duct or obstruction due to bony encroachment then a DCR is mandatory.

**Acquired epiphora** (defective drainage of tears)

*Trauma:* canalicular lacerations are not uncommon (LCW 9.11 p136). These are discussed on p468.

*Malposition of the puncta.* The puncta are normally directed backwards into the marginal tear strip but even minor degrees of medial ectropion are sufficient to allow the puncta to fall forward. This prevents effective drainage which may be aggravated by punctal stenosis (p468). The treatment consists of a lid-shortening procedure at the site of maximum laxity. In the case of medial ectropion this can be combined with excision of a diamond of tarsoconjunctiva to invert the punctum.

*Chronic canaliculitis* may result from infection by actinomyces (streptothrix). Typically there is a chronic discharge with reddening of the skin over the canaliculus and a gaping prominent punctum. Somewhat deceptively a syringing indicates patency in most cases. Treatment consists of dilating the punctum and curetting the mycelium ('sulphur granules') from the canaliculus wall along its length and the topical application of tetracycline or benzyl penicillin (p339). Canaliculitis can also be caused by the herpes simplex virus, the chicken pox virus and chlamydia.

*Dacryocystitis.* (a) *Acute dacryocystitis,* due to pyogenic organisms spreading from the lacrimal sac to involve the surrounding tissues, often on the basis of a neglected nasolacrimal obstruction, presents as redness, swelling and pain in the region of the sac and if untreated the abscess may point through the skin or rarely into the nose. It is treated with appropriate systemic antibiotics. Residual epiphora, recurrent acute attacks, persistent fistula or chronic dacryocystitis are indications for dacryocystorhinostomy (DCR) after the acute episode has resolved (Stereo plate 9.1 p213). (b) *Chronic dacryocystitis* is almost always associated with nasolacrimal obstruction but may occur in sinus disease or after trauma or rarely in the course of granulomatous diseases. The occlusion of the naso-lacrimal duct usually at the narrow lower end of the lacrimal sac occurs more commonly in women and over the age of 45. Sometimes the only symptom is epiphora. The epiphora is not constant due to the normal evaporation of the tear film. It is usually troublesome where there is corneal stimulation e.g. in windy weather. Rarely occlusion of the tear duct may occur as a presenting symptom of carcinoma of the maxillary antrum. Muco-purulent discharge from the lacrimal sac into the eye will tend to cause a relapsing conjunctivitis. A mucocoele may form if the canaliculi are blocked.

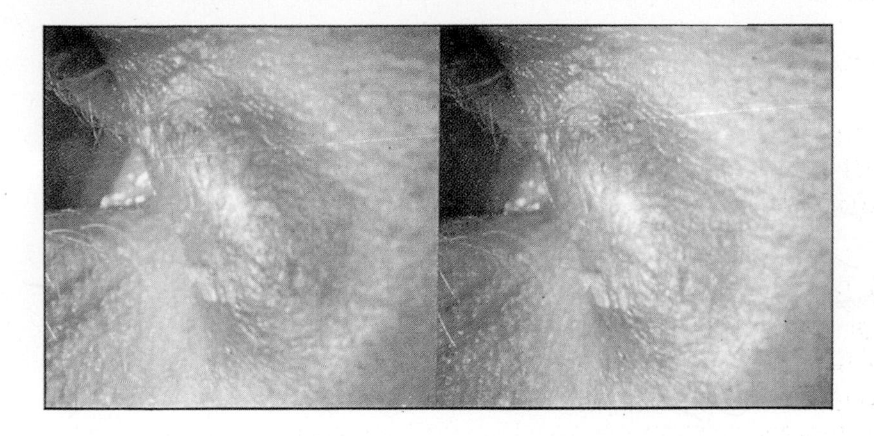

Stereo Plate 9.1 Right acute dacryocystitis. (see p4)

*Carcinoma of the lacrimal sac* may rarely have clinical features indistinguishable from chronic dacryocystitis. It may also present as a firm swelling in the region of the sac which is followed by ulceration of the overlying skin. Lymphoma or secondary spread from sinus carcinoma may also occur in this site.

**Treatment of naso-lacrimal duct obstruction in the adult**

If a patient is fit enough for surgery and considers that the epiphora is sufficiently troublesome, the operation of dacryocystorhinostomy (DCR) is carried out. It consists of removing the bony medial wall of the lacrimal fossa, opening the sac and the nasal mucosa and anastomosing them anteriorly and posteriorly, thus by-passing the naso-lacrimal duct (Figs. 9.2 - 9.7 pp214 - 215).

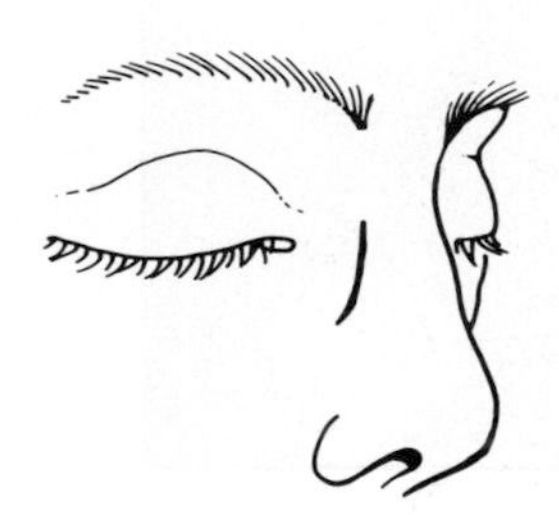

Fig. 9.2 DCR. Skin incision.

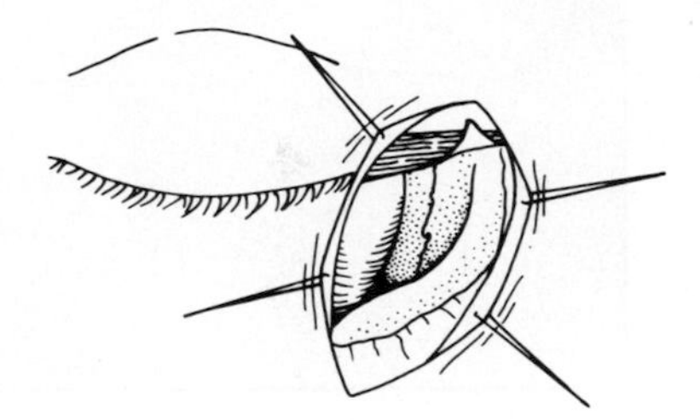

Fig. 9.3 DCR. Exposure of the lacrimal fossa.

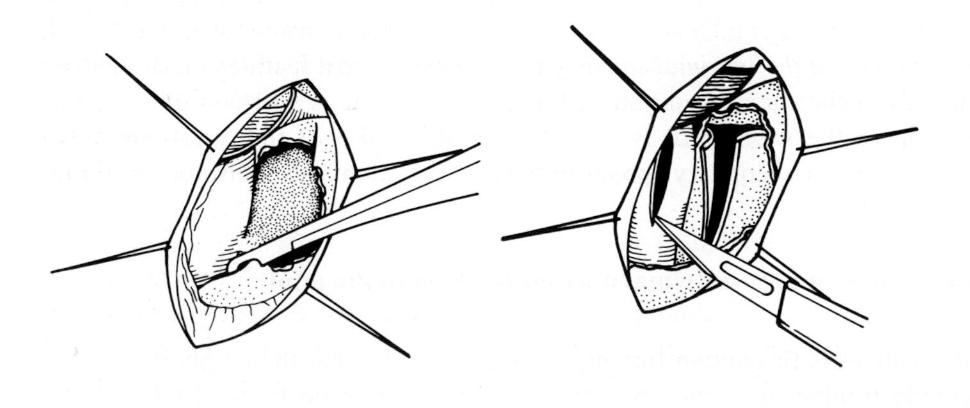

Fig. 9.4 DCR. Bone resection.

Fig. 9.5 DCR. Incising the lacrimal sac.

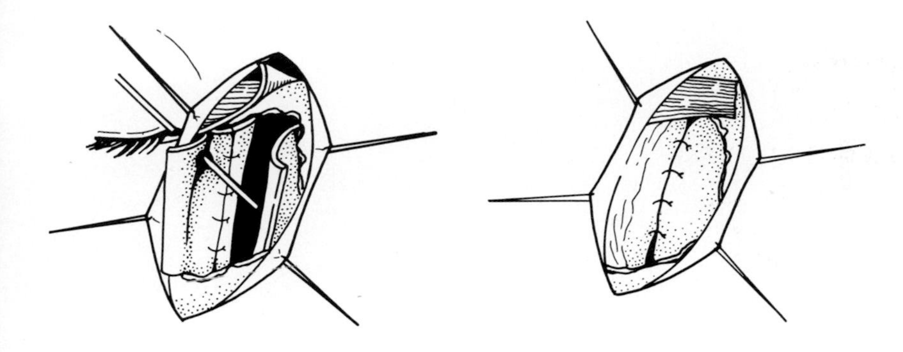

Fig. 9.6 DCR. Formation of mucosal flaps. Fig. 9.7 DCR. Closure of anterior flaps.

It is necessary to identify and deal with canalicular obstruction if present when carrying out a DCR operation. If active mucosal inflammation is found, in repeat surgery or where the common canaliculus is obstructed, silicone tubes are placed in a loop through the canaliculi and tied off in the nose to be removed when the passages have healed. When the lateral part of the common or the medial part of the individual canaliculi are blocked the canaliculi are anastomosed with the lateral sac wall and tubes are placed. If the proximal parts of the canaliculi are occluded or maldeveloped then a search can be made for them in a retrograde direction. Where little or no canalicular or sac structures exist due to congenital anomaly, inflammatory disease, trauma or radiotherapy then the DCR is combined with the passage of a permanent glass drainage tube from the medial canthus into the nose (the Lester-Jones tube). Endonasal surgery, where the ostium is created by laser from the lateral wall into the sac, is under trial.

*Carcinoma of the lacrimal sac* is treated by excision and radiotherapy.

# CHAPTER 10

# DRY EYES

## Reduction of tear flow (kerato-conjunctivitis sicca)

The tears and the tear film are described on p27. The flow varies considerably between individuals. A reduction of tear flow in the absence of ocular disease is common in the elderly. Pathological reduction of tear flow results in corneal drying and this causes the symptoms of the dry eye.

*Symptoms:* patients may say that their eyes feel dry but often they complain of a gritty feeling which is usually worse in smoke-laden atmospheres. They may present with recurrent infections of the conjunctiva or the lid margins to which patients with kerato-conjunctivitis sicca are particularly prone.

*Signs* of the reduction of tear flow may be seen with the slit lamp: thinning of the pre-corneal tear film with a faster break-up time than the normal film (LCW 6.20 p99), a slow movement of the tear film following a blink, the presence of excess mucus in the film, sometimes filaments of mucus attached to the corneal epithelium *(filamentary keratitis)* (Plate 10.1 p217, LCW 3.4 p42) and reduction of the marginal tear strip.

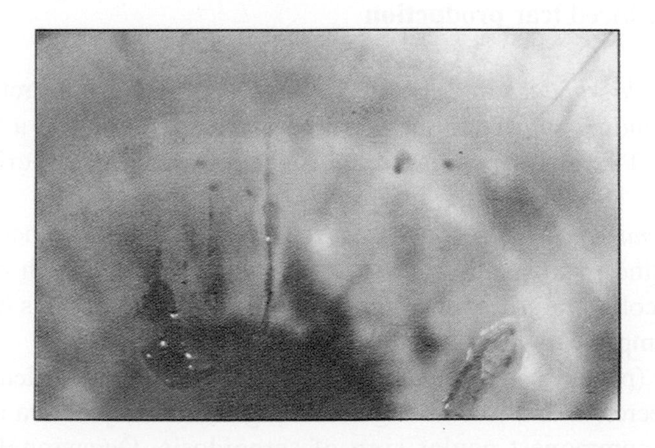

Plate 10.1 KCS filaments on cornea.

*Staining tests:* 1. *fluorescein* may be used to demonstrate more clearly the break-up time of the tear film, as the broken parts of the tear film appear darker, and Rose Bengal (p615) reveals the devitalised cells of the cornea and conjunctiva (Plate 10.2 p219). Those which are exposed in the interpalpebral fissure suffer most from desiccation. Classically, punctate staining is seen in a band across the cornea and conjunctiva in this region. However any impairment of the epithelium will result in Rose Bengal staining, so that its presence alone is not diagnostic of tear deficiency. Slight punctate staining of the conjunctiva with Rose Bengal can be regarded as a normal finding in older patients. 2. *Schirmer's test* uses standardised (Whatman's No.7) absorbent filter paper strips 5mm wide and folded at the end. The folded end is then hooked over the lower lid at the junction between the middle and nasal third of the lid margin. The papers are left in place for five minutes and the length of paper wetted by the tears is then measured and recorded. Readings in excess of 15mm are usually considered satisfactory. Falsely high readings may occur if the strips are put in immediately after the instillation of drops into the eyes because the drops absorbed by the papers are incorrectly assumed to be a part of the patient's tear flow, or because drops, e.g. Rose Bengal, stimulate increased tear secretion by irritation (stimulated Schirmer's test). 3. Lysozyme estimation can be used for research purposes.

## Causes of reduced tear production

-*Congenital alacrima* is rare. There is reduced or absent tear secretion.
-*Trauma* including surgery to the lacrimal gland may result in a loss of tear production. The gland is sometimes excised in cases of lacrimal gland tumour.
-*Sjögren's syndrome* (p359) consists of the triad of keratoconjunctivitis sicca, xerostomia and rheumatoid arthritis. It may be associated with evidence of widespread collagen disease. Juvenile chronic arthritis and Still's disease may also be accompanied by keratoconjunctivitis sicca.
-*Sarcoidosis* (p365). Sarcoid granulomas in the lacrimal glands lead to diminished tear secretion and Rose Bengal staining of the conjunctiva and cornea. This is the commonest ocular sign of sarcoidosis. Occasionally there is swelling of the lacrimal glands and the eyes usually show a severe reduction of tear secretion, Plate 10.2 p219 was such a case.

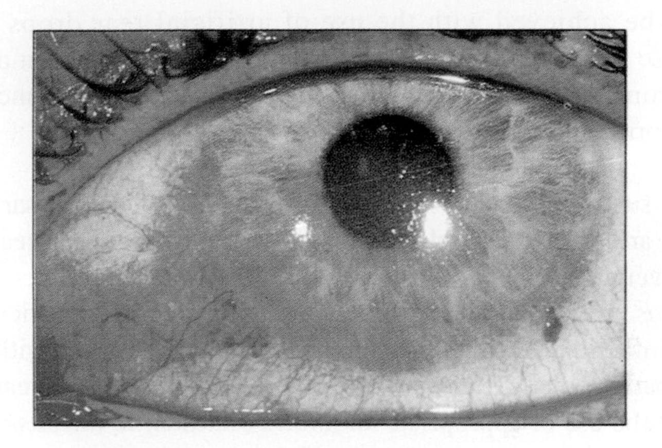

Plate 10.2 KCS Rose Bengal staining.

-*Systemic lupus erythematosis* is frequently associated with keratoconjunctivitis sicca (p363).
-*Scarring of the conjunctiva:* certain disease processes cause severe progressive destruction and scarring of the conjunctiva. This may close the ducts of the lacrimal gland. Examples include: trachoma (p195), ocular cicatricial pemphigoid (syn. benign mucous membrane pemphigoid) (p203), erythema multiforme (Stevens Johnson syndrome) (p481).

*Tumours of the lacrimal gland* (p466)

*Infection of the lacrimal gland (dacryoadenitis)* (p465)

*Treatment of kerato-conjunctivitis sicca*

Accurate diagnosis of the underlying cause is important because this may give an indication of prognosis. Sometimes the underlying cause may require treatment as in lacrimal gland tumour or sarcoidosis, but considerable symptomatic relief may be achieved with the use of artificial tear drops e.g. guttae hypromellose (p616). The patient should use the drops as often as necessary to achieve comfort and should instil them *before* the eye feels uncomfortable, i.e. before corneal epithelial drying has occurred.

Mucolytics, such as acetyl cysteine drops (p615) may be necessary especially if filaments are present but adequate hydration with artificial tears of which there are several types will, in the majority of cases, be sufficient to dissolve excess mucus. Patients with keratoconjunctivitis sicca are particularly prone to chronic infection, especially of the lashes. Inflammation with even relatively non-pathogenic organisms can upset the stability of the tear film. Lash hygiene and the use of appropriate antibiotic ointment to the *base of the lashes* can considerably improve patient comfort.

Occlusion of the puncta and the canaliculi by electro-cautery can give long lasting benefit to patients with very dry eyes. However, before occluding the puncta it is essential to be sure that there is no likelihood of recovery of tear production and that it is so reduced that epiphora is unlikely to result, so at least 12 months should elapse before considering occlusion. Temporary occlusion of the canaliculi by gelatine rods may give some indication of the risk of epiphora occurring if permanent occlusion is to be carried out. For extreme cases a variety of devices have been used to keep the eyes moist, including constant infusion of artificial tears from a reservoir. They are not in common use.

# CHAPTER 11

# SQUINTING EYES (STRABISMUS) - DISORDERS OF OCULAR MOTILITY

Eyes with full movements, no deviation and full binocular vision can be regarded as the ideal state. A disorder of motility is a broad term which is used to encompass such conditions as concomitant squint, paralytic squint nystagmus and anomalies of convergence. The terms squint and strabismus are alternatives and the simpler shorter term will be used here. Squint is due to some barrier to the reflexes which serve binocular vision and whether this is efferent, central, or afferent in type, management aims to overcome the relevant barrier. On inspection this can be judged by noting that the reflection of a light at which the patient is looking will appear either to be central (LCW 1.11 p14) or symmetrically situated with respect to the pupil, also that covering one eye leads to no movement of the other eye. (cover test p229)

A squint is defined as the inability to make co-ordinated movements of the two eyes, so that the visual axes are not directed simultaneously towards the same fixation point. The squint may be *concomitant,* where the angle of deviation remains approximately constant in all positions of gaze, or *incomitant* where the angle varies with the gaze position. Incomitant, or paralytic, squints are considered on p239. Concomitant squints may be subdivided according to whether the visual axes are deviated inwards ("eso" deviations) or outward ("exo" deviations).

## Explanation of some of the terms used in squint management

### *Manifest concomitant squint (heterotropia)*

In a manifest squint the direction of the eyes is such that with both eyes open the target image falls on the fovea of one eye but on a non-foveal part of the retina of the fellow eye. An in-turning manifest squint is called an *esotropia,* an out-turning manifest squint an *exotropia.* The squint may be constant, or only appear intermittently, and it may be present at only one viewing distance (e.g. right esotropia for near).

*Latent concomitant squint (heterophoria)*

When the stimulus to maintain binocular single vision is absent, e.g. when one eye is covered and the other views a distant object, then the eyes will take up the 'anatomical' position of rest. In these circumstances the eyes may be deviated either outwards *(exophoria,* latent divergent squint) or inwards *(esophoria,* latent convergent squint). Less commonly one eye may be elevated *(hyperphoria)* or depressed *(hypophoria).* A number of people by chance have straight eyes in the position of rest *(orthophoria).* No matter how great the deviation, provided it is instantly self-correcting at the moment binocular viewing is permitted, the term latent squint or 'phoria' is applicable.

*Binocular single vision*

Normal binocular single vision can only occur when the visual axes are correctly aligned so that the image of the object of regard falls simultaneously on the fovea of each eye. There are three degrees of capacity for binocular single vision:

1. *simultaneous perception* in which an image of an object must not only fall on the fovea of each eye but the visual cortex is able to perceive the two images simultaneously.
2. *fusion.* Not only must simultaneous perception be present but there is also the ability to fuse the two images to form one single image.
3. *stereopsis.* The person must have fusion but is also able to perceive three dimensional depth in the fused images. The image seen by each eye is slightly different due to the separation of the eyes by the inter-pupillary distance and this difference allows the object to be perceived in depth. The quality of stereopsis may be measured by assessing the smallest degree of difference between the two images which still allows depth perception.

*Diplopia*

Binocular diplopia occurs when the image of the object of regard falls on the fovea of one eye and on a non-foveal part of the retina of the other eye.

Diplopia is only appreciated when some degree of binocular vision is present. It is not appreciated when a patient is able to suppress the image of the squinting eye. Monocular diplopia is less common and is usually due to opacities in the media of the eye irregularly deviating the light rays.

*Suppression*

Suppression is the mechanism by which a patient with a squint, and therefore at risk of diplopia, is capable of ignoring the image from the squinting eye. Suppression is developed most easily in childhood. The relative inability of adults to suppress the image of a squinting eye usually causes prolonged diplopia, as is seen in acute sixth nerve palsies. One eye alone may be suppressed, but if a patient with an alternating squint uses either eye to view an object he is able alternately to suppress the image of whichever eye is squinting at the time.

*Amblyopia*

Amblyopia is a term used to describe loss of visual acuity in an eye without an apparent organic cause. Recent work suggests however that visual deprivation, at least in the developing eye, may result in histological changes in the lateral geniculate body, and the binocularly driven cells of the occipital cortex subsequently respond only to stimuli from the non-amblyopic eye. Amblyopia may be sub-classified into:

*Strabismic amblyopia.* If one eye constantly squints in a child the vision of the squinting eye is suppressed. The younger the child the more dense the suppression. This is usually associated with a fall in visual acuity as the patient does not use the fovea of the squinting eye and the consequence of this lack of use so early in the development of the visual system is that the affected eye loses the capacity for foveal vision. The degree of this amblyopia is more marked in squint of early onset (e.g. at one year) and those of long duration. Treatment is by occlusion (p230).

*Anisometropic amblyopia.* Anisometropic amblyopia occurs in patients with potentially normal binocular vision, but with a significant refractive difference between the two eyes. Usually the eye with the smaller refractive error is used for vision and the image falling on the fovea of the second eye is never clearly focused, resulting in amblyopia. It may sometimes respond to occlusion of the better eye and refractive correction of the anisometropia.

*Ametropic amblyopia.* In this situation both eyes are affected by a significant refractive error which the focusing system of the eye is unable to overcome. Due usually to astigmatism or high hypermetropia, at no time has a clear image fallen on the fovea of either eye and therefore normal foveal vision fails to develop.

*Stimulus deprivation amblyopia.* This results from the failure of a clear image to fall on the fovea of one or both eyes due to a physical obstruction to the light rays, e.g. congenital cataract, complete ptosis. This condition only occurs if the light rays are interrupted before the age of seven, because after this age the visual system is relatively mature. In general, the younger the child when stimulus deprivation occurs, the more severe is the degree of amblyopia. Removal of the obstruction to vision does not necessarily improve the sight but ptosis and cataracts should be treated appropriately.

### *Convergent concomitant squint - esotropia*

One eye of a patient may constantly turn inwards while the other eye is used for fixation, and this is called a *constant concomitant convergent squint* or esotropia (e.g. if the right eye is constantly squinting this is termed a right esotropia). Eso deviations are the most common type of concomitant squint, with the peak incidence of onset between the ages of two-and-a-half and five years. If the eyes are used alternately for fixation, this is described as an *alternating concomitant convergent squint.* The angle of squint may vary with distance, and the variation may be so great that the patient may be binocular at one distance and squint at another distance e.g. straight for distance but convergent for near.

Esotropias may be categorised as:

1. *accommodative*
2. *non-accommodative*
3. *infantile*

*Relationship between accommodation and esotropia*

A large number of concomitant convergent squints in children are associated with hypermetropia. When a normal emmetropic person views a distant object, the visual axes of the two eyes are parallel and no accommodative effort is necessary to focus a clear image on the fovea of either eye. When the object is brought closer to the patient:

-the person must *converge* the two eyes such that the image of the object falls simultaneously on each fovea
-*accommodation* must occur such that the now divergent rays from a close object are focused into a sharp image on the fovea

Although the actions of accommodation and convergence have separate aims, the actions are linked so that on accommodation a convergence movement of the eye is also made. Normally this physiological mechanism is very satisfactory because both convergence and accommodation are necessary to view a near object. However, if the patient is hypermetropic an excessive accommodative effort must be made, even to view a distant object, in which case the associated convergent movement is inappropriate. There are three possibilities in this situation. The patient may

1. have sufficiently strong binocular reflexes to overcome the converging mechanism and will remain binocular with good acuity both for near and for distance, though the binocularity is under stress.
2. prefer binocular vision to a clearly focused image and accommodation may remain relaxed so that over-convergence does not occur, but by thus 'seeing mistily' the visual acuity will be reduced.
3. allow the eye with the higher refractive error, or other cause of less good vision, to over-converge and develop a squint, the image of the squinting eye becoming suppressed.

1. *Accommodative esotropia*

If the child's squint is entirely due to uncorrected hypermetropia, prescribing the full hypermetropic correction will fully correct the angle of squint. The refractive correction is assessed by retinoscopy after paralysing the ciliary muscle (cycloplegic refraction). In such cases, if the child is fully binocular with a spectacle correction but squints without the spectacles, the squint is described as a *fully accommodative esotropia.* In many cases the spectacles only reduce the angle of squint and if a child is not binocular with the spectacles, but the angle of squint is reduced, then this is termed a *partially accommodative* or *mixed esotropia.*

A relatively uncommon form of accommodative esotropia occurs due to an inappropriate degree of convergence associated with accommodation. These patients have a normal, or low hypermetropic, refractive error and have little deviation when looking in the distance. Near vision induces significant esotropia, and this is caused by an abnormally large degree of convergence in response to accommodation. These are termed *convergence excess*, and are a difficult problem to manage successfully. Miotics may be tried, and echothiophate iodide (Phospholine Iodide) is the miotic of choice. The rationale is that the miotic causes ciliary muscle contraction so that the patient does not need to exert any accommodative effort and therefore the over-convergence does not occur. However miotics, particularly Phospholine Iodide, have side effects (p595) and prolonged use may be undesirable. Bifocals have been tried on some patients, but a child may look over the top of the bifocal segment for close work and then continue to squint. The prolonged use of bifocals may tend to weaken the child's ability to accommodate for near and might condemn the child to bifocals from an early age. Bimedial rectus surgery is usually required to correct the abnormal accommodation: convergence ratio. Convergence excess esotropias are uncommon, and a number of cases so described are in fact fully accommodative esotropias which have not had the hypermetropic error fully corrected. It is essential therefore to make sure that all children with a residual convergent squint despite cyclopentolate refraction are refracted under atropine and the full correction prescribed.

2. *Non accommodative esotropia*

Non accommodative causes of esotropia include stimulus deprivation, divergence insufficiency, consecutive esotropias and stress induced esotropia.

*Stimulus deprivation esotropia.* A squint may result from the reduction of vision in one or both eyes. If the vision is reduced severely by organic ocular or intracranial disease, the binocular reflex is lost and the eye with a poor visual acuity will drift into the anatomical position of rest (either convergent or divergent). Therefore every patient with a squint must be carefully examined to exclude any organic cause reducing the visual acuity (e.g. cataract, retinoblastoma, optic nerve disease). Regrettably cases still occur where the reduced vision in one eye is thought to be due to strabismic amblyopia, and treatment of the real cause is delayed.

*Divergence insufficiency.* This squint is the opposite to convergence excess. In this situation the patient is fully binocular for near but exhibits a manifest convergent squint for distance. This is usually due to reduced function of one or both lateral recti and it is important that any underlying sixth nerve palsy is excluded.

*Consecutive squints.* This is a term used to describe the change of a squint from either divergent to convergent or vice versa following surgery to correct the original condition. Consecutive deviations may be present immediately post-operatively or may develop some years after the surgery in a patient who does not have a sufficient degree of binocular vision to hold the eyes straight. The onset of a large consecutive squint immediately after surgery should raise the possibility that one of the muscles operated on has slipped from or become detached from the eye. The tendency is usually towards divergence with the passage of time so that it may sometimes be wise to delay operation for divergent squint in cases with poor binocularity, if cosmetically acceptable.

*Stress induced squint.* Some children who have never demonstrated any tendency to squint may develop a large esotropia acutely following a febrile illness or an accident. Once any underlying neurological defect or refractive error have been excluded, treatment is usually surgical, as there is usually a potential for good binocular function.

3. *Infantile esotropia* (LCW 8.4 p123)

This is a form of esotropia which presents in the first six months of life, usually with a large convergent squint. The baby cross fixates, using the inturning left eye to look right and vice versa, and this gives the false impression of bilateral sixth nerve palsies (p246). In addition to the esotropia there may also be latent nystagmus (p413), and an upward and outward movement of either eye on occlusion, termed dissociated vertical deviation. After correcting any amblyopia and refractive errors, early surgery is usually recommended, but long term follow up is necessary to avoid amblyopia.

**Management of a child suspected of a convergent squint**

*History*

An adequate medical history including obstetric and birth history as well as method of delivery is essential. The general progress in achieving developmental milestones, as well as the age of onset of the squint and its duration are important, as is a family history. The chance of regaining binocular vision is better:

-the later the squint developed, because the binocular reflex is more
and so more easily restored
-the sooner the squint is treated after its onset

*Examination*

All children with suspected squints should be assessed by an orth
management planned jointly. During examination of the child the
features should be noted:

-*the appearance of the child.* Much information may be gained
the patient to assess general health, apparent maturity and any
retardation. Epicanthic folds (p462) or broad nasal bridges ofte
sistible impression of convergent squint. Such an 'apparent' or
excluded by subsequent tests.
-*visual acuity and refractive error determination as far as pr*
-*ocular movements* in all directions of gaze should be as
p229, LCW 1.11-1.17 p14)
-The *cover test* and the *uncover test*

**The *cover test* to detect a manifest squint.** The patient is asked to look at an object held in the observer's hand. One eye is then covered and the other eye watched to see if it moves to take up fixation. If it moves, it must have previously been deviating. The procedure is then repeated on the other eye.

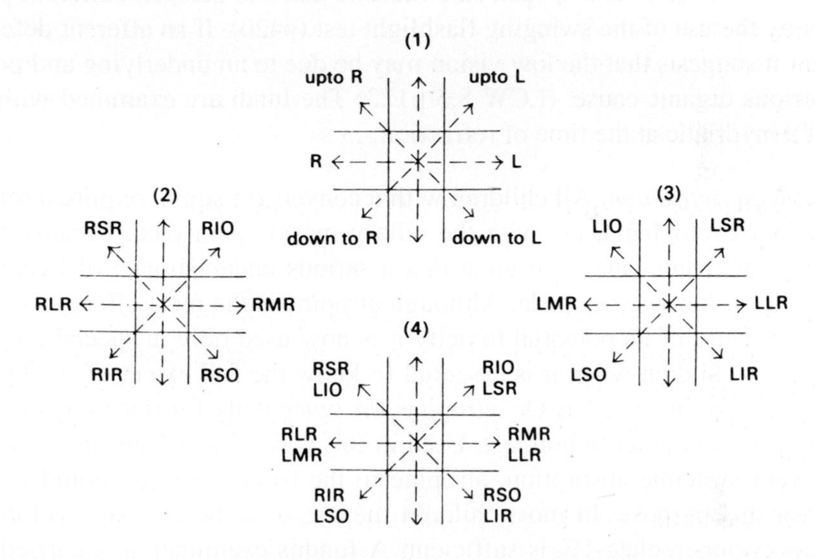


Fig. 11.1 The actions of the extraocular muscles (nine positions of gaze).

**The *uncover test* to detect a latent deviation.** One of the patient's eyes is covered and he is asked to look at an object held by the observer. The eye behind the cover takes up the dissociated position. When the eye is uncovered the movement of the eye to take up fixation is readily detected. The degree of movement reveals the angle of latent squint and the speed of recovery indicates the strength of binocular vision. This is then repeated on the other eye. Rapid movement of the cover between the eyes in the alternating cover test will detect small degrees of latent squint.

The cover test and the uncover test must be performed both for distance and near. It is common practice to do a cover test for near with a light, but a target requiring accommodation should be used, such as a small picture or letter.

Additional information is given by the Maddox Rod and the Maddox Wing Tests if practicable, which form an integral part of the tests to determine the refractive error of the eye (pp71, 72).

-a *general ocular examination* is important to exclude any abnormalities predisposing to a squint and special care must be taken to detect an afferent pupil defect by the use of the swinging flashlight test (p420). If an afferent defect is present it suggests that the low vision may be due to an underlying and possibly serious organic cause. (LCW 8.5 p123) The fundi are examined with the aid of a mydriatic at the time of refraction.

-*cycloplegic refraction.* All children with a convergent squint require a refraction under cycloplegia, i.e. with the ciliary muscle paralysed, because their ability to accommodate is so great that a serious underestimate of hypermetropia may otherwise be made. Although atropine is the most effective cycloplegic, because of its potential toxicity it is now used only in special circumstances for children when it is essential to know the full extent of the hypermetropia. It is then used as Oc. atropine 1% twice daily for three days immediately prior to the refraction test. Used in this way toxic effects are rare, but, to prevent systemic absorption, atropine in the form of drops should not be used for this purpose. In most children the use of a short-acting cyclopegic such as cyclopentolate 1% is sufficient. A fundus examination is carried out at the same time as the refraction. Very occasionally a child is unable to co-operate and an examination under an anaesthetic is necessary.

*Treatment of convergent concomitant squint (Esotropia)*

The aim of treatment is to overcome amblyopia, maximise the degree of binocular single vision and correct the angle of squint.

*Treatment of amblyopia*

Refractive error found by cycloplegic refraction requires full correction, but the only effective method of treating amblyopia is to occlude the non-amblyopic eye. In a very young child a dramatic improvement will occur with effective occlusion. Improvement in vision is usually very slow after the age of six years. Patching is the most effective method of occlusion. The patch should be worn on the face so that it totally covers the better eye. If the amblyopia is not severe or has improved with patching, then less rigorous occlusion can be used such as covering of the spectacle lens of the better eye with a patch. This type of occlusion is useless in the case of marked amblyopia as the child will simply look over the top of the spectacles. When the vision is almost equal in both eyes, the spectacle lens can be occluded with a semi-transparent material. Atropine drops or ointment may be instilled into the non-amblyopic eye to blur the vision of that eye especially for close work and encourage the use of the amblyopic eye. However in most cases atropine is only partially effective.

Occlusion is essential to improve most cases of amblyopia, but its use requires experience. The younger the child the faster the amblyopia will be improved by patching. This improvement can often be associated with a marked fall in vision of the originally better eye as the occlusion of that eye is producing deprivation amblyopia. Children who are patched must therefore be seen extremely frequently, especially when very young, to assess the progress of the amblyopic eye and of the occluded eye. Diplopia can occur if the vision of the amblyopic eye is improved. This can become a problem in children over seven when they are unable to fuse the image due to the absence of any binocular vision and sometimes cannot learn to suppress the second image.

*Correction of the angle of squint* Convergent squints which are the direct result of uncorrected hypermetropia may be completely corrected by prescribing the *full hypermetropic correction* as discovered by retinoscopy. Effective treatment of the amblyopic eye by occlusion is also necessary. The fully accommodative squint may well be cured in this way. (LCW 8.6(a), 8.6(b) p123)

If the spectacles fail to correct the squint and vision is practically equal in both eyes following occlusion, then consideration must be given to *surgery*. Care must be taken to assess whether one is treating a squint in which a binocular result can be expected or one which is solely a cosmetic problem. The potentially binocular squint is usually of recent onset and occurs in a child of two years of age and over in whom it can be inferred that binocular vision has developed to some extent. Orthoptic assessment should show the presence or absence of binocular vision.

The potentially binocular squint should be operated on as soon as possible if it persists after the refractive error has been corrected and occlusion has achieved equal vision in both eyes or has failed to produce further improvement in the squinting eye. If possible the child should be able to alternate, i.e. use either eye for fixation. This situation is ideal but is not always achieved. The child with a squint who does not show any potential binocular vision should only be operated on to achieve an acceptable cosmetic result, but the visual acuity of each eye should be tested at intervals and further occlusion used if necessary. It is not advisable to operate on a squint in a child under seven years of age until a reasonable attempt has been made to overcome any amblyopia. Children with a fully accommodative squint should not have an operation with the intention of reducing the strength of their spectacle correction because as the child becomes older his accommodative ability decreases and he will need to have his hypermetropia corrected in order to see clearly. This will then tend to cause divergence as the cause of the squint was uncorrected hypermetropia. Thus the treatment is the correction of the hypermetropia and not surgical interference with a normal physiological reflex. In children with a low hypermetropic correction who have binocular vision it is nevertheless reasonable to consider a short trial of orthoptic training to improve binocularity in an attempt to control the squint without spectacles.

*Divergent concomitant squint - exotropia*

Divergent concomitant squints do not have the same strong link with accommodation defects as convergent squints. Their aetiology is not clear even though myopic patients tend to have an *exophoria* (latent divergence) which is reduced when the myopia is corrected. The condition is much less common than convergent concomitant squint (1:14).

*Types of divergent concomitant squint*

They may be classified as either constant or intermittent.

In *constant exotropia* one eye is constantly divergent for all distances and the divergent eye is frequently amblyopic. Constant exotropia may be primary, due to breakdown of a previously intermittent exotropia, secondary to intraocular disease, or consecutive following correction of an esotropia (p227).

In *intermittent exotropia* patients have a large latent divergence (exophoria) which breaks down to a manifest squint from time to time. These patients have usually have equal vision in either eye, and amblyopia is rare. When the eyes are straight the patients have binocular vision, but when the squint becomes manifest one eye is usually suppressed, making diplopia uncommon. Intermittent exotropia is sub classified into distance exotropia, near exotropia and a non-specific or "basic" form:

-in a *distance exotropia* the patient is fully binocular for near but breaks down to a manifest divergent squint when viewing a distant object. It may only be demonstrable when a distant object further than the standard six metre target is used and for this reason a cover test should be performed using an object in the far distance as a target
-in *near exotropia* the patient is binocular for distance, but develops a manifest exotropia for near vision.
-in *non specific* or *basic exotropia* the deviation is similar for both near and distance vision. Some cases which initially appear to be purely distance exotropias, with no deviation for near, are actually non specific but appear to control for near because of over accommodation. It is important to diagnose these "pseudo distance exotropias" by orthoptic testing as the management may be different.

**Management of a child suspected of a divergent squint**

*Refraction:* any refractive error should be corrected. It is sometimes inappropriate to correct small symmetrical hypermetropic errors as this may make the squint worse if it is reasonably controlled without the spectacles.

*Amblyopia:* as with convergent squints amblyopia must be treated.

*Orthoptic treatment:* In intermittent exotropias it may be possible to improve the control of the exophoria by orthoptic means, and thereby reduce the problem.

*Surgery:* if a constant divergent squint is present then surgery should be undertaken to correct it if potential binocular function is present. If the squint is simply a cosmetic problem, the surgical decision depends on the appearance of the squint. Intermittent exotropias may require surgery, especially if they are deteriorating. True distance exotropias are often treated with bilateral lateral rectus recessions, near exotropias with bimedial resections, and non specific types with unilateral surgery, recessing the lateral rectus and resecting the medial rectus.

**'A' and 'V' phenomena i.e. concomitant squints complicated by an element of paralysis of the vertically acting muscles**

Many patients have a congenital imbalance of the vertically acting ocular muscles. The superior rectus, superior oblique and inferior oblique muscles are most commonly affected, and this may give rise to differences in the relative eye position between up gaze and down gaze, most commonly "A" or "V" patterns.

The *'V' phenomenon* This is usually associated with an underaction of one or both superior recti and as a consequence there is an overaction of one or both inferior obliques. For a patient to be described as having a 'V' phenomenon, the angle of squint when measured in the upward direction of gaze must be more divergent than when measured in a downward direction of gaze.

The *'A' phenomenon* These are usually due to inferior rectus underaction of both eyes which is often associated with superior oblique overaction. These are much more difficult to treat than 'V' phenomena.

The problem with squints that have significant "A" or "V" patterns to them is that the horizontal angle varies with upwards or downwards gaze, making symmetrical horizontal surgery unsuccessful or less effective unless the vertical imbalance is also corrected. In cases with clear oblique muscle overaction, weakening procedures such as inferior oblique recession may be required, but in other cases some vertical transposition of the horizontal recti may be needed. The judgement as to whether this type of complicated surgery is required must depend upon the extent of the 'V' or 'A' phenomenon and upon whether the patient has binocular single vision in any direction of gaze. (Fig. 11.2 p235).

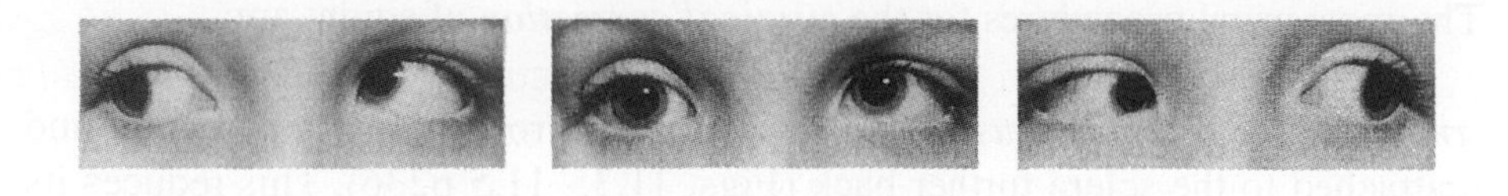

Fig. 11.2 Left concomitant convergent squint with left inferior oblique muscle overaction.

*Brown's syndrome.* Rarely children may present with impaired up gaze, particularly in adduction, but normal elevation in the primary position and in abduction, simulating an inferior oblique palsy. In order to move the eye out of elevation, and to keep binocular vision, the child may assume a head posture. This condition, which may be bilateral, is thought to reflect a congenital anomaly of the superior oblique tendon, and a click may be felt on palpation over the trochlea. The majority of cases resolve spontaneously, but in cases with a very marked head posture, surgery to the involved superior oblique muscle may be required.

## Management of concomitant squint in the older child and adult

The management of these squints (whether convergent or divergent) is similar to the management of the squint in childhood. However, occlusion over the age of seven in an effort to treat amblyopia is usually ineffective and if poor binocular function is present, over-enthusiastic treatment may induce intractable diplopia. If satisfactory binocular vision is present, surgery is often necessary to reduce the angle of squint and help the patient control it and remain binocular. If binocular vision is poor or absent and the squint is a cosmetic problem then the patient must be assessed for the risk of postoperative diplopia. This can be measured with prisms, but in many cases better information is achieved by temporarily correcting the squint with an injection of Botulinum toxin to see if diplopia is a significant problem. Diplopia can be troublesome in older patients with poor binocular vision, because the patient had learned to suppress the image falling on a particular part of the retina of the squinting eye, and when the angle of squint is corrected the image falls on a part of the retina not previously suppressed. Diplopia usually resolves in a few days, but some adults and older children may find it impossible to learn to re-suppress the new diplopia, and careful patient selection is therefore necessary before undertaking surgery.

The most usual procedures for the *surgical correction* of squint are:

-*recession of a rectus muscle* which is divided from its usual insertion and reattached to the sclera further back (Figs. 11.3 - 11.5 p236). This reduces its effective action.

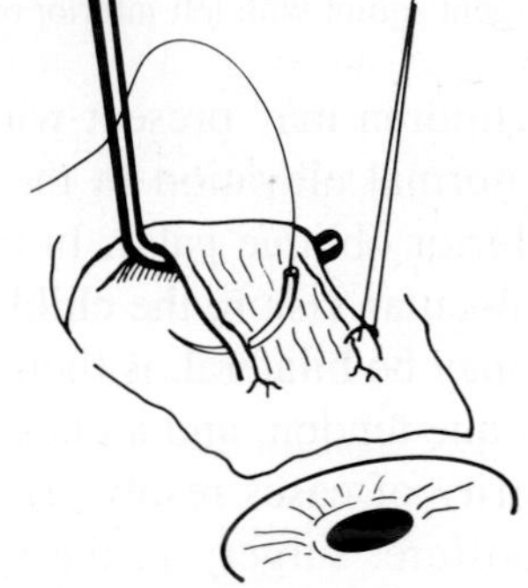

Fig. 11.3 Recession of rectus muscle (a).

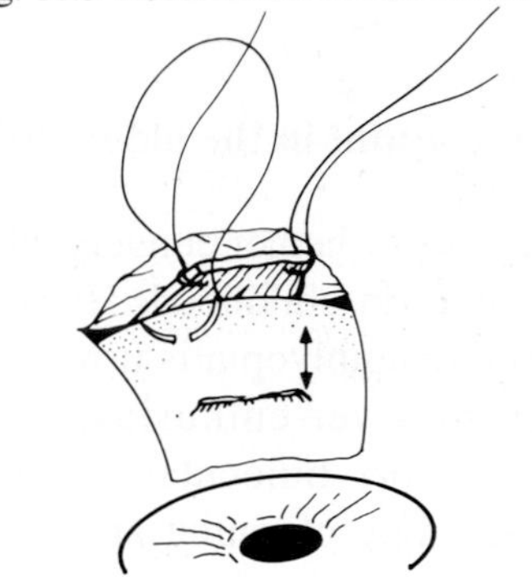

Fig. 11.4 Recession of rectus muscle (b).

Fig. 11.5 Recession of rectus muscle (c).

*-resection of a rectus muscle* which is divided and, after removing an appropriate length of muscle, is resutured to the original insertion on the sclera (Figs. 11.6 - 11.8 p237).

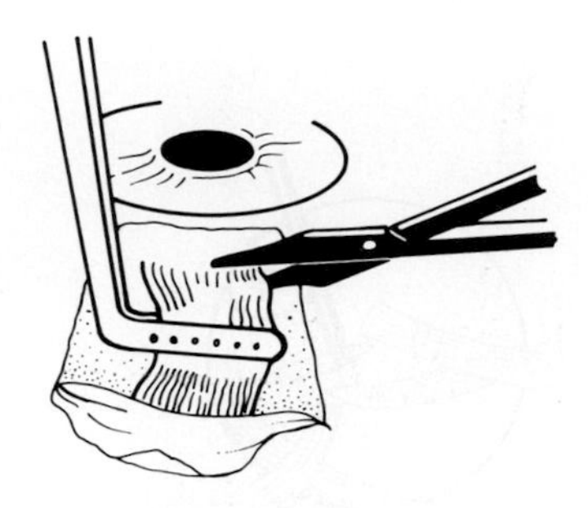

Fig. 11.6 Resection of rectus muscle (a).

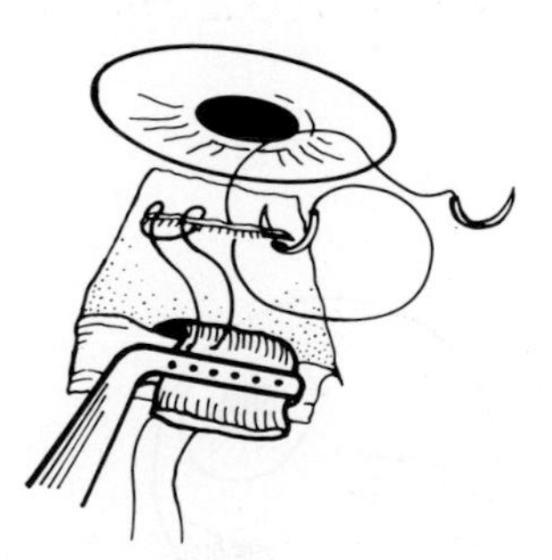

Fig. 11.7 Resection of rectus muscle (b).

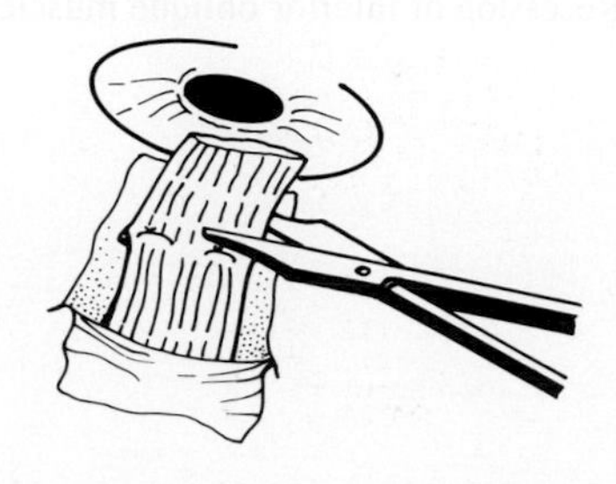

Fig. 11.8 Resection of rectus muscle (c).

*-recession of the inferior oblique muscle* to reduce its action (Figs. 11.9, 11.10 p238).

Fig. 11.9 Recession of inferior oblique muscle (a).

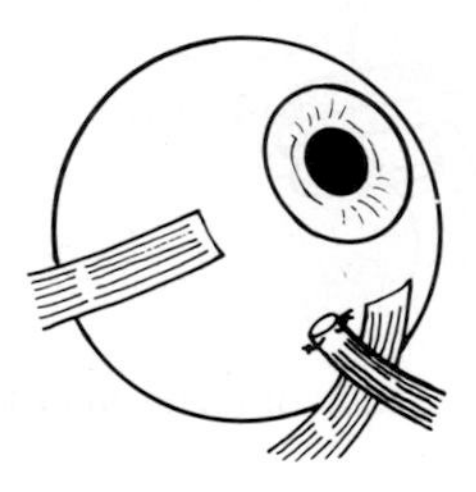

Fig. 11.10 Recession of inferior oblique muscle (b).

These indicate the principles, but many other procedures are described in texts of ophthalmic surgery.

**Paralytic squint (the usual cause of the symptom of double vision)**

A paralytic or incomitant squint, which is the usual cause of a complaint of double vision, may be defined as one in which the angle of squint varies with the direction of gaze, e.g in the case of a right lateral rectus palsy there may be a convergent squint in the straight ahead position, but the angle will increase markedly on looking to the right, i.e. in the direction of the line of action of the paralysed muscle. Conversely, the angle of squint is reduced when the eyes are turned away from the line of action of the paralysed muscle (in this case when looking to the left).

*Synergists and antagonists*

It is important to understand the effect a paralysed muscle has on the other muscles not principally affected by the paralysis. Each muscle has a *direct antagonist* and a *contralateral synergist.* For example, in the case of the right lateral rectus, which is responsible for turning the right eye to the right, its action is opposed by its direct antagonist, the right medial rectus. If the right lateral rectus is paralysed, then the action of the right medial rectus will be unopposed and therefore will relatively overact. For the eyes to be turned to the right, the right eye is moved by the right lateral rectus and the left eye by the left medial rectus, the left medial rectus being therefore the contralateral synergist to the right lateral rectus.

Fig. 11.11 Actions of extraocular muscles.

| *Muscle* | *Main action* | *Maximal in-* | *Additional actions* |
|---|---|---|---|
| Lateral rectus | Abduction | | |
| Medial rectus | Adduction | | |
| Superior rectus | Elevation | Abduction | Adduction Intorsion |
| Inferior rectus | Depression | Abduction | Adduction Extorsion |
| Superior oblique | Depression | Adduction | Abduction Intorsion |
| Inferior oblique | Elevation | Adduction | Abduction Extorsion |

*Increase in the angle of deviation when the paralysed eye is used for fixation*

It is accepted that a similar amount of nervous energy is given to each muscle responsible for deviating the eyes in any direction of gaze e.g. to turn the eyes to the right, a similar amount of energy is given to the right lateral rectus muscle and to the left medial rectus *(Hering's Law).* If the right lateral rectus is weakened then the right eye will not turn sufficiently to the right and a right convergent (paralytic) squint will be evident. However if the patient fixes with the right eye, then to turn the right eye to the right will require more energy than normal and a similar amount of energy will also be applied to the left medial rectus. Therefore the excess amount of energy going into the left medial rectus will cause it to overact. This overaction of the contralateral synergist explains why the angle of squint is always greater when a patient fixes with the eye which has a paretic muscle. If a muscle is paralysed there is thus overaction of both its direct antagonist and, even more, its contralateral synergist.

*The actions of the extraocular eye muscles*

The horizontal recti (medial rectus and lateral rectus) have a simple line of action either abducting or adducting the eye. However, the vertical recti and the oblique muscles have a much more complex action, and may cause elevation or depression, abduction or adduction, as well as torsion (rotation around its sagittal axis), either intorsion or extorsion, depending on the position of the eye. Fig. 11.11 p239, Fig. 11.12 p240 indicate the principal lines of action of these muscles.



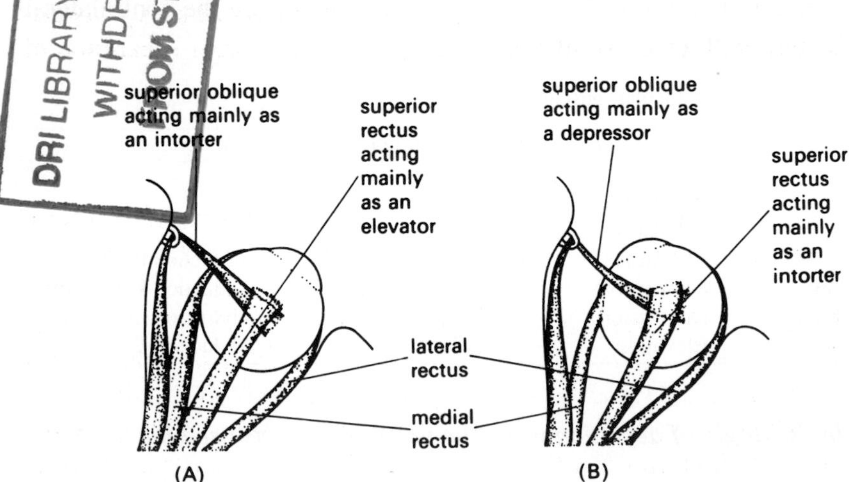


Fig. 11.12 Principal lines of action of extraocular muscles in (A) abduction (B) adduction.

Fig. 11.13 p241 gives the various antagonists and synergists of each muscle. If the muscle is paretic or paralysed, then there is overaction of both its direct antagonist and its contralateral synergist.

Fig. 11.13 Antagonists and synergists .

| Muscle (R) eye | Direct antagonist | Contralateral synergist |
|---|---|---|
| Lateral rectus | (R) Medial rectus | (L) Medial rectus |
| Medial rectus | (R) Lateral rectus | (L) Lateral rectus |
| Superior rectus | (R) Inferior rectus | (L) Inferior oblique |
| Inferior rectus | (R) Superior rectus | (L) Superior oblique |
| Superior oblique | (R) Inferior oblique | (L) Inferior rectus |
| Inferior oblique | (R) Superior oblique | (L) Superior rectus |

Once the concept of direct antagonists and contralateral synergists is appreciated then the deviation occurring in the case of a paralytic squint is easily understood. The angle of squint should be measured in the nine positions of gaze. The angle should also be measured with each eye fixing in turn. The differences in the angle of squint in the different directions of gaze reveal which muscle or group of muscles is affected and the difference in the angle of squint when fixing with each eye in turn should indicate which eye is primarily affected. ie. fixing with the affected eye gives the larger separation of images.

*The abnormal head posture*

Patients with a paralytic squint may be able to minimise the angle of squint when looking in certain directions and may be able to achieve binocular vision in this direction of gaze and avoid diplopia. These patients will move their head such that the eyes occupy a position in the orbit where the angle of squint is minimal, e.g. in the case of a right lateral rectus paresis the patient will turn his head to the right so that the eyes are deviated to the left relative to the orbit when the patient is looking straight ahead. When the horizontal recti are affected the characteristic head posture is a turn of the face to right or left; when a vertical rectus muscle or an oblique muscle is affected, a tilt of the head to the right or left with depression or elevation of the chin is adopted to reduce both the vertical deviation and the torsion.

*Causes of paralytic squint*

The following types may be encountered in practice:

-lesion of one or more of the nerves (or their nuclei) supplying the extraocular muscles (p246). This is the most common cause.
-injury or disease affecting the central control of eye movements. These gaze palsies may be further sub-divided into *supranuclear* (p401) and *internuclear* (p405)
-lesion of the extraocular muscles or their nerve endings e.g. myopathies (p406) or myasthenia gravis (p369)
-mechanical restriction of eye movements due to tethering of muscles or globe e.g. dysthyroid disease (pp309, 407) or orbital fractures (p540).

**The approach to a patient with a paralytic squint**

*History*

Most older patients with an acquired paralytic squint will complain of diplopia but if the squint occurs in young children they rapidly suppress the second image and amblyopia may supervene.

The symptom of diplopia may be of considerable help in deciding which muscle or muscles are affected. The patient is asked in which way the images are separated. If the images are only separated horizontally it is probable that either a lateral or a medial rectus is affected; when the images are separated vertically or the image is tilted *(torsion)* it is likely that one or more of the vertical recti or the obliques are affected. The variations of the separation of the images in the different directions of gaze is a further clue to the affected muscle, e.g. if the separation is greater looking to the right in a case of horizontal diplopia, then the likelihood is that either the right lateral rectus or the left medial rectus is affected. When looking in the direction of maximum diplopia the eye responsible for the further displaced image is the one with the paralysed muscle as shown in Fig. 11.14 p243. A full general history must be taken as a paralytic squint is usually due to a cranial nerve palsy or a muscle problem and may be just one manifestation of a general nervous system disorder.

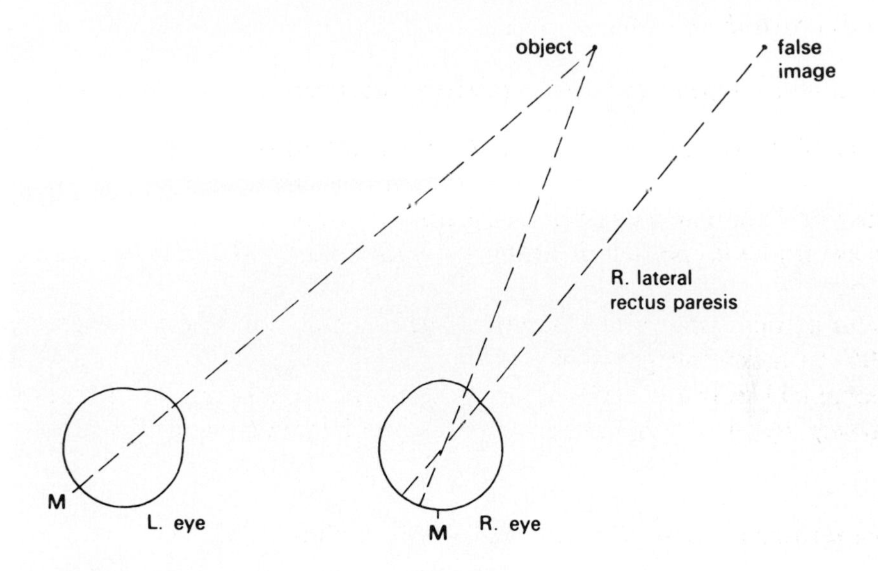


Fig. 11.14 Position of false image.

*Examination*

Any abnormal head posture is noted. A general ophthalmic examination is carried out with particular reference to ocular movements. The diplopia may be checked simply by holding a pencil in front of the patient and asking him to describe the separation of the images in the *nine positions of gaze*. The cover test should also be performed in the nine positions of gaze and the position of the eyes observed in these nine positions of gaze to allow an objective assessment. An orthoptic examination is desirable when possible. The orthoptist will measure the angle of squint and chart the ocular movements on a *Hess screen* (Figs. 11.15 p244, 11.16 p245). For this test the patient wears a red glass over one eye and a green glass over the other and is provided with a torch giving a green bar of light. A red light is shown at appropriate positions on the screen; the patient is asked to place the green bar over it. The colours dissociate the eyes and the direction and degree of deviation is revealed. Measurement of the squint is essential not only to diagnose which muscle or muscles are affected but also for a baseline measurement of the degree of the paralysis so that the patient's progress may be accurately assessed on subsequent examinations.

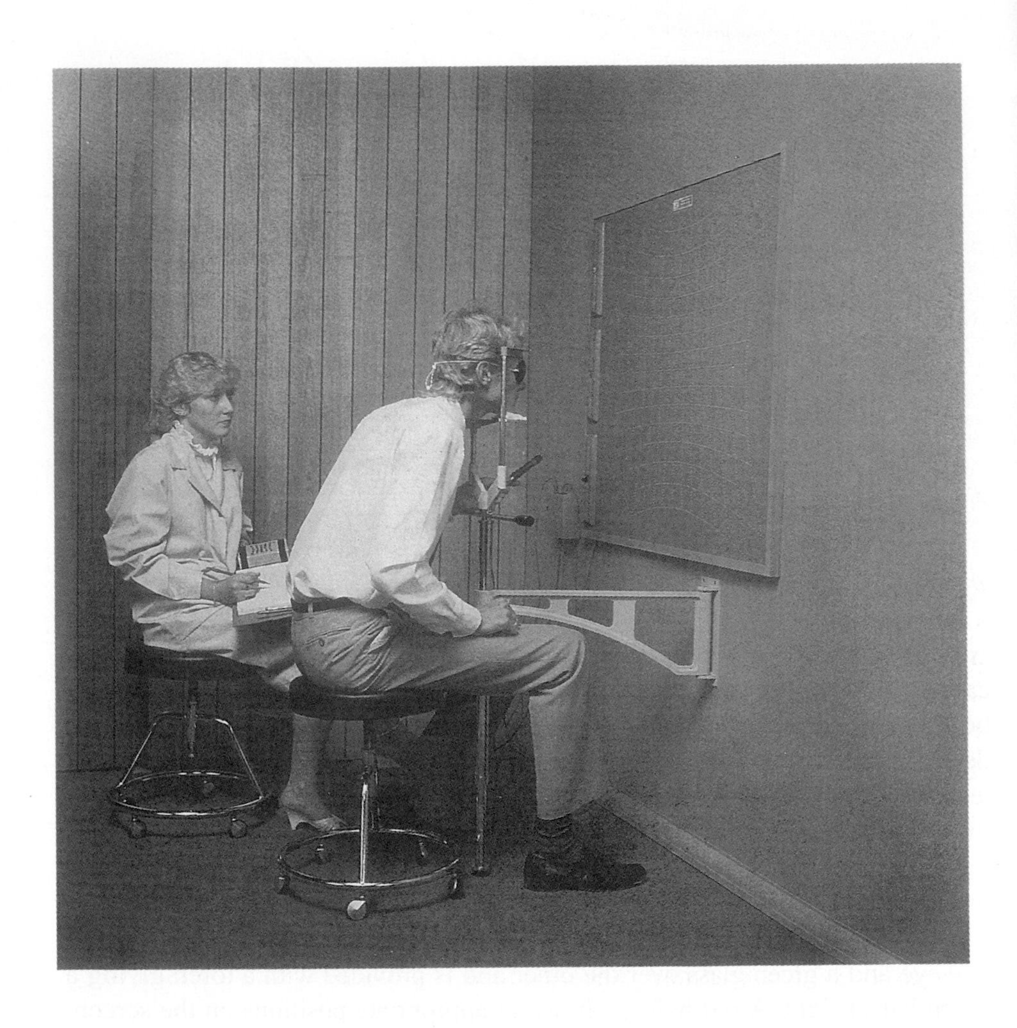

Fig. 11.15 Hess screen. (Basic equipment. Many sophisticated variants are available)

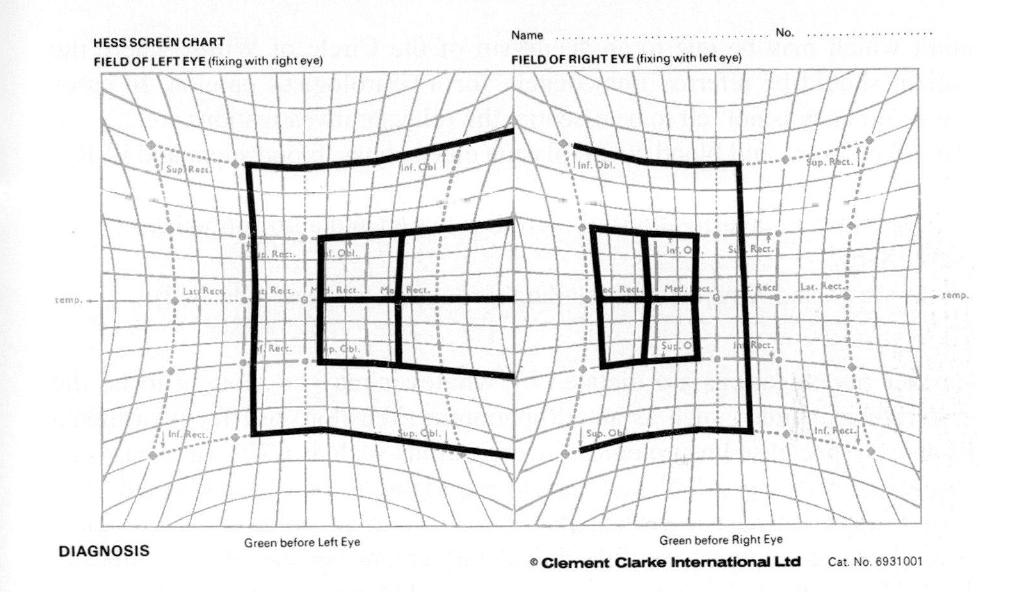


Fig. 11.16 Hess Chart showing R 6th cranial nerve palsy causing R lateral rectus paresis.

In addition to simply charting the limitation of movement of the eyes, it is also important to assess the speed of saccadic movement of the eyes both horizontally and vertically. The saccadic speed, i.e. the quick movement of the eye to take up fixation, is extremely fast, about 400° of arc per second. However, in cases of damage to the higher centres controlling ocular movements or impairment of the nerve supplying the muscle, it can often be demonstrated clinically to be much slower than normal, either in the direction of gaze controlled by the higher centre or in the line of action of the affected muscle. Attention is also paid to pupil reactions, lid movements, the absence or presence of proptosis or enophthalmos and the appearance of the fundi. A general medical and neurological examination of these patients is essential.

### *Investigations*

The degree of investigation of the patient partly depends upon the findings at clinical examination and upon the history. Some paralytic squints are due to serious neurological disorders, such as a painful third nerve palsy in a young

adult which may be due to an aneurysm of the Circle of Willis, where the patient should be referred immediately for a neurologist's opinion. In cases where urgency is not felt to be essential the relevant investigations are:
-Blood pressure, full blood count, plasma electrolytes, blood sugar and ESR
-chest X-ray
-where relevant, investigations of thyroid function and auto antibodies
-skull X-ray
-sinus views and X-ray orbit (including tomography where indicated)

Further investigations are dictated by whatever may be thought to be the underlying disorder, such as an edrophonium (Tensilon) test for myasthenia gravis, or specialised neurological investigations such as CAT scan, magnetic resonance imaging (MRI) and carotid angiography may be indicated. The degree of investigation, particularly where it involves invasive neurological investigations, requires careful clinical judgement because while a normal skull X-ray report does not necessarily exclude serious intracranial disease, such as an aneurysm or neoplasm, a number of paralytic squints of sudden onset occur in the absence of any demonstrable underlying pathology and these frequently recover spontaneously within six months.

**Neurological causes of paralytic squint (disorders of the nerve supply to the extraocular muscles)**

-Palsy of the 6th, 3rd and 4th cranial nerves or their nuclei (pp246 - 254). The anatomy of the cranial nerves is described on pp15 - 23.
-gaze palsy (supranuclear and internuclear ophthalmoplegia) (p401)
-myasthenia gravis (p369)

**Sixth cranial nerve (abducens) palsy**

The sixth nerve supplies motor fibres to the ipsilateral lateral rectus muscle, and is the most commonly affected nerve (Fig. 11.17 p247, Fig. 11.16 p245, LCW 7.16 p111). It is frequently involved in systemic disease without clinically demonstrable intracranial pathology e.g. hypertension or diabetes. A sixth nerve palsy may be the direct result of a lesion along the course of the nerve, but it can also be a false localising neurological sign. This is because the nerve is restricted at its emergence at the lower border of the pons by the posterior inferior cerebellar artery so that downward displacement of the brain stem can cause a stretching and paresis of the sixth nerve as it turns forward sharply over the edge of the petrous temporal bone (Fig. 19.5 p375). It is important to appreciate this, and that the lesion responsible for the raised intracranial pressure may be well away from the course of the sixth nerve.

*Signs and symptoms*

-*diplopia:* with the images side by side and separated maximally when looking in the direction of action of the paralysed muscle,
-*esotropia (convergent squint):* in the primary position, increasing when looking in the direction of the affected muscle,
-*abnormal head posture.* A patient who has a sixth nerve paresis and who has some degree of binocular vision will turn his face in the direction of action of the paralysed muscle to retain binocularity despite less muscle action.

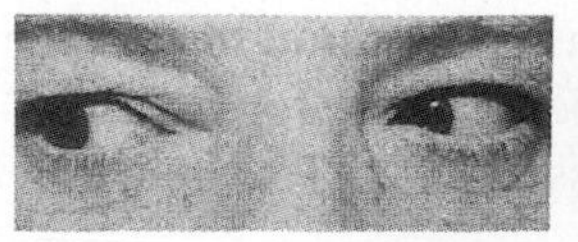 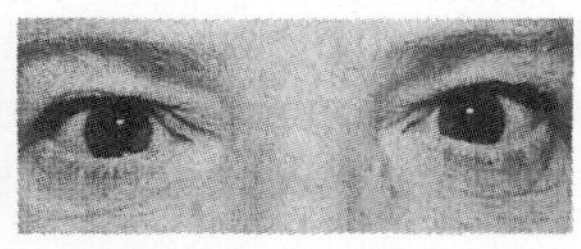 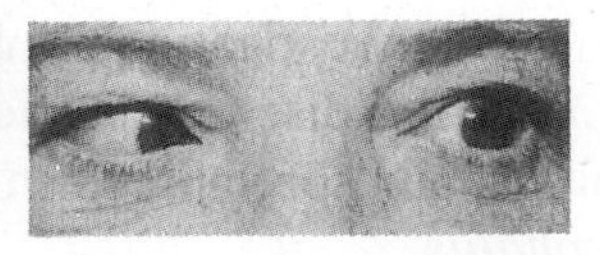

Fig. 11.17 Left sixth cranial nerve palsy (incomplete).

*Causes of sixth cranial nerve palsy*

*Congenital.* Some children are born with a sixth nerve palsy even after an atraumatic delivery. The early months may see a dramatic improvement and a few patients may eventually develop straight eyes. Care must be taken to prevent amblyopia supervening in these children. A rare condition known as *Möbius' syndrome* is characterised by bilateral sixth and seventh nerve palsies. This is due to a congenital abnormality of the nuclei of these nerves, and may also involve the ninth and twelfth cranial nerves, giving rise to feeding difficulties in these children. Children with *infantile esotropia* may appear to have bilateral sixth nerve palsy as they cross fixate, but this can be excluded by demonstrating movement of the eyes beyond the midline on spinning with the child.

*Raised intracranial pressure,* however caused, may result in a unilateral or bilateral sixth nerve palsy as described above.

*Diabetes* and *hypertension* both result in a sixth nerve palsy which often recovers and no intracranial disease can be demonstrated to cause this condition. However, the presence of diabetes or hypertension does not exclude serious unrelated intracranial pathology.

*Trauma:* the sixth nerve is particularly vulnerable to damage in closed head injuries.

*Lesion along the course of the nerve*

-*vascular, demyelinating* or *neoplastic* disease of the brain stem may affect the sixth nerve nucleus prior to affecting any other of the nuclei in the mid-brain, but often other neurological features are associated with pontine disease.
-*cerebello-pontine angle tumours* such as *acoustic neuromas* may involve the sixth nerve as it passes up in the pontine cistern, often with associated ipsilateral fifth and eighth nerve signs.
-*lesions of the petrous tip (Gradenigo's syndrome):* mastoiditis or middle ear infection may cause thrombosis of the petrosal sinus resulting in a sixth, seventh and eighth nerve lesion. The condition is associated with severe pain in the ear and may result in raised intracranial pressure (p378) *otitic hydrocephalus*
-*cavernous sinus lesions:* the sixth nerve may be the first nerve to be affected by infra-clinoid aneurysms of the carotid artery in the posterior part of the cavernous sinus (Fig. 19.5 p375). A sixth nerve palsy occurs in a cavernous sinus thrombosis but there would be severe constitutional signs and usually rapidly progressive paralysis of the third and fourth nerve and proptosis of the eye. The sixth nerve may be damaged by neoplasia anywhere along its course, e.g. meningioma, metastasis, and direct invasion by nasopharyngeal carcinoma.

*Management of sixth cranial nerve palsy*

The majority of sixth nerve palsies will spontaneously improve with time, and once a serious underlying disorder has been excluded, simple occlusion of the affected eye will relieve the diplopia. If diplopia persists beyond six months then conservative measures such as use of prisms may provide benefit. Selected cases may benefit from squint surgery, possibly with a transposition of the vertical recti laterally to aid lateral movements of the eye. Botulinum toxin injections to the ipsilateral medial rectus muscle have been found to be beneficial in selected cases.

*Duane's syndrome*

This is not due to a sixth nerve palsy, but as it is frequently misdiagnosed as such it has been included here. *Duane's syndrome* is characterised by a horizontal limitation of the movement of one or both eyes. Commonly there is a restriction of abduction (the eye may not abduct beyond the midline) associated with widening of the palpebral fissure. Adduction of the affected eye is associated with retraction of the globe and narrowing of the palpebral fissure (Fig 11.18 p249). Diplopia is extremely rare, and surgery usually not indicated except in cases where there is a significant head posture. The aetiology is uncertain but may be due to abnormal cross innervation of the lateral recti from the third nerve due to a congenital aplasia of the sixth nerve nucleus.

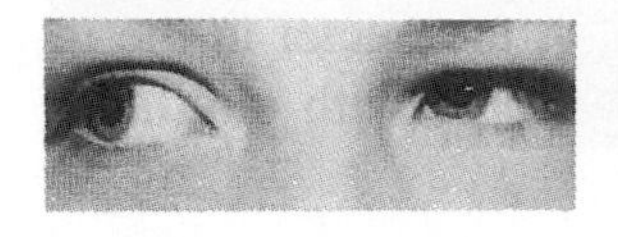
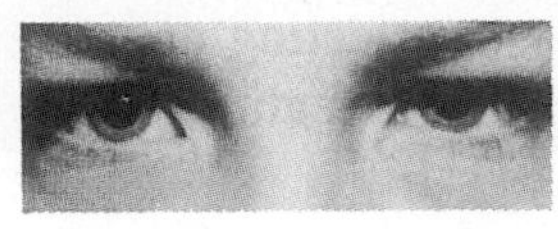
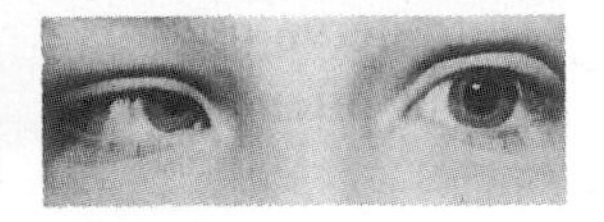

Fig. 11.18 Duane's syndrome (bilateral with failure of left abduction).

## Third cranial nerve (oculomotor) palsy

*Signs and symptoms*

The third cranial nerve supplies the superior rectus, the inferior rectus, the inferior oblique, the medial rectus and the levator palpebrae superioris. It also carries the parasympathetic fibres supplying the pupillary sphincter and the ciliary muscle which run in the branch supplying the inferior oblique muscle. In cases of a total third nerve palsy there is a complete ptosis and the eye is deviated downwards and outwards (Figs. 11.19, 11.20 p250, LCW 7.17, 7.18 p111, 7.19 p112). The pupil may be fixed and dilated indicating loss of the parasympathetic supply, producing an 'internal ophthalmoplegia'. The parasympathetic fibres run superficially in the intracranial course of the nerve, and hence are readily compromised by external compressive lesions,

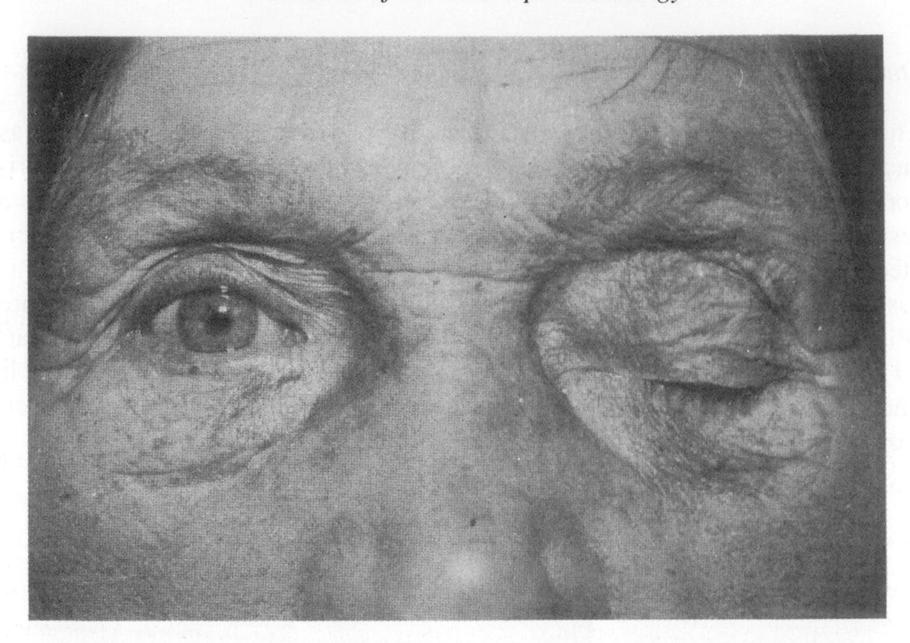

Fig. 11.19 Total third cranial nerve palsy (ptosis).

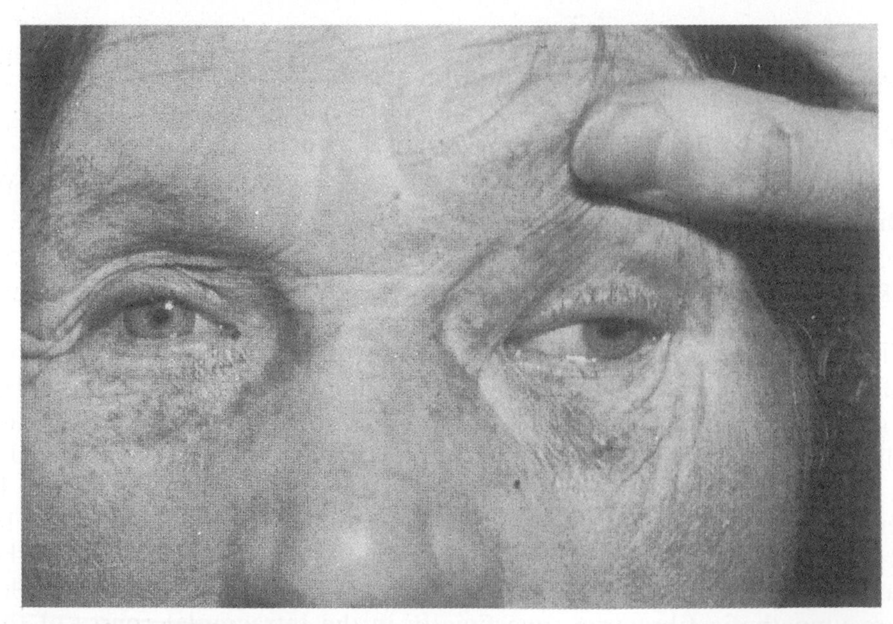

Fig. 11.20 Total third cranial nerve palsy (ocular deviation). - The ptosed L upper lid is held up to demonstrate the deviation.

but may escape injury in vascular related nerve palsies. An abnormal head posture is not present in a case with total third nerve palsy due to ptosis. However in partial third nerve palsies without ptosis the patient may be able to adjust the position of his head so that he is binocular when adopting that abnormal head posture.

*Causes of third cranial nerve palsy*

*Congenital* third nerve palsies are rare, and do not produce an internal ophthalmoplegia. External ocular muscle involvement may be associated with some ptosis.

*Trauma* either at birth or later may affect the third nerve but it is less liable to injury than the sixth nerve.

An *aneurysm of the Circle of Willis,* commonly affecting the posterior communicating artery or the supraclinoid part of the internal carotid, usually presents with severe pain around the eye and a partial or complete third nerve palsy. If aneurysm is suspected the patient should be sent as an emergency to a Neurological Centre. Aneurysms often occur in young adults. The third nerve palsy may be the first sign of an incipient subarachnoid haemorrhage. If there is an infra-clinoid aneurysm of the internal carotid artery within the cavernous sinus, the third nerve palsy may be preceded by a sixth nerve palsy.

*Diabetes* and *hypertension,* especially in the presence of arteriosclerosis, may cause either a complete third nerve palsy or just paralysis of the extraocular muscles, or it may solely affect the pupil. The presence of diabetes or hypertension does not necessarily exclude the presence of a more serious intracranial disease

*Neoplasia* may damage the third nerve by invasion of its nucleus or by, for example, meningeal, pituitary or naso-pharyngeal tumours along its pathway.

*Aberrant regeneration of the third nerve:* if the third nerve is damaged by trauma or aneurysm, it may recover by regeneration of nerve fibres following treatment of the original cause. However the growth of the axons along the neural sheaths may develop in an aberrant fashion such that some axon fibres may go to the superior rectus which were previously intended for the inferior rectus and some may go to the upper lid which were previously intended for an extraocular muscle. This may result in bizarre movements of both the lids and the eyes when looking in certain directions. One type of aberrant regeneration can be demonstrated by asking the patient to look down. On doing so the lid may elevate (the *pseudo-Graefe phenomenon).* Aberrant regeneration may also affect the pupillomotor fibres.

*Demonstration of a fourth cranial nerve involvement in a third cranial nerve palsy*

In a third nerve palsy the eye is deviated down and out so that it may be difficult to show whether the fourth nerve is also affected. The patient is asked to look down and inwards so that the superior oblique will contract if not affected. This contraction will not necessarily produce any vertical or horizontal movement of the affected eye. It will produce, however, intorsion of that eye due to the unopposed action of the superior oblique without the extorting effect of the inferior rectus to oppose it, thus demonstrating the integrity of the fourth nerve. Torsion is best detected by observing the movement of the iris pattern or limbal conjunctival vessels.

*Management of third cranial nerve palsy*

Establishing and treating the underlying cause is the principal objective. The associated ptosis usually prevents problems with intractable diplopia, and surgical correction rarely is beneficial given the combination of vertical, horizontal and torsional imbalance between the eyes.
*Partial third cranial nerve palsies* causing individual or mixed paresis of the levator palpebrae superioris , inferior oblique, superior rectus, medial rectus or inferior rectus muscles may occur with appropriate signs (LCW 7.20-7.25 p112).

**Fourth cranial nerve (trochlear) palsy**

*Signs and symptoms*

Damage to the fourth nerve will result in vertical and possibly torsional diplopia (Fig. 11.21 p253). Weakness of the superior oblique muscle causes maximal separation of the images when the affected eye looks down and in and this particularly affects reading and walking down stairs. The patient commonly adopts a head posture to maintain single vision characterised by depression of the chin, a face turn to the opposite side and head tilt away from the affected eye. (LCW 7.15 p111)

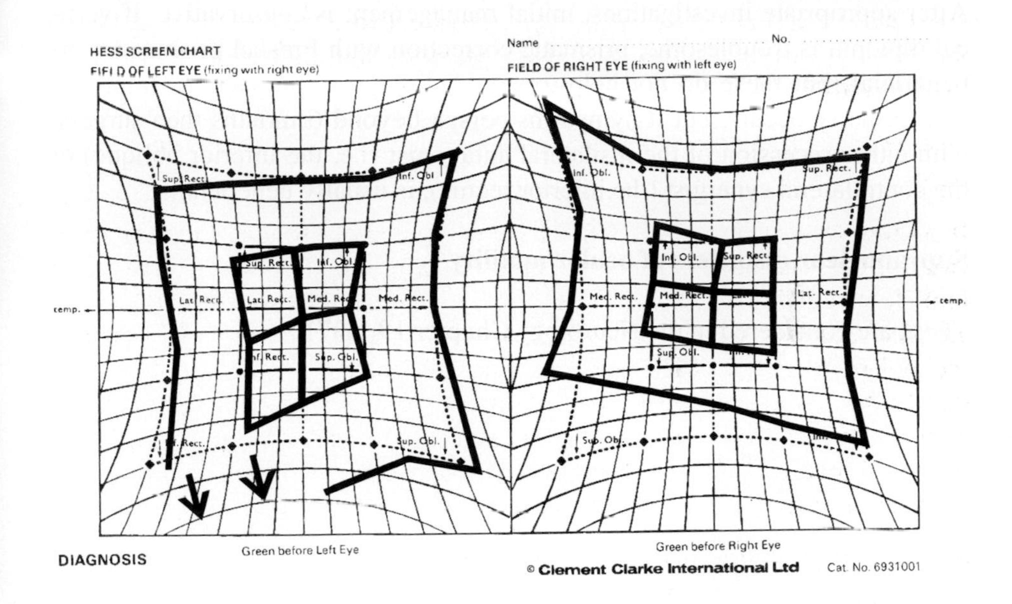


Fig. 11.21 Hess Chart showing R 4th cranial nerve palsy causing R superior oblique muscle paresis.

## *Causes of fourth cranial nerve palsy*

Many fourth nerve palsies are traumatic in origin, as the fourth nerve is vulnerable to contra-coup injury where it emerges on the dorsum of the midbrain. Vascular diseases and brainstem gliomas are other recognised causes. A fourth nerve palsy may also be congenital. (Trauma may also result in weakness of the superior oblique muscle action due to injury to the trochlear pulley through which its tendon passes).

*Treatment of fourth cranial nerve palsy*

After appropriate investigations, initial management is conservative. If vertical diplopia is troublesome, prismatic correction with Fresnel prisms may be beneficial, but these do not help torsional diplopia. Spontaneous improvement usually occurs, but if symptoms persist beyond 6 months then surgery, with either recession of the ipsilateral antagonist (i.e. the inferior oblique) or the contralateral synergist (the inferior rectus), is usually beneficial.

## Supranuclear disorders of ocular motility

These are considered in the Neurology Chapter 19, p371.

# CHAPTER 12

# EXAMINATION OF STRUCTURE

## Examination of the external eye and anterior media

*The small magnifying lens (loupe)*

Examination of the external eye and anterior segment requires magnification to detect many of the signs of disease. The use of quite simple apparatus can improve diagnostic accuracy. A powerful (x10) magnifying lens constructed to minimise distortion (sometimes called a loupe) (Fig. 12.1 p255, LCW 1.9 p13) can be used in conjunction with a focused beam of light which is conveniently provided by the ophthalmoscope or a focusing pen torch (LCW 1.5, 1.10 p13). The eyelids, cornea, iris and lens can be seen directly but a full view of the conjunctiva requires eversion of the lids. (LCW 1.6, 1.7, 1.8 p13). Keratic precipitates and even cells in the anterior chamber can be detected and when the pupil is dilated the anterior vitreous can be inspected. Binocular magnifying spectacles incorporating prisms are also very useful.

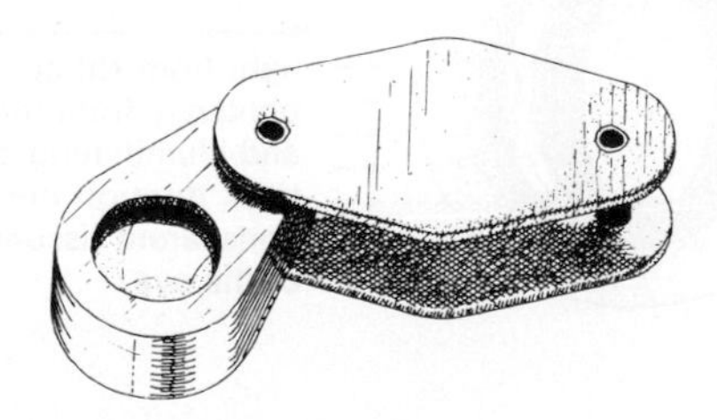

Fig. 12.1 A small magnifying lens.(x10)

*The slit lamp microscope*

A slit lamp is essential for detailed examination of the anterior part of the eye, and in combination with additional lenses allows similar views of the posterior vitreous and retina. The slit lamp is a horizontally mounted binocular microscope with an illumination beam, which passes through a variable slit aperture and illuminates a track through the semi-transparent structures of the eye. As the tissues surrounding the tract are in darkness the beam appears to cut a section through the media of the eye (optical section) and the sides of this section are viewed by the examiner through the microscope (Fig. 12.2 p256, Plate 20.1 p426, LCW 1.29 p20). The focus of the beam and that of the microscope are arranged to coincide. This relationship is maintained by coupling the mountings of the light source and the microscope so that they move together. (Fig. 12.3 p257) This leaves one of the examiner's hands free to hold the lids or manipulate apparatus which measures the pressure in the eye (tonometry) or special contact lenses for examinations such as the stereoscopic examination of the angle of the anterior chamber (gonioscopy) or the fundus (fundus contact lens). Introduction of a hand held non-contact condensing lens (+90D or +78D) into the slit lamp beam allows a highly magnified stereoscopic view of the retina which is particularly valuable in the detailed assessment of disease of the retina and optic nerve head by indirect ophthalmoscopy (p262).

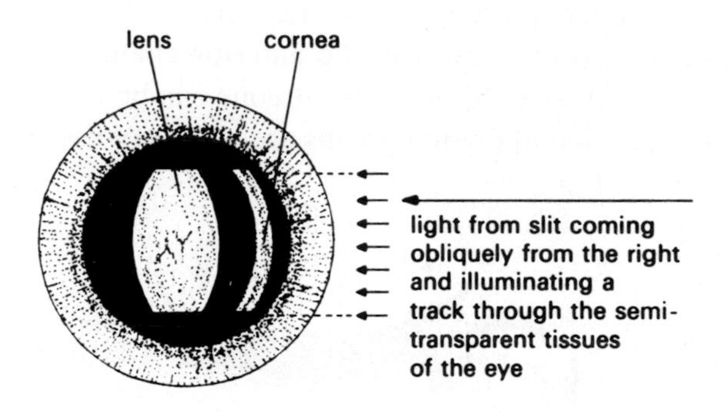


Fig. 12.2 Optical section of anterior segment as seen with slit lamp microscope. (see also Plate 20.1 p426, which also demonstrates an inferior coloboma of the iris and nuclear sclerosis of the lens of the eye).

## *The principle of the direct ophthalmoscope*

No one had observed the ocular fundus until 1847, when the mathematician Charles Babbage, and later Helmholtz in 1851, independently recognised that rays of light entering the eye and illuminating the retina retraced almost the same path on leaving the eye. They realised that this accounts for the blackness of the pupil, because little light from the retina reaches the observer's eye unless he places his eye in line with the source. Babbage solved this problem by directing light into the patient's eye by an inclined mirror which had an aperture in its silvering through which some rays emerging from the patient's eye could pass to the observer's eye. Lenses may be necessary to focus the fundus view and the strength of these will depend on the sum of both the patient's and the observer's spectacle correction (Figs. 12.4 - 12.7 pp258 - 259).

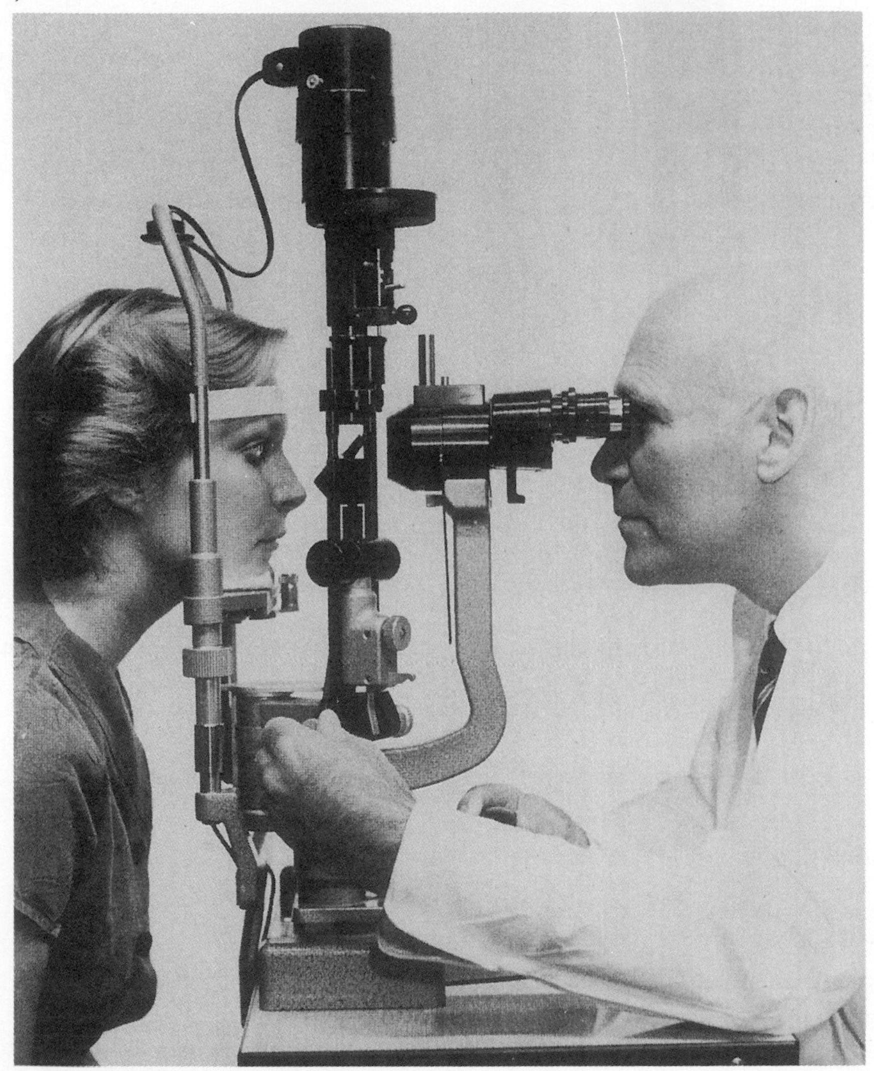

Fig. 12.3 A slit lamp microscope.

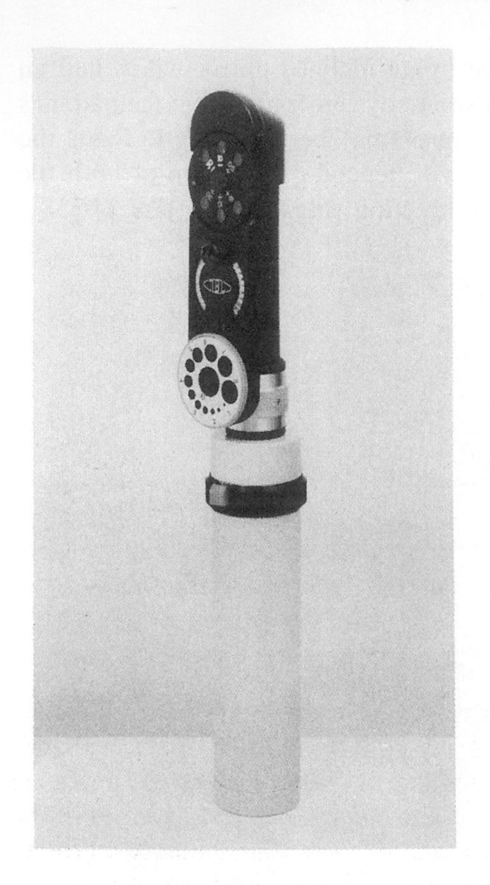

Fig. 12.4 A direct ophthalmoscope.

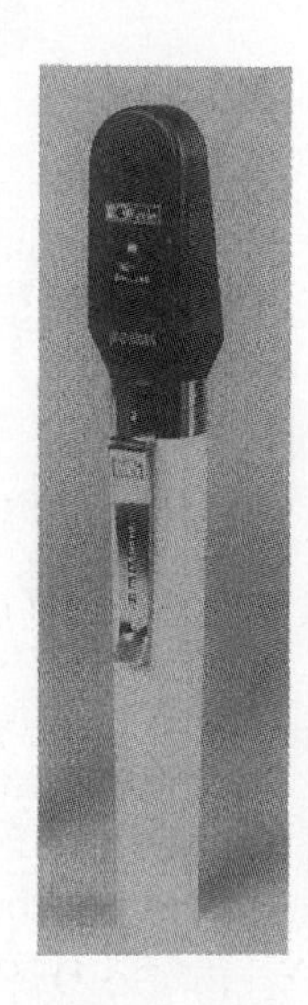

Fig. 12.5 A direct pocket ophthalmoscope.

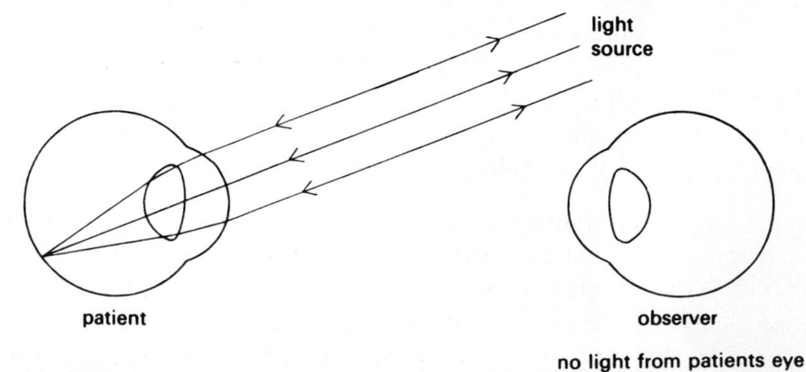


Fig. 12.6 The principle of the direct ophthalmoscope (1).

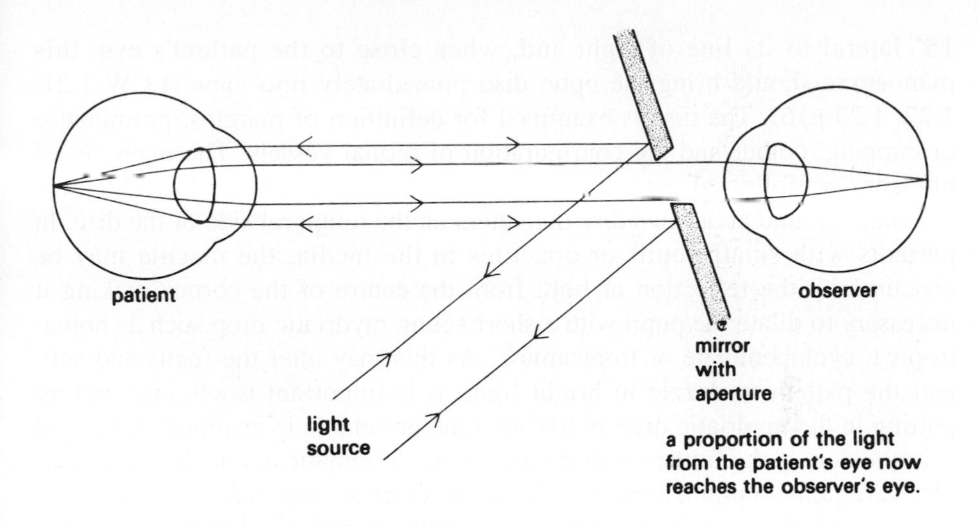


Fig. 12.7 The principle of the direct ophthalmoscope (2).

## *The use of the direct ophthalmoscope*

The direct ophthalmoscope provides a magnified view of the fundus, usually about x 15. Before examining fundus detail it is desirable to hold the ophthalmoscope at a distance of about 15 cm from the eye. In this way an unfocused reddish light is seen in the patient's pupil, the 'red reflex' , and opacities in the media show up clearly in silhouette against this even luminous background (LCW 1.18, 1.19, 1.20 p15). The depth of opacities can be detected by parallax; if the direction of view is moved, a vitreous opacity will appear to move in the same direction as the observer relative to the pupil margin, while corneal opacities appear to move in the opposite direction. Lens opacities in the plane of the pupil remain relatively stationary. If the opacity is very dense, as may occur when the lens of the eye is opaque in advanced cataract, a reduced red reflex will be obtained or may even be absent.

Having made this preliminary examination it is best to ask the patient to look at some distant object and for the observer to use his right eye for examination of the patient's right eye and left eye for the patient's left eye. He should keep his head vertical thus avoiding obstruction to the patient's other eye which, being directed at the designated fixation object, is keeping the observed eye steady. The examiner, looking through the ophthalmoscope held close to his own eye, then approaches the patient's eye at an angle of about

15° lateral to its line of sight and, when close to the patient's eye, this manoeuvre should bring the optic disc immediately into view (LCW 1.21, 1.22, 1.23 p16). The disc is examined for definition of margins, prominence or cupping, colour and the configuration of retinal vessels. The main vessel branches are followed out to the periphery. The central area of the retina, the macula, is found about two disc diameters on the temporal side of the disc. In patients with small pupils or opacities in the media, the macula may be obscured by the reflection of light from the centre of the cornea making it necessary to dilate the pupil with a short acting mydriatic drop such as homatropine, cyclopentolate or tropicamide. As this may alter the focus and subject the patient to dazzle in bright light, it is important to enquire, before putting in the mydriatic drop if driving a motor vehicle is intended. A general examination of the fundus is then carried out systematically, noting in particular variations in pigmentation or the presence of haemorrhages, 'cotton wool spots', 'exudates' or drusen (p292). It is helpful to make a sketch of what has been seen to serve as a record for other, later examiners.

*The principle of the indirect ophthalmoscope*

This method of examination was devised four years later than direct ophthalmoscopy, by Ruete (1851), who reflected light from a bright source into the patient's eye using a mirror with a central hole, but who interposed a convex lens between them. This focused the rays emerging from the patient's eye into a real inverted image which could then be observed through the sight hole in the mirror (Fig. 12.8 p260).

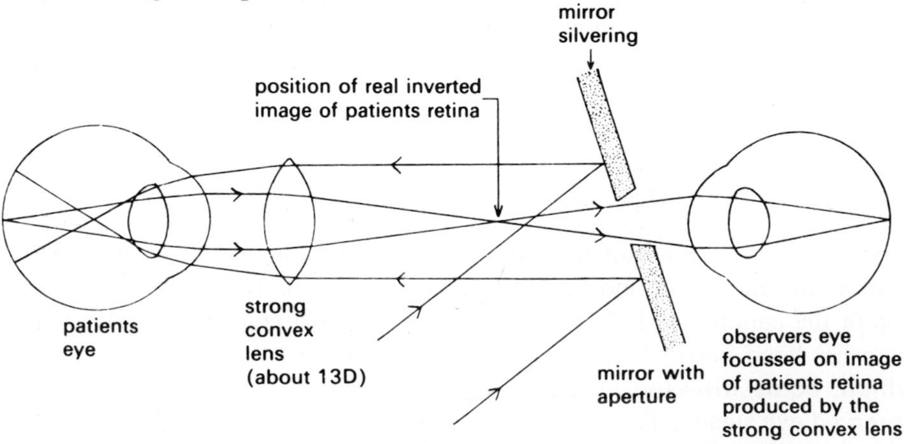


Fig. 12.8 The principle of the indirect ophthalmoscope.

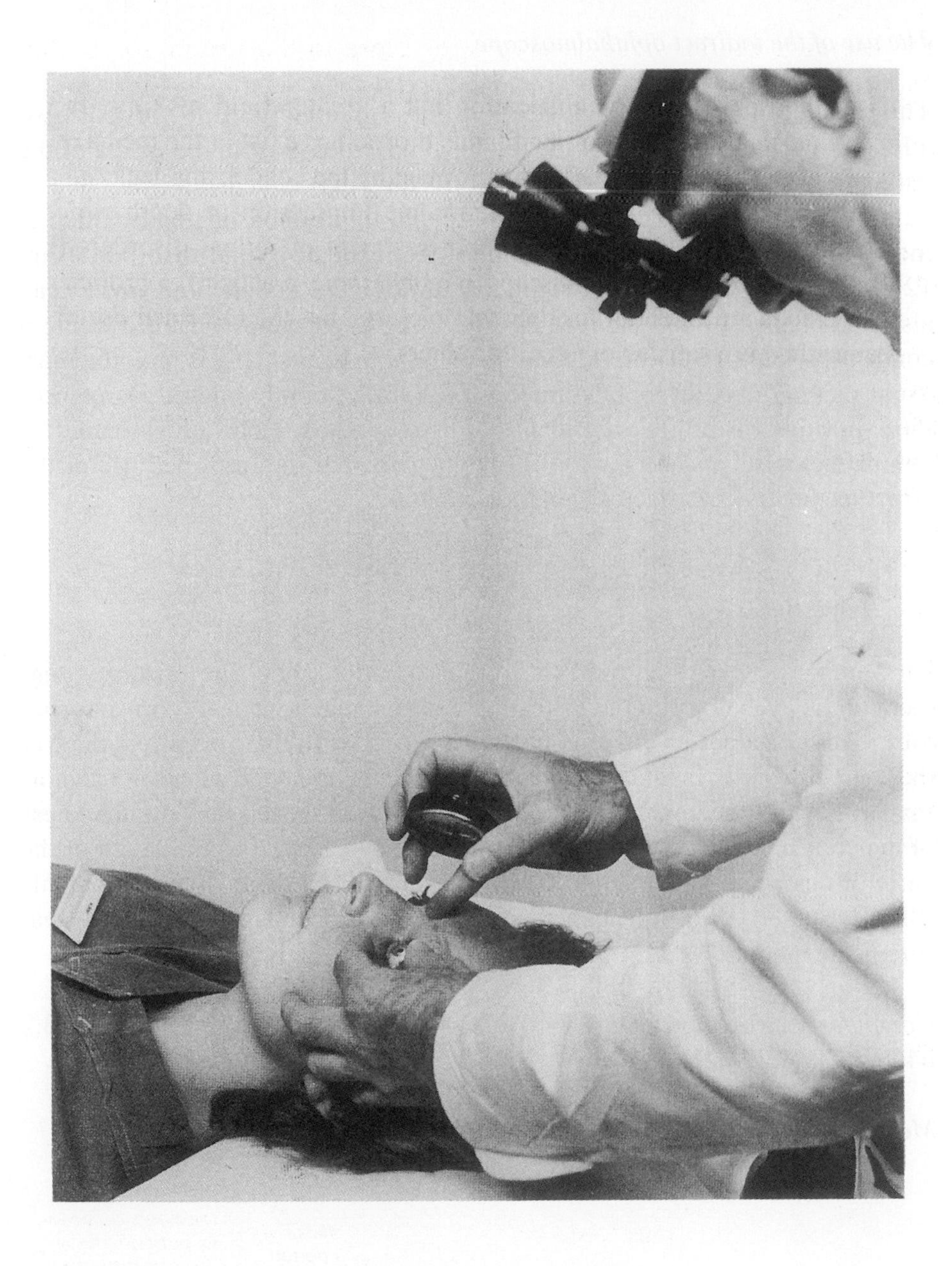

Fig. 12.9 The method of indirect binocular ophthalmoscopy.

*The use of the indirect ophthalmoscope*

This method gives less magnification but a greater field of view. It also allows a much clearer view of the fundus if opacities exist in the media of the patient's eye. The real inverted image formed by the condensing lens can also be viewed binocularly giving a stereoscopic impression of depth which is most valuable in the interpretation and treatment of retinal disorders (Fig. 12.9 p261, LCW 1.30 p20). A highly magnified stereoscopic fundus view can also be obtained by the indirect method using a slit lamp and introducing a condensing lens of 90D or 78D into the slit lamp beam. This is now the most usual method used by ophthalmologists. Indirect ophthalmoscopy requires more practice for proficiency than the direct method. Both are essential for ophthalmologists and desirable for optometrists, but the informed use of the direct method is a satisfactory goal for others.

*Ultrasonography*

Ultrasonography may be used to give information about the posterior segment of the eyeball when the media are opaque, whether due to corneal opacities, dense cataract or vitreous haemorrhage. It will also reveal cupping of the optic disc. It is invaluable in severe ocular injuries and vitreous haemorrhage prior to vitrectomy. The method is also used to measure the thickness of the cornea, lens, and the axial length of the eye as well as the depth of the anterior chamber. The calculation of the refractive power of lens implants (SRK Formula) prior to cataract extraction also depends on ultrasonic measurements (p66, LCW 1.34 p21).

*Computerised Tomography (CT)* (p371, Figs 19.10 p381, 19.12 p382, 23.2 p466, LCW 1.35 p21)

*Magnetic Resonance Imaging (MRI)* (p371)

# CHAPTER 13

# EXAMINATION OF FUNCTION

## Visual physiology

*Photoreceptors.* Electromagnetic radiation of wavelength between 400-700 nm initiates a chemical reaction in specialised retinal cells, the rods and cones, constituting a stimulus to nervous impulses which are transmitted and give rise to perception in the human visual cortex.

The foveal part of the retina is predominantly occupied by cones which contain one of the three visual pigments and subserve colour vision. Rods contain a single visual pigment (rhodopsin) and increase in number from about 2 mm from the centre, maximum density being found between 3 and 6 mm from the fovea. Cone density decreases towards the periphery but extends to the ora serrata. The rods are far more numerous than the cones, there being a total of 120 million rods and 6 million cones in each eye. There are only approximately 1 million axons in the human optic nerve, therefore there is considerable convergence of photoreceptor efferents on to ganglion cells via bipolar cells. Many more rods converge on a single ganglion cell than cones and this means that at moderate light levels most of the retinal activity is cone signalling in spite of the high rod:cone ratio. The small number of cones converging on each ganglion cell also allows the high spatial resolution (measured as visual acuity) of the cone rich fovea.

*Retinal Adaptation.* The presence of two functionally distinct photoreceptor classes allows adaptation of vision to the wide range of ambient illumination. Rods contain the visual pigment rhodopsin which is 'bleached' by bright light. The pigment is regenerated in dark conditions and the consequent slow increase in light sensitivity (extending over approximately half an hour) can be measured and plotted as the 'dark adaptation curve' (Fig 13.1 p264). The initial portion of this curve also shows the limited dark adaptation of the cones. Predominantly rod mediated night vision is achromatic and of low spatial resolution but is much more sensitive than would be possible without adaptive changes. When fully dark adapted, central (cone) vision is particularly insensitive and this explains the apparent disappearance of a faint star when viewed directly: the star 'reappears' when viewed indirectly. When illumination is high, cone activity allows high resolution central vision and colour perception.

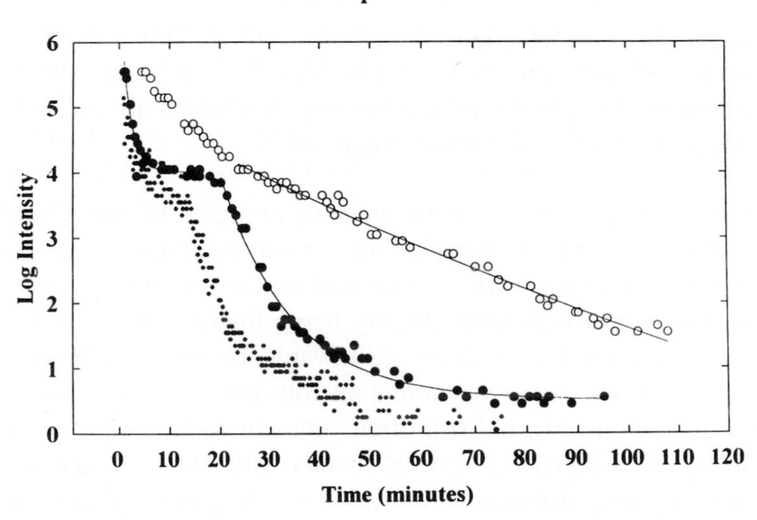


Fig. 13.1 The dark adaptation curve. The lower plot (filled circles) shows the normal dark adaptation curve. Initially the sensitivity increases followed by a pause until at about seven minutes there is again an abrupt increase in the ability to see dim stimuli,although a limit is gradually reached after about an hour. The first rapid phase in dark adaptation represents the increase in sensitivity of the cones in the dark. Following previous exposure to light which has bleached the rhodopsin on which the rods depend, their threshold is still higher than that of the cones. Regeneration of this pigment takes about half an hour. The upper plot (open circles) shows the impaired dark adaptation curve from a patient with a retinal dystrophy in which rhodopsin regeneration is delayed.

Abnormalities of rod adaptation occur in (inherited) retinal dystrophies which have in common the symptom of nyctalopia or 'night blindness'. These patients may have subjectively normal vision during the day but are abnormally incapacitated in conditions of low illumination. An example is retinitis pigmentosa in which this nyctalopia is frequently the earliest symptom (p520). Cones may also show clinically important abnormalities of adaptation: patients with macular disease may show delayed recovery of acuity after exposure to a bright light (photo-stress testing).

## Colour vision

Refracted white light forms a spectrum with increasing wavelengths from 380 nm in the violet through indigo, blue, green, yellow, orange to 700 nm in the red. Young and Helmholtz proposed the existence of three colour perceptive elements in the retina, for red, green and blue. White light is perceived when all three elements are stimulated equally and colours of various kinds when there is unequal stimulation. The trichromatic theory of colour vision is supported by histological evidence of three types of cone containing pigments with maximum absorption at 445, 535 and 570 nm. Trichromatic vision extends 20-30 degrees out from fixation and this can be assessed clinically using a coloured (red) target to confrontation.

### *Colour vision defects*

Impaired colour discrimination may be congenital or acquired. An acquired colour defect is an important sign of optic nerve dysfunction. Red/green discrimination is reported to be poor in early demyelinating optic neuritis whereas the blue/yellow axis is thought to be affected earliest in thyroid optic neuropathy and in chronic glaucoma. Colour vision may also be impaired in macular disease. A recent history of colours appearing 'washed out' or desaturated and/or asymmetry of colour vision between eyes is highly significant and excludes congenital colour defects. Congenital 'colour blindness' is genetically determined and has a complex classification:

-the *anomalous trichromat* has a weakness, not an absence, of the response to one of the colours and shows impaired discrimination between colours when compared to normals. Red or green anomaly has an X-chromosome linked inheritance and so is much more frequent in males (8%) than females(0.4%),
-the *dichromat* lacks one of the visual pigments and is called a deuteranope if the green factor is absent and a protanope or tritanope if the red or blue response is missing,
-the *monochromat* has an absence of colour appreciation. Rods or cones may both be defective. The very rare blue cone monochromats and rod monochromats have poor acuity and nystagmus.

## Tests of visual function

### *Tests for colour discrimination*

-*Lantern tests*: the Edridge-Green lantern has one aperture through which coloured lights of varying brightness can be shown successively, and the Board of Trade lantern has two such apertures. The effect of a retinal stimulus is modified by either a previous stimulus, an effect described as successive contrast or temporal induction, or an adjacent stimulus which is called simultaneous contrast or spatial induction. In this context the Edridge-Green lantern can be used to assess the first by showing successive stimuli while the double aperture in the Board of Trade lantern may indicate the effect of simultaneous contrast. These tests are used for simulating in a practical way the requirements for navigation and flying.

-*Pseudo-isochromatic plates* consist of a pattern of dots which can only be identified correctly if colour vision is normal. In some cases of colour vision defect an alternative 'wrong' answer can be given. Numbers are used in Ishihara plates and various symbols in the Hardy, Rand, Rittler test. These are valuable tests for rapid screening (Fig 13.2 p266).

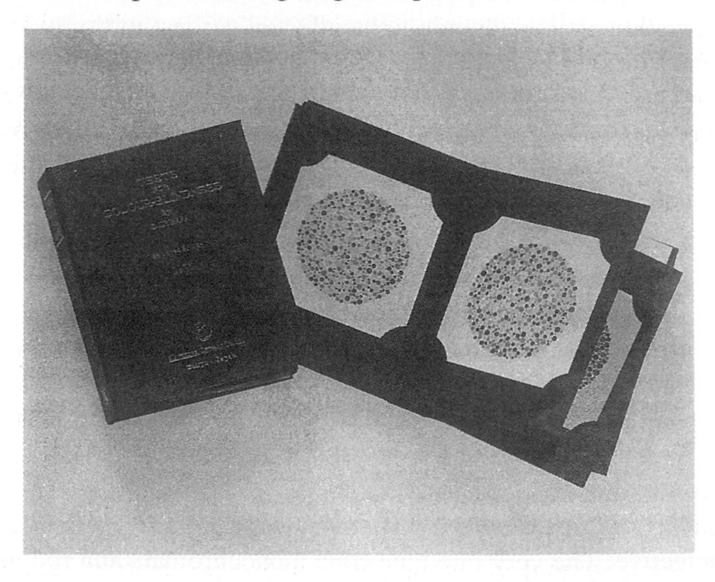

Fig. 13.2 Ishihara colour vision test.

*-Hue tests* are accurate but time consuming. They involve the arrangement in the correct order of a series of closely graded hues. An examination of this type, the Farnsworth-Munsell Hundred Hue Test, is of great value in scientific work. Automation of the test has made it more suitable for clinical work Both these tests of colour vision were developed to distinguish between congenital anomalies but are also commonly used clinically to assess acquired defects.

*Parallel visual pathways*

Recent work has suggested the presence of two functionally distinct but parallel visual pathways in the retina and optic nerve: the magnocellular and the parvocellular, arising from larger and smaller ganglion cells respectively with corresponding larger and smaller diameter axons.

Histological and psychophysical evidence indicates that the magno-pathway is phylogenetically the older and concerned with perception of depth and motion and low contrast/low spatial resolution stimuli. The more recently developed parvocellular system subserves colour vision and form perception and is maximally stimulated by high spatial resolution/high contrast stimuli. There is evidence that the earliest neuronal loss in primary open angle glaucoma may preferentially impair the magno-pathway, although as the visual system is very complex there are certain to be additional pathways. Specific tests of visual function to detect magnocellular damage are therefore being developed. Such tests may be of importance for the prevention of sight loss from primary open angle glaucoma because this disease gives rise to few or no symptoms in its early stages, which may not in fact be 'early' because the optic nerve has been shown to suffer appreciable glaucomatous damage before this is detectable by current tests.

These newer tests include 1. blue on yellow perimetry 2. motion detection 3. spatial contrast sensitivity 4. temporal contrast sensitivity 5. a combination of 3 and 4.

*Blue on yellow perimetry* - exploits the relative rarity of blue sensitive cones in the central retina. (approximately only 10% of the number of red/green cones). Perimetry using a blue light stimulus on a yellow background may detect early neuroretinal glaucomatous damage because of less 'overlap' (redundancy) between the receptive fields of the rarer blue cone ganglion cells. Hence the loss of individual cells manifests itself earlier than when using white light perimetry which stimulates the ganglion cells serving all cones. There are, however, problems involving the selective absorption of light of certain wavelengths in the presence of lens changes and with requirement of pre-test dark adaptation which tend to make the test less practical for primary ophthalmic examinations at present.

*Motion detection* - is measured by motion sensitivity perimetry in which one of a number of linear stimuli in different parts of the visual field are moved briefly on a computer screen. The subject indicates when motion is perceived and the threshold for the smallest displacement is obtained (Fig 13.3 p268). Results are influenced by the orientation of the linear stimuli but the test is largely independent of medial opacity and pupil size and refractive error up to 6 dioptres.

Fig 13.3 Fig. 13.3 Prototype VDU-based test of motion sensitivity. One of the line stimuli briefly moves from side to side at a right angle to its length. The subject presses a button when motion is perceived. A threshold for the smallest displacement that is perceived can be obtained. Patients with very early glaucoma may fail to perceive small displacements even when conventional visual field tests are normal.

*Spatial contrast sensitivity.* Some images have absolutely sharp edges whilst others do not. The ability to detect contrast under these varying conditions can be tested using sinusoidal gratings. A grating is a repeated sequence of light and dark stripes portrayed on paper or on a cathode ray tube - or television. The profile of the stripes varies as a sine wave with distance about a mean luminance (Fig 13.4 p269) and the width of one light and one dark stripe is one cycle. When brightness is plotted against distance across a sharp

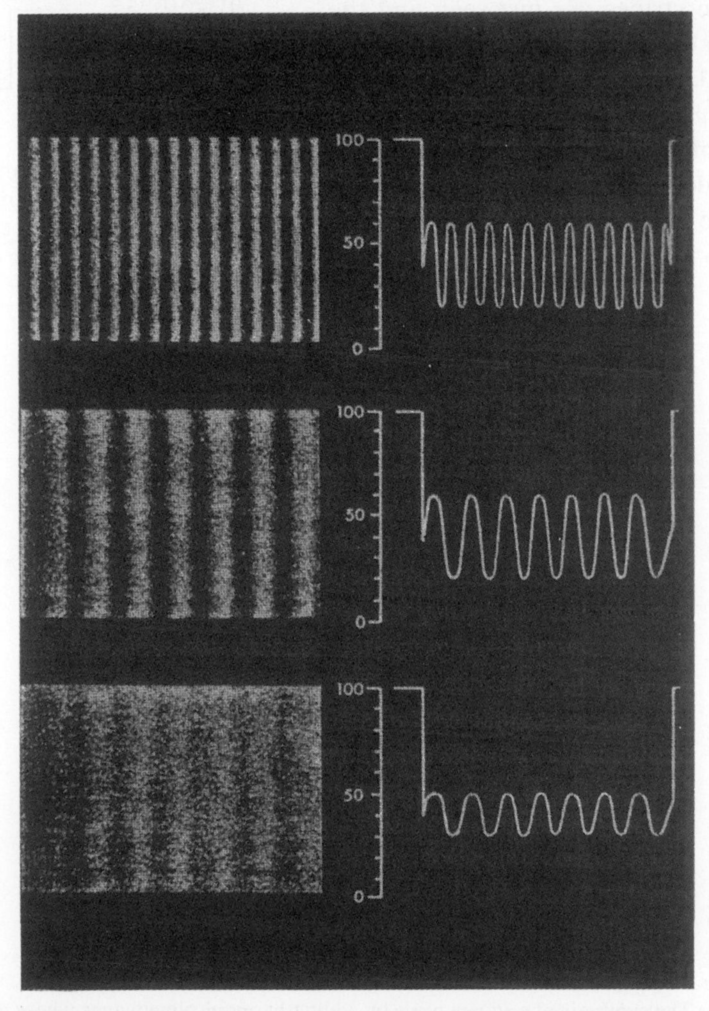


Fig. 13.4 Stripe pattern (grating) with sine wave plot of brightness against distance.

boundary between a white object and a dark background it reveals a square wave (Fig 13.5(a) p270). The square wave can be regarded (Fourier transform) as a fundamental sine wave with the addition of a series of odd harmonics (Fig 13.5(b) p270) The patient responds when the stripes are just seen and the ability to detect the stripes of gratings of varying degrees of closeness (i.e. *spatial frequency*, expressed as cycles per degree of visual angle (cpd)) and of varying degrees of luminance intensity will give a frequency response graph (Fig 13.6 p271) of the eye being tested.

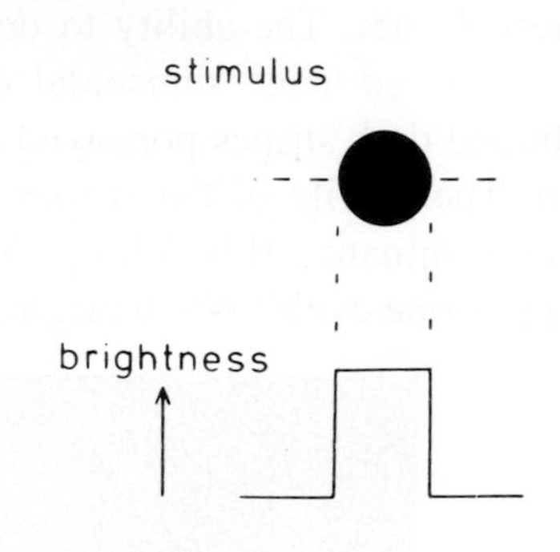


Fig. 13.5(a) Square wave pattern produced by typical perimetric stimulus; brightness plotted against distance across the diameter.

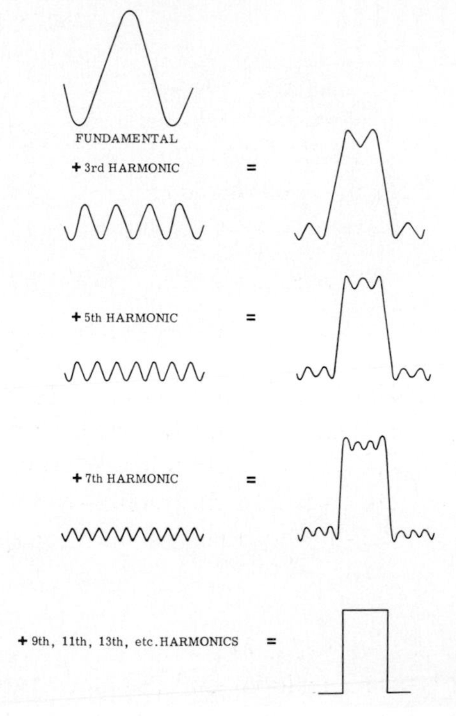


Fig. 13.5(b) The synthesis of a square wave by adding harmonic sine waves whose amplitudes may be determined by the Fourier transform.

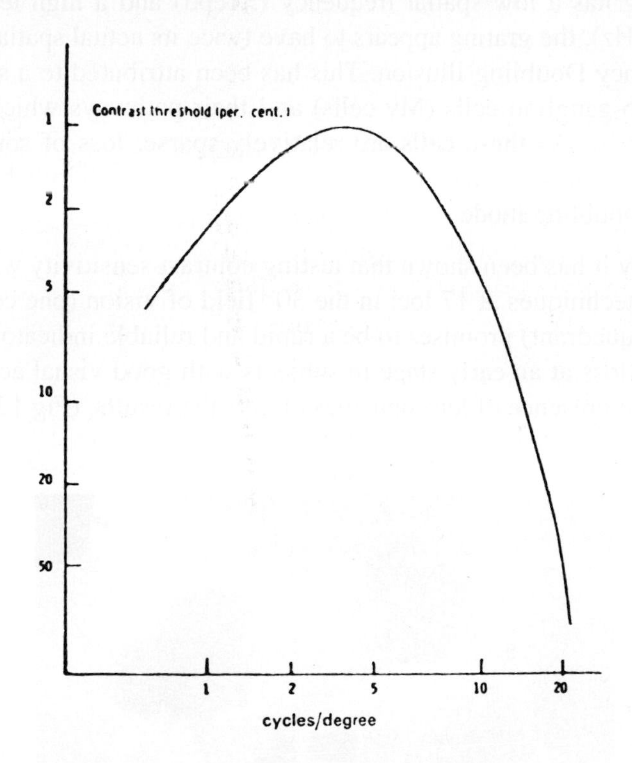


Fig. 13.6 Spatial frequency response graph.

*Temporal contrast sensitivity* - can be assessed when a grating undergoes counterphase flicker (i.e. There is a rapid contrast reversal in which light stripes become dark and vice versa, while the total luminance remains unchanged on reversal. The brightness at which flicker is just perceived is the threshold of temporal contrast sensitivity for a given spatial frequency and rate of reversal.

*Combined spatial and temporal contrast sensitivity*. For many years it has been known that both spatial and temporal contrast sensitivity are reduced in primary open angle glaucoma, especially if their results obtained separately are combined.

If the grating has a low spatial frequency (>1cpd) and a high temporal frequency (>15Hz), the grating appears to have twice its actual spatial frequency - the Frequency Doubling Illusion. This has been attributed to a small subset of the magno-ganglion cells (My cells) and their pathways which give non-linear responses. As these cells are relatively sparse, loss of some of them will reveal a disproportionate reduction of contrast sensitivity when tested in a frequency doubling mode.

More recently it has been shown that testing contrast sensitivity with frequency doubling techniques at 17 loci in the 30° field of vision (one centrally and four in each quadrant) promises to be a rapid and reliable indicator of glaucomatous field loss at an early stage in subjects with good visual acuity (6/9 or better) but the presence of lens opacities affects the results. (Fig 13.7 p272)

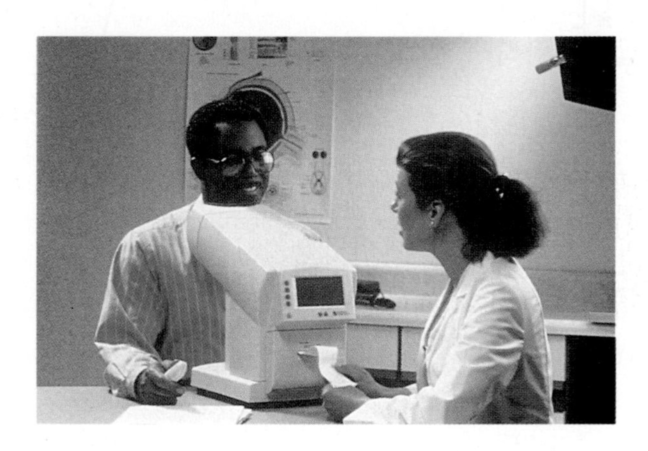

Fig. 13.7 Frequency Doubling Technology Perimeter (Carl Zeiss – Humphrey Systems).

Lens or other opacities in the media affect the results of contrast sensitivity tests but motion detection is largely independent of them, pupil size and refractive error. Consequently the results of prototype devices for testing a combination of both contrast sensitivity by frequency doubling techniques and motion detection, show a promise of rapid screening, with a satisfactory sensitivity and specificity for the detection of early primary open angle glaucoma, which is so important if satisfactory sight is to be retained for the patient's lifetime.

*Visual acuity* (p77)

Visual acuity is a clinical measure of the patient's minimum angle of resolution. The various methods for measurement of visual acuity are performed under conditions of high contrast using black figures (letters or symbols) on an illuminated background. The most familiar pattern is the Snellen chart. For a normal eye wearing the correct optical correction the resolving power is between 30 seconds of arc and 1 minute of arc. A patient with resolution of 1 minute will read the line on the Snellen chart designated 6/6 at 6 metres viewing distance or 20/20 at 20 feet. Patients with smaller resolution limits may read Snellen line 6/5. The anatomical limit on visual acuity assuming an ideal optical pathway is the density of cones at the fovea. The maximal density of cones at the fovea is two cones to each minute of arc. Further limits are imposed (at more peripheral retinal sites) by the convergence of more than one photoreceptor on to each ganglion cell.

There are visual tasks which seem to indicate a higher resolving power than the anatomical receptor density limit. These are known as hyperacuity measures. The mechanisms underlying hyperacuity are still under investigation but must involve complex neural processing. These measures can be clinically useful because they are less dependent on clear media and are resistant to refractive defocus.

*Visual field*

The visual field is the projection in space of the seeing area of the retina. An instrument designed for measurement of the visual field is called a perimeter. The purpose of testing the visual field is to obtain information about the functional status of the retina and the other visual pathways and to also allow measurement of change from one visit to the next. Taken together with the other clinical findings, the pattern and extent of any visual field loss can be of great diagnostic value.

*Methods of visual field testing (Perimetry)*

Confrontation tests (p83) if correctly carried out will detect most of the important types of field loss.

The use of perimeters to test and record precisely the extent of the field of vision is necessary in ophthalmic practice to confirm the details of any suspected defect and to provide a record for future comparison. Conventional perimetry measures the differential light sensitivity or luminance sensitivity at points throughout the expected visual field and provides a visual representation of the result. Most perimeters use a standard background illumination (31.5 apostilbs) so that the field test is conducted under mesopic conditions. This arrangement is a compromise to ensure similar contributions to the field sensitivity from both rods and cones as well as avoiding the inconvenience of the prior dark adaptation required for scotopic testing. It is important to remember that visual field tests are subjective and will depend on the performance of both the patient and the examiner. The ideal test:

1 -is rapid to avoid fatigue
2 -is simple for the patient to understand
3 -allows the patient to give a simple reply
4 -allows control of the patients fixation
5 -is carefully standardised for comparability
6 -is assessed numerically to improve facility for visual field indices follow-up
7 -in some circumstances requires for some purposes a perimeter which is inexpensive, robust, able to run on a portable electric power source and gives quantifiable results suitable for screening programmes. It should comply with the other criteria as far as possible (p278) despite less than ideal working conditions.

*Kinetic perimetry*

A target of specified colour, size and illumination is moved by the examiner into the expected field of vision both to define the limits of the field and to map the areas within these limits in which the object cannot be seen (scotomas) .

*Kinetic perimeters*

There is considerable variety in these instruments. The patient's chin is supported on a chin rest so that the eye being examined is in line with a fixation target and usually 33cm from it. The patient reports when he just becomes aware of the target, which is being slowly moved by the operator into the expected field of view from without or from within a non-seeing area or *scotoma*. As each point in the visual field is tested, a line is drawn joining points of equal sensitivity (isopter) . In some instruments the target moves mechanically along a rotatable arm in the form of a curved arc; in others, e.g. the Goldmann perimeter, it is projected as a spot of light. Care should be taken to control fixation and the level of illumination (Fig. 13.8 p275) . An important feature of the Goldmann perimeter is that it allows testing of the far peripheral field. This facility allows detection of some types of neurological field defect and the assessment of some occupational requirements.

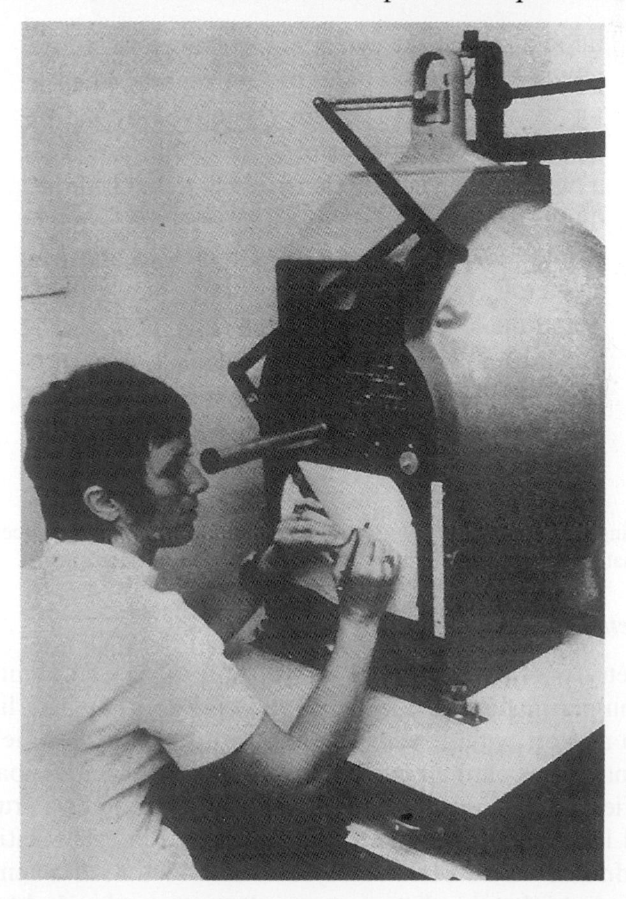

Fig. 13.8 The Goldmann bowl perimeter with projected light stimulus, rear view showing pantograph recorder and fixation monitor telescope.

*The Bjerrum or tangent screen*

The original equipment introduced by Bjerrum was a flat screen frequently used at a distance of one or two metres to examine the central 25 degrees of the visual field. A small circular disc on an inconspicuous wand or a projected spot of light is used as the target (Fig. 13.9 p276, LCW 1.26 p18) (the wall in this figure is shown bright but would be dark in practice). The Bjerrum screen requires little equipment but does require some skill to obtain reproducible measurements. It is also limited in the assessment of the far peripheral field. Care must be taken to ensure uniform illumination of the whole screen.

Fig. 13.9 The tangent screen (Bjerrum) (The surrounding wall would in practice also be black). Alternatively a projected light on a grey screen has some advantages.

*Static perimetry*

Static perimetry is often performed as part of a thorough examination using the Goldmann perimeter Fig. 13.8 p275, LCW 1.25 p18). The light stimulus is flashed on and off within suspicious areas of the expected field using different light intensities until an intensity is reached at which the patient consistently identifies the stimulus. This light intensity constitutes a crude threshold for that field location. This concept was developed in early static perimeters e.g. the Friedmann Visual Field Analyser MkI which used a semi-automated process with multiple stimuli to measure luminance thresholds at different retinal locations.

Modern (automated) static perimeters e.g. Octopus, Zeiss-Humphrey (Figs. 13.11(a) p278, 13.11(b) p279) perform a similar estimation of the differential light sensitivity at many field locations which are arranged in a regular grid pattern covering the field. Computerisation of the thresholding process ensures that the stimuli are presented at each location in a random order so there is less incentive for the patient to lose fixation.

The test typically takes longer than a Goldmann field test because it is a more detailed measurement of the visual status. Automated perimetry also requires less highly trained operators. Such detailed examinations may not be appropriate for all patients but are important in the follow-up of glaucoma patients because small changes in the size or depth of a scotoma may indicate the need for more aggressive treatment of raised intraocular pressure.

Automated and semi-automated perimeters allow the selection of different test grid patterns to adapt the field test to the individual patient. The full extent of the visual field can be tested but for glaucoma diagnosis and follow-up only the central 30 degrees of field is conventionally tested. This is because the majority of glaucoma field defects occur within the central 30 degrees and omission of the peripheral points allows time for a more detailed examination of the important central field before the patient becomes fatigued. Assessment of a suspected field defect arising from the visual path-

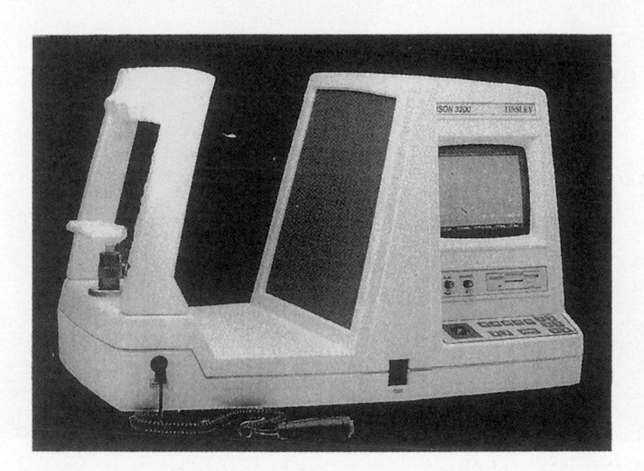

Fig. 13.10 The Henson 3200 Central Visual Field Screener and Analyser (CFA 3200).

way central to the optic nerve head does, however, require a full field test (either Goldmann or full field automated perimeter programme). Several methods of analysis of the change in serial visual fields have been devised. One of these, the *Progressor programme* (Plate 13.12 p280) provides valuable information at a glance for the clinician. This has a display which records both the degree of defect and the rate of change of sensitivity of the retina at 76 points in the 30% field for a sequence of 20 or more tests. In addition it provides pointwise analyses of its sensitivity and its rate of change together with estimates of patient reliability. A prediction capability is being developed.

It is important in many situations to be able to use a simpler instrument which fulfils requirement 7. on p274. The Friedmann Analyser MkI and the Henson CFA 3200 (Fig. 13.10 p277) go some way to satisfy this requirement and the latter is particularly suitable for surveys and screening programmes, being appreciably independent of operator expertise. A very simple card method with a centrally appearing stimulus and multiple fixation points, the Damato OKP test, has value as a screening test. Other screening perimeters are being developed from computer games to make them more user friendly.

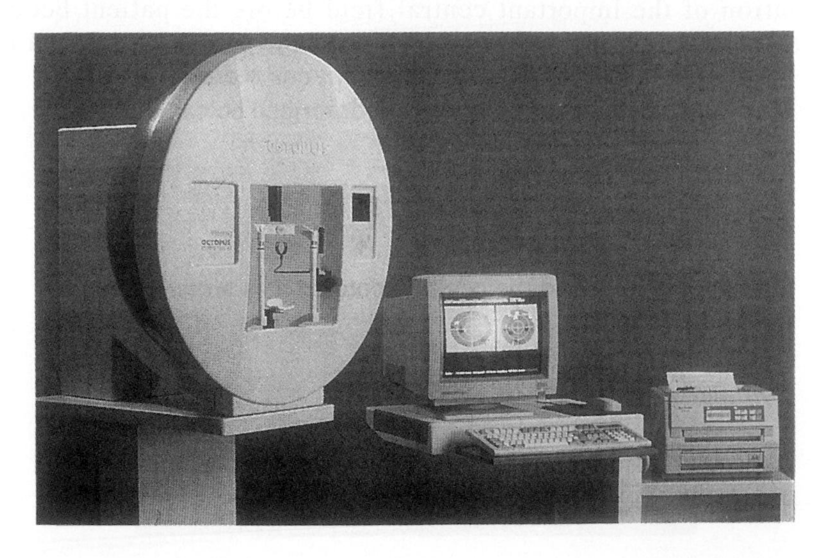


Fig. 13.11(a) The automated perimeter system Octopus 101.

*Interpretation of perimetry*

Interpretation of the results of kinetic or static perimetry requires experience and pattern recognition skills as well as an awareness of the possibility of artefact related to the patients performance or external factors. The patient's performance in correctly indicating (usually by pressing a button) when the stimulus is seen, can be affected by unfamiliarity with the task (learning) as well as fatigue. The overall extent of the measured visual field can be affected by optical factors e.g. pupil size, cataract, spectacle frames as well as the individual configuration of the periorbital tissues e.g. apparent superior field loss with ptosis. Modern automated perimeters will also indicate when a subject demonstrates unreliable testing behaviour e.g. repeated loss of fixation. Interpretation of the resulting field test must therefore be performed with knowledge of these factors as well as important clinical information e.g. optic disc appearance. (Figs. 13.13 p 281,13.12 p 280, LCW 1.28 p19)

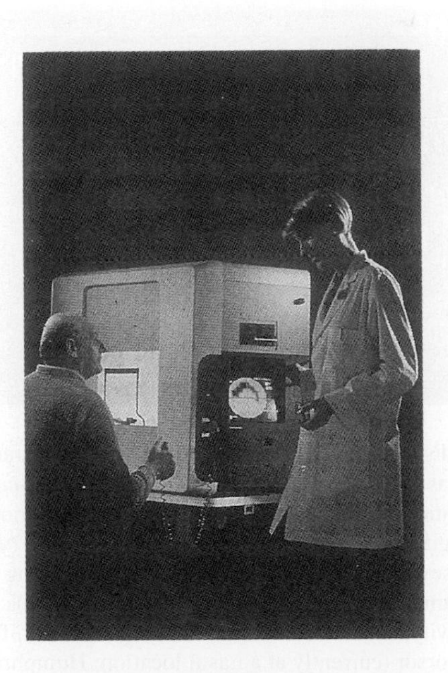

Fig. 13.11(b) A Zeiss Humphrey automated perimeter.

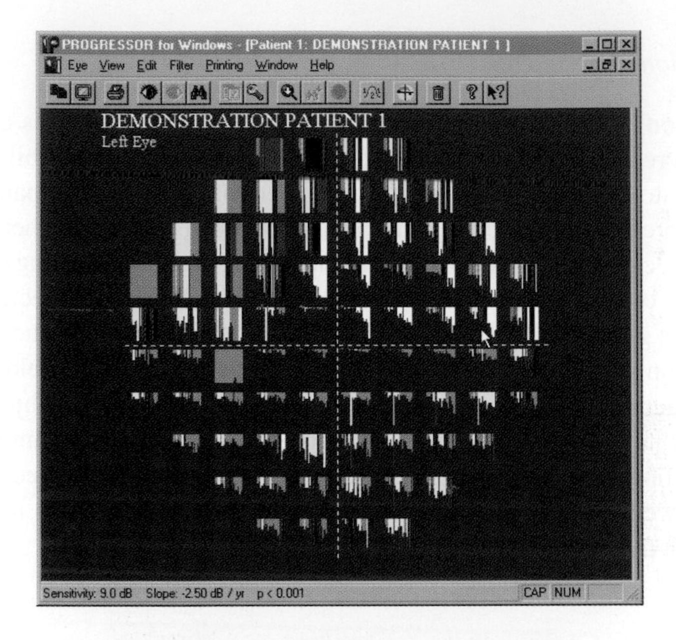


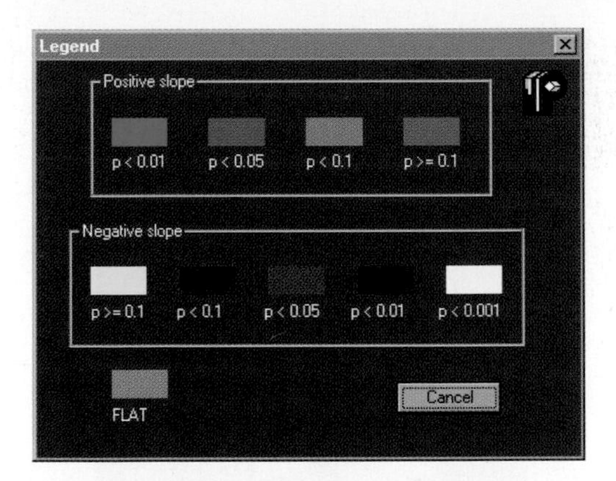


Plate 13.12 'Progressor' display. Locations in the superonasal quadrant show progressively longer bar length as well as the appearance of red and white colour-code indicating significant rate of deterioration. Locations in the inferotemporal quadrant are more stable with generally similar sequential bar lengths and non-significant regression slopes (yellow). In the superotemporal quadrant, there are several locations with generally unchanging grey bars indicating sites of absolute defects that cannot progress further. The status bar (at the bottom of the display) gives the recorded sensitivity at the last test, rate of change and significance of any location pointed at by the screen cursor (currently at a nasal location: Humphrey co-ordinates 21° , 3°, revealing a sensitivity of 9 dB and slope of -2.50 dB/yr p‹ 0.001)

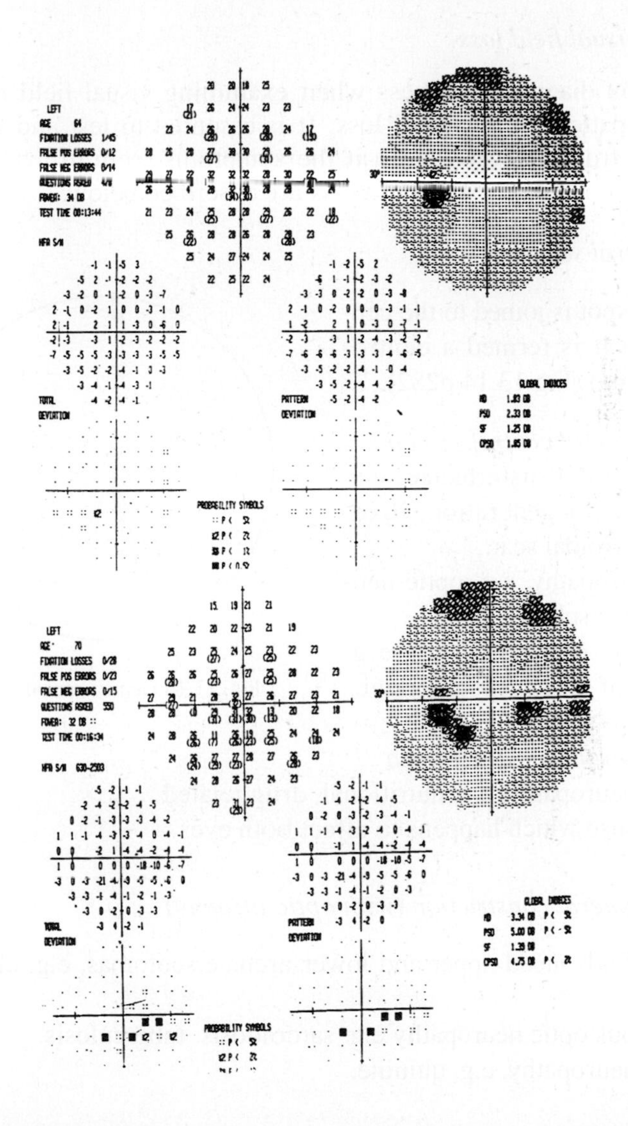

Fig. 13.13 Two visual field tests of the left eye (Humphrey automated perimeter) of a patient with optic disc appearances which raise a suspicion of early glaucoma. The upper field shows the position of the blindspot as a round dark spot on the greyscale representation. This field was judged normal with only slight diminution of sensitivity (grey shading) in the extreme superior field, within normal limits. The lower field was recorded from the same eye one year later: there is now a small area of dark shading extending from the blindspot into the inferior field. This is an early inferior arcuate scotoma. This eye is developing glaucoma and requires appropriate treatment.

*Patterns of visual field loss*

The important diagnostic process when examining visual field records is to identify the pattern of any field loss. It is essential to test and examine the visual fields from both eyes even if the symptoms seem to involve one eye only. The common patterns of field loss are described below.

1. *Central scotoma*

(If the blind spot is joined to the central scotoma it is termed a centro-caecal scotoma) (Fig 13.14 p282).

(a) *Unilateral central scotoma*
-a retinal macular disturbance, e.g. senile macular degeneration, toxoplasmosis choroidal scar.
-an optic neuropathy, e.g. optic neuritis in multiple sclerosis.
-optic nerve compression, e.g. meningioma of the sphenoidal ridge.

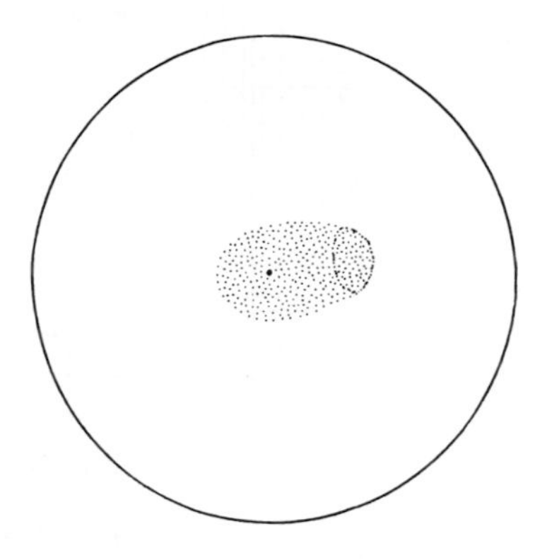

Fig. 13.14 Centro-caecal scotoma.

(b) *Bilateral central scotoma*
-toxic optic neuropathy, e.g. nutritional, drug related.
-unilateral cause which happens to affect both eyes.

2. *Peripheral constriction (with optic atrophy)*

-end stage of advanced upper and lower arcuate scotomas, e.g. chronic glaucoma
-granulomatous optic neuropathy e.g. sarcoidosis, tuberculosis.
-toxic optic neuropathy, e.g. quinine.

3. *Hemianopia*

(a) *Bitemporal hemianopia (with optic atrophy)* (Fig. 13.15 p283)
-chiasmal compression e.g. by pituitary neoplasm, carotid aneurysm. (*bitemporal hemianopic scotomas* may be present if the central crossing fibres are mainly affected.)

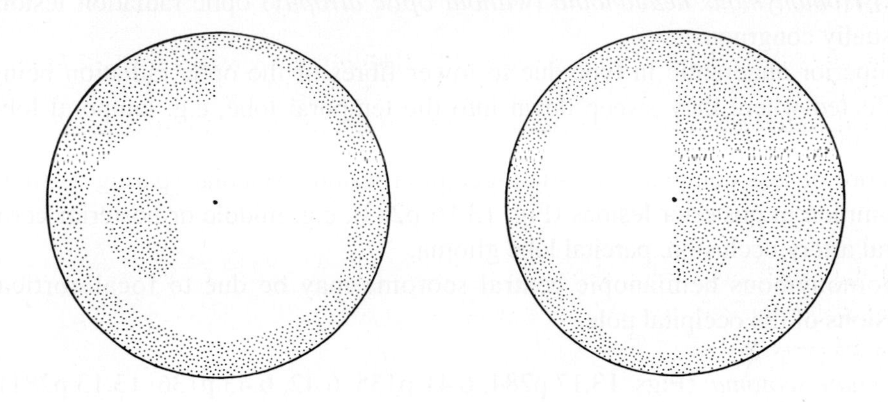

Fig. 13.15 Bitemporal hemianopia.

(b) *Homonymous hemianopia (with optic atrophy)*
-optic tract lesion, e.g. pituitary neoplasm or a craniopharyngioma extending posteriorly. It is often incongruous (i.e. the visual field defects of each eye may not be of the same size) and usually affected up to fixation point (i.e. no macular sparing). A relative afferent pupillary defect may be detectable in the eye with the worse field loss.

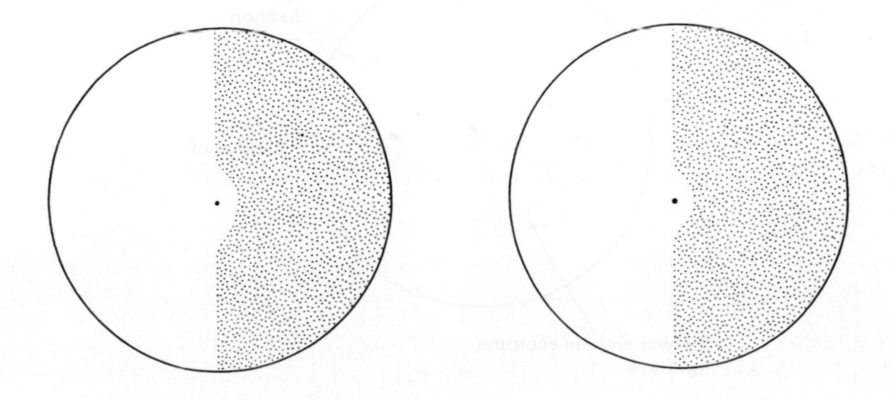

Fig. 13.16 R homonymous hemianopia with macular sparing.

(c) *Homonymous hemianopia (without optic atrophy)* optic radiation lesion, usually congruous:
-superior quadrantic in type due to lower fibres of the optic radiation being affected where they sweep down into the temporal lobe, e.g. temporal lobe glioma,
-more or less complete in parieto-occipital lesion. 'Macular sparing' is more common in posterior lesions (Fig. 13.16 p283), e.g. middle or posterior cerebral artery occlusion, pareital lobe glioma,
-homonymous hemianopic central scotoma may be due to focal cortical lesions of the occipital pole.

*Arcuate scotoma*: (Figs. 13.17 p284, 6.41 p135, 6.42, 6.43 p136, 13.13 p281)

-with glaucomatous cupping and atrophy of the optic disc, e.g. chronic open angle glaucoma,
-with pallor of segment of disc, e.g. segmental ischaemic neuropathy from giant cell arteritis,
-with a patch of choroidal scarring near the optic disc, e.g. juxta-papillary choroiditis,
-with myopic crescents or other congenital abnormality of disc, e.g. high myopia, coloboma of optic disc.

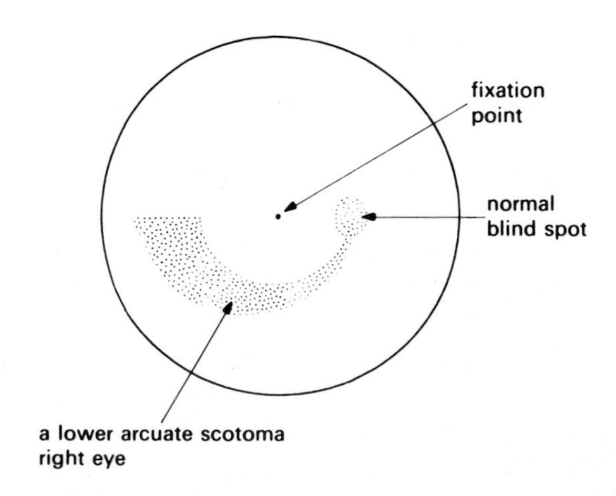


Fig. 13.17 Arcuate scotoma (right eye).

**The examination of ocular position and movements** (pp71, 85, 229, 239)

**The examination of the pupil and its movements** (pp86, 414)

**Electro-oculography, electro-retinography and electro-encephalography (visually evoked responses)**

Retinal and visual pathway functions can be objectively assessed in three further ways by these electronic methods:

*Electro-oculography (EOG)*

The retinal pigment epithelium develops a resting potential of several millivolts so that the anterior pole of the eye is electrically positive with respect to the posterior pole. The electrical potential of the light adapted eye is about twice that of the dark adapted eye. Electrodes placed at the canthi may be used to record changes in this potential when the eye is moved. The amplitude of the resting potential is recorded in light and in dark adaptation when the eyes turn a standard angle from one side to the other . The ratio of maximum amplitude in the light to the minimum in the dark should be two or more. If there is disease of the retinal pigment epithelium it will be less than two . An abnormal EOG is valuable in confirming a diagnosis of vitelliform dystrophy (Best's disease) in which the ERG may be normal.

*Electro-retinography (ERG and PERG)*

When a contact lens electrode is placed on the cornea, and an indifferent electrode on the forehead, a resting potential is recorded. When the retina is stimulated by light an action potential is superimposed and this comprises an electro-retinogram (ERG) (Fig. 13.18 p286, LCW 1.36 p21) The eye is either light (photopic) or dark (scotopic) adapted. Following the light stimulus there is a latent interval, followed by a negative 'a' wave and a positive 'b' wave. The duration is less than 250 milliseconds. The 'b' wave has usually 1.5 times the amplitude of the 'a' wave and the amplitude is greater under scotopic conditions. The ERG response has been shown to arise from retinal elements distal to the ganglion cells. The 'a' wave is generated by photoreceptor activity (relative activity of rods and cones being dependent on adaptation). The 'b' waves have been shown to reflect photoreceptor activity in primates but may also include glial cell (Müller cell) activity. The clinical usefulness of ERG testing is in distinguishing primarily retinal dysfunction from optic nerve/ganglion cell disease e.g. the ERG becomes small in retinitis pigmentosa but remains normal in optic neuropathies.

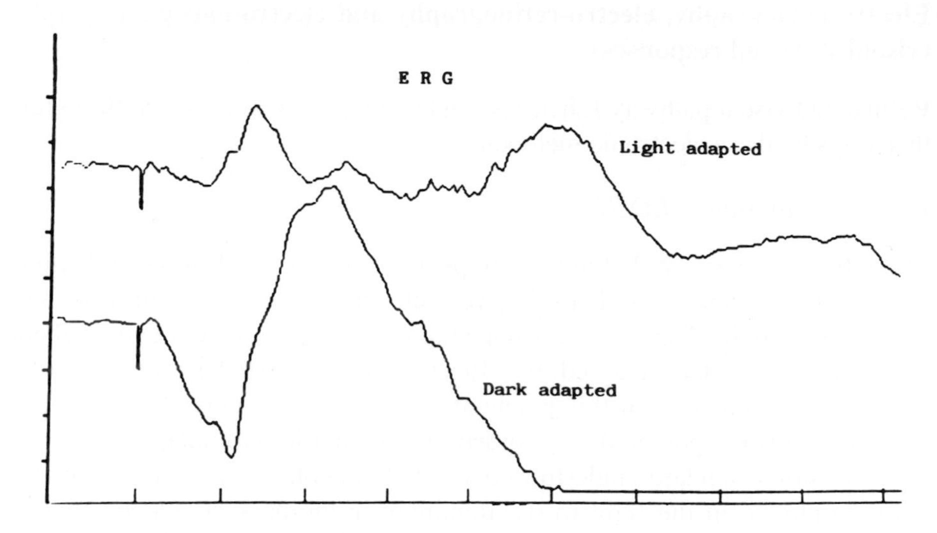


Fig. 13.18 An electroretinogram (ERG).

A pattern electroretinogram (PERG) is recorded as for the ERG but with a 100% contrast reversing chequerboard as a stimulus. It has been suggested that the PERG may reflect inner retinal activity and has been used in glaucoma suspects to detect early functional loss. The stimulus must be accurately focused on the retina and accurate interpretation of the PERG assumes functionally intact photoreceptors (Fig. 13.19 p287)

*Visually evoked response (VER)*

The electro-encephalographic responses (VER) to a flashed light stimulus recorded over the occiput gives information about the integrity the entire visual pathway (Fig. 13.20 p287). The response is a variable and complex waveform which is dominated by the foveal cones since these are disproportionately represented at the occipital pole which lies superficial closest to the recording electrodes. The VER requires clear media and accurate fixation. The VER response is reduced in optic atrophy and is reduced and delayed in optic neuritis during the acute illness.

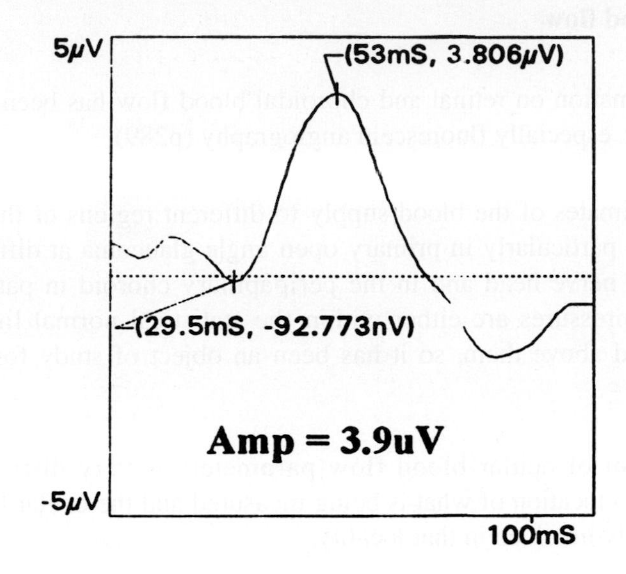


Fig. 13.19 A pattern electroretinogram (PERG).

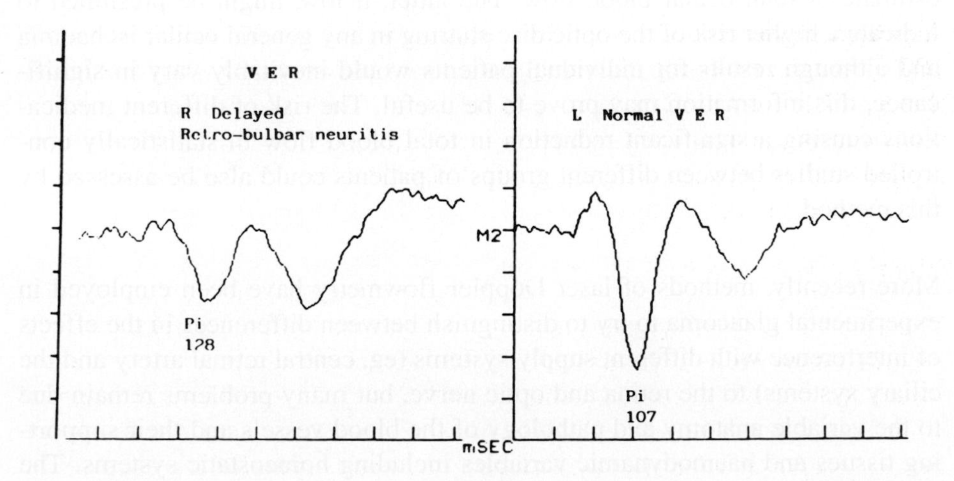


Fig. 13.20 A visually evoked response (VER).

## Ocular blood flow

Much information on retinal and choroidal blood flow has been obtained by angiography, especially fluorescein angiography (p289).

Reliable estimates of the blood supply to different regions of the eye would be valuable, particularly in primary open angle glaucoma at different depths of the optic nerve head and in the peripapillary choroid in patients whose intraocular pressures are either within the statistical normal limits or only mildly raised above them, so it has been an object of study for very many years.

Measurement of ocular blood flow parameters is very difficult, both in respect of the location of what is being measured and the ocular blood supply system mainly involved in that locality.

Pulsatile ocular blood flow and an assessment of its relevance to clinical conditions can be made by a pneumotonograph attached to a slit lamp microscope (p559). This allows the measurement of intraocular pressure and an estimate of total ocular blood flow. The latter, if low, might be presumed to indicate a higher risk of the optic disc sharing in any general ocular ischaemia and although results for individual patients would inevitably vary in significance, this information may prove to be useful. The risk of different medications causing a significant reduction in total blood flow in statistically controlled studies between different groups of patients could also be assessed by this method.

More recently, methods of laser Doppler flowmetry have been employed in experimental glaucoma to try to distinguish between differences in the effects of interference with different supply systems (eg. central retinal artery and the ciliary systems) to the retina and optic nerve, but many problems remain due to the variable anatomy and pathology of the blood vessels and their supporting tissues and haemodynamic variables including homeostatic systems. The method is also in clinical use in trying to assess the influence of blood supply in ocular disease especially in chronic simple and 'normal pressure' glaucoma, but the results for individual patients have to be interpreted with caution.

# CHAPTER 14

# CARDIOVASCULAR CONDITIONS

## General considerations of the retinal effects of vascular disorders

### *The retinal circulation*

The retinal circulation is unique in being the only microcirculatory bed in the body accessible to direct inspection. The arterial and venous vessels around the disc are about 150μ in diameter and when using the ophthalmoscope, vessels as small as 10μ in diameter can be seen. With the aid of fluorescein angiography, abnormalities can also be visualised within the capillary bed itself.

### *Fluorescein angiography*

Sodium fluorescein (p614), 3ml of 20% solution, is given intravenously (LCW 1.33 p21). The dye passes through the pulmonary circulation and then to systemic arteries including the internal carotid. It reaches the retina via the central retinal artery and is brought to the choroid and the vascular circle around the optic nerve head via the short posterior ciliary arteries. The dye arrives at the fundus after 10-15 seconds and its passage through the retinal circulation takes about 10 seconds. The dye is visualised by exposing it to blue light which it re-emits as green light. This is photographed through a yellow filter which eliminates blue light so that only the fluorescein emitted green light is recorded on the film. Several photographs are taken during the initial transit at one second intervals and subsequently every minute, or it may be recorded continuously by film or video camera. The essential feature is that, due to 'tight cell junctions', the normal retinal vessels are impervious to fluorescein and areas of capillary abnormality are revealed by deformities in their pattern, or by their absence or because they leak fluorescein. In contrast the dye penetrates the choroidal vessels readily, giving a uniform background glow which will be screened to a varying extent by the overlying retinal pigment epithelium (Fig. 19.9 p379). Late pictures will reveal leakage of dye from abnormal retinal vessels or the take-up of dye by structures such as drusen. Haemorrhages will appear as black areas. Stereoscopic photographs will afford a dramatic revelation of the exact position, depth and the character of the changes in hypertension, diabetes and other diseases affecting small vessels which are of great interest to the physician (Stereo Plates 6.38 p127, 19.14 p385).

*The retinal vessels*

Different diseases may specifically involve different types of vessel. Hypertension for instance primarily affects arterioles, while diabetes mellitus particularly involves capillaries and veins.

*The retinal arteries: Arterial occlusion* is considered on p118. *Arterial changes occurring in hypertension* are discussed in the section on hypertension, p294 and briefly on p126 and *giant cell arteritis* on p124.

*The retinal veins. Dilatation* of retinal veins may be seen in many different conditions. The most important of these are:

-*Diabetes mellitus:* irregular venous dilatation, sausaging or beading, may be one of the earliest changes to be seen in the diabetic fundus.
-*Partial or complete obstruction of retinal venous flow:* generalised venous distension can occur in central retinal vein obstruction and in raised intracranial pressure. Arterial insufficiency may also reduce flow and cause ischaemia and venous dilatation.
-*Blood viscosity:* severe venous dilatation and irregularity may occur in several conditions in which the common factor appears to be an elevated blood viscosity. These include macroglobulinaemia (p303), polycythaemia rubra vera (p300) and secondary polycythaemia associated with chronic respiratory disease and congenital heart disorders.
-*Retinal vasculitis* (p518).

*Retinal vein occlusion* is considered on p121.

*The retinal capillaries*

*Microaneurysms.* Normal capillaries are too small to be seen with the ophthalmoscope unless the media are extremely clear; when dilatations occur within the capillary system they become visible as microaneurysms. The larger microaneurysms can easily be seen with the ophthalmoscope. Fluorescein angiography which allows visualisation of normal capillaries will often demonstrate many microaneurysms not otherwise seen. They generally occur at the venous end of capillaries. Although most commonly found in diabetic retinopathy, microaneurysms may also occur in the various types of stasis retinopathy and various congenital vascular abnormalities. The pathogenesis of microaneurysms is not completely understood but damage to capillary pericytes is thought to be a factor.

*New vessel formation and proliferative retinopathy.* New vessels are abnormal vascular channels which arise from the retinal circulation. They will eventually be accompanied by a visible fibrous support; it is the association of new vessels with fibrous tissue formation to which the name *proliferative retinopathy* is given. It is useful to consider the processes involved as two stages, early and late, which have different ophthalmic appearances and which are liable to different complications. At first endothelial tubes grow forward from the retinal veins accompanied by connective tissue which is ophthalmoscopically invisible. They usually lie flat in the plane of the retina or may grow forwards on the posterior surface of a detached vitreous and when visualised with fluorescein angiography they usually fill more slowly than surrounding vessels and leak profusely. It is at this stage that these abnormal vessels are most likely to bleed, and if they do, they will either form a haematoma between retina and vitreous *(pre-retinal or subhyaloid haemorrhage)* or they will bleed into the vitreous *(vitreous haemorrhage).* In the late stages of proliferative retinopathy, atrophy of new vessels with extensive fibrous tissue formation occurs, and dense white sheets are seen in the fundus. Contraction of fibrous tissue causes retinal traction and detachment, which is a major cause of visual disability at the last phase of proliferative retinopathy. The mechanism of new vessel formation is unknown but it appears to be induced by the presence of ischaemic living tissue. It may occur at some distance from the ischaemic focus, as in rubeosis of the iris (p316, Plates 6.33 p122, 16.11 p313, LCW 4.15 p63), The processes involved are complex. Several vascular endothelial growth factors (VEGF's) have been identified and their concentration in the aqueous has been found to correlate with neovascular activity and falls after retinal photoablation. Inhibitors of VEGF's have also been found and this has given hope that eventually neovascularisation can be prevented or reduced in circumstances where its presence can lead to so much damage.

*Retinal haemorrhages* can occur from either the superficial or the deep capillary plexuses of the retina, as well as from abnormal vessels and may either localise pre-retinally or within the retina.

*Superficial bleeding* from the superficial capillary plexus occurs into the nerve fibre layer. The nerve fibres run radially out from the disc and haemorrhages into these fibres will follow their course giving the appearance of linear 'flame shaped' haemorrhages. Haemorrhages of this type occur in hypertension, retinal vein thrombosis and blood disorders such as leukaemia, where white leukaemic deposits may be surrounded by haemorrhage.

*Deep bleeding* from the deep capillary plexus occurs at the level of the bipolar cells so that the blood is narrowly confined within this layer. Such small round irregular haemorrhages may easily be confused with large microaneurysms and are often known as 'blot haemorrhages' which are most commonly found in diabetes.

*A pre-retinal haemorrhage (subhyaloid),* is contained between the vitreous and the internal limiting membrane of the retina, and may break through into the vitreous: becoming a *vitreous haemorrhage* (p111). A *subhyaloid* haemorrhage is easily recognised by its rounded lower border and upper fluid level. It occurs in all forms of proliferative retinopathy. It may also be seen following a subarachnoid haemorrhage (p398); it is possible that in this particular association bleeding is the consequence of an acute rise of intracranial pressure compressing the central retinal vein in the subarachnoid space.

*Retinal 'white lesions'*

White lesions in the fundus are often confused. The commonest lesions are 'cotton wool spots' , exudates and 'drusen'.

*'Cotton wool spots'* are white retinal patches with indistinct edges, and in diabetes can be quite faint. They range in size from 1/2 to 1/10 disc diameter. They are in fact acute retinal infarcts, the white material being the accumulation of axoplasmic material in the ischaemic nerve fibres. They resolve completely over a period of six to twelve weeks. Fluorescein angiograms of 'cotton wool spots' demonstrate that they often appear in relation to areas of capillary closure. The conditions in which 'cotton wool spots' are most commonly seen are malignant hypertension, diabetes mellitus, systemic lupus erythematosis and polyarteritis nodosum.

*'Exudates'* are well demarcated white or yellow patches in the retina, sometimes referred to as 'hard' exudates. Histologically they are seen to be collections of phagocytes swollen with lipid, situated usually in the deeper layers of the retina. They arise in areas surrounding oedematous retina and are commonly seen in the macular region as a 'macular star'. Diabetic exudates tend to surround areas of retinal oedema or microaneurysms. They are associated with visual field defects and their slow resolution is not necessarily accompanied by better vision, despite the improved fundus appearance.

*Arteriosclerosis*

This term embraces those metabolic disturbances of the blood vessel walls which result in various deposits and other changes in their tissues.

*Age changes.* Senile vascular sclerosis as seen in the healthy elderly patient involves the replacement of active muscle and elastic tissue in the vessel walls by fibrous tissue which is less compliant and which then tends to contract. The blood vessels thus become narrowed and straighter.

*Atheroma.* This affects both large arteries such as the aorta and the carotids and small ones like the cerebral arteries, the central retinal arteries and their larger branches. It may be associated with hypercholesterolaemia. Yellow streaks of lipid form in the intima which is raised over these deposits by nodular thickening of its connective tissue which may become hyaline. The internal elastic lamina is broken up by the lesion and the media in its vicinity become thinned causing local vessel dilatation. The deposits may undergo calcification or may ulcerate and discharge atheromatous material or adherent platelet aggregations into the bloodstream which may embolise the central retinal artery or cause transient cerebral ischaemic attacks. The nodules in small vessels like the central retinal artery may imperil the lumen directly or by precipitating thrombosis.

*Arteriolar sclerosis.* Hyaline change is the characteristic arteriosclerotic lesion in the arterioles and small arteries *(the resistance vessels).* Arteriolar sclerosis involves hyaline thickening of the subintimal tissue, hyperplasia of the elastic tissue and sometimes proliferation of the endothelial cells. The hyaline and elastic changes are at first confined to the intima, but later involve the entire wall. The condition is common in essential hypertension and it may be found in the hypertensive experimental animal, but similar lesions may appear in the elderly normotensive patient and they are not uncommon at an earlier age in certain organs such as the spleen or choroid in the absence of a generally raised blood pressure.

*Fibrinoid arteriolar necrosis.* This occurs when high blood pressure is associated with spastic constriction of the arterioles as in malignant hypertension. The arteriolar lesions, though of the same type as arteriolar sclerosis are more serious and in parts the whole thickness of the vessel wall becomes structureless. The necrotic area allows the escape of fluid leading to retinal oedema or haemorrhage.

**Hypertension** (p126)

Hypertension damages the heart and the large arteries as well as the resistance vessels (the small arteries and arterioles). The nature and degree of these changes depend primarily upon the extent of the rise in blood pressure. Whereas hypertension in the majority of patients pursues a relatively benign course, in a minority, probably less than 1% of all hypertensives, it may pursue a rapidly fatal malignant course.

*Benign hypertension.* (LCW 6.9, 6.10, 6.11 p96)

The arterial change responsible for the excess morbidity and mortality of benign hypertension is arteriosclerosis, a combination of atherosclerosis and arteriolar sclerosis. Atherosclerosis is a disease of large arteries, the effects of which are particularly felt in the coronary, cerebral and retinal vessels. About two thirds of patients with benign hypertension die as a consequence of their hypertension, and in the majority of these the fatal event is either a myocardial or cerebral infarction. As a result of high pressure, aneurysms may develop on the small cerebral perforating arteries. Rupture of such aneurysms is the cause of cerebral haemorrhage in hypertensive patients.

*Malignant hypertension* (Stereo plates 6.36 p126, 6.37, 6.38 p127)

The pathological change which characterises malignant hypertension is fibrinoid necrosis of the arterioles (Plate 14.1 p295) with resulting oedema and haemorrhage. This focal necrosis of the vessel wall with subsequent scarring involves arterioles in several organs, especially in the kidney, leading to rapidly progressive renal failure which is the cause of death in the majority of such patients. With treatment, the outlook for patients with malignant hypertension has improved but this primarily depends upon the degree of renal damage which has occurred before treatment can be commenced.

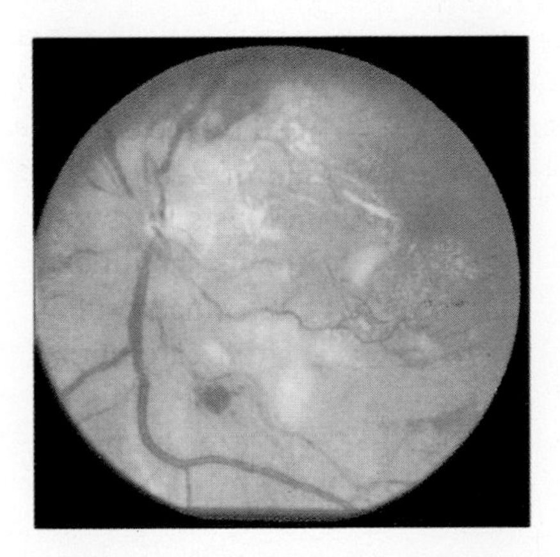

Plate 14.1 Hypertensive retinopathy (malignant KWB Grade III).

As the retinal vessels are the only part of the microcirculation which can be observed directly, the retinal changes seen in hypertension are commonly used as a guide to the severity of the disease and therefore to its prognosis. The Keith, Wagener and Barker (KWB) ophthalmoscopic grading of hypertensive changes requires experienced interpretation due to the presence of appearances resulting from atherosclerosis and occlusions of retinal veins and arterioles superimposed on those due to hypertension itself (p126).

# CHAPTER 15

# BLOOD DISEASES

There are a number of haematological abnormalities that give rise to ocular signs. Recognition of these may assist in the diagnosis of the general condition and influence its treatment. The more important blood diseases to consider apart from acute blood loss are sickle cell disease, polycythaemia, Vitamin B 12 deficiency (pernicious anaemia), the leukaemias and lymphomas, multiple myeloma and Waldenstrom's macroglobulinaemia. Histiocytosis-X is also considered in this section.

*Fundus appearances of anaemia.* (LCW 6.12 p97) In severe anaemia from any cause, retinal veins may become dilated. The retina and also the blood column of the retinal vessels are pale. There may be retinal oedema, disc swelling, 'cotton wool' spots and a liability to retinal haemorrhages, occasionally subhyaloid in type. Sometimes retinal haemorrhages with white centres occur, and are called Roth Spots, classically associated with subacute bacterial endocarditis. In acute blood loss, the retinal vessels are narrowed and in conditions of increased blood viscosity such as polycythaemia the veins are markedly dilated predisposing to venous occlusion.

*Acute blood loss.* This may be traumatic or result from internal haemorrhage. It can lead to transient loss of vision in one or both eyes which may recover rapidly with energetic treatment, although all degrees of optic atrophy and visual field defect may result, including sometimes even total blindness. Nerve fibre bundle defects causing non-progressive arcuate scotomas may later lead to a mistaken diagnosis of 'low tension' glaucoma so that a history of a haemodynamic crisis should always be sought in patients suspected of this condition.

## Sickle cell disease

This is practically confined to Africans and is inherited as a Mendelian dominant. The essential factor is an abnormal haemoglobin which causes the red cells to assume a sickle shape when exposed to the lower end of the physiological range of oxygen tension. These sickle cells do not pass easily through the small vessels so that there is a tendency to thrombosis and haemorrhage. Haemoglobin contains a protein, globin, combined with an iron porphyrin compound. The globin has two pairs of amino acid chains. Normal adult haemoglobin HbA contains a pair of alpha and a pair of beta chains. Foetal haemoglobin of which there are traces in the adult, contain a pair of alpha and a pair of gamma chains. Alterations in the sequence of amino acids in the beta chains may cause changes in the responses of the haemoglobin molecule. Thus HbS has valine substituted for glutamic acid in the sixth position of the beta chain and HbC has lysine similarly substituted.

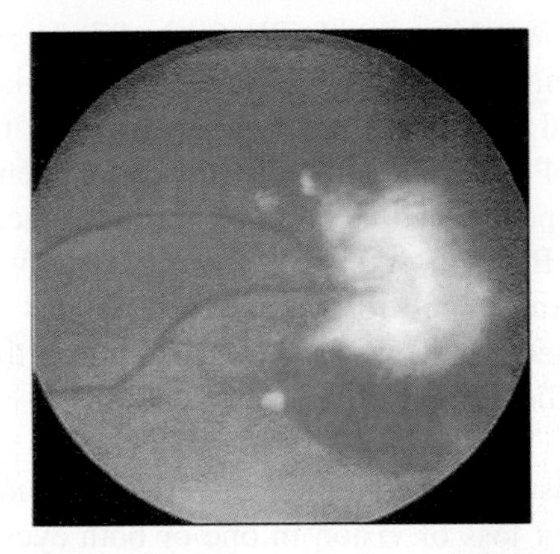

Plate 15.1 Vascular fans and haemorrhage in sickle cell disease.

*Clinical Signs.* Clinically, sickle SS disease has homozygous HbS haemoglobin with some HbF. Sickle cell trait, SA disease, has HbS and HbA. SC disease has HbS and HbC and leads to a particularly severe retinopathy. Homozygous HbC is not a form of sickle cell disease and only causes haemolytic anaemia as does thalassaemia in which there is a deficiency of alpha or beta chains. In homozygous sickle cell disease an episode of low oxygen tension may result in sickled red cells being trapped in the small vessels, causing infarction. Retinal lesions are found in the periphery where arteriolar occlusions are followed by arteriovenous anastomoses and the formation of new vessels which grow forwards in a fan-like pattern on the posterior surface of the vitreous, forming a "sea-fan". These delicate vessels are liable to bleed on eye movement, causing vitreous haemorrhage and sometimes lead to the development of a tractional retinal detachment. Areas of atrophic chorio-retinal scars occur due to ischaemia and form localised, pigmented lesions called "sunburst spots". Saccular dilatation of the conjunctival vessels may also occur (Plate 15.1 p298, Fig. 15.2 p299, LCW 6.13 p97).

*Investigations, prophylaxis* and *treatment.* Haemoglobin electrophoresis is an important investigation in those patients of Mediterranean or African ancestry who show signs of retinal haemorrhage. *It is also obligatory before general anaesthesia in such patients so that special care can be taken to maintain good oxygenation.* Unfortunately, the results of treatment of established retinal lesions with photocoagulation or cryotherapy of the ischaemic peripheral retina, have been poor. However, repair of a detached retina may be possible, though if an encirclement operation is required there is a risk of anterior segment ischaemia.

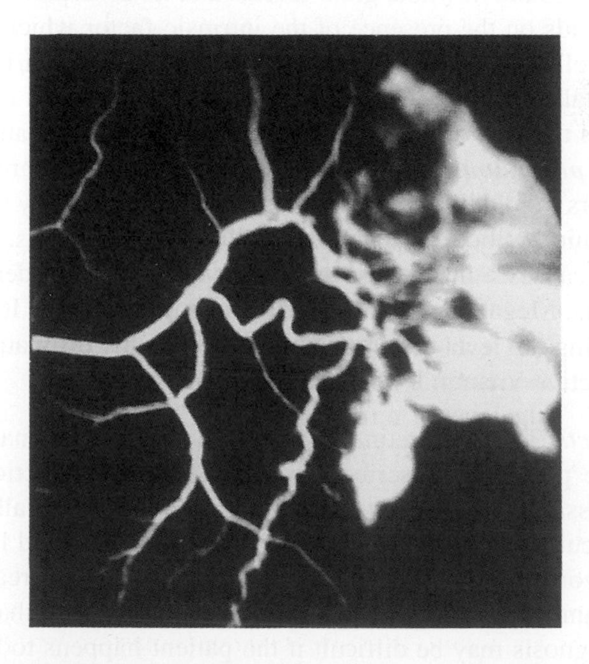

Fig. 15.2 Fluorescein study of sickle retinal fan.

## Polycythaemia

In this condition there is an absolute increase in the number of red blood corpuscles. In *primary polycythaemia* overactivity of all haematopoiesis takes place. *Secondary polycythaemia* is a response to excessive erythropoietin due to other causes such as low oxygen tension in mountain dwellers, in congenital heart disease, chronic lung disease or renal disease.

The *ocular effects* are due to hyperviscosity with retinal vein dilation proceeding to retinal vein occlusion or in some cases to subhyaloid haemorrhage (LCW 6.14 p97).

*Treatment* is by regular venesection, or chemotherapy with oral busulphan, oral hydroxyurea or intravenous radioactive phosphorus ($P^{32}$)

## Vitamin B12 (cyanocobalamin) deficiency syndrome p355

Vitamin B12 is found in whole grain cereals and its absorption from the small intestine depends on the presence of the intrinsic factor which is secreted by the parietal cells of the gastric mucosa. Deficiency may arise rarely from inadequate intake or from mucosal abnormalities of the small intestine but the major cause is the lack of parietal cell activity as a result of atrophic gastritis *(Addisonian pernicious anaemia)* or following gastrectomy. Pernicious anaemia occurs in middle age and in families with a tendency to develop gastric cytoplasmic antibodies which destroy the parietal cells. Deficiency of Vitamin B12 causes a macrocytic anaemia due to the disordered maturation of the red cells. Megaloblasts are found in the bone marrow. It can be detected by measuring the level of serum Vitamin B12 or by estimating the absorption of radioactive Vitamin B12 (Schilling test).

*Clinical Aspects.* The usual retinal changes associated with anaemia from any cause may be present as described on p297. Lifelong injections of vitamin B12 are necessary to control the disorder and must not be allowed to lapse because subacute combined degeneration of the posterior and lateral columns of the spinal cord as well as optic atrophy may supervene if treatment is irregular. Significant recovery of this optic atrophy is unusual so that prevention is essential. Diagnosis may be difficult if the patient happens to have been taking preparations of folic acid because these may correct the blood changes but do not alleviate the neurological complications. As there is evidence that cyanide intoxication may be a factor in the optic atrophy of tobacco amblyopia (p383) due to the inhalation of extra cyanide with the tobacco smoke, and as cyanide may play a part in some other conditions with optic atrophy where detoxication of cyanide by the liver is impaired, it is best to substitute hydroxycobalamin injections for cyanocobalamin in routine therapy

## Leukaemia

Leukaemia is a disease characterised by abnormal proliferation of the leukopoetic tissue. Acute lymphocytic leukaemia mainly affects children, and chronic myeloid and lymphocytic types affect adults. In all types, widespread cellular infiltration with enlargement of the spleen and liver, anaemia and platelet deficiency are features. Marked splenomegaly and infiltration of the skin and pruritis are features of chronic myeloid leukaemia while in the lymphoid type, enlargement of the lymph nodes is pronounced. As a result there is general debility and a liability to intercurrent infection as well as the local effects of infiltration and anaemia including haemorrhages from mucous membranes, the orbit, skin, periosteum and kidney.

*Ocular effects.* The ocular effects are similar in all types of leukaemia and more common in the acute disease. Ophthalmoscopic appearances are as in any severe anaemia (p297). The conjunctiva, sclera and choroid may be infiltrated and thickened and similar infiltration may obstruct the lacrimal passages and predispose to dacryocystitis. Infiltration of the optic disc may occur and give the appearance of a swollen disc. This must be differentiated from indirect infiltration of the CNS and meningitis causing increased intracranial pressure and papilloedema. Cranial nerve palsies can also occur with extraocular muscle involvement.

*Treatment.* In addition to local palliative ocular therapy the treatment is that of the general condition by a combination of radiotherapy, steroids and chemotherapy. The latter includes: 1 Antimetabolites which compete directly for substances essential for cell growth, e.g. methotrexate which prevents the uptake of folic acid. 2 Alkylating agents which inactivate cellular proteins, nucleic acids and amino acids by covalent bonding of alkyl groups (e.g. chlorambucil and cyclophosphamide) to them, thus preventing chromosomal division . The mechanism of steroid action is not fully understood.

## Lymphoma, lymphosarcoma, reticulum cell sarcoma and Hodgkin's disease

These are closely related neoplasms affecting lymphoid tissue and are treated similarly. They may cause local effects in the eye and other tissues due to infiltration or compression. The lids may become thickened from infiltration or a smooth, whitish subconjunctival tumour may be found especially in the lower fornix. 2% of all malignant systemic lymphomas involve the conjunctiva. Orbital or lacrimal gland masses will produce pressure effects with diplopia and proptosis, with retrobulbar tumours causing extraocular muscle palsies. Orbital lymphomas are usually of B-cell origin and need biopsy for diagnosis. Treatment is with fractionated radiotherapy. Cerebral lymphoma may present with a uveitis or vitreous deposits, and often the only method for definitive diagnosis is by a vitreous biopsy.

## Multiple myeloma and Waldenstrom's macroglobulinaemia

*Types of lymphocyte*

Some lymphocyte stem cells from the bone marrow may mature under control of the thymus *(T-lymphocytes)* and others under control of gut associated lymphoid tissue such as Peyers patches *(B-lymphocytes)*. T-lymphocytes mediate cellular immunity and B-lymphocytes are responsible for the production of immune globulins which comprise antibodies. Different kinds of B-lymphocytes produce one of the four main types of immune globulins.

1. *IgA immunoglobulins* are secreted into mucosal surfaces (tears, saliva) preventing gastro-intestinal and secretory gland infection.
2. *IgE immunoglobulins* are responsible for immediate hypersensitivity reactions such as atopic dermatitis or allergic asthma, and are elevated in protozoal diseases.
3. *IgG immunoglobulins* are relatively small molecules so that they can diffuse into the interstitial body fluids with ease and most antibacterial and antiviral antibodies are of this type. Some can activate complement enzyme systems which while usually valuable in defence may cause unwanted tissue damage in immune disorders such as systemic lupus erythematosis (p363).

4. *IgM immunoglobulins* have large molecules and thus mostly remain in the vascular compartment and are chiefly concerned in combating bacteraemia. They can activate complement. Some autoimmune antibodies are of this type such as rheumatoid factor (p357).

*Plasma Cells.* The immune globulin-producing B-lymphocytes mature to plasma cells which also produce and secrete immune globulins. Malignant or abnormal deviation can occur at any stage of maturation. Multiple myeloma and Waldenstrom's macroglobulinaemia are each considered to be such a deviation.

*Multiple Myeloma*

This is a neoplasm of plasma cells producing an excess of predominantly one type of immune globulin. Anaemia occurs with osteolytic lesion in the bones. The abnormal myeloma proteins or their subunits, Bence-Jones proteins, may be found in the urine. In the blood the proteins cause hyperviscosity effects of the cryoglobulin type. These precipitate with reduced temperatures as when exposed in the conjunctiva causing dilation of these vessels. The main ocular effects are those arising from the local orbital or intracranial presence of the osteolytic lesions or from the effects of increased viscosity such as retinal haemorrhage or vascular occlusion. Cysts of the pars plana containing protein are common and the cornea may be engorged with abnormal cells. *What is apparently a corneal dystrophy with deposits in the stroma may occur some years before the overt disease.*

*Waldenstrom's macroglobulinaemia*

This involves plasma cells secreting IgM immune globulin. It is a disease of late middle life with severe anaemia and enlarged lymph glands, liver and spleen. Increased proteins cause hyperviscosity and widespread vascular occlusions. In the eyes dilated veins, haemorrhages, 'cotton wool' deposits and even exudative detachments occur. As in multiple myeloma an apparent corneal dystrophy may antedate the other signs. *If electrophoresis is carried out at the stage when only the corneal dystrophy is present, early detection is possible.* Treatment aims at reducing the production of protein by the abnormal primitive plasma cells by means of cytotoxic agents such as melphalan and by the removal of the abnormal protein by cell separation and plasmaphoresis which may lead to a remarkable temporary improvement in the condition.

*Histiocytosis X*

It is probably best to include this lipid abnormality here because there is no metabolic deficiency as in the disturbances of lipid metabolism described under Genetic Diseases (p423). The condition includes eosinophilic granuloma, Hand-Schuller-Christian disease, Letterer-Siwe disease and juvenile ocular xanthogranuloma. Lipid is liberated as a result of local inflammatory destruction of tissue. Transitional forms between the different types are seen, all of which are characterised by a diffuse reticuloendothelial hyperplasia.

In *eosinophilic granuloma* one or more foci of osteolytic activity containing histiocytes and eosinophils are found.

*Hand-Schuller-Christian Disease* shows multiple lipid granulomas particularly affecting the skull causing diabetes insipidus, exophthalmos and otitis media. Ophthalmoplegia, papilloedema and loss of vision may result.

*Letterer-Siwe Disease* is rapidly fatal due to progressive anaemia and hepatosplenomegaly as well as skull involvement.

*Juvenile ocular xanthogranuloma* is an ocular sign of the disease naevoxanthoendothelioma which presents as orange coloured plaques on the head and trunk in early childhood. When the eye is involved, spontaneous hyphaema may occur which may proceed to secondary glaucoma but as the iris and ciliary body are packed with histiocytes, it is not surprising that glaucoma may also occur independently of hyphaema. The typical skin lesions should distinguish the condition from other possible causes of anterior chamber haemorrhage or juvenile glaucoma. The various types of histiocytosis-X are treated with irradiation or corticosteroids together with local therapy appropriate to the particular lesions.

# CHAPTER 16

# ENDOCRINE DISORDERS

## Ocular manifestations of thyroid disease

*Hyperthyroidism (Thyrotoxicosis)*

*Thyrotoxicosis* is the clinical syndrome caused by excess secretion of thyroid hormones. Thyrotoxicosis is a feature of two quite distinct disease entities: toxic nodular goitre and Graves' disease.

*Toxic nodular goitre* (LCW 6.15, 6.16, 6.17 p98)

In this condition, thyroid hormones are secreted by a thyroid adenoma independently of the normal pituitary controlling mechanism (Figs. 16.1 p305, 16.2 p306). Multinodular disease occurs in females over 60 years old, and the less common singular nodule is seen in younger patients. Unless the tumour

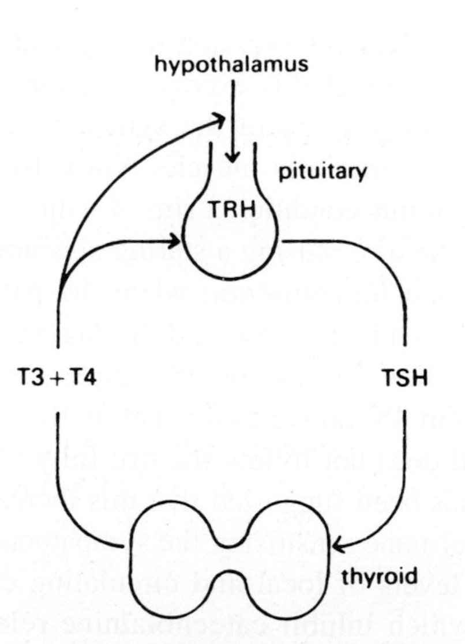


TRH - thyroid releasing hormone
TSH - thyroid stimulating hormone
$T_4$ - thyroxine
$T_3$ - tri iodo thyronine

Fig. 16.1 Normal thyroid hormone control. In health the thyroid produces mainly $T_4$ and a small proportion of $T_3$. $T_3$ is about 3 times more potent than $T_4$ has an immediate effect and is quickly metabolised.

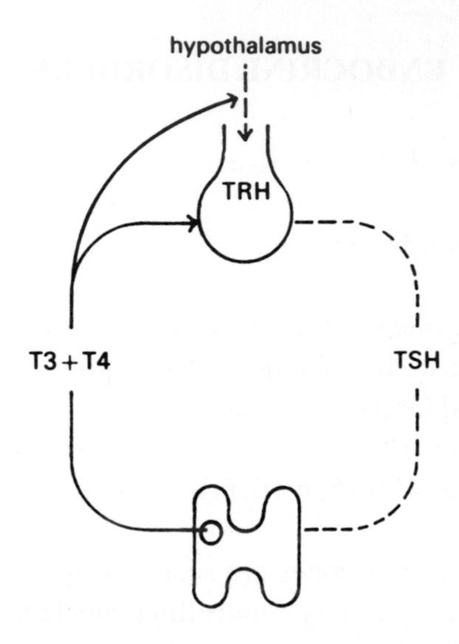


Fig. 16.2 Toxic thyroid nodule effect.

is sufficiently large to cause symptoms by local compression the clinical features of the disease are the consequence solely of this excess secretion. The only ocular abnormality to be found is over activity of the sympathetically innervated fibres of the levator palpebrae superioris muscle. Normally the upper eyelid covers some of the iris, but in this condition a rim of white sclera is exposed above the upper margin of the iris, causing a staring appearance with widening of the palpebral fissure and *lid retraction* when the patient looks ahead (Fig. 16.3 p307). If lid retraction is not observed, lid lag may be demonstrated by asking the patient to follow the examiner's finger held at arms length from him through an arc from 45° above to 45° below the horizontal. If *lid lag* is present, the upper lid does not follow the eye fully and a rim of white sclera will be exposed. It has been suggested that this increased muscular tone is due to excess thyroid hormone sensitising the sympathetically innervated muscle fibres to normal levels of local and circulating catecholamines. Guanethidine eye drops which inhibit catecholamine release from adrenergic nerve endings can sometimes provide relief from lid retraction. It is very important to appreciate that patients with a toxic nodular goitre, unlike those with Graves' disease, do not have forward protrusion of

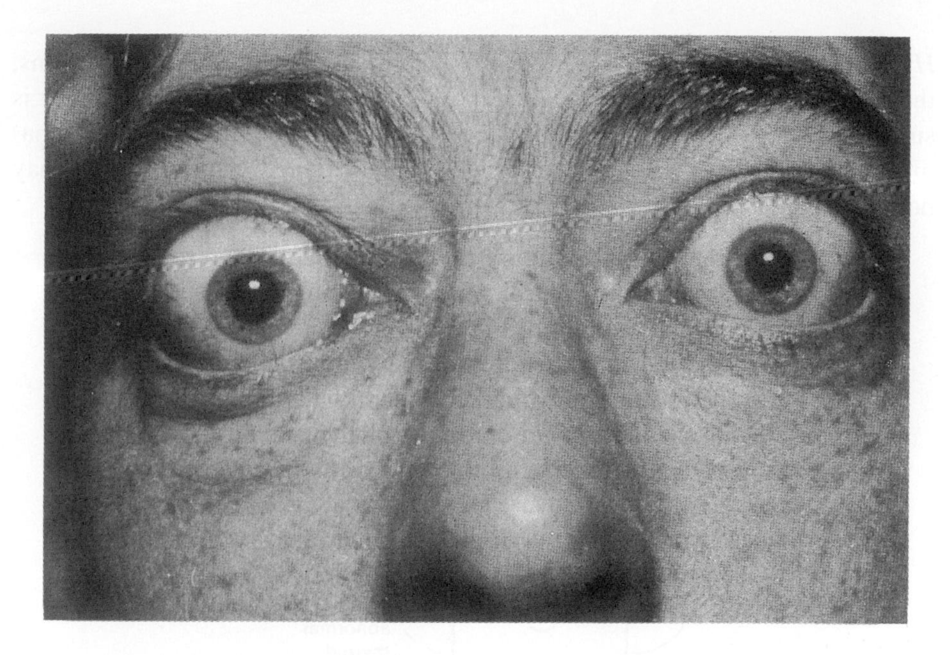

Fig. 16.3 Lid retraction in thyrotoxicosis.

the eyeball *(proptosis)* but only lid retraction which can cause an illusion of proptosis unless the two features are carefully distinguished. Superior limbic keratitis is frequently associated with hyperthyroidism (p480, Plate 25.2 p480). Diagnosis is made with thyroid function tests revealing elevated levels of $T_4$ and $T_3$ and reduced TSH levels. Radioisotope scans and uptake tests are also helpful. Treatment is with anti thyroid drugs such as carbimazole and propyluracil, radioactive iodine $^{131}I$ , or partial thyroidectomy.

*Graves' disease*

Graves' disease is an autoimmune, systemic disease affecting many tissues which may include the thyroid gland and the eye. Unlike the toxic nodular goitre whose clinical manifestations are only those of excess thyroid hormone secretion, in Graves' disease thyrotoxicosis is usually, but not always, present and is only part of the disorder, and the ocular changes, except for lid retraction are due to other still unknown causes. Graves' disease is due to the presence of IgG antibodies to the TSH receptor of the thyroid follicular cell. The features of Graves' disease are:

*Hyperthyroidism,* when present, is due to excess thyroid hormone secretions, the production of which is stimulated by ill understood factors whose effect is superimposed on the normal pituitary and hypothalamic controlling mechanisms. It is the nature of the control which has changed and the patient may be either hyperthyroid or euthyroid (Fig. 16.4 p308).

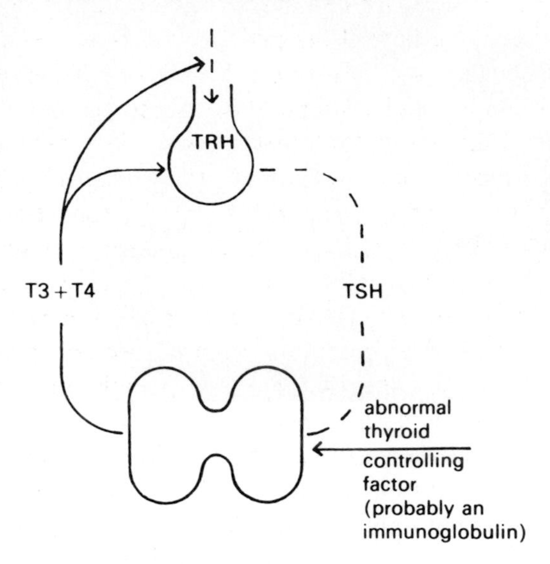


Fig. 16.4 Graves' disease. Abnormal thyroid control.

*Pre-tibial myxoedema* is characterised by bluish brown plaques of mucopolysaccharide infiltration which are found on the feet and lower legs. It can occur in association with finger clubbing, phalangeal periosteal new bone formation and overlying oedema, a triad known as thyroid acropachy.

*Ophthalmopathy including exophthalmos.* The factors underlying the development of exophthalmos remain unknown and it is important to appreciate that the various manifestations of this disease can occur separately and may progress or regress quite independently of each other. Exophthalmos (or proptosis) is forward displacement of the eye. The term endocrine exophthalmos is, however usually applied to all the ocular manifestations of Graves' disease. The major pathological abnormality to be found within the orbit is infiltration of the orbital tissues including the extraocular muscles with fat, mucopolysaccharide and lymphocytic oedema. Infiltration within the limited volume of the orbit will inevitably increase the intraorbital pressure. The clinical features are the outcome of these changes within the orbit. These may be summarised as follows:

*-proptosis:* the eye is displaced forwards (LCW 2.16 p32). This may be demonstrated by standing behind the patient who is seated. Normally when his head is tilted slowly backwards, the first structure to be seen in line with the supraorbital ridge is the cheekbone. If the globe and not the cheekbone is seen in line with the supraorbital ridge, proptosis is indicated as this would be most unusual otherwise. Fullness of the upper lid especially on the temporal side is characteristic. A rim of sclera will often be exposed both *above and below* the iris when the patient looks directly forward. If found this assists differentiation from a retro-orbital tumour where the forward displacement of the eye is often accompanied by the finding of an exposed rim of sclera *below* but not above the iris. In endocrine exophthalmos the proptosis may be accompanied by a sensation of grittiness in the eye with or without photophobia due to exposure of the cornea. If proptosis is severe the patient may be unable to close the affected eye, and the globe is liable to severe keratitis with ulceration and permanent visual impairment if treatment is not energetic.

*-conjunctival oedema (chemosis):* the conjunctival vessels are engorged due to the compression of orbital veins and in severe cases the conjunctiva may become so oedematous that it becomes a fluid filled bag hanging over the lower lid.

*-papilloedema and visual loss:* compression of the optic nerve will cause optic atrophy, and compression of the central retinal veins and other veins draining the eye will cause dilatation of the retinal veins with retinal oedema and papilloedema. Unrelieved papilloedema may progress to optic atrophy and irreversible visual loss. The findings of papilloedema or central retinal oedema with a fall in visual acuity due to central visual field loss in a patient with endocrine exophthalmos demands urgent action.

*-extraocular muscle palsy:* infiltration of the extraocular muscles can dramatically increase their bulk and may be sufficient to cause their weakness and impair ocular movement, with consequent diplopia. Characteristically in the early stages the movements most affected are elevation and abduction but as the condition progresses all movements may become restricted (p407).

*Diagnosis and differential diagnosis of endocrine exophthalmos*

The presence of bilateral exophthalmos if associated with a goitre and the other manifestations of Graves' disease should present little difficulty in diagnosis. Where necessary the diagnosis may be confirmed by immunological tests for thyroid peroxidase or microsomal antibodies. The main differential diagnostic problem is from an orbital tumour, and CT and MRI scanning is very useful in determining the involvement of extraocular muscles. Thickened muscle bellies, particularly of the inferior rectus and medial rectus are characteristic of this disease. In addition to these tests there are some helpful clinical pointers, such as the presence of both upper and lower lid retraction and asymmetry of exophthalmos which is rarely greater than 6mm in endocrine exophthalmos, whereas it can be more than this in many orbital tumours.

*Management of endocrine exophthalmos*

It is important to follow the progress of any type of proptosis by serial exophthalmometer measurements. This is a simple instrument which measures the distance forward from the lateral bony margin of the orbit to the anterior convexity of the cornea. It can be done simply by a millimetre scale but more accurate measurements are possible with one of the special instruments of which Hertel's exophthalmometer, in which an inclined mirror superimposes the corneal image on a millimetre scale, is one of the most useful (Fig. 16.5 p310).

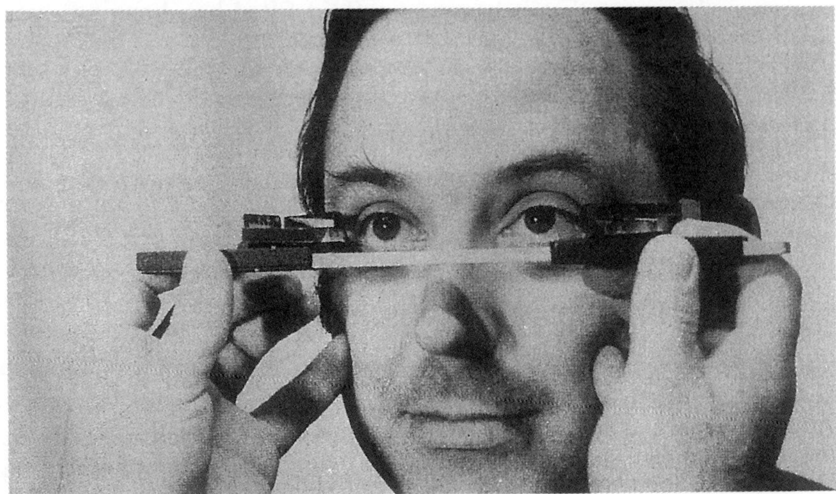

Fig. 16.5 Hertel's exophthalmometer.

The endocrine exophthalmos may threaten vision by exposure keratitis, retinal oedema, papilloedema and optic nerve compression. Exposure of the cornea in mild cases is prevented by methyl cellulose drops or eye ointments and vaseline gauze dressings. Tarsorrhaphy is necessary if the condition is progressive and should not be long delayed as it may become difficult to accomplish when the orbital pressure is tending to separate the lids. The tarsorrhaphy should be adequate (a mere approximation with a suture is not enough). Union of the lateral four-fifths of the bared posterior half of the lid margins with four mattress sutures will usually be secure and allow measurement of visual acuity and a fundus view at the medial end when the eye is adducted (Fig. 16.6 p311, LCW 6.18, 6.19 p98)

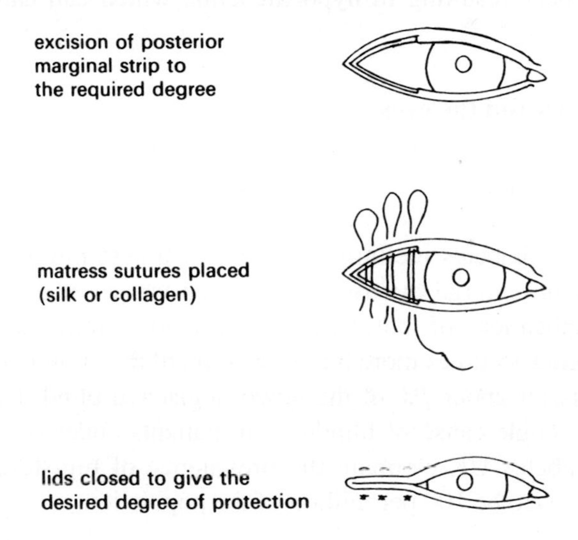


Fig. 16.6 Tarsorraphy (lateral tarsorraphy which can be extended medially as is necessary for protection. This allows a view of the cornea on adduction).

When optic nerve function is threatened, treatment is urgent and is aimed at shrinkage or escape of the orbital soft tissue. This may be done medically using high dose oral corticosteroids, or with orbital radiotherapy, or both. If, despite treatment, the vision deteriorates, surgical decompression of the orbit carried posteriorly to include the apex is necessary. The bony orbital walls are removed to relieve pressure. The original operation was laterally into the temporal fossa. but now this is either upwards and laterally into the temporal fossa, or via a transantral approach removing the inferior and medial orbital walls to release the orbital contents into the antrum below and the ethmoids medially. The patient with this demoralising condition requires much support and assurance that the disorder is ultimately self-limiting. Residual diplopia due to fibrosis in affected extraocular muscles may eventually require surgical correction. Serial examinations using the Hess screen is valuable in assessing progress and indicating arrest of muscle involvement.

*Parathyroid glands*

Although functionally unrelated, the parathyroid glands may be injured during thyroid surgery resulting in hypocalcaemia which can cause tetany and cataract.

## Diabetes mellitus and the eyes

*The size of the problem*

Many features of diabetes mellitus are those of a systemic immune disease. Although renal and arterial disease are the main causes of death in diabetics, the ocular complications of the disease are a major determinant of disability. Diabetics are about 15 times more likely to go blind than non-diabetics and this disease accounts for about 7% of the newly registered blind. Diabetes is now the commonest single cause of blindness in patients under 65 years in Great Britain and Diabetes UK estimate the prevalence of registerable blindness caused by diabetes to be 100 per million of the population.

*Diabetic eye disease*

The main causes of blindness in diabetic patients are retinal disease, accounting for about 80%, and cataract formation which accounts for the majority of the remaining 20%. Other ophthalmic conditions to which diabetics are liable include optic neuritis and extraocular muscle palsy, both of which usually have a favourable prognosis. Diabetes causes widespread changes in the tissues of the eye including degenerative changes in the epithelium of the iris and ciliary body. Although the lens and retina may be affected together they will be considered separately here for convenience.

*The lens (cataract formation in diabetes)* (p98)

*The retina in diabetes*

General retinal vascular changes, including those occurring in diabetics, have been considered on p290. The most important factors affecting the incidence of diabetic retinopathy are age and the duration of the disease. More than 90% of insulin-dependent diabetics have some form of retinopathy after 20 years. Established retinopathy is adversely affected by co-existent hypertension. The visual symptoms may be gradual when there is slow encroachment on the macula of a ring of exudates surrounding an area of oedema or microaneurysms (Plate 16.7 p313), or rapid when a vitreous haemorrhage occurs. It is important to classify the grades of diabetic retinopathy because the grade of retinopathy not only gives an indication of the visual prognosis and further management, but it may also be a guide to the life expectancy of the patient. The onset of proliferative retinopathy is associated with a median survival of 5.4 years, and death is usually due to ischaemic heart disease and renal failure. The presence of microaneurysms which are tiny outpouchings from blood vessel walls, vascular tortuosity, venous irregularities, "dot and blot" haemorrhages, exudates and new blood vessels suggests the possibility of diabetic retinopathy, which may be divided into the following subsections:

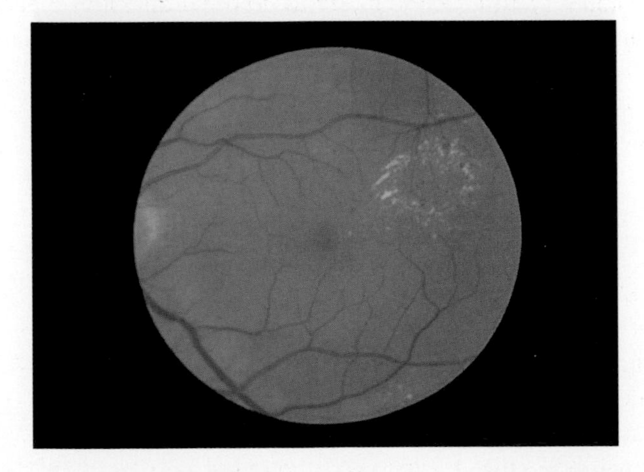

Plate 16.7 Background diabetic retinopathy - ring of hard exudates (circinate) which may later encroach on the macular area.

*-background diabetic retinopathy.* This benign stage is characterised by microaneurysms (p291), dot and blot haemorrhages, a few hard exudates, venular dilatation and tortuosity, with relative absence of these changes in the macular area (Plates 16.8, ,16.9 p314, LCW 6.1, 6.2 p91).

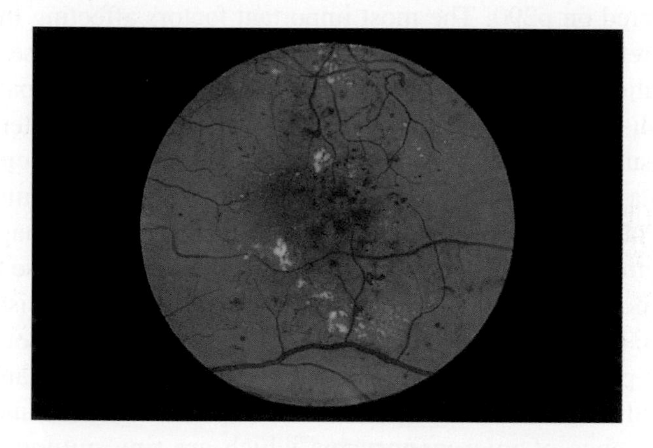

Plate 16.8 Background diabetic retinopathy, exudative type, showing microaneurysms, haemorrhages and 'hard' exudates.

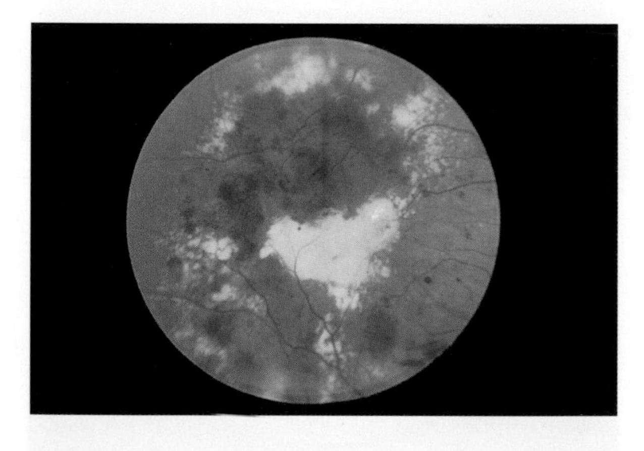

Plate 16.9 Background diabetic retinopathy, exudative type, showing ring of exudates surrounding an oedematous area of retina.

*-maculopathy.* Exudation at the macula results from microvascular leakage and are lipid deposits. These lipids diffuse out from an oedematous area in rings of increasing diameter (Plate 16.9 p314, LCW 5.12 p78). When these lipids encroach on the macula, they distort the cones and cause visual distortion and loss of acuity. Exudates normally occur in the form of circinates, but occasionally are deposited in Henle's nerve fibre layer of the retina to create a radial star pattern (macular star)(p293). Circinates may be treated with laser photocoagulation to stop areas of vascular leakage (p76). The Early Treatment Diabetic Retinopathy Research Group confirmed that treating exudative diabetic maculopathy with focal laser treatment reduced the risk of blindness by 50% at 5 years.

*-pre-proliferative diabetic retinopathy.* Fundal changes of this stage in the disease process are characterised by the presence of cotton-wool spots (p293), deep and superficial haemorrhages, intra-retinal microvascular anomalies, venous dilatation, looping and irregularities. By definition, no new vessels are present,

*-proliferative diabetic retinopathy.* Retinal ischaemia in diabetes causes neovascularisation (Plate 16.10 p315, LCW 6.3 p92, 6.5 p93, 6.6 p94). New ves-

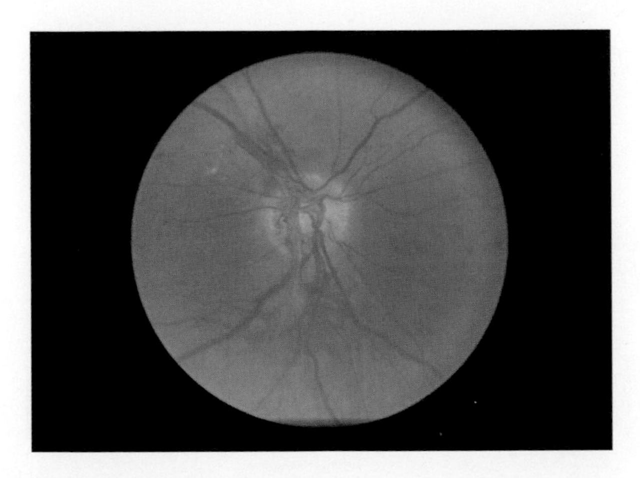

Plate 16.10 Diabetic retinopathy, proliferative type, showing new vessels especially in the optic disc.

sels may be identified by their abnormal position, small calibre and tendency to grow forwards into the vitreous cavity (p291). Treatment of proliferative retinopathy is with pan-retinal photocoagulation with the argon laser, and this has been demonstrated by the Diabetic Retinopathy Study to reduce severe vision loss by 50% at 5 years. (LCW 6.7, 6.8 p95) Blood vessels may also grow on the iris in diabetes, and it is important to assess the anterior segment during ocular examination with the ophthalmoscope or microscope, and to note the presence of *rubeosis iridis* (p581, Plate 16.11 p316, LCW 4.15 p63). Secondary rubeotic glaucoma is usually intractable and though both photocoagulation and surgical methods of treatment can be attempted and may be effective for a time they usually only delay the progress of the condition. Bleeding from new vessels causes vitreous and subhyaloid haemorrhages, and fibrosis may lead to the development of tractional retinal detachments (Plate 16.12 p317, LCW 6.4 p92). Vitreous haemorrhage even when quite dense may absorb spontaneously with good visual recovery. However, if blood persists in the vitreous, a vitrectomy with a suction-cutting vitrector may be successful in restoring reasonable clarity. Dense pre-retinal membranes can sometimes be carefully separated with advantage and tractional detachments treated using advanced techniques.

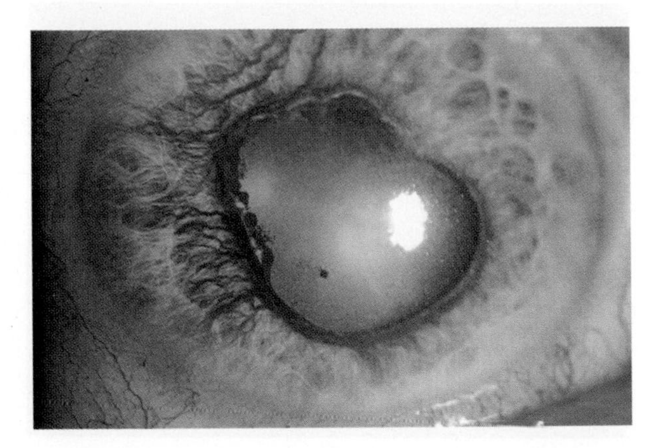

Plate 16.11 Diabetic rubeosis of the iris.

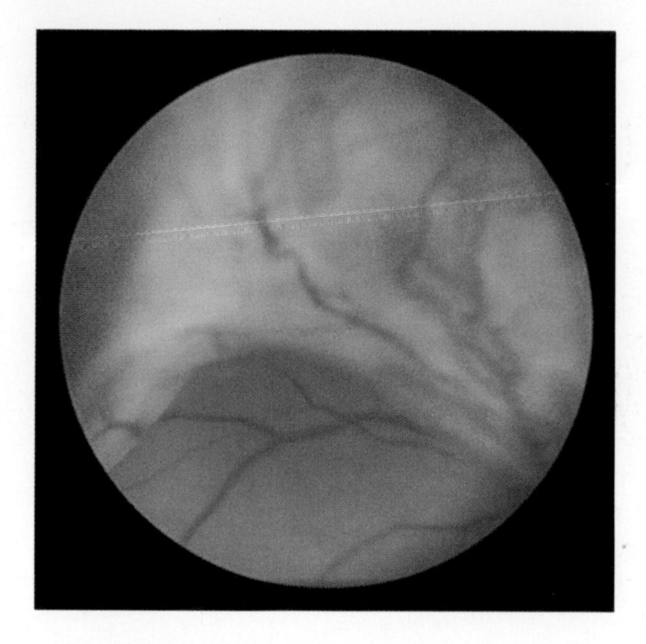

Plate 16.12 Diabetic retinopathy - fibrosis of neovascular tissue - 'retinitis proliferans' predisposing to retinal detachment.

## *Screening*

The European Consensus Document recommended that diabetics should be screened at least 2 yearly after puberty. Screening is extremely important in detecting early changes of treatable lesions of diabetic retinopathy that may then be managed by an ophthalmologist. It is recommended that patients with proliferative disease or those with exudation close to the fovea or individuals with recent onset of severe diabetic eye disease are referred urgently by physicians or optometrists to ophthalmologists.

# CHAPTER 17

# INFECTIONS AND INFESTATIONS AND NUTRITIONAL DEFICIENCY

The eyes may be involved in infection from without or from the blood stream. The lids, conjunctiva, and cornea will be subject to involvement in infections of the skin. Corneal abrasions are liable to secondary infection even though immune globulins and the lysozyme of the tears and the intact epithelium form an effective barrier to most organisms. Useful immune responses may be complicated by troublesome hypersensitivity reactions. Blood borne infections are particularly liable to affect the highly vascular uveal tract. The aim in this chapter is to present a clinically orientated guide to the more important infections and infestations which affect the eye. Figs. 17.1 p320, 17.2, 17.3 p321 summarise the range of bacteria, viruses and other organisms involved. They will be studied approximately in this order. This chapter includes the more important tropical diseases and a summary of the nature, effects and problems of nutritional deficiency.

## Bacterial infections

### *Staphylococcus*

The pathogenic *S. aureus* (coagulase positive) and the opportunistic pathogen *S. epidermidis* (hitherto *S. Albus*)(coagulase negative) commonly infect eye-lash follicles, their associated glands and the meibomian glands causing a chronic staphylococcal blepharitis (Plate 17.4 p322). Infections of the glands of the lash follicles cause styes. It is probable that infection of the meibomian glands is associated with changes in sebum leading to obstruction of their ducts and the accumulation of sebaceous material which excites a foreign body reaction. This results in a meibomian or tarsal cyst, which is an infected cyst with considerable redness, swelling and pain in the lid (p454). Staphylococcal blepharitis may lead to chronic conjunctivitis and keratitis with marginal infiltrates, the nature of which is frequently unrecognised (pp194, 206 and Plate 17.5 p322). Rosacea is frequently complicated by a staphylococcal blepharitis (p206). A perforating wound or corneal ulcer infected with staphylococci rapidly leads to panophthalmitis if treatment is not immediate and energetic (pp158, 166).

Fig. 17.1 Bacteria (organisms described - •).

| ***Cocci*** | | | |
|---|---|---|---|
| *Gram +ve* | *Gram -ve* | | |
| •Staphylococcus<br>•*S. aureus*<br>•Streptococcus<br>•*S. pneumoniae* | •Neisseria<br>•*N. Meningitidis*<br>•*N. Gonorrhoeae* | | |
| ***Bacilli*** | | | |
| *Gram +ve* | | *Gram -ve* | |
| *Aerobic* | *Anerobic* | *Aerobic* | *Anerobic* |
| •Mycobacterium<br>•*M. Tuberculosis*<br>•*M. Leprae*<br>•Corynebacterium | Clostridium<br>Lactobacillus | Enterobacteria<br>-Escherichia<br>-Klebsiella<br>-Salmonella<br>-Proteus<br>-Shigella<br>Pseudomonas<br>Parvobacteria<br>-Haemophilus<br>-Brucella | Bacteroides<br>Fusobacterium |
| ***Spirochaetes*** | | | |
| •Treponema<br>•*T. pallidum*<br>•Leptospira | | | |
| ***Filamentous bacteria*** *(described with fungi below)* | | | |
| •Actinomyces<br>•Streptomyces | | | |

*Streptococcus*

The alpha haemolytic *Streptococcus pneumoniae* (pneumococcus) may be responsible for conjunctivitis, endophthalmitis and dacryocystitis and is increasingly resistant to penicillin which makes drug susceptibility testing important. Beta haemolytic streptococci are fortunately very sensitive to penicillin and other antibiotics because untreated they may give rise to devastating inflammation. They may cause a membranous conjunctivitis and corneal invasion may lead rapidly to pus in the anterior chamber. The skin of the lids may become involved in erysipelas, an acute streptococcal infection of the skin which is not common but usually affects the face and head of the elderly or debilitated.

Fig. 17.2 Viruses (organisms described - •).

| *DNA Viruses* | | | |
|---|---|---|---|
| *Pox* | *Herpes* | *Adeno virus, causing-* | *Papova* |
| •Molluscum | •Simplex<br>•Varicella - zoster<br>•Cytomegalovirus | •Epidemic kerato-conjunctivitis<br>•Follicular conjunctivitis<br>•Pharyngo conjunctival fever | •Papilloma |

| *RNA Viruses* | | | | |
|---|---|---|---|---|
| *Picorna* | *Paramyxo causing-* | *Orthomyxo* | *Toga* | *Retro* |
| Entero | •Measles<br>•Mumps | Influenza | •Rubella | •HIV |

Fig. 17.3 Other organisms.

| *Chlamydia* | | *Rickettsiae, causing-* | |
|---|---|---|---|
| •Trachomatis<br>•Lymphogranuloma venereum | | •Q fever<br>•Rocky Mountain Fever | |
| *Fungi* | | | |
| *Moulds* | *Yeasts* | *Yeast-like* | *Dimorphic* |
| •Tinea<br>•Aspergillus | •Cryptococcus | •Candida | •Histoplasma |
| *Protozoa* | | *Soil transmitted helminths* | |
| •Toxoplasma<br>•Acanthamoeba<br>•Entamoeba histolytica<br>•Malaria<br>•Trypanosoma | | •Toxocara<br>•Ascaris<br>•Ankylostoma<br>•Trichina | |
| *Filarial worms* | | *Parasitic cysts* | |
| •Onchocerca volvulus<br>•Loa loa<br>•Wucheria bancrofti<br>*Myiasis*<br>•Myiasis | | •Taenia solium<br>•Taenia echinococcus (hydatid) | |

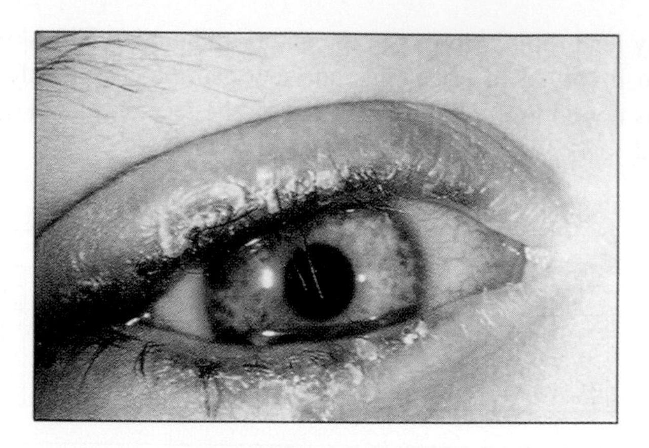

Plate 17.4 Staphylococcal blepharitis.

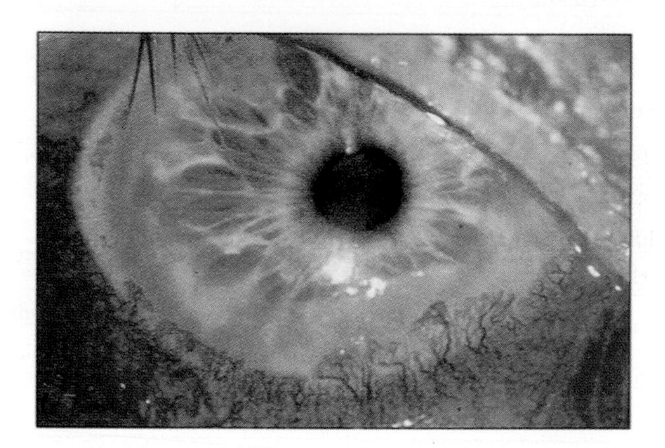

Plate 17.5 Staphylococcal keratitis.

### *Neisseria*

Neisserial gram negative organisms include the gonococcus and the meningococcus, both of which are important ophthalmologically.

*N. Gonorrhoeae* causes inflammation of mucous membranes particularly of the genito-urinary tract which may be followed by bacteraemia, and sometimes there may be a severe acute exudative uveitis. In purulent conjunctivitis during the first 21 days of life *(ophthalmia neonatorum),* gonococcal infection should always be excluded (p198) and although in the UK it is now responsible for fewer cases, decisive intervention is both possible and important.

*N. Meningitidis* infection enters through the nasopharynx, causes a bacteraemia and leads to lesions in the meninges with the risk of loss of vision due to involvement of the visual pathways or paralytic squint due to lesions of the oculomotor nerves. Sometimes metastatic infection of the uveal tract causes endophthalmitis.

*Mycobacteria*

Mycobacteria cause chronic infectious granulomas. *M. tuberculosis* and *M. leprae* are the most important.

**Tuberculosis**

Tuberculosis in man is caused by human and bovine strains of the tubercle bacillus. Primary infection in children is characterised by the local formation of tuberculous follicles and a similar but greater reaction in the draining lymph nodes. Depending on the degree of immunity and hypersensitivity acquired as a result of the primary episode, re-infection may lead in the immune non-allergic patient to a fibrotic reaction. Re-infection in the hypersensitive patient leads to varying degrees of caseation which may present as a circumscribed tuberculoma when immunity is good or, when it is poor, as diffuse tuberculous granulation tissue with some caseation. In the non-allergic patient a bacillaemia may lead to acute miliary tuberculosis when immunity is low or to chronic miliary tuberculosis if it is high.

*Primary disease.* The conjunctiva and its preauricular nodes may be the site of a primary tuberculous complex (p194) presumably caused by droplet infection. This may spread locally into the eye or by the blood stream into the lungs. It is one cause of *Parinaud's Oculo-Glandular Syndrome* which embraces a group of conditions which are characterised by a unilateral necrotic conjunctival lesion with preauricular adenitis. The syndrome also includes lymphogranuloma venereum, chancre, tularaemia and 'cat scratch fever'.

*Phlyctenulosis.* Primary ocular tuberculosis is rare but tuberculoprotein from primary infection in the lung or elsewhere may be a frequent cause of allergic phlyctenular disease in countries where tuberculosis is common. This is a condition mainly affecting the conjunctiva in children (p205). Symptoms may be slight or there may be a complicating mucopurulent conjunctivitis. When the cornea is involved, photophobia and lacrimation are often intense. The phlyctenule is a small pinkish-white elevation often situated near the corneo-scleral junction with an associated leash of conjunctival vessels. If the cornea is affected, nodules appear in the superficial layers followed by connective tissue formation and deep and superficial vascularisation. Eventually in the absence of treatment, corneal scarring may be severe. The phlyctenule consists of an exudation of polymorphonuclear and mononuclear leucocytes into the deeper layers of the conjunctiva. Untreated, it may resolve but more often it sloughs and heals by granulation. The tubercle bacillus is not found in the phlycten which is a Type IV hypersensitivity reaction and is readily controlled by local steroid applications with antibiotics as necessary. Phlyctenular disease may indicate active pulmonary tuberculosis but increasingly in the UK it results from allergy to other organisms.

*Miliary disease.* The eye is commonly affected in miliary tuberculosis by direct haematogenous infection. Characteristic tubercles are seen in the choroid and sometimes in the iris. The choroidal lesions appear first as indefinite white patches under a zone of slight vitreous haze, later becoming more defined and slightly pigmented

*Adult tuberculosis. Ocular tuberculosis* is seldom seen in patients with active post-primary tuberculosis but those affected are often healthy individuals with apparently inactive lesions. It is presumably due to an occasional bacillaemia. It is usually difficult to establish the tuberculous nature of *uveitis.* It has been shown that ocular tuberculosis with a high local sensitivity to tuberculo-protein is compatible with a low skin sensitivity. However, a high skin sensitivity in conjunction with clinical features and after exclusion of other known aetiological factors is some evidence in favour of the tuberculous nature of an eye lesion. A non-specific uveitis is usual but a granulomatous response may result in nodules in the iris, ciliary body and choroid.

*Retinal tuberculosis* is usually caused by direct extension from choroidal tubercles but may take the form of retinal periphlebitis (p518). Optic neuritis may occur (p383). The sclera may develop a chronic, dusky red, brawny congestion which can be seen beneath the bulbar conjunctiva anteriorly and which heals eventually as a white scar (p494). When Tenon's capsule is involved there may be proptosis or immobility of the eye with oedema of the lids. There is always some underlying uveitis. Episcleritis is a more superficial condition, being an allergic reaction to a variety of allergens of which tuberculo-protein may be one (Plate 25.15 p493). The cornea may be affected by extension from the sclera or as a sectorial patch of interstitial keratitis, vascularised by both superficial and deep vessels. It is also frequently unilateral and leaves heavy scarring. In contrast, syphilitic interstitial keratitis is almost always bilateral, involves all parts of the cornea and scarring is often less dense. The treatment of ocular tuberculosis is prolonged and normally involves isoniazid, rifampicin and pyrazinamide for 2 months followed by isoniazid and rifampicin for 4 months. Ethambutol or streptomycin is included in the initial regime until sensitivity results are available. If there is no improvement in the condition after 1 month of therapy the diagnosis must be questioned. In uveitis suspected of tuberculous origin and under treatment, atropine drops and the careful use of steroids locally may reduce the inflammatory response.

*Leprosy (Hansen's disease)*

*Mycobacterium leprae* was discovered as the cause of leprosy in 1873 by Hansen. The optimum temperature for growth of this slowly dividing organism is below 37°C hence the heaviest infiltrations are in the cooler areas of the body. Humans are the only natural host and the immune response to infection is extremely variable. There are two polar forms of leprosy; tuberculoid, in which cell mediated immunity (CMI) is high, and lepromatous, in which CMI is low. Intermediate forms have been recognised between these extremes which are unstable and may change with time. The disease primarily affects the skin, peripheral nerves and anterior ocular structures.

*Tuberculoid leprosy* - (high CMI, paucibacillary) more typically results in involvement of the 5th and 7th cranial nerves which may give corneal anaesthesia and lagophthalmos with consequent exposure keratopathy and secondary infections. A type 1 reaction in a facial patch has been shown to be predictive of subsequent facial nerve damage. *Lepromatous leprosy* (low

CMI, multibacillary) may result in loss of the lateral portions of the eyebrows and eyelashes (madarosis) and is typically associated with lepromatous keratitis with corneal vascularization, uveitis and scleritis. The uveitis may be pathognomonic if iris pearls (small white lepromas) are present and low intraocular pressure, synechiae, and cataract are frequent complications. Selective atrophy of the dilator pupillae results in a pin-point pupil.

*Treatment.* Because ocular leprosy is a significant threat to vision, early diagnosis and treatment is critical. Diagnosis beyond clinical detection, still often relies on skin scrapings or biopsies. A simple, reproducible screening test is required along the lines of the protein-glycolipid-1-antigen ELISA. Polymerase chain reaction methods are also being developed. Dapsone used to be the only therapy required for this disease 10mg/kilo body weight per week over 2 or 3 years or longer but resistance to this drug means that multidrug therapy is now preferred involving dapsone, rifampicin, and clofazimine 100mg on alternate days. Therapy may need to be taken for up to 2 years but episodes of ocular inflammation may recur. Many of the allergic effects can be alleviated by the judicious use of corticosteroids.

## Spirochaetes

### *Syphilis*

Syphilis is a specific infectious disease caused by *Treponema pallidum.* It is either acquired by intimate body contact, usually coitus, or through the placenta of a syphilitic mother in the case of the congenital disease. The tissue reaction to the organism is by perivascular infiltration and obliterative endarteritis, phlebitis and lymphangitis. A gumma is the characteristic lesion consisting of a necrotic central area surrounded by giant and epithelioid cells with a peripheral mantle of small lymphocytes and plasma cells.

*Acquired syphilis.* There are three stages. The *primary* external lesion is the chancre usually appearing 2-4 weeks after contact, which sloughs centrally to give a punched-out appearance. Examination of the slough may reveal the treponema on dark ground illumination. After the chancre has been present for two or more weeks, mucous membrane ulcers or skin rashes appear and the serological tests become positive, constituting the *secondary* stage. This stage may be complicated by conjunctivitis with a marked papillary reaction, scleritis and later a severe granulomatous iridocyclitis, choroiditis and optic neuritis (Plate 17.6 p327). If basilar meningitis occurs it may lead to papilloedema, optic atrophy or cranial nerve palsies. The *tertiary* stage occurs ten or more years after the chancre in the form of tabes dorsalis or general paralysis of the insane (GPI). There are few organisms found but gumma formation is destructive and may occur almost anywhere including the lids and uveal tract. In the CNS degenerative changes may occur in the posterior columns of the spinal cord with ataxia and loss of vibration and muscle and joint sense. Deep reflexes are lost and posterior root 'lightning pains' may occur. Argyll-Robertson pupils which are small, irregular and react briskly to accommodation and convergence but poorly or not at all to light, are diagnostic of tabes. The lesion is considered to be in the posterior part of the midbrain. Primary optic atrophy (p388, Plate 19.15 p389) is the next most common sign of tabes causing peripheral field restriction which proceeds to tubular vision and finally loss of sight. Serological tests on blood and cerebrospinal fluid may be positive in tabes but not as invariably as in general paralysis, which is a slowly progressive organic psychosis due to continued inflammatory reaction in the CNS. In the advanced state there is pyramidal tract involvement giving a positive Babinski sign or exaggerated deep reflexes as well as slurred speech and tremors. Ocular involvement in tertiary syphilis usually occurs in those who have had inadequate therapy. The absorbed fluorescent treponemal antibody test (FTA-ABS) is specific. Active infection can be distinguished by the presence of IgM antibodies, when penicillin therapy is highly effective.

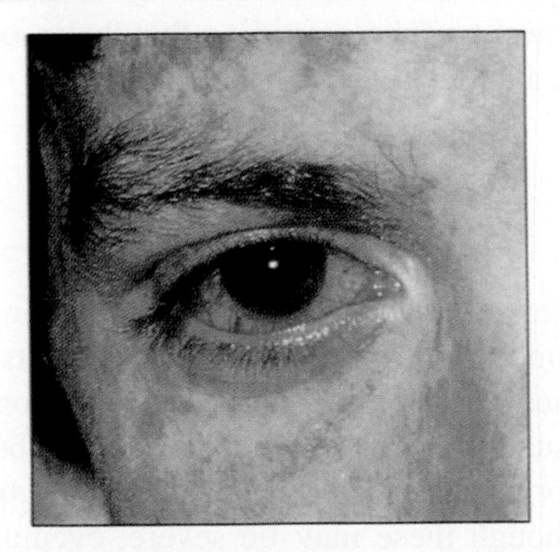

Plate 17.6 Secondary syphilis, uveitis and skin lesions.

*Congenital syphilis.* In congenital syphilis the foetus is infected *in utero*. Infants may also suffer the acquired disease due to infection after birth, and the condition then follows the usual adult course. Symptoms of congenital syphilis may be present at birth with a fatal bullous eruption but they usually appear after about four weeks as a papillo-macular rash with a snuffly nose and fissures around the mouth. There is a failure to thrive and later Hutchinson's triad of notched incisor teeth, nerve deafness and interstitial keratitis may be found. Saddle bridge of the nose due to collapse of the nasal septum, splenomegaly, periostitis of the (sabre) tibiae and exostosis of the cranial bones (hot cross bun head) are also characteristic signs.

*Congenital Syphilitic Interstitial Keratitis (IK)* (LCW 3.23 p49) occurs at about the age of 10 years (5-20 years). It commences as anterior uveitis. The cornea then develops oedema with circumcorneal injection causing blurred vision, pain, photophobia and watering. Vessels then invade the hazy swollen stroma from the periphery. Untreated, the inflammation tends to settle down with some clearing although the residual empty blood vessels (ghost vessels) are visible indefinitely with the microscope, and degenerative changes such as calcareous deposits may eventually appear (band opacity). The iridocyclitis and anterior choroiditis may be severe and can be complicated by secondary glaucoma. Fortunately, the active keratitis, an immune reaction in the stroma, is amenable to treatment by local corticosteroids in addition to general antibiotic therapy. Should corneal scarring have occurred, this usually responds well to corneal grafting providing that the associated uveitis has not led to severe complications.

*Leptospirosis*

Leptospirosis is due to a motile spiral organism occurring worldwide. Human infection results from contact with animal hosts. In its severe form (Weil's disease), with jaundice and intense conjunctival injection and haemorrhage, it may be fatal. It is usually mild but after recovery may be followed by a granulomatous anterior or posterior uveitis and sometimes by an acute exudative type of uveitis. Although these may be severe, eventual recovery is usual. The condition is treated by penicillin in high dosage.

## Viral infections

Viruses are obligatory intraocular parasites which contain a central core of nucleic acid responsible for infectivity and genetic characteristics surrounded by protein which is concerned with antigenic properties. Recovery of the virus from infected cells by tissue culture is the most definite method of diagnosis but increasing antibody titre to autogenous specific virus indicates an immunity reaction to the virus although this may be slight in isolated ocular infection. The eye can be affected by almost all viruses. As bacterial disease is often amenable to antibiotics, virus infections are a major problem, especially those affecting the eye externally. Fortunately, antiviral agents are proving effective in herpes simplex, which is one serious cause of corneal disease and to some extent in AIDS. Viruses affecting the eye include:

-*Molluscum contagiosum.* (pp201, 456) This is a DNA virus affecting the skin in which the virus forms large eosinophilic inclusions in epidermal cells giving rise to a pearly white umbilicated nodule (Plate 17.7 p329) Man is the only host and the eyelids of children are often affected. In AIDS patients the lesions tend to be multiple and extensive.

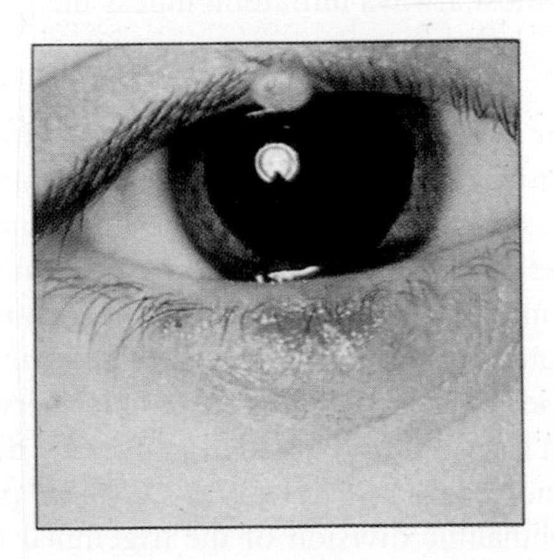

Plate 17.7 Molluscum contagiosum.

*Herpes viruses*

These are a group of DNA viruses which include herpes simplex virus, the herpes zoster virus which causes varicella and herpes zoster (shingles), and the Epstein-Barr virus, which is implicated in both mononucleosis and Burkitt's lymphoma.

*Herpes simplex virus* (HSV). Primary infection in previously unaffected individuals, usually in early childhood, may give rise to adenopathy, fever and malaise, but the symptoms are usually minor and unidentified. When eczema is present a dangerous widespread herpetic disease with severe toxaemia and Kaposi's varicelliform eruption may occur. Inoculation against smallpox with vaccinia in eczematous subjects can also have the same result. Meningo-encephalitis may also be a complication of herpes simplex. Infection results in antibody formation, and in these patients subsequent re-infection leads to recurrent vesicles on an erythematous patch (cold sore) often at mucocutaneous junctions on the lip or nose with enlargement of the draining lymph glands. When the eye is affected, which occurs frequently after exposure to ultraviolet light or during fever for any cause, this takes the form of a follicular conjunctivitis with preauricular adenopathy, punctate corneal staining and dendritic ulceration (p168). While the conjunctivitis may be bilateral, corneal herpetic disease is almost always unilateral unless the patient is immunocompromised.

*Varicella zoster virus* (VZV). This virus causes both chicken pox and herpes zoster (shingles). Chicken pox affects mainly the young, although it may occur at any age. Vesicles may appear on the lids, conjunctiva and cornea. Herpes zoster is more common after middle age and, in a patient of less than 40 years of age, should raise a suspicion of immunocompromise . It tends to attack those debilitated by other conditions. The virus produces its effect mainly at the posterior root or homologous cranial nerve ganglia leading to severe neuralgic pain followed by vesiculation over the distribution of the corresponding sensory nerves.(Plate 7.20 p172, LCW 2.3 p28). In herpes zoster ophthalmicus the ophthalmic division of the trigeminal nerve is affected and the eye is very liable to be involved especially if vesicles are found at the site of the termination of the nasociliary nerve on the side of the nose because the nasociliary is the main sensory nerve of the eye (Hutchinson's sign). The cornea, uvea and the optic nerve are frequently sites of inflammation and secondary glaucoma is common (p171). Motor nerves may also be affected in about 10 % of cases, causing extraocular muscle pareses and diplopia. A severe degree of meningo-encephalitis is not rare. Both simplex and zoster may result in acute retinal necrosis. This is a confluent peripheral necrotising retinitis, arteritis and periphlebitis with moderate vitritis. Patients are usually elderly and fit, presenting with red painful eyes and blurred vision. It may also occur in immunocompromised individuals. Diagnosis is by viral titres or raised IgM antibodies and treatment consists of high dose intravenous aciclovir.

*Cytomegalovirus* (CMV) is a widespread virus disease resulting in large cells with inclusion bodies which are found in secretions or tissues. Pregnant women may show a raised antibody titre and about 2% of newborn excrete the virus but few show signs of the disease. In those who do, many tissues may be involved causing jaundice, splenomegaly, haemolytic anaemia and encephalitis with its sequelae. It is one cause of intracranial calcification. The eyes may be affected by choroidoretinitis varying from a single focus to severe disorganisation. An acute retinal vasculitis may occur in patients with AIDS and adults following organ transplants involving immuno-suppressive treatment. Anti-viral treatment is partially effective as described in the therapy of AIDS but there is a serious risk of retinal detachment.

*Adenovirus* (p200) comprises a group of over 30 immunologically distinct types of DNA virus. Adenovirus 8 and 19 cause epidemic kerato-conjunctivitis (p200) and adenovirus types 3, 4 and 7 may cause simple follicular conjunctivitis but they may also cause pharyngoconjunctival fever (p200). The latter is an acute disease affecting all age groups but occuring mainly in children. It may be transmitted by bathing in infected swimming pools. The incubation period is about a week. it is infectious for the first 10 days and persists for three weeks although an associated superficial keratitis may continue for several months. There is usually a mild nasopharyngitis with lymphadenopathy and fever.

*Infective mononucleosis.* This troublesome disease, believed to be due to the Epstein-Barr virus may give rise to lacrimal gland swelling and inflammation, conjunctivitis, uveitis, retinitis and optic neuritis. It is associated with antibodies which will agglutinate sheep erythrocytes (the Paul Bunnell test). Usually the ocular effects are not severe and treatment is non-specific.

*Burkitt's lymphoma.* This is the most prevalent malignant neoplasm among children living in tropical Africa; orbital involvement with proptosis occur in about 20% of cases. The disease is not confined to Africa, although exposure to malaria in infancy may predispose to it. A neoplastic proliferation of mainly B lymphocytes is believed to occur in response to the Epstein-Barr virus in the form of closely packed lymphoblasts interspersed with pale staining macrophages. Convalescent sera containing antibodies are of some value in treatment.

*Verrucae (warts).* The virus causes papillomatous excrescences which may occur on the lid margin causing chronic keratoconjunctivitis which usually resolves when the wart is excised.

*Measles* is due to an RNA virus and causes a conjunctivitis with a mucous discharge and a superficial keratitis with punctate erosions causing photophobia. Recovery is usually complete provided secondary bacterial infection is prevented by topical antibiotic therapy. Nutritional deficiency, especially involving Vitamin A is particularly serious in measles (p354).

*Mumps* is an RNA virus infection and may cause lacrimal gland as well as salivary gland inflammation and swelling. Uveitis can occur and is usually mild, although meningitis may cause optic neuritis and corneal nerve palsies.

*Rubella (German Measles).* During the course of an attack of rubella there is mild conjunctivitis with reddening of the bulbar conjunctiva, but when rubella is contracted by a pregnant woman it is the severe effect on the foetus which makes the disease of great ophthalmic importance. The foetus is particularly liable to developmental defects from maternal rubella during the early weeks of pregnancy when microphthalmos with cataract formation (p428) may occur. There may be uveitis and failure of anterior chamber development resulting in iris atrophy and a miosed pupil and sometimes congenital glaucoma. The development of the pigment epithelium of the retina may be disturbed to cause pigmentary mottling but, although there is diffuse hyperfluorescence on angiography due to injury to the pigment epithelium, this impairment does not usually reduce vision appreciably if the eye is otherwise unaffected. Extensive neural damage may cause deafness and mental retardation and cardiac malformation may also occur. The virus can persist in the child for 2-3 years after birth and may be a source of risk for non-immune pregnant women who should avoid contact with such a child. Blood tests can indicate the presence of immunity to rubella and immunisation of both boys and girls with rubella vaccine has been recommended with a view to reducing the risk of exposure to the disease and of its being contracted during pregnancy. As vaccination could affect the foetus this can only be carried out in women when the possibility of pregnancy can be excluded and when continued non-pregnancy for three months can be assured. The treatment of cataract due to congenital rubella is described on p96 and p100.

**HIV / AIDS**

*Epidemiology.* In 1981 the first cases of acquired immunodeficiency syndrome (AIDS) were reported. By 1994 the millionth case was reported to the World Health Organisation (WHO) and the pandemic is now affecting all continents. Case definitions and laboratory facilities for diagnosis vary between countries. Africa bears the biggest burden of infection, however the numbers in Asia are rising fast (Fig. 17.8 p334).

The three principal modes of transmission of the virus are:

1. heterosexual or homosexual intercourse 2. transfusion of infected blood or injections with infected needles (drug related) 3. mother and infant transmission in utero or by breast feeding.

The WHO global estimate of HIV/AIDS sufferers at the beginning of 2000 was 34.3 million distributed as in Fig. 17.8 p334. A survey by the Joint United Nations Programme on HIV/AIDS (Unaids) at the beginning of 2000 estimated the death toll from HIV/AIDS related diseases during 1999 to be 2.6 million. Of these 95% are in the developing world, 70% of whom are in sub-Saharan Africa, and most of these will die during the next ten years to join the 13.7 million Africans who have already died from the epidemic. Unaids also reports that HIV infections have doubled in the last two years in the former Soviet Union mainly due to the use of infected syringes for drug taking. Half of all people infected by HIV were less than 25 years old and tragically they typically died by the age of 35. The report emphasised that the disease remains fatal and that the decline in deaths due to anti-retroviral therapy is tapering off.

The 1991 WHO estimates of risk per single exposure are given in Fig. 17.9 p334. World-wide 75 - 85% of HIV infections in adults have been transmitted by unprotected sexual intercourse, with heterosexual intercourse accounting for more than 70%.

Fig. 17.8 WHO estimates of HIV/AIDS at the beginning of 2000.

| **Area** | **Estimated HIV/AIDS (millions)** | **Prevalence** | **%children** |
|---|---|---|---|
| Sub Saharan Africa | 24.50 (19.00 - 41.80) | 8.6% | 4.1% |
| South and South East Asia | 5.60 (3.60 - 6.60) | 0.50% | 3.7% |
| Latin America | 1.3 (1.00 - 1.60) | 0.50% | 2.1% |
| Established market economies | 1.4 (1.10 - 1.70) | 0.20% | 1.0% |
| Caribbean | 0.36 (0.26 - 0.47) | 2.10% | 2.7% |
| Eastern Europe-Central Asia | 0.42 (0.31 - 0.53) | 0.21% | 3.6% |
| East Asia-Pacific | 0.53 (0.39 - 0.68) | 0.06% | 1.0% |
| North Africa-Middle East | 0.22 (0.14 - 0.30) | 0.12% | 3.6% |

Fig. 17.9 1991 WHO estimates of transmission of HIV infection.

| **Type of exposure** | **Risk per single exposure** |
|---|---|
| Blood transfusion | >90% |
| Perinatal | 30%* |
| Heterosexual | 0.1-1.0% |
| Homosexual | 0.1-1.0% |
| Injecting drug use | 0.5-1.0% |
| Health care | <0.5% |

* More recent data has shown this risk to be 15%. Breast feeding, however, carries a risk of about 15%, hence the original figure of 30% is still valid in the majority of developing countries.

*Virology.* The T-lymphocyte tropic RNA retrovirus which causes this condition has two isolates to date which have been called human immunodeficiency virus (HIV) 1and 2. Research has progressed at such an impressive rate that the entire genome of the virus has been mapped. Diagnosis 3-6 months after infection can be made by highly sensitive and specific serological tests for antibodies to the virus. Before individuals become seropositive, culture, polymerase chain reaction, and antigen tests must be used.

*Natural history of infection.* The natural history of HIV-1 infection can be viewed in three consecutive stages: an acute stage corresponding to primary infection with HIV-1; a chronic stage representing a period of clinical but not necessarily virological latency; and a crisis stage where profound immunodeficiency exists, manifest by opportunistic infections and other pathological conditions. The acute stage may have fever, headache, lymphadenopathy, myalgias and rash associated with transiently high levels of virus in the plasma.

The chronic or latent stage frequently lasts for 7-11 years with low serum levels of virus. The function of T helper cells may be impaired even at this stage in the face of normal CD4+ counts (>500 CD4+ cells/ml). An increase in viral burden heralds the commencement of the crisis stage during which severe depletion in CD4+ cells occurs with concomitant opportunistic infections. There are many systemic infective patterns which may occur with the acquired immunodeficiency syndrome. Use of the drug AZT (zidovudine) lengthens survival in patients with AIDS.

*Ocular manifestations.* Syphilis must be considered as a possible aetiology in any HIV-related ocular inflammatory process.

As a result of immunodeficiency the number of organisms that can become pathological in patients with AIDS is very large. The more common patterns are:

*Lids*

*Molluscum contagiosum* (p329) is a common cause of eyelid and conjunctival disease, especially in children. In AIDS patients the lesions tend to be multiple and widespread.

*Kaposi's sarcoma* is a multicentric malignant tumour derived from endothelial cells and is second only to *Pneumocystis carinii* as a presenting manifestation of AIDS. It occurs mainly in homosexuals, affecting 15% to 24% of patients with AIDS. Single or multiple lesions have been described on the eyelids and conjunctiva following a histopathological and clinical staging. Stage I and II lesions are patchy, flat (less than 3mm in height), and of less than 4 months duration. Stage III lesions are more nodular (>3mm height and >4 months duration). Conjunctival lesions have been treated by injection with interferon alpha or local excision whilst eyelid lesions may be treated with cryotherapy or radiotherapy.

*Lymphomas* have been reported to occur in the orbits of AIDS patients causing proptosis, lid swelling and ocular motility disturbance.

*Neuro-ophthalmic abnormalities*

*Primary HIV infection* causes an encephalopathy with progressive dementia. Later, gaze palsies and nystagmus and cranial nerve palsies may occur.

*Opportunistic infections* causing meningitis, and both diffuse and focal encephalopathy result in papilloedema in addition to focal lesions resulting in gaze palsies, cranial nerve palsies, nystagmus, and hemianopia. *Cryptococcus* is the most common cause of meningitis whilst *Cytomegalovirus* is the most common cause of encephalitis. The focal encephalopathy of toxoplasmosis may be the presenting feature of AIDS. Other common associations include tuberculosis, herpes simplex and herpes zoster.

*Intracranial neoplasms* may result in neuro-ophthalmic symptoms and signs. Examples include primary CNS lymphoma and metastatic Kaposi's sarcoma.

*Anterior segment.* Herpes simplex keratitis and blepharitis may occur in AIDS patients and are more common than the recently recognised microsporidial keratoconjunctivitis caused by *Encephalotizoon bellem.* Topical fumagillin C has been used in the long term treatment of this condition.

*Posterior segment.* Due to the depletion of CD4+ cells, opportunistic infections affecting the posterior segment in patients with AIDS are seldom associated with a marked inflammatory response. As a result the patients are often asymptomatic despite extensive tissue destruction.

The most common ocular manifestation of AIDS is a *noninfectious retinal microangiopathy* that clinically and pathologically resembles that of diabetes mellitus, hypertension, and collagen disease. The cotton wool spots, with or without associated retinal haemorrhage, may be difficult to distinguish from early CMV retinitis, however they fade over several weeks in distinction to CMV which almost invariably progresses. They are mainly distributed at the posterior pole near the main retinal vessels. These lesions probably reflect systemic vascular disease as evidenced by concurrent cerebral impairment and abnormal cerebral blood flow. They are possibly due to immune complex deposition.

It is estimated that about 25% of patients with AIDS will develop *cytomegalovirus (CMV) retinitis.* It is the most common ocular infection in AIDS and is responsible for at least 90% of cases of infective retinitis. CMV infection is fortunately becoming much less frequent in populations where HIV sufferers are receiving prophylactic antiviral therapy, particularly protease inhibitors. The clinical appearance is of coalescent areas of whitish retinal necrosis, especially around the vascular arcades with haemorrhage at their borders. They may be confused with cotton-wool spots.

Treatment not only preserves vision but may also prolong life by reducing systemic CMV infection. Ganciclovir sodium and foscarnet sodium intravenously are both effective in the initial treatment of CMV retinitis but chronic maintenance therapy is required to delay recurrent disease. An improved survival has been shown with foscarnet compared to ganciclovir in the presence of good renal function, the latter drug being superior when renal function is impaired. Intraocular drug delivery is used either by regular intravitreal injection or by surgical implantation of sustained-release devices in order to decrease systemic drug toxicity. Viral resistance to these drugs has been reported as high as 11% in some areas.

After one year up to 50% of patients with CMV retinitis develop retinal detachments which are often bilateral. These detachments are difficult to treat, often requiring vitrectomy and silicone oil or gas tamponade. With improved survival the complications of these procedures become increasingly important.

Many *other infective agents* may affect the posterior pole both as a primary event and as a result of secondary spread. Herpes zoster can cause a rapidly progressive outer retinal necrosis in AIDS patients which is usually bilateral and poorly responsive to any therapy. Toxoplasmosis is a rare cause of chorioretinopathy despite being a common intracranial pathogen in people with AIDS. Unlike the disease in immunocompetent individuals the lesions do not typically arise from pre-existent scars. The lesions are single, multiple, or diffuse, and have ill defined edges. Serum antitoxoplasma IgG titres are present making this a useful screening test in suspected cases. Most patients respond to sulphadiazine and pyrimethamine and steroids are unnecessary. Flat, yellow, round, irregular lesions at the level of the choroid and posterior to the equator may represent *Pneumocystis carinii* infection of the choroid. This may be the first manifestation of extrapulmonary systemic dissemination. Finally fungal infections may occur. Cryptococcus is the most common, mainly in the setting of cryptococcal meningitis. Creamy choroidal lesions are the most commonly described sign with a variable response to systemic antifungal therapy.

## Chlamydia

These organisms resemble bacteria in many ways, but like viruses they are obligatory intracellular parasites. They are responsible in the eye for *adult and neonatal inclusion conjunctivitis* and for *trachoma* which is one of the major causes of blindness in the developing countries (pp173, 195). Also in this group is *lymphogranuloma venereum.* This is a contagious venereal disease causing a vesicle which breaks down to form an ulcer with a sharply defined edge followed by suppurative involvement of the regional lymph glands which may require aspiration. Primary infection of the lid and conjunctiva may occur with preauricular adenitis. Uveitis and sclerokeratitis may result from haematogenous spread. Diagnosis is by microscopy and immunodiagnosis as with the antigen ELISA. The condition is treated with a combination of tetracycline and sulphonamides.

## Rickettsiae

Rickettsiae are obligatory intracellular parasites which are visible with the light microscope. They are transmitted to man either directly from animals in the case of Q fever, or by the larvae of mites in the case of scrub typhus (Tsutsugamushi). Ticks are the source of Rocky Mountain Spotted fever in which patients show a high temperature with skin rash and the ocular changes are part of the generalised vasculitis which is the hallmark of Rickettsial infections. Ocular complications include conjunctival haemorrhage and retinal venous engorgement, sometimes with optic disc oedema, vitreous haemorrhage, arterial occlusion and neuro-ophthalmic complications. Prophylaxis is by vaccines and avoidance or elimination of animal hosts. Treatment with tetracyclines and chloramphenicol is effective, the latter being used in younger children.

**Fungi**

Fungi are single celled parasitic or saprophytic organisms characterised by the formation of filaments, hyphae which intertwine to form a mycelium and by the production of spores. For diagnosis special stains and culture media are necessary and occasionally skin tests may be useful. Treatment with cortico-steroids and immunosuppresive agents predispose to fungus infection and its possibility should always be considered in extensive indolent corneal ulceration, especially if the adjacent skin or the nasal sinuses are also abnormal. Three widely distributed fungi which may involve the eye are actinomyces, histoplasma and candida. Blastomyces and apergillus are mainly found to cause eye lesions in North and South America. In addition many other fungi such as coccidioides, cryptococcus and sporothrix produce lesions somewhat similar to those of actinomyces.

*Actinomyces* is bacterial and not a true fungus even though it grows in colonies and forms hyphae. In the tissues it tends to break up and form free-living bacterial organisms. The ophthalmic effects may be:

-lid and orbital infection spreading from the sinuses,
-lacrimal canalicular infection with a troublesome inflammation of the conjunctiva and a watering eye,
-indolent corneal ulceration which may be associated with a hypopyon and corneal perforation.

In the case of infection of the lacrimal passages, colonies of the fungus are seen as small yellowish ('sulphur') granules in the discharge if the affected canaliculus is gently curetted. This will frequently relieve the local symptoms especially when aided by penicillin therapy (p212).

*Aspergillus* appears microscopically as a round mass of black spores and may cause a dark brown discharge from a canaliculus with associated conjunctivitis.

*Candida* (monilia) may complicate injury or dendritic corneal ulceration and may produce florid lesions in AIDS patients. It tends to have poor invasive powers but is, nevertheless, the most frequent ocular fungal pathogen and may cause endogenous uveitis.

*Histoplasma* is seen as small spherical bodies inside histiocytes *(histoplasma cells)* and causes a systemic infection resulting in pulmonary inflammation or localised granulomas of the skin. Recovery may be spontaneous but it can be a fatal condition by blood spread to the viscera and brain. It may cause severe ocular inflammation, appearing as multiple yellowish white patches in the fundus with symptoms of patchily blurred or distorted vision. (LCW 5.9 p76)

*Blastomyces* produces effects similar to actinomycosis which may be severe and additionally may cause a granulomatous uveitis.

*Cryptococcus* is the most common cause of choroidal whitish inflammatory lesions in the course of AIDS complicated by cryptococcal meningitis.

*Treatment* for fungal infections is unsatisfactory due to the resistance of the organisms and the scarcity, expense and toxicity of the drugs. Much treatment has to be merely symptomatic.

Examples are:

-*Antibiotics* such as tetracycline orally or penicillin G in high doses topically, subconjunctivally or systemically are usually effective but dapsone orally or natamycin (Pimafucin) locally as drops or ointment may be given for actinomyces
-*long acting sulphonamides* may help in blastomycosis
-*Nystatin* locally, e.g. as drops or ointment, is suitable for candida or aspergillus but only for superficial lesions as it has poor penetration. It is fungistatic rather than fungicidal and is *toxic when given parenterally.*
-*Amphotericin-B* when administered systemically as an intravenous infusion may be effective in many types of fungus infection but must be used with great care and only when the diagnosis is certain because rigors and impaired renal function may result. 5-flucytosine may be given orally for blastomyces, candida, histoplasma and coccidoides or as drops for candida.

## Protozoan infections

The eye may be involved in diseases caused by certain unicellular organisms, the most important of which are toxoplasmosis, malaria, amoebic dysentery, acanthamoeba infection, and trypanosomiasis.

*Toxoplasmosis* infection has been considered with uveitis (p182 and Stereo Plate 7.26 p184)

*Malaria* has relatively few ocular complications and anti-malarial treatment will usually prevent their occurrence although increasing drug resistance is a problem. The high fever may activate herpes simplex of the lids and the cornea may be affected with dendritic ulceration and its complications. Capillary embolism at the height of the attack, complicated by anaemia may lead to retinal haemorrhages. Cranial nerve palsies may also result from embolism. Sensory nerve neuralgia, especially of the trigeminal nerve, may be troublesome.

*Acanthamoeba infection* is becoming more widely recognised as a cause of serious disease and while it can cause a granulomatous encephalitis by haematogenous spread, in the eyes it is important as a cause of keratitis by the contact of a combination of a foreign body, (including hard or soft contact lenses) and contaminated fluid with the cornea. The contact lens may have been stored in contaminated fluid derived from a well or spring water or may have been worn while the patient swam in contaminated water. The amoeba inhabits soil and water. It feeds on bacteria and releases enzymes that allow tissue invasion. It also forms resistant cysts which can necessitate prolonged treatment. The cysts may also become air-borne. The resulting keratitis follows the course of the corneal nerves with a dendritic spread of a non-suppurative character and may result in a disciform lesion. Diffuse scleritis may also occur. Diagnosis is difficult but more recently greater awareness has resulted in earlier treatment and systemic therapy can be given in the form of ketoconazole by mouth. Locally several agents have been used especially 0.15% dibromopropamidine which hitherto has been the most effective medication but drops of polyhexamethylene-biguanide (PHMB) have recently shown promise. The timing and rôle of penetrating keratoplasty in this condition is controversial. Because this can be a devastating eye infection, relatively unresponsive to current therapy, it is important that contact lens practitioners emphasise the dangers and advise that contact lens patients use only sterile fluids to store or rinse contact lenses and avoid swimming with them in water which could be contaminated. The use of contact tonometers and gonioprisms requires scrupulous attention to disinfection to avoid keratitis caused by these and other organisms. Disposable tonometer prisms are now increasingly employed.

*Amoebiasis.* While a mild uveitis may be encountered in amoebic dysentery, due to *Entamoeba histolytica* it is occasionally very severe with hypopyon and is then probably similar to the abscesses which may affect the liver and brain. In the chronic ill health associated with liver abscess, night blindness and fatigue of accommodation are frequently reported.

*Trypanosomiasis* is due to flagellated protozoans *Trypanosoma gambiense* or *T. rhodesiense* carried to humans in the bite of the tsetse fly, *Glossina palpalis.* Urticarial swelling of the lids may occur in the early stages of the disease as in other parts of the body. Invasion of the cornea causes a deep keratitis, and the iris and choroid may sustain a haemorrhagic uveitis. A mild optic neuritis may occur and papilloedema can result from raised intracranial pressure. Arsenical therapy, which is very effective in clearing the corneal lesions, can itself cause a toxic amblyopia so great care is required in treatment.

## Metazoan infestation

Most infections are due to simple organisms, but a few result from infestation with multicellular animals, either worms or insect larvae. In myiasis the larval stage of the insect is harboured by man, the adult fly not being parasitic. In the case of worms the parasites spend part of their life cycle in two hosts. When man is involved he is usually the definitive host of the mature worm, but he may be also the intermediate host of the larval stage. Worm diseases of the eye are of two main types, either:

-a mobile worm, which may be either adult or a larva, affects the eye or surrounding tissue, e.g. onchocerciasis
-the larva is sessile and forms a cyst, e.g. cysticercosis.

## Nematodes (onchocerciasis, filariasis)

### *Onchocerciasis*

Onchocerciasis is an important cause of world blindness after cataract, trachoma, and glaucoma. Onchocerciasis has always been limited to equatorial

Africa with a relatively very small number (<1% total) of affected communities in the Sudan and Yemen and also Guatemala and Mexico. This is because the disease is spread by a small black fly of the genus *Simulium*. Subspecies of this fly have either tropical forest or Sudan and Guinea savannah as their natural habitat. All require fast flowing,turbulent water for their larvae hence infection occurs in communities living near breeding sites and the illness has become known as river blindness. Up to one in ten people may be blind in affected communities.

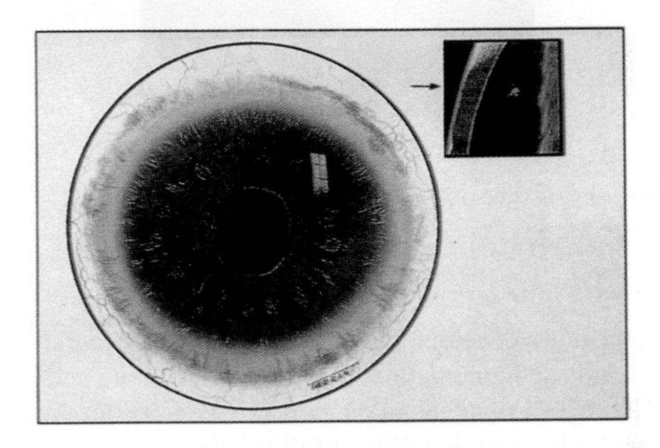

Plate 17.10 Onchocerciasis - anterior segment changes and (inset) microfilaria in anterior chamber. (painting)

The disease itself is caused by a filarial worm *Onchocerca volvulus.* After mating a female simulium takes a blood meal for the maturation of her eggs. If this meal is taken from an infected host then she also ingests microfilariae which subsequently develop in her to infective larvae. The cycle is completed when these enter the next human she feeds upon. Inside the human host the infective lavae develop into adult worms which may be found in nodules over bony prominences. These worms produce millions of microfilariae which cause the symptoms and signs of onchocerciasis.

The cardinal symptom of onchocerciasis is intense pruritis which disturbs sleep and often results in people using sticks for scratching. Microfilariae may be found widely, especially in the skin, cornea and anterior chamber in large numbers with no pathological reaction (Plates 17.10, p343, 17.11 p344).

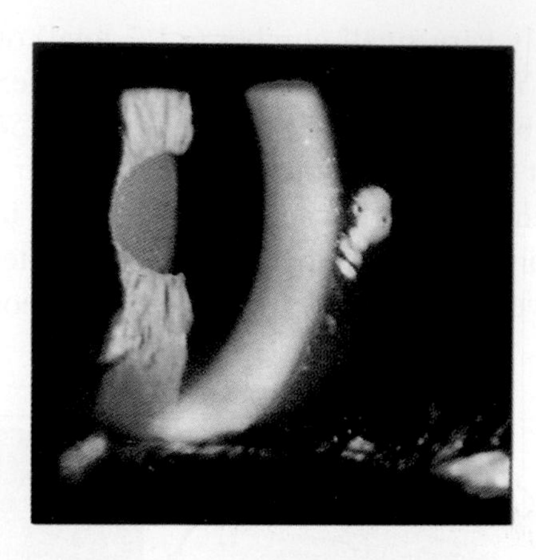

Plate 17.11 Onchocerciasis - microfilaria in anterior chamber. (photograph)

If pathological changes develope they are predominantly in the skin and eyes. Skin lesions include depigmentation of the shins and acute and chronic maculopapular rashes. Eye lesions include a punctate keratitis around dead corneal microfilariae and sclerosing keratitis, a full thickness fibrovascular change in the cornea continuous with the limbus. These corneal changes occur typically at the three and nine o'clock positions. The typical uveitis of onchocerciasis is flare without cells, intraocular pressures are lower in infected populations and peripheral anterior synechiae are related to infection. The pupil may become pear shaped. In the posterior segment, optic atrophy is common and may be the only clinical sign. If there is other tissue involvement then the retinal pigment epithelium is the first to be damaged just temporal to the macular area. Active inflammation most commonly involves the optic nerve but may occur elsewhere. The end-stage fundal appearance is termed a Hisset-Ridley fundus after those who gave the first detailed descriptions of it (Plate 17.12 p345). There is advanced optic atrophy with sheathing of the peripapillary vessels and extensive choroidoretinal atrophy of the entire posterior pole leaving only attenuated major retinal vessels, large choroidal vessels and some clumps of pigment covering the sclera (Plate 17.13 p345). Diagnosis is by skin snip. When placed in saline the microfilariae emerge from the skin specimen and can be visualised using a darkground illuminating microscope.

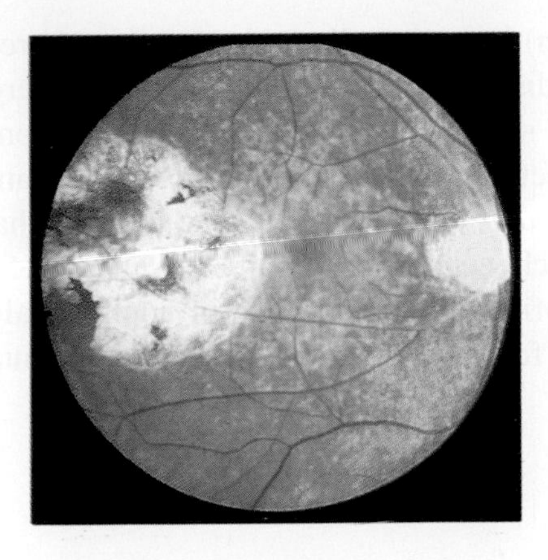

Plate 17.12 Onchocerciasis - optic atrophy, choroido-retinal atrophy.

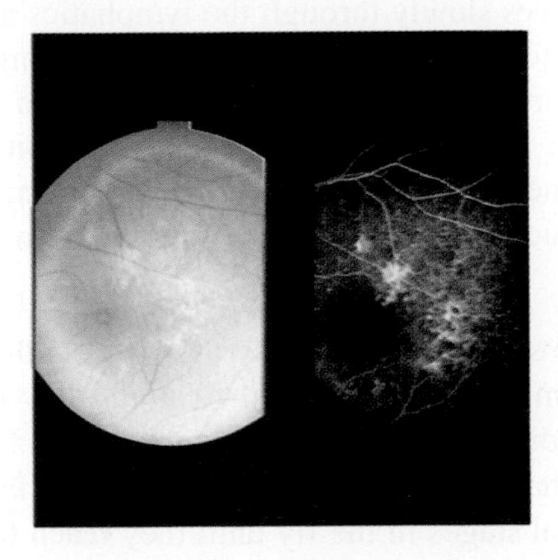

Plate 17.13 Onchocerciasis - choroido-retinal atrophy, retinal vasculitis.

*Prevention and treatment*

In a large area of West Africa the Onchocerciasis Project (OCP) has successfully controlled the disease by larvicidal treatment of the Simulium larvae in the rivers. This, however, is very expensive and suffers from the problem of Simulium reinvasion as soon as spraying stops. Removal of nodules never removes all the worms from the body because they do not all reside in nodules. There is no effective and safe macrofilaricide at present although some new products are currently being evaluated.

Ivermectin (Mectizan) is a microfilaricide that has offered much hope for the management of this disease. It is very safe and has undergone extensive clinical trials which have shown it effective in the prevention of progressive ocular disease. The 'Mectizan' committee monitors programmes for ivermectin distribution to affected communities because the drug has been donated free for human use in onchocerciasis. A major advantage of ivermectin is that it affects unborn microfilariae in the uterus of adult female worms and consequently has an effect for between 6 and 18 months. Annual treatment is therefore practicable.

*Loiasis*

Infection with the filaria Loa Loa (the African eye worm) rarely leads to severe eye trouble. The adult is about three to five centimetres in length and has a long life. It moves slowly through the lymphatics and subcutaneous tissues and is attracted to warmth. It may enter the subconjunctival tissue causing severe pain and irritation of the eye. The condition is also characterised by 'calabar swellings' which may affect the lids or orbit. These are transient oedematous lumps one or two inches in diameter. This may be due to the sudden production of microfilariae, a toxic excretion from the worm or a local allergic reaction.

Loa loa microfilariae spread widely. They may be found in midday blood and can be aspirated from calabar swellings. At night they retreat to the deeper viscera. The intermediate host is the female mangrove fly, *Chrysops.* It flies by day but prefers shade. As in the case of onchocerciasis the microfilariae develop through larval stages in the fly until they reach the region of the proboscis and are transmitted when man is next bitten.

*Treatment.* Ivermectin offers promise for control of the disease. Steroids and antihistamines may help to control allergic reactions. When the adult worms are seen under the surface of the conjunctiva, they can be removed after instilling drops of local anaesthetic and *seizing and holding the worm with toothed forceps*. (Indecision may result in failure as the worm rapidly retreats.) The conjunctiva is then incised and the worm extracted slowly and progressively with a second pair of forceps so that it is removed whole. Other adult worms may however still be present. Preventive measures resemble those for onchocerciasis.

*Wuchereria Bancrofti*

Filariasis due to *Wuchereria bancrofti* causes elephantiasis by lymphatic obstruction. The lids or intraocular structures may be involved. Treatment is as for other filarial conditions.

*Ascariasis*

Ascariasis is due to the giant nematode which mainly infests the intestine in children and has no intermediate host. It may cause colic, perforation or allergic reactions. Ingested eggs hatch in the bowel. The migrating larvae spread widely causing iridocyclitis or even endophthalmitis, sometimes with a violent tissue reaction and marked eosinophilia as well as general illness, such as pneumonia and encephalomeningitis. From the lungs the larvae break through into the bronchi and reach the pharynx where they gain access to the oesophagus passing down to become mature worms in the intestine.

*Ankylostomiasis*

*Necator americanis,* the hookworm, is common in the Southern parts of North America. It also has no intermediate host. The eggs hatch in the soil into hooked larvae which penetrate the skin and migrate, sometimes causing ocular inflammation. They ultimately reach the lungs and find their way to the intestine like ascaris larvae to become adult worms.

*Toxocara canis or cati*

The nematode worm of the dog or cat may infest puppies and kittens particularly, discharging eggs in large numbers. Children, especially due to inefficient hygiene, may ingest them. They hatch in the gut and the larvae migrate throughout the body via the bloodstream and lymphatics. The adult worm cannot develop in humans, but the larvae may give protean symptoms with fever and eosinophilia by being distributed widely. In the eye, a larva may appear in the fundus as a whitish lesion about the size of the optic disc. A stereoscopic view (Stereo Plate 17.14 p348) reveals their rounded shape and the larva within. They may be multiple, vitreous traction bands may be attached and endophthalmitis may occur. While the visceral larval migration illness may have occurred at between two and five years of age, the eye lesion may develop about five years later and by that time the blood count may have returned to within normal limits. Skin testing to toxocara antigen due to the presence of IgE antibodies (Type I hypersensitivity) may give help in diagnosis but as these may persist, their presence does not necessarily indicate active infection. The serum ELISA test for antibodies is considered positive at a dilution of 1 in 8. The diagnosis in most cases still depends on the history of contact and clinical findings.

*Treatment.* Prophylactic hygienic measures to prevent contact of children with animal excreta, particularly that of puppies and kittens, are extremely important and frequently neglected. Treatment is symptomatic using steroids to minimise allergic reactions. Surgical treatment for retinal traction detachments with vitrectomy may be indicated in some cases.

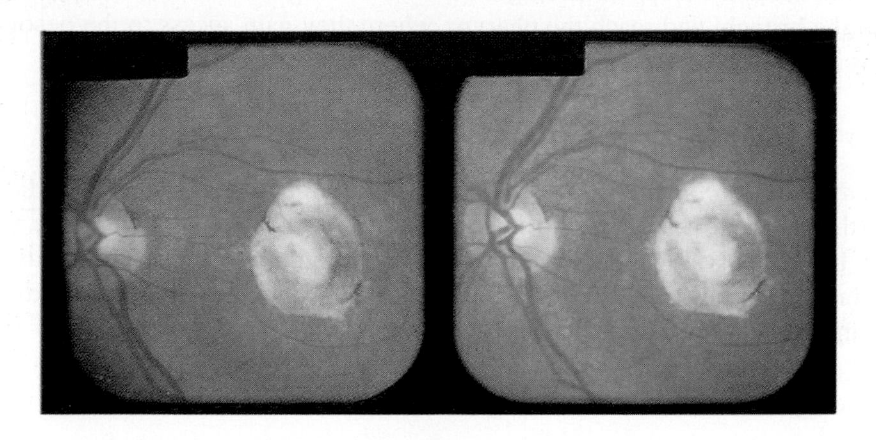

Stereo Plate 17.14 Toxocara choroido-retinal lesion. (see p4)

### *Trichinosis*

This is due to the nematode, *Trichina spiralis,* whose larvae infect many animals but pigs are the most dangerous reservoir for man. Encysted larvae are ingested when undercooked pork is eaten and become adult worms in the intestine. The female discharges eggs in large quantities which hatch into larvae. These penetrate the gut and enter the circulation about a week after the infected food has been eaten. They are distributed widely in the body with generalised fever and muscle weakness and pain accompanied by eosinophilia. Oedema may affect the orbit, particularly the upper lid, and ocular movement may become painful. Subconjunctival haemorrhage may occur. Larvae may be found in biopsy specimens and skin testing may be suggestive.

*Treatment.* This is non-specific, steroids being used to suppress allergic reactions. Prophylactically, undercooked meat should be avoided in areas affected.

## Parasitic cysts

### *Cysticercosis*

Man is usually the definitive host of *Taenia solium* but only becomes the intermediate host when eggs are ingested (sometimes by auto-infection). The cyst stage then occurs in human tissue. They are 0.5-1 mm in diameter and may appear in the skin, heart, lung and brain; sometimes the eye and orbit are affected. In the orbit it may cause fluid swelling and proptosis, with little eosinophilia, the cause frequently being obscure. In the eye a cyst is usually subretinal, leading to a solid looking detachment, but may occur in the anterior chamber or conjunctiva. These cysts may become calcified and if the larva dies, a foreign body reaction may cause inflammation. The cysts are not affected by chemotherapy although this will be necessary to treat the intestinal worm if one co-exists. Surgical treatment for ocular cysts has a poor prognosis and cyst contents may cause intense inflammation.

### *Hydatid*

Man may be one of the intermediate hosts of *Taenia echinococus,* the minute tape worm of dogs and cats. The eggs can contaminate food eaten by pigs, cattle and man, and penetration of the gut wall results in larval spread by bloodstream or lymphatics so that larval cysts are formed in most tissues including the orbit and the eye. The cyst grows slowly but in some situations may reach a diameter of 10 cm. In the eye it may appear as a small white sphere in the vitreous or as a retinal detachment. In either case progression leads to destruction of the eye and excision without rupture must be attempted. Chemotherapy is not effective.

*Ocular myiasis*

Certain species of fly breed in the nasal sinuses of animals such as the Bot fly of sheep *(Oestrus ovis)* and occasional infection of the human eye and adnexa by insect larvae occurs in conditions of poor hygiene and abundant flies. The ox warble fly *(Hypoderma bovis)* and the flesh fly *(Wohlfahrtia magnifica)* may deposit larvae near the eye which burrow inwards to cause severe foci of inflammation. Most cases of myiasis however, result in subretinal hypopigmented migratory larval tracks which are usually asymptomatic.

## Nutritional deficiencies

In the developed world specific nutritional deficiencies are occasionally seen in individuals who adopt rigorous dietary restrictions, or who suffer from malabsorption syndromes and alcohol abuse. The most common of these is a deficiency of vitamin B12 with pernicious anaemia or in vegans. More recently evidence is emerging of macrobiotic diets resulting in impaired growth of children less than 5 years old, important deficiencies being vitamins B12 and D.

On a global scale such isolated deficiencies are rare, but poverty, failed harvests, wars and natural disasters account for the development of most nutritional deficiencies. The four most prevalent and potentially serious forms of malnutrition are protein-energy malnutrition, vitamin A deficiency, iodine deficiency and iron deficiency. Once the body (in particular the immune system) is weakened by nutritional deficiency then the chances of infection increase and the effects of many infections are more profound.

## Criteria for nutritional deficiency disease

Assessment of nutritional deficiency in any organ is complicated by the possibility that the particular deficiency may affect several tissues and that a tissue may suffer from several deficiencies at the same time which often affect it in a similar manner.

*Criteria:*for a condition to be regarded as a deficiency disease the requirements are:

-evidence of inadequacy of an essential nutrient,
-symptoms and signs compatible with a deficiency,
-improvement where the deficiency is remedied.

*Evidence of inadequacy:* this may be a primary dietary deficiency or the result of abnormal diets dictated by religious or other beliefs, or the artificial deprivations of prison camps or devastation from war. It may also be a secondary deficiency due to:

-defective absorption as in chronic intestinal diseases,
-increased utilisation from physical work, pregnancy or disease,
-the presence of dietary anti-vitamins especially the maize factor affecting nicotinic acid and predisposing to pellagra.

*Symptoms and signs compatible with deficiency:* optic neuropathy associated with other eye signs suggests that they too may be deficiency effects. Animal experiments may help to elucidate the connection but animals may react differently, e.g. rats develop cataract after ascorbic acid deficiency. Only exceptionally is one symptom or sign conclusive of deficiency.

*Therapeutic test:* this is often difficult to carry out on ethical grounds and also because a specific nutrient may have a curative action on a condition not due to its deficiency. In addition the nutritional defect may be irreversible, as in some cases of optic neuropathy, which although caused by Vitamin Bl deficiency cannot be helped by Vitamin Bl treatment.

*Multiple deficiencies:* many proteins and vitamins are present in larger quantities in relatively expensive foods so that several deficiencies can affect the poor simultaneously. Often, particular nutrients occur in similar foods so that they may be missed entirely if the diet is limited. Multiple deficiency must be cured by multiple sustained therapy using rich sources of nutrients such as meat, milk, fresh fruit and vegetables, yeast and liver.

## Diseases due to nutritional deficiency

### Vitamin A deficiency

This fat soluble vitamin is mainly derived from fish oil and liver and its precursor (carotene) is found in plants, vegetables and cream. It is stored in the liver, hence severe liver disease may result in deficiency. Diseases such as measles result in a much increased demand for vitamin A and may precipitate deficiency.

Lack of this vitamin has a profound effect on both childhood morbidity and mortality such that up to 75% of children blinded as a result of vitamin A deficiency die within a few months of the blinding episode. This is because deficiency results in an impaired immune response and in particular the squamous metaplasia of epithelial surfaces giving rise to pulmonary infections. Finding one severely affected child in a population means that there are many more deficient in the same population.

The photopigment rhodopsin in the retina consists of an opsin bound to the aldehyde of vitamin A, consequently one of the first symptoms of deficiency is night blindness. Unfortunately the most common group to be affected by vitamin A deficiency are pre-school children who have been weaned from their mothers' breast milk. These children are too young to report night blindness.

The squamous metaplasia of epithelial surfaces affects the conjunctiva and cornea causing xerophthalmia (Plate 17.15 p353). The temporal bulbar conjunctiva is the first to become affected, loosing its normal sheen. Bitot's spots are silver-grey in colour with a 'foamy' or 'cheese-like' surface quality. They are usually bilaterally located near the temporal limbus whilst larger ones become more triangular or elliptical. These spots are colonised by the saprophytic bacillus *Corynebacterium xerosis* and may not disappear with vitamin A treatment in older children.

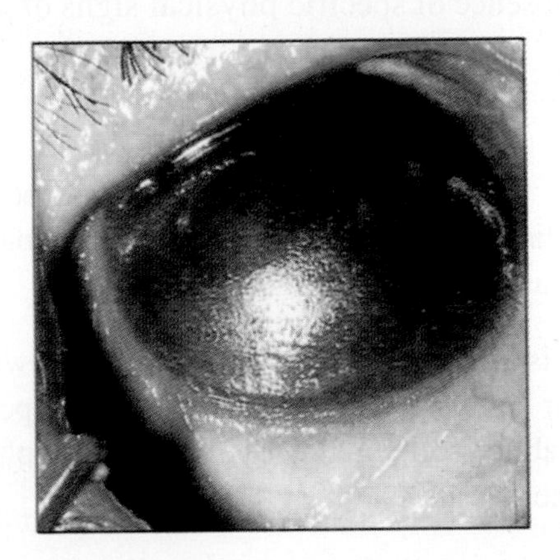

Plate 17.15 Xerophthalmia.

With a severe decrease in vitamin A levels the cornea may melt (keratomalacia) in a few hours. Those children fortunate enough to survive have a white corneal scar frequently centrally unless secondary infection has occurred causing phthisis bulbi.

Plasma retinol (vitamin A alcohol) levels are by themselves a poor indicator of vitamin A status but two tests are helpful. The first is a relative dose response test, in which a low dose of vitamin A is given and the plasma retinol levels measured just before and 5 hours after administration. In the presence of a deficiency the levels fail to rise. The second test is conjunctival impression cytology, which if positive shows loss of goblet cells and keratinisation of epithelial cells.

Treatment is currently 200,000 IU of vitamin A every 3-6 months to pre-school age children in deficient areas. In the long term, prevention involves improvement of infant feeding with preformed vitamin A since this is more accessible for absorption and use. Price subsidies, fortification and nutritional education are some of the strategies currently employed. High dose Vitamin A supplements are recommended for all children with measles who may be at risk of vitamin A deficiency and for all children with moderate or severe measles even in the absence of specific physical signs of xerophthalmia.

**Vitamin B deficiency**

Vitamin B deficiency states are usually multiple and occur especially when there is increased utilisation as in physical stress, pregnancy and malaria, or decreased absorption as in gastroenteritis.

*Thiamine* deficiency is primarily involved in beri-beri with either peripheral neuritis ('dry' type) or right sided heart failure ('wet' type). Three quarters of patients have ocular abnormalities, dry eyes, optic atrophy with centrocaecal scotomas and oculomotor palsies.

Lack of *riboflavine* leads to the *orogenital syndrome* in which fissures and excoriation occur around the mouth and keratotic scrotal lesions may be associated with ocular signs of limbal vascularisation, lacrimation and photophobia.

*Nicotinic acid* deficiency is responsible for the dermatitis, diarrhoea and dementia of *pellagra*. Investigation of patients in prison camps revealed that an initial difficulty in focusing is followed by defective vision with or without papilloedema and eventually some degree of optic atrophy with central or paracentral scotomas.

*Treatment.* A good general diet with supplements of thiamine, riboflavine, nicotinic acid and yeast and liver extracts is given, and although improvement in early cases may be expected following treatment, in the other aspects of these vitamin B deficiency conditions, once degenerative changes have occurred in the nerves little visual improvement or recovery in other nervous function is to be expected.

*Vitamin B12 and folic acid deficiency* (p300). Deficiency of either of these normal dietary constituents leads to a *megaloblastic anaemia.* Lack of folic acid is mainly responsible in developing countries. Additional iron deficiency may, however, lead to a normocytic blood picture. A megaloblastic anaemia may be associated with protein energy malnutrition which is not improved by either vitamin B12 or folic acid therapy. Increased dietary protein in necessary. Usually, peripheral neuritis and subacute combined degeneration of the spinal cord are absent in pure folic acid deficiency so that vitamin B12 in the form of hydroxocobalamin must be given when neurological lesions are present. Laboratory facilities are rarely present where such malnutrition exists so that iron preparations, folic acid and hydroxocobalamin are best used together in most cases.



## Vitamin C deficiency

Vitamin C (ascorbic acid) found in fresh fruits and green vegetables has an important rôle in collagen formation, consequently scurvy is a disease of connective tissues with haemorrhages into the skin, mucous membranes, and body cavities. These haemorrhages may be into the lids, subconjunctival space, anterior chamber, vitreous cavity, or retina. Relative vitamin C deficiency may occur in alkaline chemical burns of the eye due to increased tissue demands. It has been shown that topical and oral administration of vitamin C in these cases enables healing and reduces subsequent ocular damage.

## Vitamin D deficiency

This deficiency leads to hypocalcaemia which may result in cortical cataract formation.

# CHAPTER 18

# SYSTEMIC IMMUNE DISORDERS (COLLAGEN DISEASES)

In this group of inflammatory diseases ocular lesions may be a direct result of the immunological disorder or they may occur, directly or indirectly, as a consequence of immune vasculitis involving the deposition of antigen-antibody complexes in the vessel walls.

## Rheumatoid arthritis

Rheumatoid arthritis is a chronic systemic disease affecting predominantly women in middle age. It arises insidiously with an inflammatory and exudative alteration in connective tissue. Joints are affected and fasciae may develop nodules of necrosis surrounded by fibroblasts and granulation tissue. Typically, the synovial joints become intermittently hot, swollen, painful at rest and associated with restricted movements. The metacarpophalangeal joints are most commonly affected (95%) followed by the wrist, proximal interphalangeal joints and knees. General effects such as enlargement of lymph glands and the spleen accompanied by pyrexia (Felty's Syndrome), leucocytosis and a raised ESR may occur. Rheumatoid nodules are present in 30% of patients, usually occurring on the extensor surfaces. Other extra-articular features are: vasculitic leg ulcers, pericarditis, alveolitis, pleurisy and Sjögren's syndrome. The origin of the condition may be viral or bacterial infection followed by the development of autoimmune antibodies of the IgM type, known as rheumatoid factor, which act against the Fc portion of the IgG molecule. (p302). 70% of patients with rheumatoid arthritis are sero-positive for rheumatoid factor while almost all have antiglobulin antibodies of some type, nevertheless rheumatoid factor may occur in other conditions and in the apparently healthy. The role of rheumatoid factor is not certain but the arthritis tends to be worse in those with a high titre. In addition synovial production of antiglobulin antibodies by plasma cells has been demonstrated. It is possible that complement is activated with the release of enzymes which erode the articular surfaces.

*Ocular effects*

Keratoconjunctivitis sicca, or dry eyes, is a common problem in rheumatoid patients (LCW 6.20 p99) and may occur in secondary Sjögren's disease (p359). About 4% of patients develop uveitis but even when it occurs it is probable that it is mainly associated with the more common scleritis (LCW 6.21, 6.22, p99) which may cause a severe 'boring' pain and in severe cases result in extensive staphylomas (p493 and Plate 18.1 p358, LCW 6.23 p99).

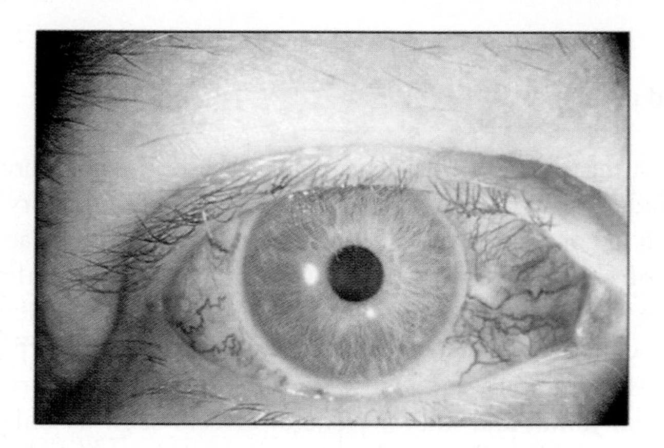

Plate 18.1 Scleritis in rheumatoid arthritis.

*Management*

No single test provides a diagnosis of rheumatoid disease. Inflammatory parameters such as elevated ESR and CRP are useful, as are immunological tests. Radiology and the analysis of joint effusion aspirates are also helpful. The condition usually responds to systemic steroids but resistance and complications may make therapy difficult. Poor prognostic indicators include female patients with high titres of rheumatoid factor and anti-nuclear antibody, evidence of radiological erosions and finally the presence of extra-articular disease.

**Juvenile chronic arthritis (JCA) and Still's disease** (p178)

This is defined as an arthritis occurring in patients less than 16 years old and of greater than 3 months duration. There are three groups: 1. polyarticular disease with an associated poor prognosis; 2. systemic disease with signs of lymphadenopathy and pyrexia; and 3. oligo or pauciarticular disease, Still's disease, with a good articular prognosis but with an increased liability to chronic uveitis, band keratopathy and cataract (pp99,178, 487). Still's disease is associated with antinuclear antibody in 30-80% and rheumatoid factor in 10%. It is important for an ophthalmologist to examine regularly patients with JCA as the onset of ocular involvement is insidious and progressive.

**Sjögren's syndrome** (p218)

Primary Sjögren's disease mainly affects middle aged women and is characterised by kerato-conjunctivitis sicca and xerostomia, or dry eyes and dry mouth. Secondary disease is associated with rheumatoid arthritis, progressive sytemic sclerosis, dermatomyositis and systemic lupus erythematosis (see below). Rheumatoid and antinuclear factor are present in many patients and some show circulatory antibodies to salivary duct antigens. Primary disease is associated with human leucocyte antigens HLA-B8, -DR3 and -DRW52. B-cell hyperactivity is present with Ro (SSA) and La (SSB) antibodies. Epithelial cells of the ducts of the lacrimal glands proliferate and lymphocytic infiltration is marked, leading to acinar atrophy and reduced tear secretion. This in turn causes areas of hyaline conjunctival and corneal epithelial degeneration which stain with Rose Bengal and are associated with tenacious mucus and corneal filaments (Plate 10.1p217 and 10.2 p219). Dryness of the eyes can be relieved partially by frequent application of artificial tear drops and sometimes by occluding the lacrimal puncta. The pain of mucus filaments can be relieved by mucolytics such as acetyl-cysteine. Dryness of the mouth may be extreme and in some cases there may also be pancreatitis, interstitial nephritis and hepato-biliary or thyroid disease. In these organs lymphocytes and plasma cells may proliferate causing impairment of function.

**Ankylosing spondylitis**

This disease of unknown aetiology predominantly affects young men (male to female ratio 2.5:1). There are marked genetic factors and spondylitis has been found in relatives many times more frequently than in the general population.

The histocompatability antigen HLA B27 is present in about 90% of patients suffering from ankylosing spondylitis compared with a figure of 6-14% in the general white population. It is associated with an elevated ESR in 50% of cases, but rheumatoid factor is negative. It causes a sacro-ileitis and lower back pain and ultimately leads to the formation of a "bamboo spine", so described due to the inflammatory changes occurring at the site of the insertion of the annulus fibrosus producing a characteristic x-ray appearance. Uveitis develops in about 40% of patients with ankylosing spondylitis and is a recurrent problem. It is often associated with a severe inflammatory response, and hypopyon may develop. It appears that both the spondylitis and the uveitis are independent expressions of an underlying common aetiology. Treatment of the uveitis is described on p181.

**Reiter's disease**

This is also a syndrome mainly affecting young men. It is of unknown aetiology and characterised by a triad of non-gonococcal urethritis, arthritis and conjunctivitis. Uveitis may also occur in a small proportion of cases. It may also be complicated by balanitis, stomatitis, buccal ulceration and carditis. The disease may appear following dysentery especially in women. There is a strong link with HLA B27 which is found in 70%-80% of patients. It may be associated with ankylosing spondylitis but there is no direct immunological evidence that Reiter's disease is a form of this condition. Chlamydial infection has been found to be associated with urethritis in a proportion of Reiter's disease patients.

**Behçet's syndrome**

This is more common in Mediterranean countries and in Japan and is believed to be higher in populations around the ancient Silk Road. It is linked with HLA B5 (found in 80% of cases in Turkey) and occasionally C4 antigens. The disease is characterised by recurrent uveitis with hypopyon (Fig. 18.2 p361) and buccal and genital ulceration (Fig. 18.3 p361). The basis of these manifestations is an obliterative vasculitis (Plate 18.4 p362). Retinal vessels may be affected and can be reduced to thread-like proportions associated with optic atrophy (Fig. 18.5 p362). Thrombo-phlebitis of the legs and aneurysms of large arteries may occur. Antibodies to mucosa from several sites including the mouth have been demonstrated but tests for rheumatoid and antinuclear factor are usually negative. There is also evidence of cell mediated delayed hypersensitivity. Treatment is with immuno-suppressants such as corticosteroids, colchicine, thalidomide and cyclosporin.

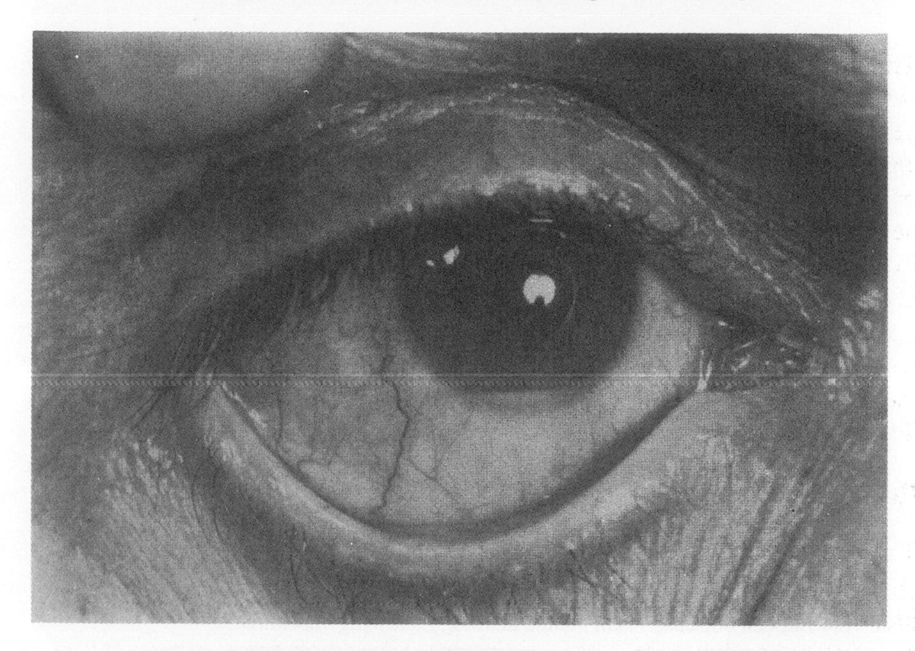

Fig. 18.2 Behçet's disease - vasculitis affecting the uvea resulting in hypopyon.

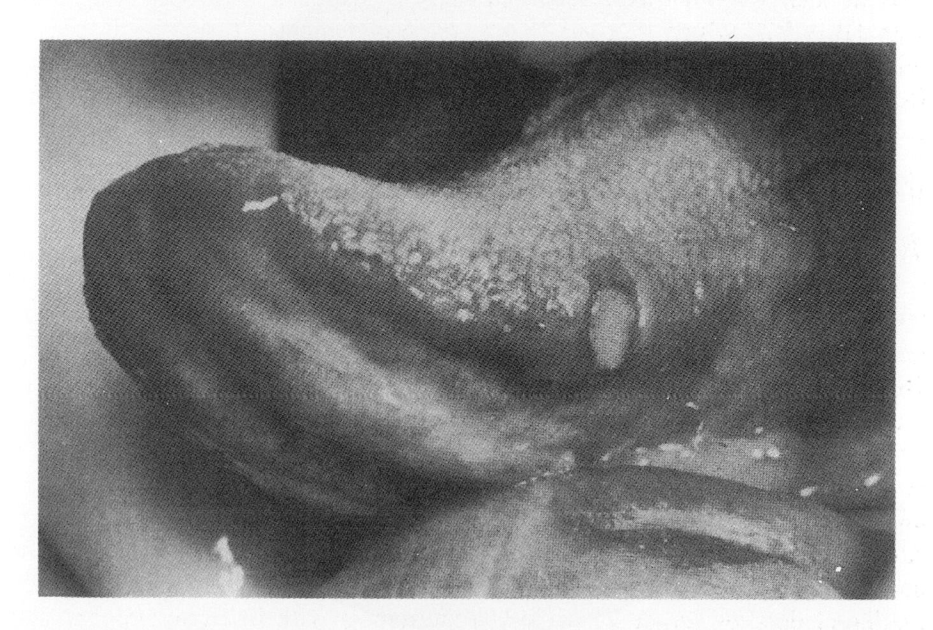

Fig. 18.3 Behçet's disease - buccal ulcers.

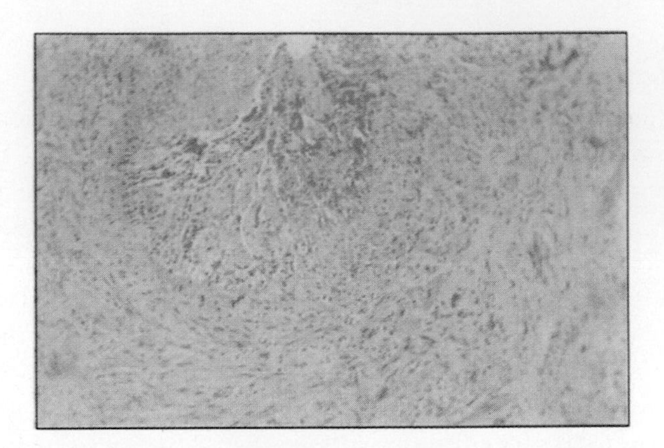

Plate 18.4 Behçet's disease - skin nodule histology showing vasculitis.

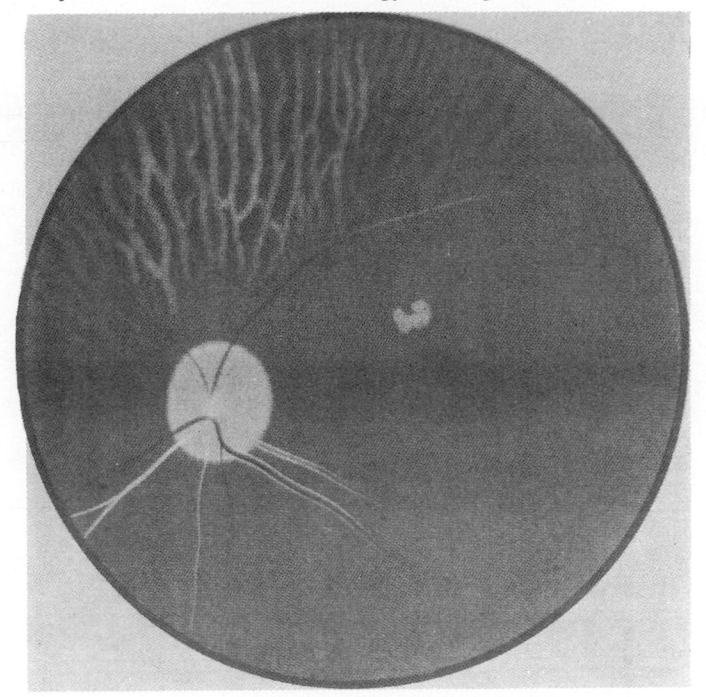

Fig. 18.5 Behçet's disease - obliterative retinal vasculitis and optic atrophy.

## Systemic lupus erythematosus

Systemic Lupus Erythematosus (SLE) is a multisystem, inflammatory disease characterised by the presence of autoantibodies leading to the deposition of immune complexes in the small blood vessels with resulting local fibrinoid necrosis due to complement activation. Its prevalence is 4-250/100,000 and is more common and severe in Asians and black Americans, affecting women predominantly (female to male ratio 10:1). Diagnostic features include a macular and discoid skin rash, photosensitivity, oral ulcers, arthritis, serositis, CNS disease such as diplopia, nystagmus and psychosis, haemolytic anaemia, leucopenia, thrombocytopenia, and immunological disorders involving the LE cell, anti-dsDNA, anti-Sm and antinuclear antibodies. Renal involvement makes the prognosis poor. The tissue damage mainly results from activation of complement (p302). Antinuclear factors can be demonstrated by immunofluorescence in almost all patients. The ESR is raised and serological tests for syphilis may be falsely positive.

Kerato-conjunctivitis sicca is frequent and retinal vasculitis with 'cotton wool' spots and haemorrhages is characteristic (Plate 18.6 p363). A retinal arterial vasculitis can occur with evidence of arterial obstructions. Fluorescein angiography may reveal retinal ischaemia and neovascularisation. Treatment is dependent on the severity of the disease, and includes in mild cases hydroxychloroquine, and later mepacrine, azathioprine and cyclophosphamide.

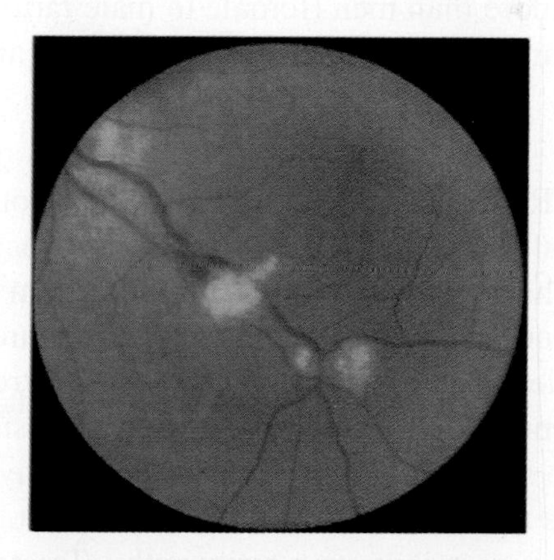

Plate 18.6 Retinal effects of systemic lupus erythematosus.

**Polyarteritis nodosa** (p518)

This is a widespread necrotising vasculitis affecting the small and medium sized arteries predominantly in middle aged men, The lesion is a fibrinoid necrosis. There is evidence that it may be a virus induced immune complex disorder and Hepatitis-B surface antigen has been found in 20-40% of cases. Fever and arthralgia with marked leucocytosis, eosinophilia and a raised ESR may be present. Myocardial infarction, pulmonary infiltration and renal disease with hypertension may be present, and mortality is high from complications such as aneurysms, thromboses and infarctions. About a third of patients show the cotton wool spots and haemorrhages typical of retinal vasculitis and in addition hypertension secondary to renal involvement may contribute to the fundus picture. Uveal, scleral and corneal necrosis may occur. Subarachnoid haemorrhages can complicate the clinical picture and there may be cranial nerve palsies and lesions of the visual pathways which may lead to visual field loss. Treatment is with steroids with or without cyclophosphamide, and plasmaphoresis.

**Progressive systemic sclerosis (scleroderma)**

This is a chronic disease mainly affecting women, with sclerosis of the skin which becomes leathery. It has an incidence of 18/million/year in the UK. Women are affected more than men (female to male ratio 3:1). There is widespread vascular disturbance, associated with fibrosis and activation of the immune system.

Patients may exhibit Raynaud's phenomenon, or cutaneous systemic sclerosis which may be generalised or limited. Complications include interstitial lung and renal disease. There is a moderately raised serum immunoglobulin in many patients and rheumatoid and antinuclear factors are frequently present. In particular, antibodies to scleroderma-70 and anti-centromere antibodies are present. Tissue systemic arterioles show fibrinoid necrosis with little lymphocytic infiltration. The systemic arterioles appear to be hyper-reactive. Due to sclerosis the lids are tight. Tear secretion is deficient. Hypertension from renal involvement may contribute to a retinopathy. Treatment is with D-penicillamine, interferon, cyclosporin and methotrexate.

## Dermatomyositis

This is a rare disease of middle life affecting the skin and causing inflammation of muscles. In many cases it occurs in association with a neoplasm. There is oedema of the lids which exhibit a reddish brown erythema (heliotrope). Diplopia and a retinopathy due to vasculitis have been described.

## Sarcoidosis

This systemic disease of unknown aetiology with widespread granulomatous lesions shows a uniform histological picture of epitheloid cell non-caseating follicles which undergo resolution by hyalinisation followed by fibrosis. (Fig. 18.7 p365)

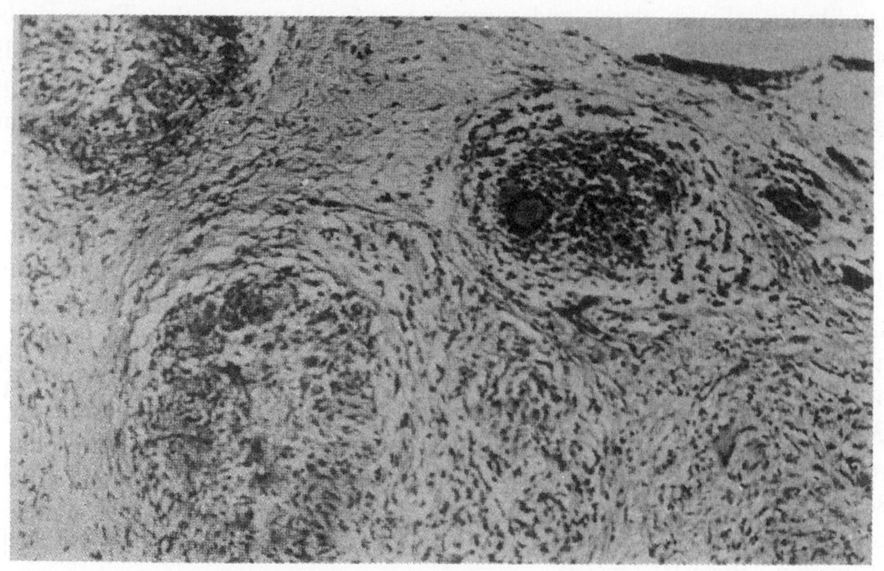

Fig. 18.7 Sarcoidosis - conjunctival histology. Non-caseating aggregations of epithelioid cells, some with giant cells.

Sarcoidosis is believed to be a malformation of T-lymphocytes and patients with sarcoidosis have long been known to give a negative Mantoux reaction. It is more common in females over the age of 15 years who may present with lung disease due to pulmonary infiltrates, bilateral hilar lymphadenopathy and skin or eye lesions.

*Uveitis* occurs in about a third of patients (Plate 7.23 p177) which has no special features in many cases but characteristically shows cells in the aqueous and translucent keratic precipitates of assorted sizes in both eyes. About 12% of uveitic cases have sarcoid nodules in the iris (Koeppe's and Bussaca's nodules) which may obstruct the angle of the anterior chamber. Acute uveitis is more common, and is frequently associated with bilateral hilar lymphadenopathy and erythema nodosum. Chronic uveitis can occur with lupus pernio, bone cysts and pulmonary fibrosis. The appearance of the eye is frequently almost normal despite an appreciable degree of uveitis and sometimes even when accompanied by secondary glaucoma. Posterior uveitis may occur, the lesions being mainly in the periphery. Ocular sarcoidosis may present as uveoparotid fever (Heerfordt's disease) in about 10% of cases in which a severe uveitis is associated with enlarged, infiltrated parotid and submandibular salivary glands and sometimes a facial nerve palsy. There is also lacrimal gland involvement with kerato-conjunctivitis sicca and occasionally swelling of the gland. The salivary and lacrimal gland swelling and the facial palsy usually subside after a few weeks but the uveitis can persist, requiring minimal steroid treatment over a long period. Such patients tend to have severe pulmonary and other symptoms.

The eyes may also be affected in sarcoidosis in respect of:

-*conjunctival follicles* of characteristic appearance especially in the lower fornix fold which may be a useful source of tissue for biopsy (Figs. 18.7 p365, 18.8 p367).
-*calcifications of the cornea and conjunctiva* due to a raised serum calcium associated with calciferol sensitivity (Plate 18.9 p368)
-*kerato-conjunctivitis sicca* from tear deficiency resulting from lacrimal gland infiltration. (Plate 10.2 p219)
-*vasculitis retinae* mainly affecting veins, causing irregularity in calibre and perivascular cuffing and pigment epithelial defects, both of which may be revealed by fluorescein angiography even when not clinically obvious.
-*orbital masses* which may cause proptosis, and paralytic squint.
-*meningeal sarcoidosis* where granulomas may also cause extraocular muscle palsies, diabetes insipidus, and affect the optic nerves. There may be raised intracranial pressure and papilloedema.

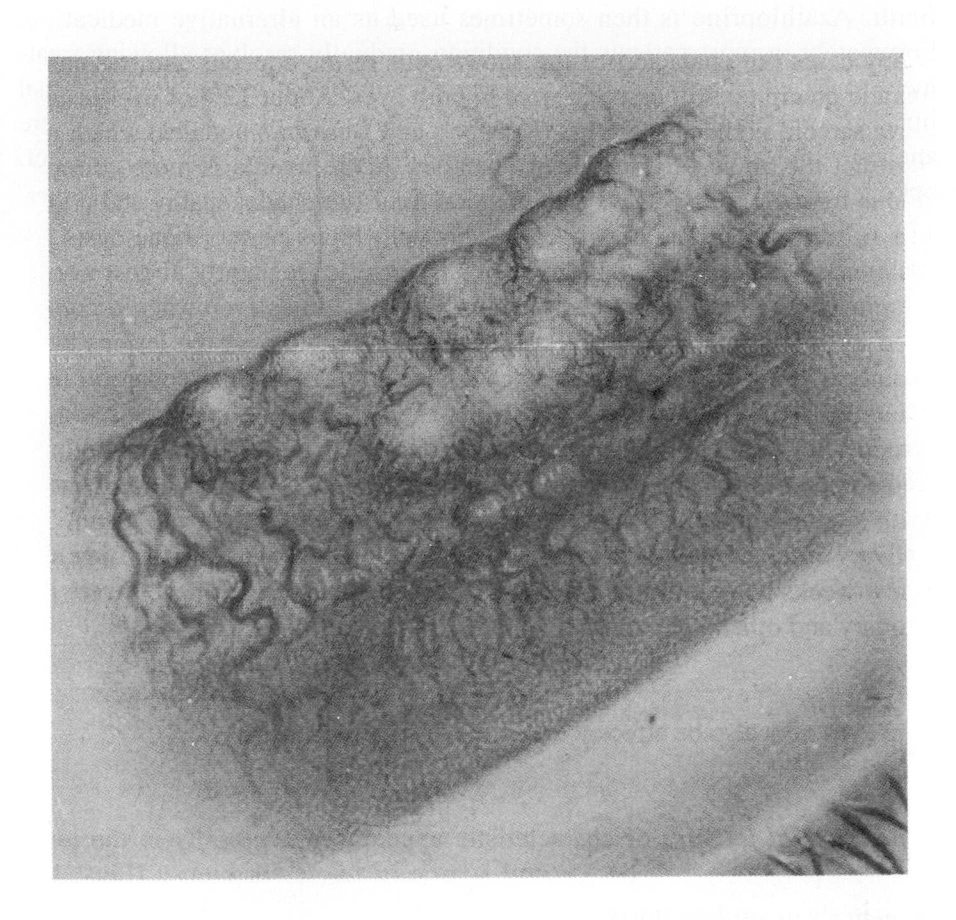

Fig. 18.8 Sarcoidosis - conjunctival follicles in lower fornix.

### *Treatment of ocular sarcoidosis*

Posterior uveitis and anterior uveitis of marked degree and those with vasculitis require systemic steroids. Topical steroids and mydriatics are used in mild anterior uveitis cases. Steroid therapy will have to be continued in the absence of resolution, although the side effects of steroids have to be weighed carefully against the degree of ocular inflammation and decisions can be difficult. Azathioprine is then sometimes used as an alternative medication. Fortunately in most patients the condition gradually resolves allowing treatment to be reduced progressively so that most patients emerge with minimal impairment of ocular function; very rarely the inflammation may be severe and fail to respond to treatment leading to secondary glaucoma, cataract, optic nerve involvement and loss of vision.

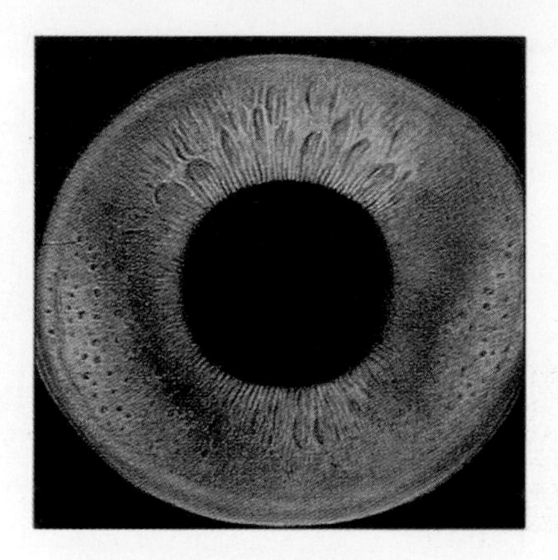

Plate 18.9 Sarcoidosis - chronic corneal 'band opacity'.

## Wegener's granulomatosis

This is a rare chronic disease in which a widespread necrotising vasculitis with granuloma formation affects the upper respiratory tract, lungs and kidneys. Men are affected more than women. Orbital granuloma, scleritis and retinopathy have been reported. There may be a widespread vasculitic rash. Diagnosis is based on the demonstration on cANCA antibodies which are found in more than 90% of patients. Immunosuppressive treatment, e.g.cyclophosphamide or azathioprine as well as corticosteroids may control the systemic signs and symptoms but the orbital granulomas are often resistant to all treatment.

**Stevens-Johnson syndrome** (pp203, 481)

**Giant cell arteritis (syn. temporal arteritis)** (p124)

## Myasthenia gravis

This condition is characterised by fatiguability of voluntary muscle due to failure of neuromuscular transmission. It occurs with a bimodal distribution in cases not associated with a thymoma, affecting patients aged between 10-30 and 60-70 years. Patients with a thymoma are usually middle aged women. Of the total number of cases, 10% have a thymoma present, and 75% are found to have an abnormal thymus. Acetylcholine receptor site antibodies are present and they lead to a reduction in the end plate potential of the neuromuscular junction. The disease may be divided into four groups: 1. Ocular myasthenia - where fluctuating extraocular muscle weakness causes diplopia and ptosis (Fig. 18.10 p370, LCW 7.18 p111) with evidence of fatigue., Ocular muscles may be the first to be involved in 65% of patients, and 90% of cases have ocular myasthenia at some stage of their disease. 2. Generalised myasthenia affecting all striated muscle. 3. Acute severe generalised myasthenia. 4. Chronic severe disease.

Anticholine receptor antibodies are found in 90% of generalised disease. Patients with a thymoma have anti-striated muscle antibodies in 90% cases as opposed to 30% in other patients. Diagnosis is confirmed by the *edrophonium (Tensilon) test.* This is given intravenously in a dosage of 2-10 mg. In positive cases improvement occurs in about one minute and lasts for two or three minutes. A Hess chart (Fig 11.15 p244) for eye movements should be recorded before and after Tensilon. Hyperthyroidism occurs in a small proportion of myasthenics and should be excluded as should bronchial carcinoma which may co-exist and give rise to the Eaton-Lambert syndrome. This is a presynaptic disorder due to the presence of IgG antibodies to the pre-synaptic nerve terminal. It is non-fatiguable and only responds to immuno-suppression. The Eaton-Lambert syndrome is associated with small cell lung carcinoma in 60% of cases.

Treatment of myasthenia gravis is by anticholinesterases (neostigmine, pyridostigmine) which are graded to produce the appropriate degree of effect, this will vary because the condition is subject to remissions. In moderate disease steroids, azathioprine and plasmapheresis are used. Surgical thymectomy is performed on patients with abnormal thymus glands.

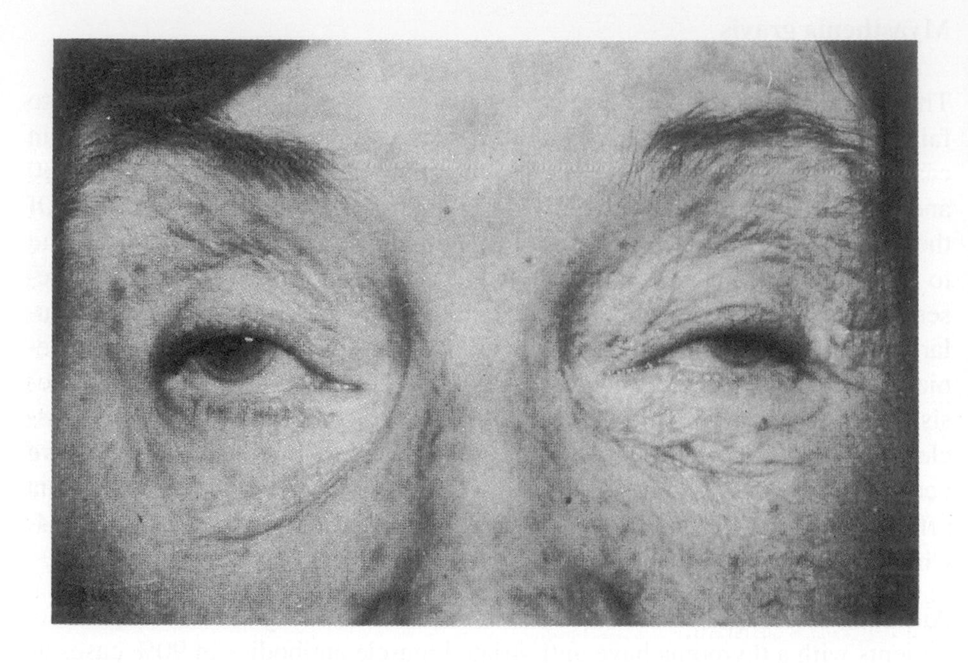

Fig. 18.10 Myasthenia gravis - ptosis and ocular deviation.

**Graves' disease** (p307)

**Diabetes mellitus** (p312)

# CHAPTER 19

# NEUROLOGY

*The investigation of neurological disease*

The progress in technology which has made possible new forms of investigation has been of particular advantage in neurology. The advent of magnetic resonance imaging (MRI) and ultrasonography, combined with computerised tomography (CT) and other improvements in radiography and angiography, now provide the physician with the means of revealing hitherto inaccessible soft tissue lesions in the central nervous system and their relationship to adjoining structures. Previously the presence of these could only be inferred and the advance has brought great benefit to patients and relief to their medical advisers, both neurologists and ophthalmologists. Specifically in ophthalmology, the laser scanning ophthalmoscope now allows optic disc and retinal conditions to be investigated and monitored with great accuracy. This includes the thickness of the retinal nerve fibre layer and the topography of the optic disc in the assessment of open angle glaucoma. The study of visual function is also leading to practical applications, such as contrast sensitivity and colour vision tests, blue on yellow and motion displacement perimetry and electrophysiological tests. It promises to improve both ophthalmic investigation and treatment.

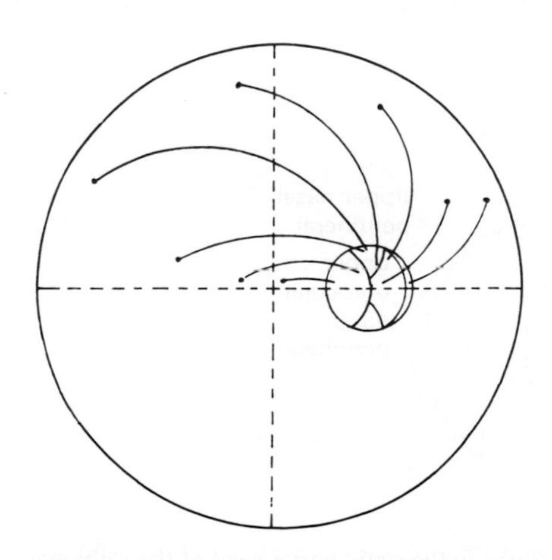

Fig. 19.1 The retinal nerve fibres of the right eye showing the course of a few selected ganglion cell axons. (Note the crescent of fibres on the nasal side of the optic disc subserving the unpaired temporal field)

*The practical anatomy of the optic nerve*

The retina contains the sensory cells for vision, the rods and cones, which generate nervous impulses in response to light (p263). These are transmitted by the bipolar cells to stimulate the ganglion cells. The response to light is modified by a previous stimulus *(successive contrast or temporal induction)* and, by the horizontally connected cells, it influences the response in the neibouring retina *(simultaneous contrast or spatial induction)* (pp267, 269). The axons of the ganglion cells pass in a regular pattern to the optic nerve (Fig. 19.1 p371). It will be noted that the central retinal fibres occupy the greater part of the temporal side of the optic disc, hence temporal pallor of the disc is seen when these fibres are destroyed by inflammation or degeneration. A variety of conditions affecting the retina will also interfere with its nervous function. These are considered together in Chapter 29 (p517).

The optic nerve contains visual fibres passing to the lateral geniculate bodies and pupillary fibres which form the afferent part of the pupillary reflex arc running to the tectum of the mid-brain. There are other nerve fibres the function of which is being intensively studied. The fibres lie in a glial framework supported by connective tissue septa. The optic nerve is described anatomically in four parts:

1. The *intraocular part* (Fig. 19.2 p372) begins at the optic disc towards which the axons of the ganglion cells converge to leave the eye through holes in a fenestrated layer of sclera (the *lamina cribrosa)* . An understanding of the

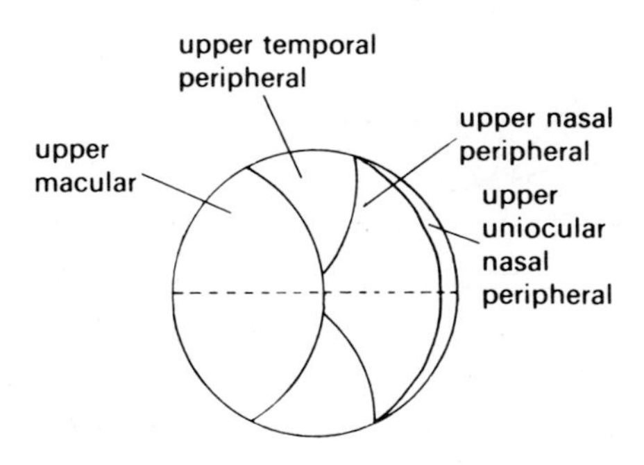


Fig. 19.2 The disposition of the retinal nerve fibres in the optic nerve head of the right eye.

*blood supply of the optic nerve head* as shown in Fig. 19.3 p373 is very important. The supply is almost entirely derived from the posterior ciliary arteries and only slightly if at all, from the central retinal artery via controversial recurrent branches, with the result that the retina and the optic nerve may be affected quite differently in vascular disturbances. There is a variable arterial plexus surrounding the nerve head which gives off branches to the nerve in a centripetal fashion. Certain segments of the disc may be more subject to ischaemia than others and this may result in the segmental or nerve fibre bundle type of visual field defect.

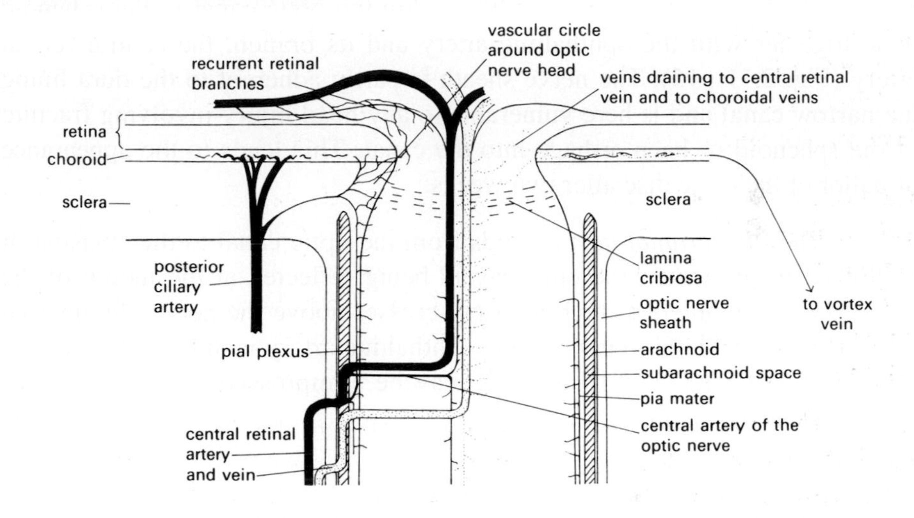


Fig. 19.3 The blood supply of the optic nerve head.

The centre of the optic disc is usually depressed, the *physiological cup*, so that the whiter lamina cribrosa may be seen at its centre. (Plate 31.8 p561, Fig 31.11 p563) This whiteness should not be interpreted as implying atrophy and attention must be directed to the pinkish white colour given by capillaries to the ring of disc tissue, composed of the non-medullated nerve fibres and neuroglia, which surrounds the central cup. The central retinal vessels and their main divisions are seen entering and emerging from the physiological cup, usually towards its nasal side. The intraocular part of the optic nerve is surrounded by the intermediary tissue, a glial condensation which separates it from the retina and choroid and by a connective tissue border layer between the nerve and sclera.

2. The *orbital part:* the optic nerve fibres normally have myelin sheaths proximal to the lamina cribrosa and the nerve runs a slightly sinuous course to the optic foramen. The central retinal artery and vein enter and leave the nerve by passing through the meninges and the cerebrospinal fluid space about 1cm behind the globe.

3. The *intracanalicular part* lies within the optic canal in the sphenoid bone together with the ophthalmic artery and its branch, the central retinal artery (Fig. 2.11 p15). The nerve sheath is partly adherent to the dura lining the narrow canal and is here vulnerable to any head injury involving fracture of the sphenoid or haemorrhage into the canal. This leads to the appearance of pallor of the optic disc after two weeks.

4. The *intracranial part* extends from the optic canal to the chiasma. It is clothed in pia mater, the other layers being reflected at the mouth of the optic canal. The anterior cerebral artery crosses above the nerve, the internal carotid artery is lateral to it while the ophthalmic artery is infero-lateral (Figs. 19.4 p374, 19.5 p375, 2.10 p14). It may be compressed by aneurysms of these vessels.

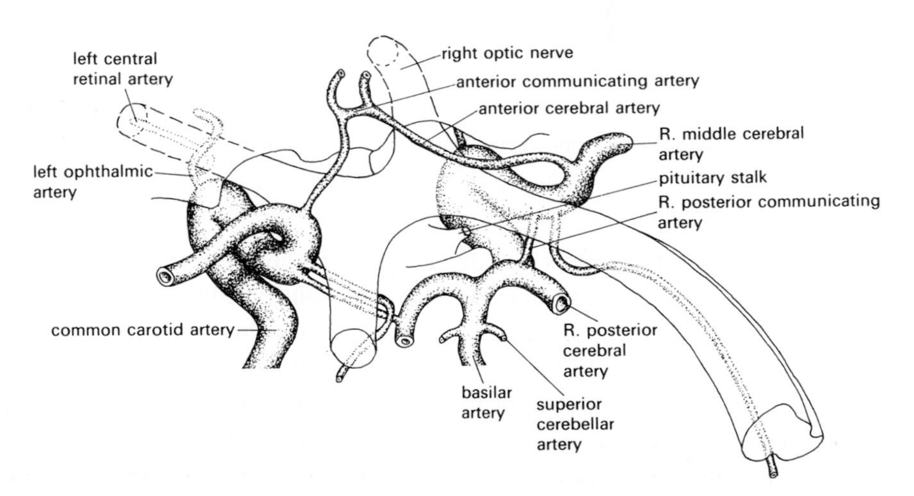


Fig. 19.4 Arteries in relation to the optic chiasma - redrawn after John Patten.

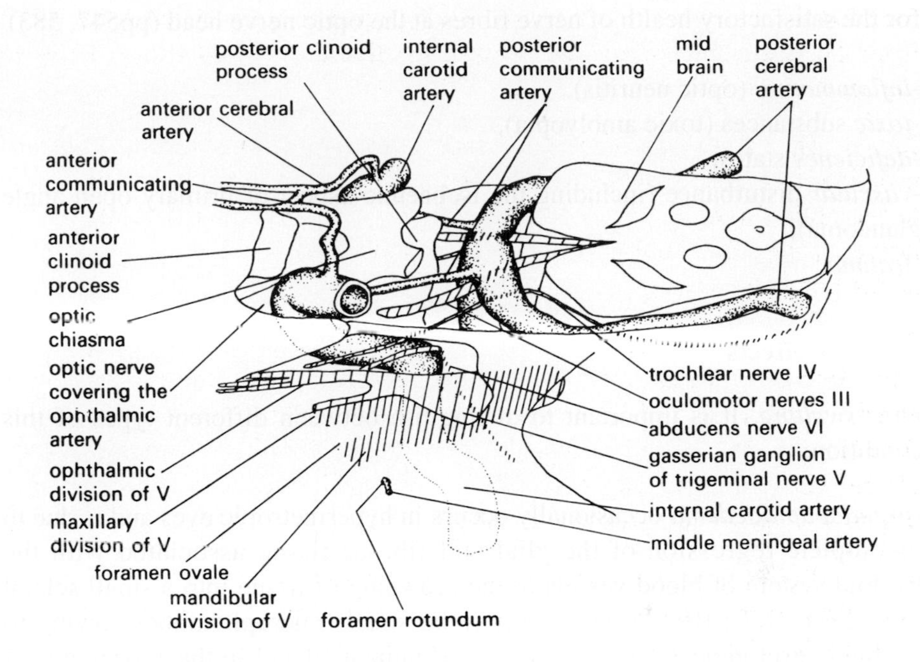


Fig. 19.5 A general view of the chiasmal region and related structures from the left side.

## Optic nerve disease

This is revealed by visual impairment accompanied on ophthalmoscopic examination by one or more signs of swelling and cupping of the optic disc and pallor of its neuro-retinal rim. The thickness of the retinal nerve fibre layer is revealed by the laser scanning ophthalmoscope.

*Types and causes of optic nerve disease (optic neuropathy)*

These may be classified as (p145):

*-pressure effects:* the basic lesion, e.g. tumour, may affect the optic nerve indirectly by causing raised intracranial pressure and thus papilloedema, or by direct optic nerve compression or by a combination of the two. In addition the intraocular pressure may be too high (glaucoma) or too low (hypotony)

for the satisfactory health of nerve fibres at the optic nerve head (pp547, 583)

-*inflammation* (optic neuritis),
-*toxic* substances (toxic amblyopia),
-*deficiency* states,
-*vascular* disturbance (including the ischaemic aspect of primary open angle glaucoma),
-*trauma.*

**Pressure effects**

*Disc swelling.* It is important to distinguish between different types of this condition:

-*pseudopapilloedema* occasionally occurs in hypermetropic eyes and is due to incomplete regression of the glial and fibrous tissue associated with the hyaloid system of blood vessels or the crowding of tissue into a small scleral ring (LCW 7.7 p108). Such eyes may be small (microphthalmos) owing to arrested development. There is no loss of intimate detail in the appearance of the disc as in true oedema and there are no haemorrhages. The retinal vessels appear normal and fluorescein angiography will confirm that there is no leakage of dye from the capillaries of the optic disc (p289).

-*drusen of the optic disc* are hyaline bodies occurring on or just deep to the surface of the disc (LCW 7.6 p108). They may be seen on the surface or may be buried, giving the appearance of general disc swelling, frequently with some irregularity of the surface. Drusen may occur as an irregularly dominant hereditary condition. It is often difficult to distinguish with confidence between drusen and optic disc oedema. Drusen-like bodies may also form in cases of extremely long-standing disc oedema. The diagnosis can be made with certainty only by fluorescein angiography when no leakage of dye from the vessels can be detected and the drusen themselves fluoresce in the late phase.

-*swelling of the nerve head due to raised intracranial pressure (papilloedema):* (Stereo Plates 19.6, 19.7 p377, LCW 7.3, 7.4 p107) raised intracranial pressure is transmitted by the cerebro spinal fluid in the optic nerve sheath to the optic nerve. It is hypothesised that here it exerts pressure on the small

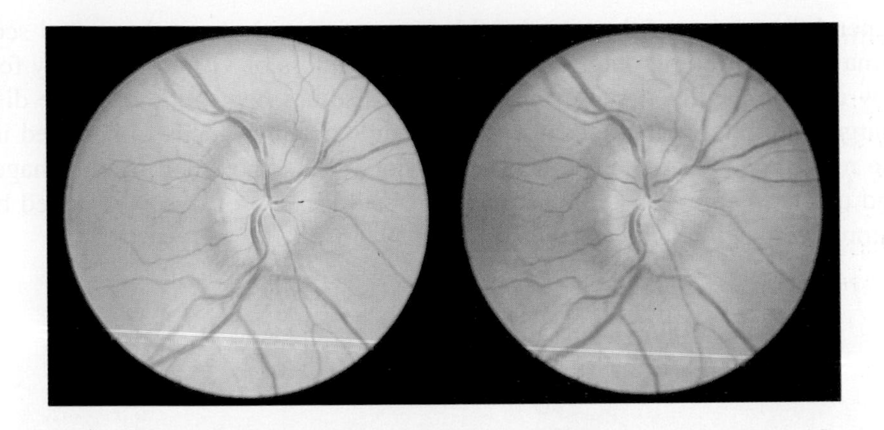

Stereo Plate 19.6 Papilloedema due to benign intracranial hypertension. (see p4)

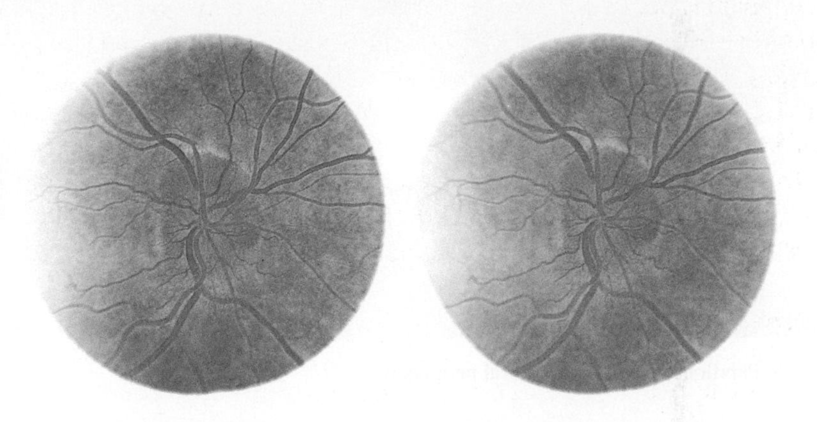

Stereo Plate 19.7 Papilloedema - fluorescein angiogram.
Same patient as Stereo Plate 19.6 (see p4).

vessels in the vascular circle at the nerve head and on the central retinal vein, thus causing a combination of vascular stasis and ischaemia which results in oedema at the nerve head, as well as the accumulation of substances involved in axonal flow. This swelling is called 'papilloedema' or 'plerocephalic papilloedema' to distinguish it from oedema from other causes. Vision is materially affected by papilloedema itself only when it has been present for several weeks or longer and then takes the form of transient obscurations of vision or general depression of the visual field, contrasting with the early central scotoma of disc swelling due to optic neuritis. Late loss of vision may sadly follow chronic papilloedema even after the cause has been relieved. The disc margins are blurred and elevated, the physiological cup tends to be filled in, the retinal veins which are congested do not pulsate and small haemorrhages and exudates appear at the disc margins. Late leakage of dye is revealed by fluorescein angiography (Figs 19.8 p378, 19.9 p379).

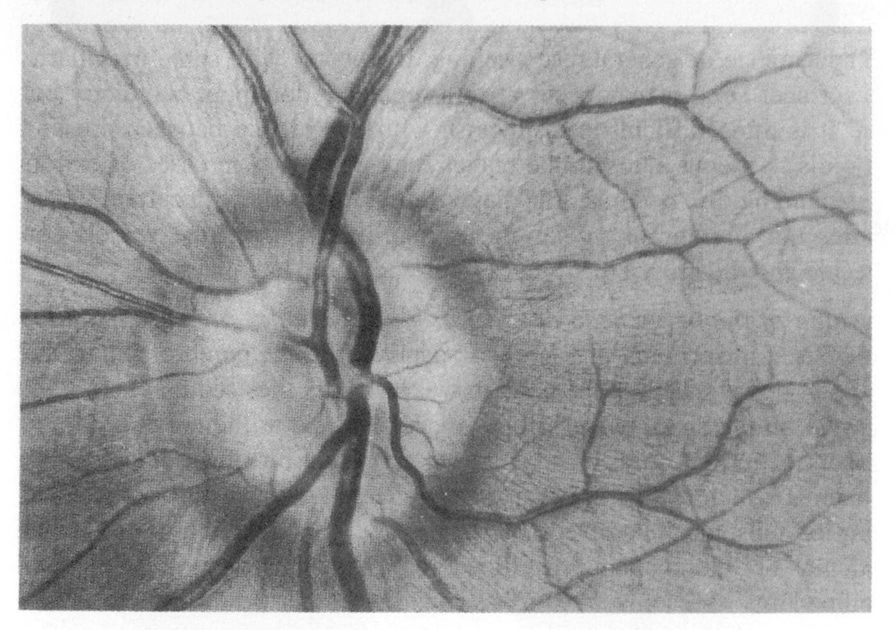

Fig. 19.8 Papilloedema due to cerebral neoplasm.

Disc swelling may be very pronounced in papilloedema whereas it is usually moderate in optic neuritis. The most common cause of raised intracranial pressure is a cerebral tumour but papilloedema may be a sign of benign intracranial hypertension. This may be caused by venous sinus thrombosis sometimes resulting from middle ear disease *(otitic hydrocephalus).*

*-raised intraorbital pressure* may produce similar signs of disc swelling in the case of Graves' disease (endocrine exophthalmos), orbital cellulitis, pseudotumour p535, trauma and nasal sinus mucocele or tumour, although the cause will usually be obvious. Signs of optic nerve involvement in these causes of proptosis are indications for immediate measures to reduce pressure in the orbit.

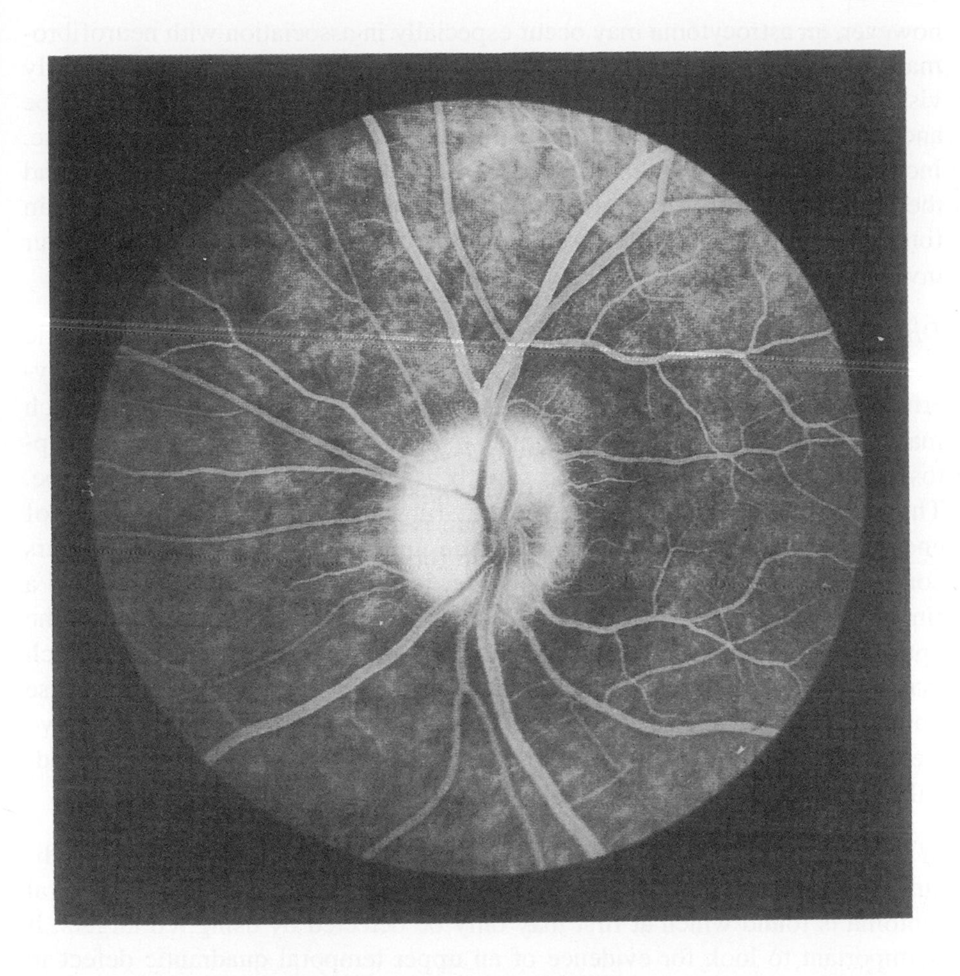

Fig. 19.9 Papilloedema due to cerebral neoplasm - fluorescein late stage of the fundus appearance of the disc shown in Fig. 19.8.

-*optic nerve tumours* may also be accompanied by disc swelling. The most important are tumours of the nerve sheath and the glioma of the optic nerve. Tumours of the optic nerve sheath *(meningiomas)* in the orbit also cause proptosis and loss of visual field and acuity although this may be only slowly progressive. Gliomas of the optic nerve usually lead first to optic atrophy but disc swelling may occur and in adults they carry a poor prognosis. In children however, an astrocytoma may occur especially in association with neurofibromatosis and can involve any part of the optic nerve or chiasma with early visual loss. It produces a fusiform swelling with axial protrusion of the globe and radiological enlargement of the optic foramen and does not metastasise. Increase in size may make excision of the nerve and tumour necessary and there is only a small recurrence rate. If possible the eye is allowed to remain for cosmetic reasons. Radiotherapy may also be used especially for a tumour involving the chiasma.

*Direct pressure on the optic nerve.* In the orbit this may result from optic nerve glioma or meningioma of the optic nerve sheath. Neurofibromas or cavernous haemangiomas may also rarely occur. A meningioma, (p538) which may be associated with neurofibromatosis, will cause appreciable axial proptosis before vision is affected and usually is situated close behind the globe. The absence of upper lid retraction distinguishes it from many cases of endocrine exophthalmos. By contrast the upper lid in optic nerve tumours comes forward with the eye but the lower fails to do so and so may reveal a rim of while sclera below the cornea. Trauma may cause a fracture across the optic canal associated with haemorrhage, oedema and thus pressure which may be difficult to demonstrate radiologically. Optic nerve damage may cause immediate loss of sight but pallor does not appear at the disc for almost two weeks. Pressure due to Paget's disease or a meningioma is usually very gradual. Craniosynostosis may also cause optic atrophy (p543).

In the case of *intracranial optic nerve compression* the onset is usually insidious. Optic atrophy may be present when the patient is first seen and a central scotoma is found which at first may only be detected by using red targets. It is important to look for evidence of an upper temporal quadrantic defect in the field of the other eye which, if present, is due to involvement of the lower crossing fibres from this eye as they loop forwards into the termination of the affected optic nerve, sometimes called the *anterior chiasmal syndrome.* The usual causes are a forward extension of a chromophobe pituitary adenoma (Fig. 19.10 p381), a meningioma of the sphenoid or the olfactory groove (Plate 19.11 p382, Fig. 19.12 p382) or an aneurysm of the carotid, anterior cerebral or ophthalmic arteries. The *superior orbital fissure (or orbital apex) syndrome* may be present with, in many cases, pupil signs and impairment of ocular movement indicating involvement of the 3rd, 4th and 6th cranial nerves and pain and anaesthesia over the distribution of the ophthalmic division of the 5th nerve. Horner's syndrome may also affect the pupil and upper lid due to a sympathetic lesion. Meningiomas of the sphenoid are diagnosed radiologically and by MRI and in some cases can be removed. They are radioresistant. The treatment of aneurysms and pituitary tumours is also a neurosurgical matter. Optico-chiasmal arachnoiditis may occasionally give rise to diagnostic difficulties and may respond to steroid treatment.

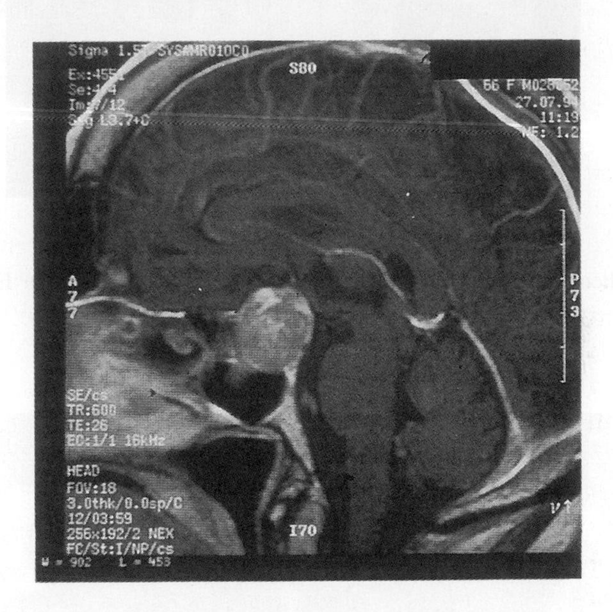


Fig. 19.10 Pituitary tumour - C.T. scan.

The *Foster Kennedy Syndrome* is characterised by optic atrophy in one eye and papilloedema in the other. The lesion is usually an olfactory groove meningioma, compressing one optic nerve and causing papilloedema on the other side because of raised intracranial pressure (N.B. papilloedema does not occur in a previously atrophic nerve head).

It is not possible to over-emphasise that *any central visual field depression which does not readily improve should be investigated thoroughly as a possible case of optic nerve compression.*

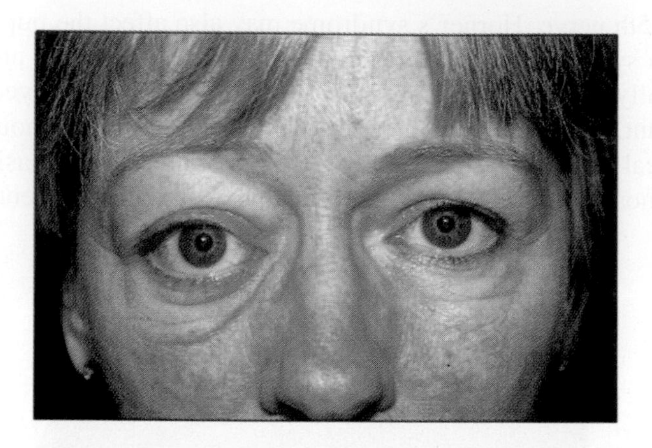

Plate 19.11 Sphenoidal meningioma - external (same patient as in C.T. scan Fig 19.12).

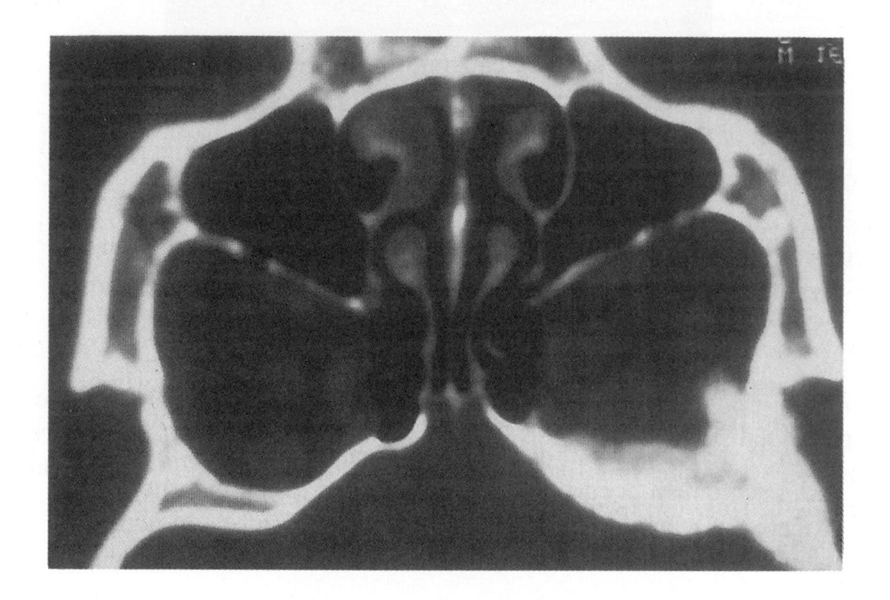

Fig. 19.12 Sphenoidal meningioma - C.T. scan (Same patient as Plate 19.11).

**Inflammation**

*Demyelinating disease* in its common form of acute optic neuritis is described on p146. Rarely it presents as neuromyelitis optica or encephalitis periaxialis diffusa. *Neuromyelitis optica* is a severe bilateral optic neuritis sometimes combined with a transverse myelitis (Devic) which is thought to be an unusual form of multiple sclerosis. The optic nerve lesions may follow but usually precede the myelitis. Considerable recovery may take place despite initial field loss. In *encephalitis periaxialis diffusa* (Schilder) loss of vision is due to both cerebral and optic nerve demyelination.

*Other inflammatory and infective causes of optic neuritis*. The optic disc may share in vasculitis or posterior uveitis whether of unknown aetiology (p179) or due to toxoplasmosis (p182), tuberculosis (p324), syphilis (p327) or associated with sarcoidosis (p366), Behçet's disease (p360) or virus infection such as measles (p332) or herpes zoster (p331). Infection may spread to the orbit from the paranasal sinuses causing optic nerve compression from the raised orbital pressure due to orbital cellulitis as well as the direct effect of inflammation of the nerve.

**Toxic causes of optic neuropathy (toxic amblyopia)**

*Toxic agents* may produce a bilateral optic neuropathy with early visual loss and these are often grouped together as the toxic amblyopias:

-*Tobacco:* usually strong tobacco smoked for many years by poorly nourished patients who frequently drink alcohol to excess. Many cases may be related to Vitamin B deficiency although the subject is controversial. Central and paracentral scotomas occur. It has been ascribed to the inability of the liver to detoxicate the cyanide ingested with the tobacco smoke in addition to the normal endogenous cyanide. It is treated by avoidance of tobacco and alcohol and by injections of hydroxycobalamin which combines with cyanide to form Vitamin B12 (cyanocobalamin).

-*Methyl alcohol* causes acute disc oedema and central loss of sight which is frequently irreversible and followed by optic atrophy.

-*Lead poisoning* may cause a toxic optic neuropathy with disc swelling and a central scotoma. It may also produce lead encephalopathy with raised intracranial pressure which in turn may cause papilloedema.

-*Leber's disease* (hereditary optic neuropathy) is an hereditary condition of severe bilateral optic neuropathy usually affecting young adults leading to optic atrophy and impaired central vision. It is included in this section because it has been suspected that it is due to the inherited inability to detoxicate endogenous cyanide, and hydroxycobalamin has been used in treatment.

Some *important therapeutic agents* may cause optic disc oedema:

-*Steroids* (p609): sight is little affected until post-papilloedema optic atrophy supervenes. Children receiving steroids systemically over long periods are particularly at risk and should be carefully watched for disc oedema. The mechanism is uncertain and it may appear on withdrawal of steroids making decisions regarding treatment very difficult,
-*tetracycline* (p603),
-*ethambutol* (p325, p622),
-*arsenic compounds,*
-*quinine,* even in small dosage in susceptible patients, may cause peripheral visual field loss associated with attenuation of the retinal arterioles (p621),
-Very rarely *salicylates* may also cause peripheral field loss.

**Nutritional deficiency conditions** (p351)

Optic neuritis may result from avitaminosis which is frequently a multiple deficiency. Vitamin B and especially thiamine are important. Thus it may accompany beri-beri and also occur in pellagra and pernicious anaemia (p354).

**Vascular causes of optic neuropathy** (p290)

*Ischaemic optic neuropathy* Atherosclerosis (p293) or giant cell arteritis (p125) (temporal arteritis) may cause occlusion of the small vessels at the disc derived from the posterior ciliary circulation. The resulting ischaemia causes disc pallor and mild oedema, which may be sectorial, often accompanied by small haemorrhages (Stereo Plates 19.13, 19.14 p385). The retinal vessels may appear normal except that they too show atherosclerotic changes. The field of vision may be grossly affected by sectorial or nerve fibre bundle types of defect.

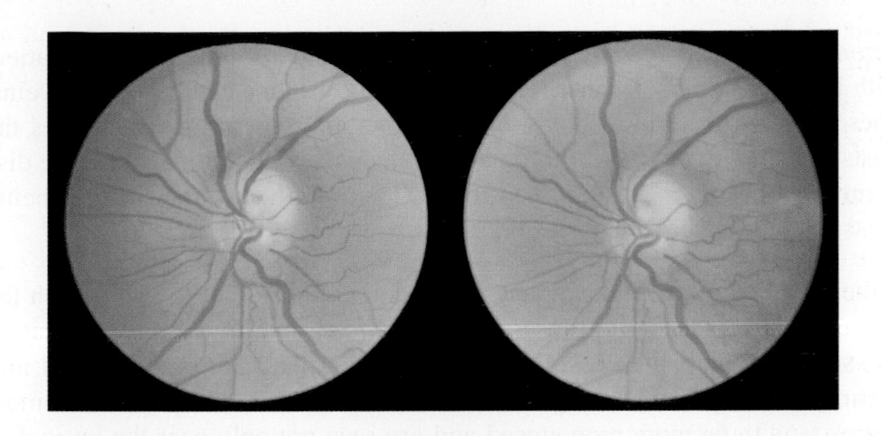

Stereo Plate 19.13 Ischaemic optic neuropathy. Giant cell arteritis. (see p4)

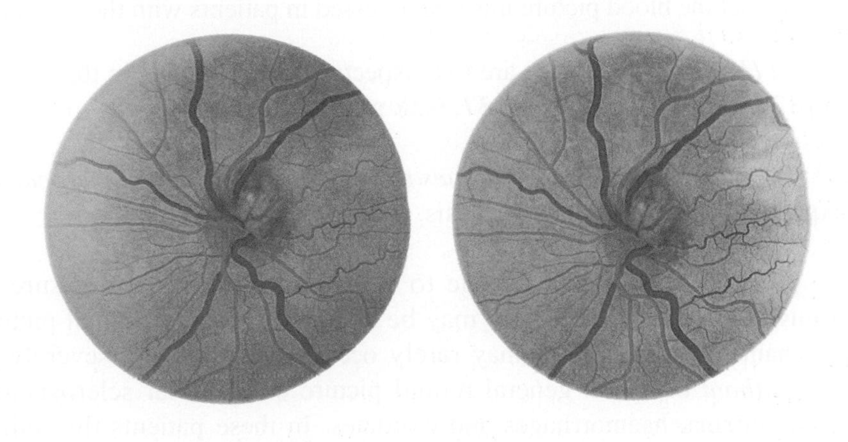

Stereo Plate 19.14 Ischaemic optic neuropathy. Giant cell arteritis - fluorescein angiogram (see p4).

*Retinal venous stasis or occlusion* (p121)

*Retinal arterial insufficiency*, especially in the elderly atheromatous patient with hypertension, may cause reduced blood flow in the capillaries and veins. These vessels dilate in response to ischaemia and this further intensifies the stasis which may proceed to thrombosis. Oedema of the retina and optic disc occurs and haemorrhages are seen in relation to the veins, the changes being gross in total retinal vein occlusion.

If the vein occlusion is incomplete it is sometimes difficult to distinguish the fundus appearances from those of papilloedema due to raised intracranial pressure in which the vessels on or near the disc may be engorged and accompanied by haemorrhages. In vein occlusions, however, the haemorrhages tend to be more pronounced and are seen not only near the optic disc but also along the course of the vein as far as the retinal periphery.

*Inflammatory changes* in the retinal veins with intimal proliferation, venous obstruction and thrombosis may give similar appearances in younger patients.

*Polycythaemia* may be accompanied by dilated retinal veins, haemorrhages and oedema so that the blood picture must be assessed in patients with these signs.

*Malignant Hypertension.* There are two aspects of disc swelling in this condition (pp127, 295) (Stereo Plates 6.37, 6.38 p127, LCW 7.2 p107).

1. It may be due to the *disc tissues sharing in the general retina oedema* consequent on retinal arteriolar necrosis.

2. It may be *papilloedema* due to the raised intracranial pressure of hypertensive encephalopathy. This may be associated with the retinal picture of malignant hypertension but may rarely occur acutely in very severely ill patients *without* the usual general retinal picture of arteriolar sclerosis and necrosis, oedema, haemorrhages and exudates. In these patients the retinal vessels are extremely attenuated in tonic contraction and acute renal failure is likely. Similar appearances may be encountered rarely in severe toxaemia of pregnancy.

**Trauma** (pp153, 540)

## Optic atrophy

### *General considerations*

The end result of optic nerve disease is a varying degree of optic atrophy as revealed by pallor of the optic disc. The quality of this appearance may indicate the cause of the atrophy. The normal very slightly pink colour of the optic disc, ignoring the white appearance of the lamina cribrosa at the bottom of the physiological cup, may be lost in a variety of conditions in which atrophy of the nerve fibres occurs. (Stereo plate 31.8 p561) When this is due to local ischaemia, it may be accompanied by widening of the normal disc cup owing to the collapse of the glial supporting tissue. Advanced cases are easily recognised; (LCW 7.8 p109) in others there may be difficulty in deciding whether the appearances are within the normal range. Visual field examination is always necessary. Sometimes slight disc pallor suggestive of optic atrophy may be seen when in fact none is present. This may be found in myopia in which the optic disc may also have a white scleral crescent, usually on the temporal side, where the retina and choroid do not quite extend to the disc margin. The physiological pit in the centre of the normal disc may be deep, revealing the white lamina cribrosa which may be interpreted as atrophy, but in these cases the pinkish sides of the cup will give a true indication of the absence of atrophy. A normal value for the thickness of the nerve fibre layer may also be revealed by the scanning electron microscope, taking into consideration the size of the optic disc. Hyaline granules (drusen) on or near the surface of the disc may give the impression of atrophic pallor, as may opaque medullated nerve fibres of characteristic white whiskery appearance radiating from the disc.

If the optic nerve fibres are involved suddenly in a lesion which destroys them, there is degeneration of the fibres and overgrowth of new glia in an irregular fashion and this fills up the gaps left by the degenerated fibres. It partly covers up the reduced fine capillary network to create the appearance of a flat white disc. If the original process is associated with oedema which may have led the disc to appear to overlap its usual boundaries, the pale disc will tend to have an indefinite margin. If the degenerated axons are replaced in an orderly manner by astrocytes then a pale disc of normal configuration results. But if due to apoptosis (p553), the degenerated axons are not so replaced, vacuoles are produced which tend to collapse and produce cupping of the disc *(cavernous atrophy)*. If in addition the intraocular pressure is raised, the pressure outwards on the walls of the cup tends to enlarge it, creating a sharper edge to the optic cup but the same level of pressure may have a varying effect in different discs.

*The clinical classifications of optic atrophy by ophthalmoscopic appearance*

It has been usual to describe optic atrophy as primary or secondary depending on ophthalmoscopic evidence. A pale disc unaccompanied by other abnormal signs was called 'primary' atrophy even if for example it was known to be secondary to tabes dorsalis or optic nerve compression. Otherwise the atrophy was called 'secondary'. If the disc margin was indefinite, suggesting former oedema, it was sometimes called 'post-neuritic' or if there was a fundus condition visible, such as pigmentary degeneration of the retina, it might be called 'consecutive atrophy'.

As there has been considerable confusion over the use of these terms it is proposed that a simple descriptive classification of optic atrophy is adopted. The colour of the neuro-retinal rim of the disc is always pale but the *margin* may be abrupt or indefinite, the appearance of the *retina* may be normal or abnormal and a *glaucomatous* type of cupping may or may not be present .

*Ophthalmoscopic types of optic atrophy*

*Abrupt margin, normal retinal appearance, no glaucomatous cupping* (Plate 19.15p389) as in optic nerve compression; following *retrobulbar* optic neuritis or neuropathy or optic nerve trauma; as part of a central nervous system disease such as a demyelinating condition, herpes zoster, tabes and G.P.I., cerebellar ataxia, peroneal muscular atrophy and in genetically determined optic atrophy. (This has often been called the 'primary' type).

*Abrupt margin, normal retinal appearance, glaucomatous cupping* as in: primary chronic open angle or primary chronic closed angle glaucoma (p560) (Plates 31.12 p563, 31.13 p564, LCW 4.11 p61) and some cases of late stage anterior ischaemic neuropathy (e.g. temporal arteritis) (p124) (This has often been called the 'glaucomatous' type).

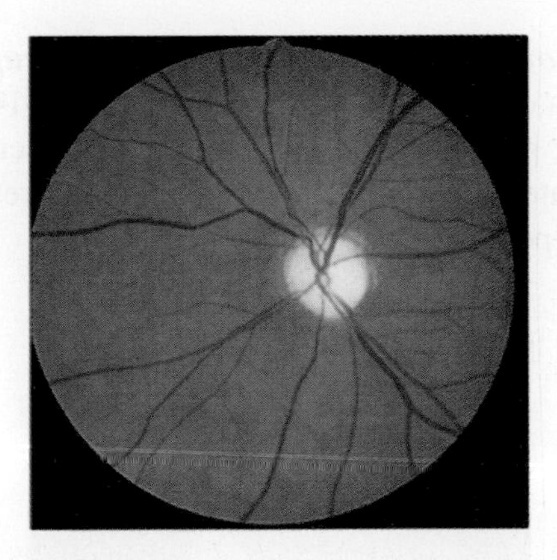

Plate 19.15 'Primary' optic atrophy.

*Variable margin, abnormal retinal appearance, no glaucomatous cupping* as in: scarring from choroidoretinitis (p179) or commotio retinae (p153), pigmentary degeneration of the retina *(retinitis pigmentosa)* with characteristic 'bone corpuscle' pigment deposits, waxy pallor of the disc and attenuation of the retinal vessels (p521) (Plate 19.16 p389, LCW 7.11 p109). Old central retinal artery occlusion with attenuated retinal arterioles of irregular calibre (Plate 6.30 p119, LCW 7.9 p109) (This has often been called the consecutive type).

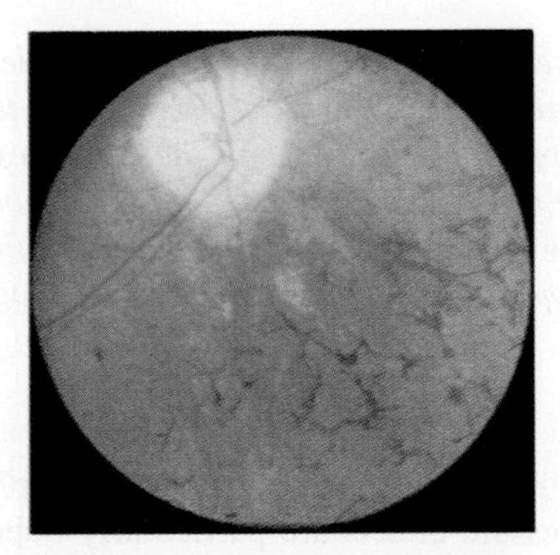

Plate 19.16 'Consecutive' optic atrophy (pigmentary degeneration of the retina).

*Indefinite margin, normal retina, no glaucomatous cupping* as: following anterior optic neuritis or neuropathy at the optic disc (p146, p384, LCW 7.10 p109) and following papilloedema due to raised intracranial or intraorbital pressure (p378) (Plate 19.17 p390) or hypertensive retinopathy (This has often been called the post-neuritic type).

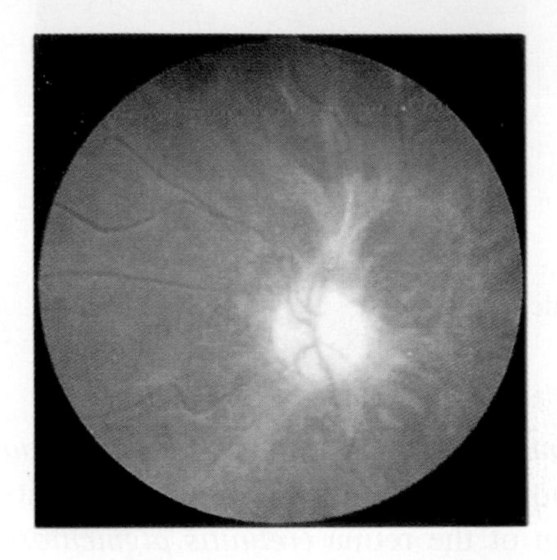

Plate 19.17 'Post neuritic' optic nerve atrophy (following papilloedema).

*Thus in a routine assessment of an optic disc, having decided that it is paler than normal in the part of the disc which is significant, the margin is inspected, glaucomatous cupping considered, and the fundus examined with special reference to the state of the retinal vessels and fundus background. A system of this type together with an examination of the visual field will rapidly indicate the category of optic atrophy and sometimes the diagnosis.*

*The optic chiasma*

The anatomy of this region is indicated in Figs 19.4 p374, 19.5 p375, 2.10 p14, 19.18 p391. Pressure effects are particularly important in this region although the chiasma may also be affected in demyelinating disease and chronic meningitis.

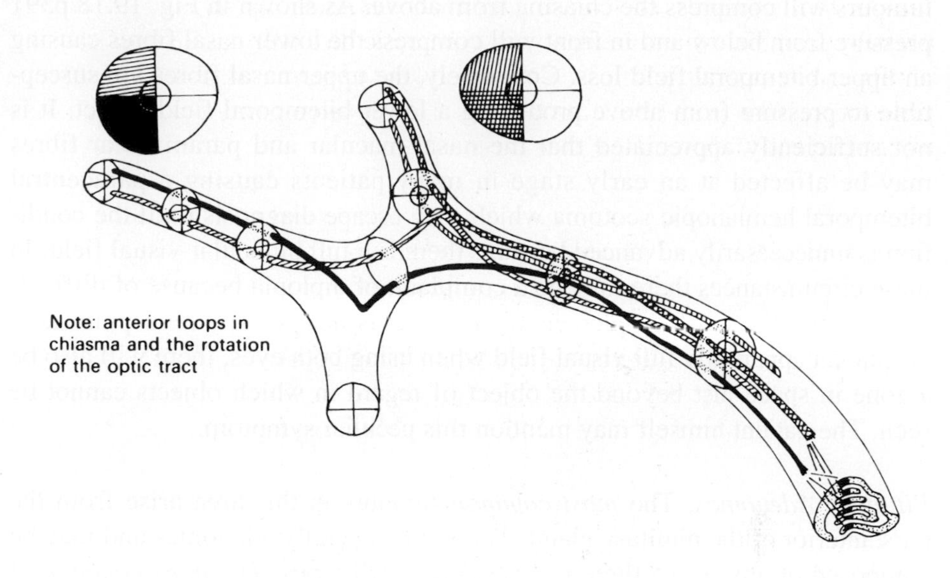

Note: anterior loops in
chiasma and the rotation
of the optic tract


Fig. 19.18 The axons of the retinal ganglion cells subserving the left homonymous hemifields, and their course through the optic nerves, optic chiasma and the right optic tract to the right lateral geniculate body showing the decussation of the nasal fibres, the anterior and posterior loops and the rotation of the optic tract. Redrawn after John Patten.

## *Chiasmal conditions of clinical importance*

The commonest lesion of the optic chiasma is due to *compression* by an expanding condition from within or above the pituitary fossa. The nature of the visual field defect so caused will depend upon the site of compression. As such lesions generally arise in the midline they characteristically compress the nasal retinal fibres as they decussate, causing a bitemporal hemianopic type of visual field defect. *Tumours*, however, which are the most common of such lesions, seldom arise symmetrically and some evidence of pressure on the outer temporal fibres is often found as well. In addition lateral extension of pituitary tumours may involve structures in the cavernous sinus with associated oculomotor and sensory disturbance. Tumours arising within the pituitary fossa will compress the chiasma from below, whereas suprasellar tumours will compress the chiasma from above. As shown in Fig. 19.18 p391 pressure from below and in front will compress the lower nasal fibres causing an upper bitemporal field loss. Conversely, the upper nasal fibres are susceptible to pressure from above producing a lower bitemporal field defect. It is not sufficiently appreciated that the nasal macular and paramacular fibres may be affected at an early stage in many patients causing a paracentral bitemporal hemianopic scotoma which may escape diagnosis until the condition is unnecessarily advanced because there is a full binocular visual field. In these circumstances there may be a complaint of diplopia because of difficulty in fusing the two complementary half fields near the fixation point and, despite an apparently full visual field when using both eyes, there will also be a zone in space just beyond the object of regard in which objects cannot be seen. The patient himself may mention this peculiar symptom.

*Pituitary Adenomas*. The *most common* tumours in this area arise from the pars anterior of the pituitary gland. These are generally adenomas and may be composed of any of the three main pituitary cell types. The *most common* of these is the *chromophobe adenoma,* which may secrete prolactin causing amenorrhoea and galactorrhoea in women and hypogonadism in men. It may also cause effects by compression of adjacent pituitary tissue (Fig. 19.10 p381). Less common is the *eosinophil adenoma* which may secrete excessive growth hormone, producing gigantism in children before fusion of the epiphyses, and acromegaly in adults after epiphysial fusion. Eosinophil adenomata may also eventually compress adjacent pituitary tissue with consequent hypopituitarism. The least common pituitary tumour is the *basophil adenoma* which can secrete excess ACTH, producing the hyperadrenocorticalism of Cushing's disease. The basophil tumours do not expand sufficiently to cause either chiasmal or pituitary compression.

*Craniopharyngiomas*. Tumours which compress the chiasma may arise from structures other than the pituitary gland. The craniopharyngioma is a carcinoma arising from remnants of Rathke's pouch, generally from above the sella turcica. They commonly occur in children and may not only compress the chiasma but may also compress the pituitary gland producing growth retardation and delayed puberty. They may calcify and some become very large causing hypothalmic effects, raised intracranial pressure, hydrocephalus and dementia.

*Other tumours*. Chordomas of the dorsum sellae arising from remnants of the notochord may invade the pituitary fossa and cause a clinical picture indistinguishable from a pituitary tumour as may an *aneurysm* arising anteriorly from the carotid which expands to occupy the sella. Other expanding lesions in this area include meningiomas and secondary deposits.

*Granulomas*. Rarely granulomas may occur in conditions such as tuberculosis, sarcoidosis and histiocytosis.

*Aneurysms* (p397). Chiasmal compression may be caused by supraclinoid aneurysmal dilatation of the carotid artery at the point of bifurcation into middle and anterior cerebral arteries. The visual field defects are variable but usually have bitemporal features.

*Investigations*

The objectives of investigations in a patient with a field defect suggestive of chiasmal compression are:

-The demonstration of expansion of the pituitary fossa including where possible:

-the nature and extent of such a lesion,

-the assessment of pituitary function.

*Skull X-rays* . Plain lateral views of the skull may be sufficient to demonstrate expansion of the pituitary fossa. Such views may also show suprasellar calcification which is suggestive of a craniopharyngioma. If expansion of the pituitary fossa is not apparent on a lateral skull X-ray, the finding of a double floor to the pituitary fossa in antero-posterior or lateral skull films or in a CT scan of the pituitary fossa region is a more sensitive index of an expanding pituitary lesion. The appearance of the ‘double floor’ is due to the asymmetrical enlargement of the tumour, with subsequent distortion of the floor of the pituitary fossa. Magnetic resonance imaging (MRI) may reveal the size, situation and character of soft tissue masses.

*Carotid angiography* . Carotid angiography may also be required in these patients to exclude an aneurysm.

*Pituitary function* . The assessment of pituitary function is necessary to investigate the possibility of hypopituitarism, and where appropriate, of acromegaly and hyperprolactinaemia. Although gonadotrophin function is lost first in hypopituitarism, impaired pituitary function is where appropriate, conveniently demonstrated by the response of serum growth hormone (GH) to insulin induced hypoglycaemia. In an individual with normal pituitary function, lowering the blood sugar to less than 40mg% or 2.2 mm/l, and sufficiently to cause the patient to sweat, will elevate serum GH levels to above 10u/ml. In hypopituitarism elevation of serum GH to such levels is not achieved. Conversely, the normal response to an oral glucose load (such as the standard oral glucose tolerance test) is a fall in serum GH to less than 5u/ml. Failure of serum GH to fall, or alternatively, the finding of a paradoxical rise in serum GH during an oral glucose tolerance test is diagnostic of acromegaly.

The management of pituitary dysfunction both medical and surgical is changing rapidly and reference to a neurologist and endocrinologist is necessary once the ophthalmologist has detected the problem.

**The optic tracts**

Optic tract lesions will produce homonymous hemianopic field defects. In the optic tracts, fibres from corresponding points in each retina are not adjacent so that defects do not affect exactly the same part of the visual field of each eye. Hemianopic field defects of this type are described as being incongruous. The tracts also undergo an inward rotation as they pass backwards. This makes it likely that pressure inferomedially on the tracts will produce lower incongruous homonymous hemianopic field loss and not upper defects as would otherwise be expected because, as can be seen in Fig 19.18 p391, the lower crossed fibres rotate away from inferomedial pressure while the upper uncrossed fibres become even further from this pressure. Tract pressure is uncommon but may occur with both pituitary tumours and craniopharyngiomas if extending posteriorly.

## The lateral geniculate bodies

In the lateral geniculate body the nerve fibres from corresponding points come to lie in the same sector but are represented in all the layers, the distribution being as shown in Fig. 19.18 p391. Lesions from the lateral geniculate body onwards will produce more nearly congruent field defects although in the optic radiations, particularly the lower part, there is some separation of corresponding fibres producing incongruity. This becomes less as the occipital cortex is approached and cortical lesions are exactly congruous.

## The optic radiations

As shown in Fig. 19.19 p396 there is a separation of the paths of the upper and lower relayed fibres of the optic radiation. The lower fibres loop forward into the temporal lobe and then pass backwards to join the more direct upper fibres in the parietal region before terminating in the visual (calcarine) cortex on the medial side of the occipital lobe above and below the calcarine fissure. These anatomical arrangements influence the symptoms and signs of intracerebral lesions whether vascular, neoplastic or of other causation. Frontal tumours may be silent for a long time with evidence eventually of raised intracranial pressure and character change without localising signs. A temporal lobe tumour may cause formed visual hallucinations and disturbances of taste and smell *(uncinate fits)* Theoretically it should cause a homonymous superior quadrantanopia but a hemianopic defect is more common, as it is with a parietal lobe tumour, although an occasional inferior homonymous quadrantic defect may be found due to the involvement of upper radiational fibres. Disturbances of spatial recognition and dyslexia may appear first. Occipital neoplasms may show an exactly congruous homonymous hemianopia with unformed visual hallucinations. They may cause raised intracranial pressure and papilloedema.

## The visual cortex

This cortical area (striate cortex, Brodman area 17) extends above and below the calcarine fissure and is identifiable macroscopically by a layer of white fibres dividing the fourth layer of cells, the *line of Gennari*. The posterior two thirds represents the macular area. The large central representation results

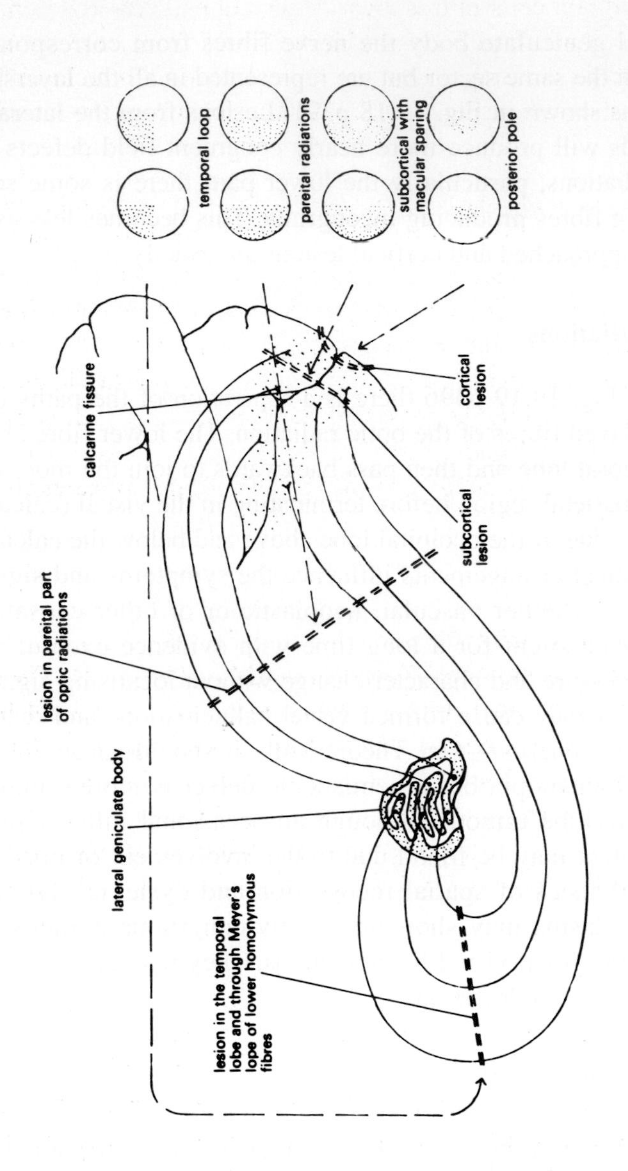


Fig. 19.19 The optic radiations showing the visual field loss corresponding to lesions in the temporal, parietal and occipital lobes.

from the density of cones at the fovea and the one to one relationship of cones and ganglion cells in this area. Associational para-and peristriate areas surround area 17 which also links with the frontal centre and the superior colliculi. Each striate area represents the homonymous half field of the opposite side, the part above the calcarine fissure representing the lower half field and vice versa.

Small lesions involving the posterior pole of the occipital lobe as in vascular disturbance or gunshot wounds may cause small central hemianopic scotomas. *Sparing of the macula* which is so often seen in lesions of the radiations or the occipital cortex is unexplained although it undoubtedly occurs. The presumption is that the lesion does not in fact destroy the entire radiation, as there is no evidence of special tracts which would duplicate the macular representation, and it may be in some cases a result of overlapping blood supply from the occipital and middle cerebral arteries. Anterior lesions of the radiations, the lateral geniculate bodies and the optic tracts cause more complete hemianopias.

**Cerebral aneurysms**

Cerebral aneurysms mainly arise from the circle of Willis and its branches or from the internal carotid artery (Figs. 19.4 p374, 19.5 p375). They are sometimes multiple. Many aneurysms remain undetected throughout life but they may suddenly rupture causing subarachnoid haemorrhage or they may present with focal signs. Most are congenital 'berry' aneurysms found at the bifurcation of the arteries. Atheroma is responsible for some of those affecting the internal carotid artery. Traumatic and infective aneurysms and those resulting from the lodgement of infected emboli (mycotic) are much less common.

*Carotid artery aneurysms*

*Supraclinoid aneurysms* arising from the terminal portion of the internal carotid, the anterior cerebral and the anterior communicating arteries may press on one or both optic nerves and the structures passing through the superior orbital fissure to cause visual field loss, sensory disturbances in the distribution of the trigeminal nerve and oculomotor paresis, i.e. the 'superior orbital fissure syndrome', similar to sphenoidal ridge meningiomas, although characteristically with a sudden painful onset (Fig. 2.11 p15). Occasionally there may be proptosis. Anterior cerebral and anterior communicating aneurysms may cause chiasmal compression with bitemporal field loss. A posterior cerebral aneurysm may rarely cause a homonymous defect. Papilloedema is a very unusual result of unruptured aneurysms.

*Infraclinoid or intracavernous aneurysms* give rise to sudden severe pain in or above the affected eye. The third nerve, including its pupillary fibres, the fourth and sixth nerves are also usually affected. Bruits may occasionally be heard by the patient or by the examiner. Posterior communicating aneurysms may involve the third and sixth nerves as may vertebral, basilar and posterior cerebral aneurysms though these are usually accompanied by brain stem signs.

Erosion of one or both clinoid processes, the carotid canal or the superior orbital fissure may also be seen (Fig. 19.4 p374). Arteriography allows the visualisation of the majority of aneurysms. Treatment is by carotid ligation (having ascertained that adequate collateral circulation is available) or by other neurosurgical techniques including aneurysm clipping.

*Subarachnoid haemorrhage* is most commonly caused by rupture of an intracranial aneurysm. The onset is abrupt with severe occipital headache and neck rigidity. Consciousness may be lost and death ensue. In those who do not lose consciousness photophobia and mental confusion may be marked. Papilloedema is frequent but variable in degree and retinal haemorrhages near the disc are common, sometimes being subhyaloid and these may break through into the vitreous (p292). Haemorrhage into the sheaths of the optic nerves may cause a varying degree of visual defect and lead to optic atrophy. The oculomotor nerves may be affected and the pupils dilated.

*Arteriovenous fistula* may result from rupture of an intracavernous aneurysm of the internal carotid artery. Mainly traumatic in origin, this may also result from congenital weakness. Occasionally gradual, the onset is typically acute with pain, loss of vision, pulsating proptosis and gross oedema of the orbit and lids with a loud bruit diminished by carotid compression on the side of the lesion. A sixth nerve palsy may occur but the other oculomotor nerves can also be affected. Having ensured that the alternative blood supply is adequate, ligation of the internal carotid artery in the neck, sometimes with simultaneous clipping of the internal carotid and ophthalmic artery intracranially, may improve the situation. Occasionally spontaneous thrombosis may occur but the fistula can also be blocked by balloon embolisation or other interventional procedures.

*Intracerebral haemorrhage* may occasionally result from rupture of an aneurysm.

**Migraine**

Migraine, which is often referred to as 'sick headache', occurs in about 10% of the population. It is more common in women and usually commences before the age of 20. There is a strong hereditary liability and possibly a tendency to a self-driving personality. A migrainous episode often follows a period of stress, exasperatingly just when the patient is starting to enjoy a relaxed interlude. Typically the aura of an attack appears as a blank hemianopic or quadrantanopic scotoma which is seen to be surrounded by a fine coloured zigzag of light. This shimmers at the rate of about 10 per second and expands for about 10-15 minutes until it appears very bright and white with large zigzags in the periphery, following which vision in the more central part recovers again. There may be aphasic symptoms and nausea and vomiting. The headache then follows which is characteristically throbbing and unilateral on the side opposite to the visual effects, although this may be a minor factor in older patients. It may be transient or last for some days. The scalp is tender over the side of the headache with dilated vessels and the ipsilateral conjunctiva may be suffused. There are many variants. Occasionally it may be bilateral with or without macular sparing and curious parietal disturbances of space perception may occur, with objects being transferred to other parts of the visual field, perhaps inverted.

A permanent scotoma or hemianopia has been recorded and a similar catastrophe may extremely rarely affect the eye itself with occlusion of one or both central retinal arteries. The attack may occasionally be associated with paresis of the third, fourth and sixth nerves with a dilated pupil, designated *ophthalmoplegic migraine*, and possibly also transient facial palsy, hemiplegia or monoplegia. These complications require full investigation for possible organic causes of secondary migraine such as cortical angioma, aneurysm or tumour.

Prophylactic treatment by sympathetic beta blocking agents is helpful and analgesics can be given during the attack. Ergotamine tartrate will reduce the vasodilation causing the migraine headache and may be used by mouth or suppository. Its use must be carefully regulated with due attention to the general cardiovascular state of the patient. It should not be given concurrently with sympathetic beta blocking drops or with sumatriptan.

*Sumatriptan* is a serotonin agonist and acts by the constriction of dilated intracranial vessels in migraine and has been a notable therapeutic advance, but again there are many cardiovascular conditions which contraindicate its use. Anti-emetics, such as the dopamine antagonist, domperidone, have also an important rôle in migraine therapy.

Sometimes a precipitating factor can be identified and then avoided, and for those who are too conscientious the calculated avoidance of stressful circumstances is sometimes possible. *Cluster headaches* affect mainly men in middle life, in phases recurring for several days. They may last for up to 2 hours and may cease quite quickly. As in migraine the headaches are unilateral and pain is felt in the eye and temporal region.

**Cerebrovascular insufficiency**

Occlusion of the carotid and vertebro-basilar systems account for a considerable proportion of cerebrovascular accidents. Incomplete occlusion leads to focal cerebral ischaemic episodes which abate after a matter of minutes. Such patients have a considerable risk of serious stroke. Early recognition of the situation may allow treatment to be given in time to prevent this.

*The carotid systems* supplies the internal capsule, the basal ganglia, the frontal, parietal and part of the temporal lobe. The chief symptoms of carotid artery ischaemia is transient loss of vision in one eye in the form of a sudden field loss varying from hemianopia to total loss. Vision gradually returns. Repeated attacks may cause cotton wool patches in the eye on the affected side due to small thrombi in retinal vessels. There may be showers of cholesterol emboli or platelet emboli from the affected vessels. All patients with carotid ischaemia who may be suitable for surgical treatment, should have carotid artery imaging with Duplex studies and MRI angiography when available.

*The vertebro-basilar system* supplies part of the temporal lobe, the occipital lobe, the cerebellum and the brain stem. It is linked to the carotid system by the posterior communicating arteries. Inadequacy of the basilar artery tends to produce bilateral signs and of the vertebral artery ipsilateral effects. Headache, vertigo, nausea, deafness, tinnitus and hemiparesis may occur. Episodes of blurred vision, diplopia, transient homonymous hemianopia and a migrainous type of shimmering scotoma may be premonitory signs of complete occlusion which when it supervenes gives rise to homonymous hemianopia, paresis of gaze movements, internuclear ophthalmoplegia and nystagmus. It can also cause Horner's syndrome. These transient ischaemic attacks may be confused with migraine, epilepsy, and Meniere's disease. A sudden decrease in blood pressure or change in pulse rate may be associated. Obvious cause such as temporal arteritis, hypertension, neuro-syphilis, systemic lupus erythematosus or polyarteritis should be excluded. Medical treatment to control hypertension or anticoagulants or aspirin to reduce platelet stickiness can be beneficial. Surgery to affected segments of vessels may be indicated in certain conditions. In some cases of stenosis of the first part of the subclavian artery, there may be reverse flow in the vertebral artery on the same side giving rise to vertebro-basilar symptoms on using the arm (the '*subclavian steal*' syndrome).

### Disorders of ocular motility

'Disorder of ocular motility' is broad term which is used to encompass such conditions as concomitant squint, paralytic squint, nystagmus and supranuclear movement disorders. Eyes with full movements, no deviation and binocular vision can be regarded as the ideal. Squint is due to some barrier to the reflexes which serve binocular vision and whether this is efferent, central, or afferent in type, management aims to overcome the barrier.

**Concomitant squint** (p221)

**Paralytic squint** (p239)

### Supranuclear gaze palsies

The final common nervous pathway for ocular movements begins at the third, fourth and sixth cranial nerve nuclei and its disturbances have already been considered (p246). Lesions affecting the motor pathway before these nuclei are reached are termed supranuclear or gaze palsies. The term gaze palsy is used because lesions at this level usually result in an inability to direct both eyes in a particular direction, e.g. 'upwards' or 'to the right'. These are called conjugate movements of gaze. The control of conjugate gaze, to the right or left, is under control of the frontal and occipital cortex via the pontine gaze centre. The saccadic (fast) movements are controlled by the frontal cortex and the pursuit or following movements by the occipital cortex. The frontal cortex on either side drives the eyes to the opposite side whereas the occipital lobes drive the eyes to the same side. The pathways in the control of gaze movements are (Fig. 19.20 p402, 19.21 p403):

-frontal cortex,
-occipital cortex,
-vestibular system (labyrinth, vestibular nucleus, cerebellum and stretch receptors in the neck muscles).

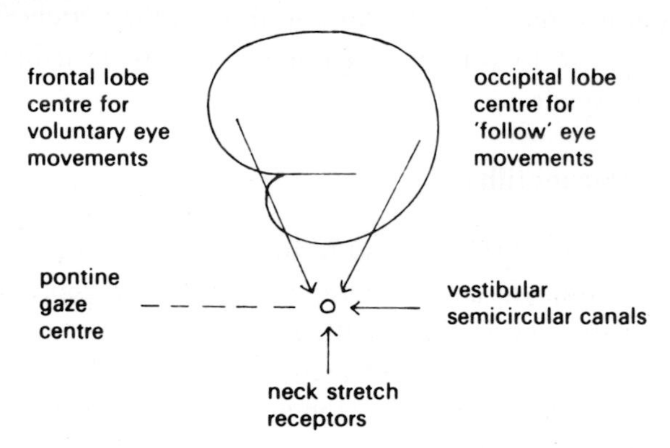


Fig. 19.20 Pontine gaze centre inputs.

It is believed that the tracts for conjugate lateral gaze arise from the frontal cerebral cortex. If one considers voluntary conjugate lateral gaze to the right, turning the eyes to this direction is initiated by the left cerebral cortex. The fibres pass from the cerebral cortex to the brain stem, crossing the mid-line to pass to the centre for lateral conjugate gaze situated in the pons. The pontine lateral gaze centre for right gaze is situated in the right side of the mid-line, i.e. the pontine gaze centres are responsible for lateral gaze to the same side.

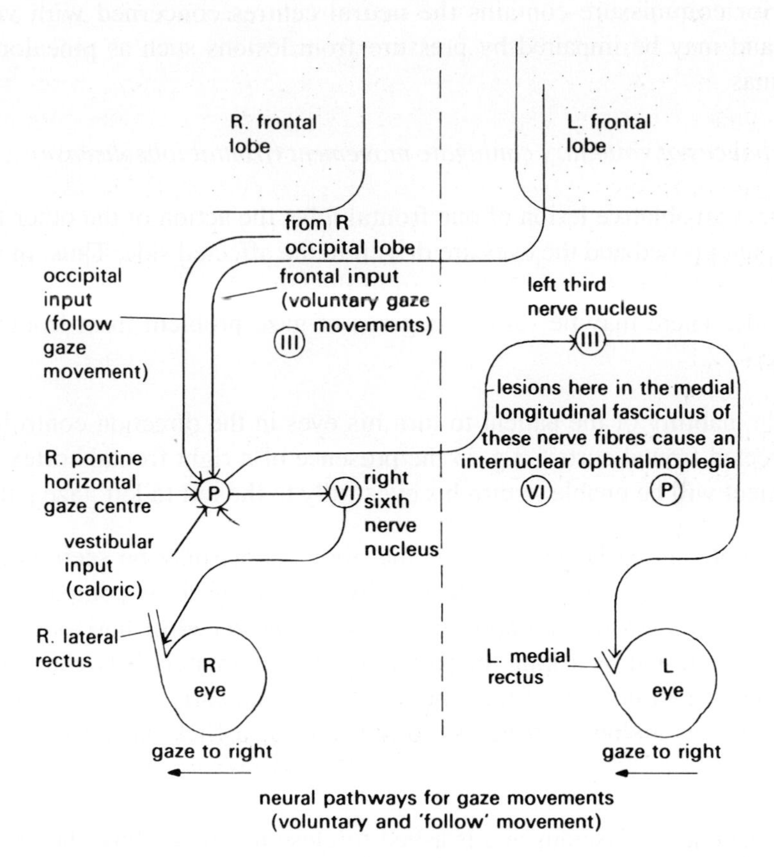


Fig. 19.21 Neural pathways for gaze movements (voluntary and 'follow' movements). The connections of the right pontine centre for horizontal gaze.

There are presumably differences between the pathways from the frontal and occipital cortex. Fibres must pass from the right pontine gaze centre to the right sixth nerve nucleus and then via the right sixth nerve to the right lateral rectus, so that the right eye can be turned to the right. Fibres must also pass to the left third nerve so that the stimuli can also reach the left medial rectus causing the left eye to move to the right. The relay between the pontine gaze centres and the third nerve nuclei on the opposite side is via the *medial longitudinal fasciculus.* The pathways concerned in vertical gaze and convergence are less well understood and do not pass through the pontine gaze centre. The posterior commissure contains the neural centres concerned with vertical gaze, and may be impaired by pressure from lesions such as pinealomas or teratomas.

*Disturbances of voluntary conjugate movement (frontal lobe disease)*

If there is an ablative lesion of one frontal lobe, the action of the other frontal lobe is unopposed and the eyes are driven to the affected side. Thus, in a right frontal lobe lesion the eyes fail to turn to the left side and so are driven to the right side. There may be various degrees of gaze problem from frontal lobe disease:-

-Simple inability of the patient to turn his eyes in the direction controlled by the affected frontal cortex, e.g. in the presence of a right frontal cortex lesion the patient will be unable to turn his eyes fully to the left (a left gaze palsy).

-More subtly in early gaze palsies the above signs may be seen as a *slow movement* of the eyes to the affected direction of gaze as compared to the turning of the eyes to the opposite direction. The speed of horizontal movement can be tested by asking the patient to look alternately between two fixation objects held in front of him about 40° apart; an early left gaze palsy may be detected as a slow drift across to the left but a flick back rapidly to the right.

-If the patient is conscious and is asked to close his eyes whilst the examiner holds the eyelids open, the normal *Bell's phenomenon* (elevation of both eyes with divergence) is *altered* in that the patient will also have a horizontal deviation of the eyes towards the affected frontal cortex.

-In the unconscious patient with a *serious ablative lesion of one frontal cortex,* it is found on examination, in addition to the associated neurological signs that are to be expected, that the eyes are driven in the direction of the lesion.

-Corollary: some lesions of the cortex may be irritative rather than ablative in which case the opposite situation occurs in that the eyes are driven away from the side of the lesion.

*Disturbances of following "pursuit" movements (occipital lobe disease)*

Normally an object can be tracked quite smoothly if it is moving slowly across the visual fields at less than 40° per second and these smooth following (pursuit) movements are mediated by the occipital cortex. Tracking an object to the right depends on the right occipital cortex so that *failure in this is almost always accompanied by visual field defect*, in this case a left homonymous hemianopia. If the following movement is defective, it breaks up into a series of jerks. The significance of a normal following movement in the presence of a defect of voluntary movement is that the lesion must involve the frontal cortex and not the pontine gaze centre.

*The tonic neck reflexes*

The stretch receptors in the neck provide another input to the pontine gaze centre via the tonic neck reflexes and they contribute to the maintenance of steady fixation in the presence of head movements. In the 'doll's head' test the patient's head is turned rapidly to one side. When turning the head to the right the eyes should turn to the left. It can be elicited in the unconscious patient. The 'doll's head' movement test (like the caloric test) helps to distinguish between hemisphere and pontine gaze palsies. In defects of vertical gaze when caloric testing (p411) is unsatisfactory, the 'dolls head' test will indicate the integrity of the final common pathway from the third and fourth nuclei.

*Internuclear ophthalmoplegia*

Internuclear ophthalmoplegia (INO) is a disorder of conjugate horizontal gaze due to a brain stem lesion interrupting the medical longitudinal fasciculus (MLF) or other pathways between the nuclei of the nerves which supply the extraocular muscles. The MLF transmits impulses from the pontine centre and sixth nerve nucleus to the contralateral third nerve nucleus. Lesion in the MLF therefore give rise to an ipsilateral failure of adduction (impaired signal to third nerve nucleus) with a contralateral abducting nystagmus. The cause of the nystagmus is not clear, but is thought to represent an error in the pulse-step control of the antagonist muscles. *The convergence response is usually preserved,* as these fibres run more cranially, *confirming the presence of normal medial rectus function.* Large pontine lesions may involve both MLF's giving rise to a bilateral INO, and this is often associated with demyelinating disease in young patients, whereas unilateral INO's are usually vascular in origin and occur in a more elderly population. Subtle INO's can be seen more readily by making the patient perform repeated rapid horizontal saccades, making the ocular imbalance or "dysmetria" more obvious.

*Variations of internuclear ophthalmoplegia*

Lesion of the medial longitudinal fasciculus close to the third nerve nuclei may extend to affect directly the nuclei themselves. In this case not only will adduction on attempted horizontal gaze be affected, but the convergence movements of the medial recti may be impaired, and a manifest divergent squint may be seen.

Large lateral pontine lesions may involve one pontine gaze centre and both MLF's resulting in no gaze to the side of the injured gaze centre, and only abduction of the contralateral eye on attempted gaze to the opposite side as a result of the INO, the 'one and a half syndrome'.

## Myopathies and disease or injury of muscle causing squint

*Myasthenia gravis* (p369)

*Myopathies*

Ocular myopathies such as chronic progressive external ophthalmoplegia, a mitochondrial transmitted myopathy, may result in grossly impaired eye movements. Diplopia is uncommon however, as there is usually symmetrical involvement of the eye muscles. The movement of the eye becomes more and more restricted and is often accompanied by a ptosis. There may be associated cardiac conduction defects and pigmentary retinopathy. In the majority of cases the only treatment possible for these patients is conservative measures to relieve the ptosis (if necessary), because surgery on the ocular muscles has little beneficial effect. Great care is required in treatment of the ptosis to prevent corneal exposure as orbicularis action is often weak. Sometimes raising the lid only on one side is desirable as diplopia may be precipitated by bilateral ptosis operations.

*Thyroid disease and squint*

Dysthyroid eye patients with Graves' disease often present with squint in addition to their proptosis and lid retraction (p309). Sometimes the squint can be the principal presenting symptom. The diplopia is usually worse in the morning, possibly due to orbital oedema from lying flat all night. Diplopia when it occurs is usually vertical, and it can be shown that the affected eye has restriction of elevation. The restriction of movement is due to infiltration and oedema of the muscles which subsequently become fibrotic and contracted. Surgery for the squint associated with dysthyroid eye disease is difficult and should be delayed until the degree of squint has become stable. Meanwhile prisms may improve the range of binocular single vision.

## Nystagmus

Nystagmus is an involuntary rhythmical oscillation of the eyes. It may be subdivided into two groups: jerk nystagmus and pendular nystagmus depending on the periodicity of the movement, and may involve horizontal, vertical and rotatory components.

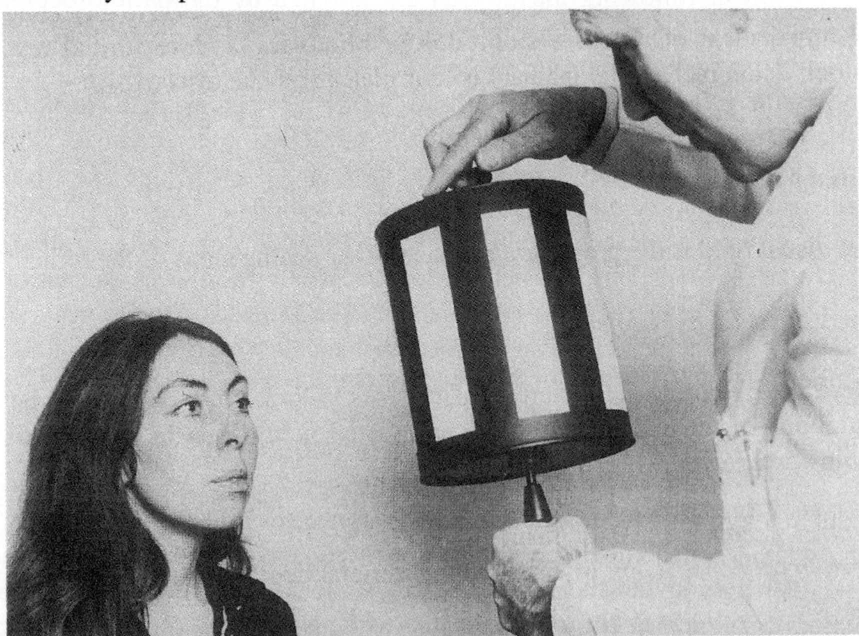

Fig. 19.22 A general purpose optokinetic drum.

*Jerk nystagmus*

Jerk nystagmus is an oscillating movement of the eyes which consists of a slow and a fast movement in opposite directions. To understand the type of movement involved one can consider the type of nystagmus known as *optokinetic nystagmus.*

*Optokinetic nystagmus* can be observed when looking at the eyes of a person viewing a series of similar objects passing, e.g. watching telegraph poles from the window of a moving train. The eyes fix on one pole and follow it as the train moves *(slow pursuit movement)*, but as the pole reaches the limit of vision, the eyes make a rapid re-fixation movement *(saccadic movement)* in the opposite direction to the previous pursuit movement to fixate on the pole coming into the field of vision. Thus, as long as the person continues to fix on the passing regularly spaced objects, the eyes will move rhythmically with slow movement in one direction and a fast movement in the opposite direction. Pursuit or following movements are initiated by the parieto-occipital cortex and are used to maintain the image of moving object on the fovea. The re-fixation movement is a saccadic movement initiated by the frontal cortex; it is extremely fast, with speeds of about 400° of arc per second. Optokinetic nystagmus can be demonstrated clinically by asking the patient to view a revolving drum painted alternately with black and white stripes. The test is useful in assessing ocular movements in both horizontal and vertical directions of gaze. Absence of optokinetic nystagmus, in one direction of gaze only, suggests a unilateral cortical lesion or an interruption in the pathway from the cortex to the brain stem nuclei (Fig. 19.21 p403). Demonstration of optokinetic nystagmus can also be used to indicate visual acuity (p80) and functional blindness. (p441and Fig 19.22 p407).

*Pendular nystagmus*

Pendular nystagmus differs from jerk nystagmus in that the oscillations in each direction are of equal speed, i.e. there is no fast or slow phase. The nystagmus is nearly always horizontal, but sometimes changes to a jerk pattern on extremes of gaze. It is often associated with a severely reduced visual acuity, e.g. as in albinism or one of the cone dystrophies.

*Nomenclature of nystagmus*

It is now accepted that the direction of nystagmus is described on the basis of the direction of the fast phase, i.e. if the slow movement is to the left and the correction fast phase is to the right, then this is described as nystagmus to the right or a right-beating jerk nystagmus. The amplitude of nystagmus increases when the patient looks in the direction of the fast phase, e.g. in a right-beating nystagmus the nystagmus increases in amplitude when the patient looks to the patient looks to the right and the nystagmus may even be absent on looking to the left or straight ahead.

Nystagmus may be congenital or acquired. Congenital nystagmus may occur as an isolated condition or in association with severe ocular or neurological disease. Acquired nystagmus is due to interruption of either the sensory afferent or the motor efferent pathways, most commonly the vestibular or cerebellar pathways.

*Disease processes which may cause acquired nystagmus*

*Labyrinthine disease:* the labyrinth may be damaged as a result of injury or inflammatory disease. The resulting nystagmus is usually horizontal or rotatory and is transient (see below).

*Central vestibular disease:* intrinsic lesions may directly affect the vestibular nuclei and these include multiple sclerosis, vascular accidents, gliomas and syringomyelia. The resulting nystagmus persists longer than that induced by labyrinthine disease and as the pathways from both sides may be affected, the direction of the horizontal nystagmus may alter with the direction of gaze.

*Cerebellar disease:* such as multiple sclerosis, infarcts and metastasis produce ipsilateral nystagmus, worse on gaze to the affected side (see below).

Tumours such as acoustic neuromas growing into the *cerebello-pontine* angle might be expected to cause vestibular nystagmus due to destruction of the pathways from the ipsilateral labyrinth, resulting in a fine nystagmus with the fast phase directed away from the lesion. However, in practice, the extending tumour also causes compression of the brain stem and the cerebellum, and the vestibular nystagmus is usually superceded by a coarse cerebellar nystagmus with the direction of the fast phase directed *towards* the side of the lesion. Then the amplitude of the nystagmus increases when the patient looks *towards* the side of the tumour.

*Localising nystagmus.* Some rarer patterns of nystagmus are associated with specific sites of neuroanatomical disease. *See-saw* nystagmus, with one eye deviating upwards whilst the other moves downwards, together with associated torsional movements, is often an indicator of chiasmal disease. *Periodic alternating nystagmus* is a form of nystagmus where the direction and speed of the horizontal movement repeatedly reduces with time and then reverses direction. This cycle may be punctuated by periods of relative stability, and may be associated with craniocervical disease. *Opsoclonus* describes a rare pattern of eye movements with seemingly chaotic conjugate saccadic movements in all directions. This may be associated with neuroblastoma in children, and metastatic visceral carcinoma in adults.

*Lesions in the region of the foramen magnum* involving distortion of the brain stem, e.g. meningiomas and Arnold Chiari malformations, cause a typical down-beating nystagmus.

*Lesions of the cerebral cortex,* particularly of the frontal cortex, cause defects of horizontal conjugate gaze. Nystagmus may sometimes be seen when the patient looks in the affected direction of gaze. This is the result of weakness of the horizontal gaze in that direction, the eyes tending to drift back towards the mid-line followed by a saccadic movement in the opposite direction in an attempt to maintain fixation. This is called *paretic nystagmus.*

*End-point nystagmus* is a normal finding. It occurs when a patient is asked to look in the extremes of horizontal gaze. Attempts to hold the eyes in this uncomfortable position are associated with a slow drift off the target followed by a corrective movement. It is important to recognise this as a normal physiological response. Recorded falsely as nystagmus it may confuse the diagnosis.

A *pinealoma* may cause compression of the mid-brain resulting in gross abnormalities of the eye movement and typically a paralysis of upward gaze *(Parinaud's gaze syndrome).* It may be associated with vertical nystagmus and in addition to the vertical beating of the eyes there may be a rhythmical retraction of the globes into the orbit on attempted upward gaze (convergence retraction nystagmus).

*Vestibular nystagmus*

*The labyrinth*

The horizontal semi-circular canal, which is not truly horizontal but is inclined backwards at an angle of 30° to the horizontal, exerts a tonic effect on the horizontal position of the eyes. The left semi-circular canal drives the eyes to the right. If the left semi-circular canal were suddenly destroyed then the eyes should be driven to the left by the unopposed action of the normal right labyrinth. The patient, however, would try to rectify this abnormal position of the eyes by making a fast corrective movement in the opposite direction towards the undamaged side. This forms the basic component of jerk nystagmus. Nystagmus due to labyrinthine destruction is usually short-lived, is maximal at the time of the labyrinthine destruction and gradually disappears over a period of days or weeks. The nystagmus is most marked when the patient is unable to fix clearly on an object, e.g. during periods of low illumination or with the vision impeded by frosted lenses. The amplitude of nystagmus increases when the patient turns his eyes in the direction of fast phase.

*Caloric test*

If the patient lies back at an angle of 60° so that the horizontal semi-circular canals are vertical and cold water is syringed into one ear (20cc of iced water), convection currents in that canal inhibit its action as though it had been partially ablated and result in the eyes deviating towards the side of the syringed ear. This slow drift is corrected by a jerk to the opposite side, producing a jerk nystagmus. The value of this test is to reveal the slow movement which demonstrates the integrity of the pontine gaze centre and helps to distinguish between hemisphere and pontine gaze palsies. The test can also be performed in the unconscious patient. Caloric testing of vertical eye movements is unsatisfactory.

*The eighth nerve and the vestibular nuclei*

Strictly unilateral destructive disease of the eighth nerve or the vestibular nuclei will produce a horizontal nystagmus with the fast phase directed away from the lesion (i.e. the direction of the fast component is the same in the disease of the eighth nerve or nucleus as that in labyrinthine disease). However strictly unilateral disease is not the rule and many patients have damage to the vestibular nuclei on both sides and therefore have horizontal nystagmus which varies in the direction of the fast phase as the patient changes direction of gaze. When this happens, the direction of the fast phase is the same as the direction of gaze, i.e. when the patient looks to the right the fast phase of the nystagmus is to the right. Disease of the vestibular nuclei may also result in vertical or rotatory nystagmus. Nystagmus resulting from damage to the vestibular nuclei tends to be more persistent than nystagmus resulting from labyrinthine disease.

*Cerebellar nystagmus*

The mechanism by which cerebellar disease causes nystagmus is not well understood. The essential point to note is that the direction of the nystagmus is the reverse of that caused by diseases of the labyrinth or vestibular nuclei i.e. damage to the right cerebellar hemisphere causes drift of the eyes to the left (away from the damaged side) and therefore the direction of the corrective fast phase is to the right, that is *towards* the damaged side. Nystagmus due to unilateral cerebellar disease tends to be coarser than the relatively fine nystagmus of vestibular disease.

*Congenital nystagmus*

Congenital nystagmus occurs within the four months of birth, and may be an isolated relatively benign condition, or reflect severe ocular or neurological disease. Typical *idiopathic congenital motor nystagmus* is purely horizontal in all directions of gaze and is symmetrical in both eyes. There may be an associated head posture, and often there is an astigmatic refractive error, but the rest of the ocular and neurological examination is normal. The visual prognosis in such cases is reasonably good. Atypical features such as incomitance of the nystagmus, or associated vertical and rotatory components suggests *secondary congenital nystagmus* and should prompt specialised electrophysiological and neuroradiological investigations for conditions such as cone dystrophy, albinism and optic nerve or chiasmal tumours.

Patients may find one position of gaze where the nystagmus is least, and this is referred to as the *null point.* They may adopt an abnormal head posture so that the position of the eyes in the orbits is as close to the null point of the nystagmus as possible. If the null point is situated in the left gaze, the patient will turn his face to the right so that the eyes occupy a left gaze position within the orbits when the patient views an object directly in front of him.

*Prismatic and surgical treatment for nystagmus*

Prisms have been used in an attempt to correct the abnormal head posture by moving the null point, but the thickness of the prism required may make this impractical. Symmetrical surgery to all four horizontal muscles in an attempt to move the eyes in both orbits such that the null point occurs when the patient looks straight ahead can be rewarding. This abolishes the abnormal head posture and often improves the vision further (possibly due to the reduction of tone in the neck reflexes and their effect upon nystagmus). However, the surgery must be precise and the pre-operative position of the null point accurately calculated.

*Latent nystagmus*

This congenital jerk nystagmus is only manifest when one eye is occluded, the nystagmus disappearing when both eyes are uncovered. It is a benign condition but if the examiner is unaware of its existence, it may give a false impression of reduced visual acuity when the vision is tested monocularly because the resulting nystagmus may severely reduce the visual acuity. These children should have their binocular visual acuity tested and then each eye tested in turn with a high convex lens before the other eye so that although the vision on that side is blurred light can enter the occluded eye and so tend to prevent the nystagmus. It may occur in cases of infantile esotropia.

*Alternating sursumduction (syn. dissociated vertical divergence)*

This peculiar ocular movement is often seen in patients with latent nystagmus. When one eye is occluded, the eye under cover moves upwards and when the occlusion is removed, the eye moves slowly down to regain the point of fixation. It is the slow speed of this movement which indicates the diagnosis in comparison with the rapid re-fixation movement occurring in a hyperphoria. Alternating sursumduction is benign condition, again found in infantile esotropias.

*Spasmus nutans*

This is a rare acquired pendular nystagmus which is characterised by dissociated nystagmus, head nodding and an abnormal head posture. The rate of nystagmus may vary between the two eyes and may even occur monocularly. It usually starts in the first two years of life and lasts for a few weeks or months before fully resolving. Because of the atypical features, it is important to exclude chiasmal and suprachiasmal compressive lesions as a cause, and the diagnosis of spasmus nutans should only be made retrospectively after serious underlying pathology has been excluded.

## Disorders of the pupil

Abnormalities of the pupil may result in a dilated pupil (mydriasis) or small pupil (miosis). The pupil may not only be abnormal in size, but also be irregular in shape. Disease processes can directly affect the iris, either its musculature or any part of the nervous pupillary reflex arcs. The pupillary reflexes are in response to light and dark and to 'near', which involves both convergence and accomodation. When assessing a patient with unequal pupils it is obviously important to decide which is the abnormal pupil, i.e. one has to decide whether the pupil on one side is abnormally dilated or whether the pupil on the other side is abnormally miosed.

*Anatomy and physiology*

The *sphincter pupillae* consists of smooth muscle arranged in a ring around the pupil margin. The sphincter is controlled by the parasympathetic nervous system. The *dilator pupillae* is a less well defined muscle arranged radially in the iris and this is supplied by the sympathetic nervous system (pp24, 418).

## The pathways of the pupillary reflexes

**The light reflex** (Fig. 19.23 p415)

*Afferent.* The sense organs are the rods and cones which stimulate certain ganglion cells, whose axons travel in the visual pathway and partially decussate at the chiasma to enter the optic tracts. They then separate from the visual fibres and pass via the superior brachium to a pretectal centre on each side.

*Central.* Here they relay to both Edinger-Westphal nuclei of the 3rd cranial nerve. *Efferent* fibres arise in these nuclei and travel with the branch of the 3rd cranial nerve supplying the inferior oblique muscle which they leave to relay in the ciliary ganglion. Postganglionic fibres enter the ciliary nerves to supply the sphincter pupillae.

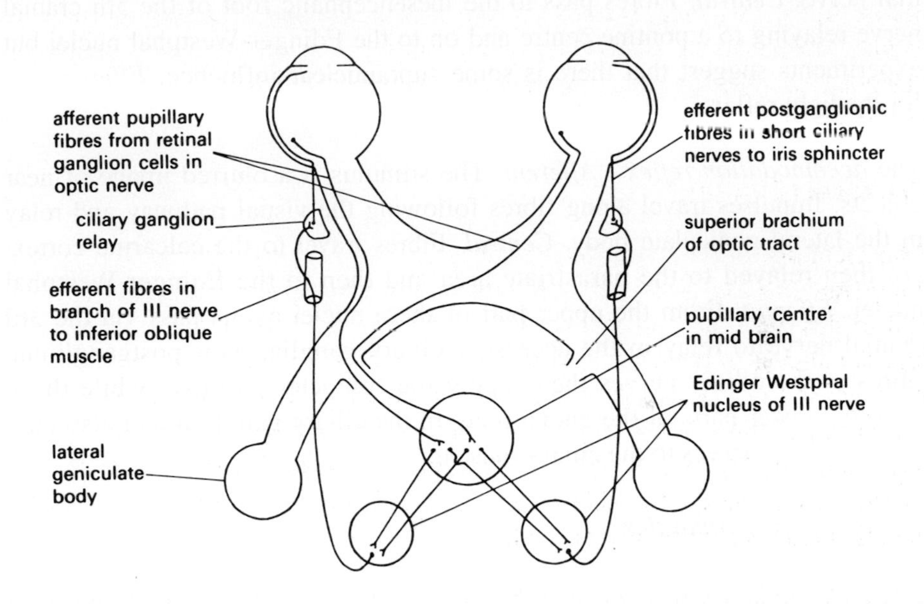


Fig.19.23 The path of the light reflex to the sphincter muscle of the pupil.

## The dark reflex

*Afferent.* As for the light reflexes as far as the superior brachium. *Central.* The fibres then follow a pathway, largely unknown, to relay in the ciliospinal dilator centre in the grey matter of the spinal cord at the upper thoracic level. *Efferent.* Fibres pass to the superior cervical ganglion to relay and postganglionic fibres are conveyed in the sympathetic plexus on the carotid artery and its branches and the 5th cranial nerve via the ciliary nerves to the dilator pupillae.

**The near reflex**

This has two components which usually operate together:

The *convergence reflex - Afferent.* Proprioceptive impulses are initiated by the contraction of the medial rectus muscles and pass along the 3rd or 5th cranial nerve. *Central.* Fibres pass to the mesencephalic root of the 5th cranial nerve relaying to a pontine centre and on to the Edinger-Westphal nuclei but experiments suggest that there is some supranuclear influence. *Efferent.* As for the light reflex.

The *accomodation reflex - Afferent.* The stimulus is a blurred image of near objects. Impulses travel along fibres following the visual pathway and relay in the lateral geniculate body. *Central.* Fibres travel to the calcarine cortex, are then relayed to the parastriate area and then to the Edinger-Westphal nuclei. *Efferent.* From the upper part of these nuclei axons pass via the 3rd cranial nerve to relay in the accessory ciliary ganglia, their postganglionic fibres in the ciliary nerves then supply the sphincter pupillae, while those from the lower parts of the nuclei relay in the ciliary ganglion and also pass via the ciliary nerves to the ciliary muscle.

*Congenital Abnormalities*

Unequal pupils *(anisocoria)* may be found at birth and they usually have no serious significance. There may be a complete absence of the iris *(aniridia)* (p427) and eccentric pupils or more than one pupil *(polycoria)* may occur.

*Acquired Mydriasis*

*Local ocular cause.* In acute *angle closure glaucoma* the pupil is semi-dilated and does not react to light or accommodation either directly or consensually. The other symptoms and signs of acute glaucoma are of course present.

*Blunt trauma* to the eye can result in a dilated pupil due to damage to the iris or its nerve supply. Slit lamp examination may also reveal single or multiple ruptures of the sphincter pupillae. In some cases the pupil may recover its normal size. Recent injury will be obvious but in other cases a history of trauma may be elicited.

In ***myotonic pupil (Holmes-Adie pupil)*** the patient presents with sudden dilatation of one pupil which may be associated with blurred vision, partly because of the pupil dilatation, but also because of the paresis of accommodation on that side. The underlying cause of the disease is unknown. The site of the lesion is probably the ciliary ganglion. The parasympathetic fibres both for accommodation and the pupil sphincter are impaired. This condition occurs most commonly in women in the second or third decade and may be associated with loss of knee jerks, but there are no other neurological complications. It is a troublesome but benign peripheral lesion and must not be confused with a partial third nerve palsy. On examination; the pupil is found to be dilated. Examination of pupil reactions usually show a sluggish and incomplete response to light. The dilatation of the pupil is also much slower than that of the fellow eye.

It is obviously important to distinguish between the benign Holmes-Adie pupil and a more serious compressive lesion of the third nerve, e.g. aneurysm of the circle of Willis. Accommodation and pupil reactions are affected in both cases but a third nerve lesion is usually associated with paresis of the extraocular muscles. The important difference between the effects of these two is that *the third nerve lesion affecting the parasympathetic nerve supply is preganglionic whereas the lesion of the ciliary ganglion results in post-ganglionic denervation causing an increased sensitivity of the receptor sites of the sphincter muscle to acetylcholine.* This denervation sensitivity of the pupil can be demonstrated by instilling one drop of the acetylcholine analogue, metacholine chloride 2.5% (Mecholyl), into the conjunctival sac of the affected eye; the myotonic pupil will constrict, while normally or in preganglionic paresis of the sphincter iridis it will not.

*Third cranial nerve damage.* Pressure on the third nerve will normally produce a combination of extraocular paralysis and a dilated pupil. Sometimes only the extraocular muscles are involved but the fibres carrying the parasympathetic supply to the pupil are situated on the outer aspect of the nerve trunk and are more often the first to be affected. Thus a dilated pupil may be the only sign of pressure on the third nerve (see causes of third nerve palsy - p249).

*Coning of the temporal lobe* due to raised intracranial pressure sometimes develops in patients with severe head injury or intracranial neoplasms. This herniation of the temporal lobe through the tentorium compresses the third nerve and will cause a dilated pupil on the same side as the temporal lobe herniation and is an early warning sign of coning. When dilatation of the pupil is associated with contralateral hemiparesis and is followed by dilatation of the pupil on the other side, this is a sign of severe coning.

*Lesions of the superior colliculus.* The superior colliculus may be compressed by tumours of the adjacent structures, e.g. a pineal gland tumour, or by direct infiltration. This results in dilatation of the pupils which may be associated with defects of upward gaze *(Parinaud's syndrome).*

*Drug induced mydriasis.* Pupil dilatation may be induced by the local effect of sympathomimetic drugs, such as phenylephrine, or of anticholinergic drugs like atropine. Accidental instillation does occur and the patient will sometimes deny using any drops, making the diagnosis difficult. Systemic absorption of either sympathomimetic or atropine-like drugs in sufficient quantities will produce pupil dilatation e.g. some bronchial or intestinal relaxants.

*Acquired miosis.* Iritis, inflammation of the musculature of the iris, usually results in spasm. The more powerful sphincter muscle will overcome the effect of the dilator muscle and the pupil is therefore constricted. Adhesions may subsequently form between the iris and the lens resulting in an irregular pupil margin. The eye is usually red and on examination the other signs of iritis will be found (p178). *Blunt trauma* to the eye usually results in a dilated pupil. However this is not invariable and sometimes the pupil on the damaged side is constricted due to an irritative lesion or associated traumatic iritis.

*Lesions of the sympathetic nerve supply to the pupil:* ***Horner's syndrome.*** Any lesion of the sympathetic pathway of the eye results in Horner's syndrome with the following signs and symptoms (Fig. 19.24 p419):

1. a *miotic pupil,* which retains its light reaction - the difference in size of the pupils is most obvious when the patient is viewed in reduced illumination. 2. *ptosis of the upper eyelid*, which is usually mild in degree (1-2 mm), due to relaxation of the superior palpebral muscle of Müller. 3. enophthalmos is classically described but is not usually demonstrable. It may be merely simulated by the ptosis. 4. if the sympathetic supply to the facial sweat glands is involved, there may be dryness on the same side of the face and *hyperaemia* of the conjunctiva and face on the side of the affected eye due to loss of vasoconstrictor tone.

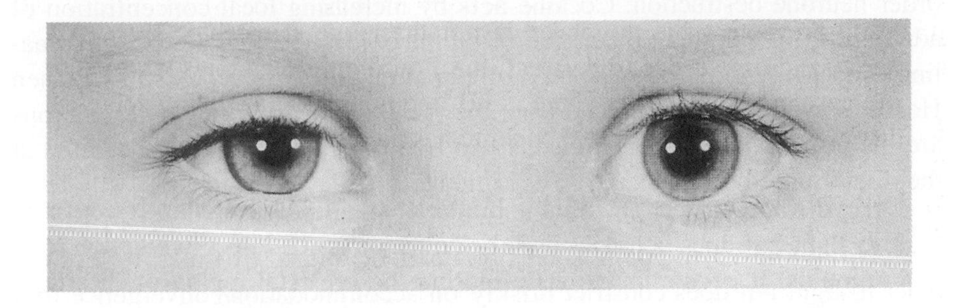

Fig. 19.24 Horner's syndrome (R).

*Causes of Horner's syndrome.* 1. *Massive cerebral damage* may result in an ipsilateral Horner's syndrome. 2. *Brain stem* and *cervical cord lesions* may interrupt the sympathetic pathway, e.g. multiple sclerosis, neoplasms, vascular lesions. The proximity of the sympathetic pathway to the central canal of the spinal cord renders it vulnerable to damage by such conditions as syringomyelia. 3. *Apical lung disease*. Classically the Pancoast bronchogenic carcinoma tumour causes destruction of parts of the brachial plexus and may involve the sympathetic pathway as it leaves $T_1$. 4. *Cervical chain lesions*- may be caused by many diseases of the neck, e.g. malignancies or trauma and the chain may be divided during major surgery to the neck. 5. *Carotid artery aneurysm* can damage the intracranial pathway of the sympathetic system.

*Pharmacological tests* may be used to distinguish between damage to the third order sympathetic neurones whose axons are the post ganglionic fibres and damage to the first or second order neurone. 1. *adrenaline* in a concentration of 1/1000 when instilled into the conjunctival sac will not usually dilate the normal pupil (p591). However the pupil affected by destruction of a third order neurone whose axons are postganglionic fibres will dilate in response to this weak concentration of adrenaline due to denervation sensitivity. This response will be absent in first or second order neurone destruction which does not cause denervation sensitivity. 2.*cocaine 4%* will dilate the normal pupil (p591and p612) as well as Horner's pupil when due to first or second order neurone destruction. Cocaine acts by increasing local concentration of adrenaline, by inhibiting the action of amine oxidase (which destroys adrenaline) and preventing the re-uptake of noradrenaline at receptor sites. When Horner's syndrome is due to destruction of third order neurones whose axons are the postganglionic fibres, there is no free noradrenaline being secreted at the dilator pupillae and therefore cocaine has no dilating effect.

***The Argyll-Robertson pupil.*** This is bilateral, small and irregular. It does not react to light but does constrict briskly on accommodation/convergence to a near stimulus. It occurs in lesions affecting the pretectal region of the midbrain and so may be found in encephalitis and in vascular and traumatic lesions but it is characteristic of neurosyphilis. The irregular shape and miosis and the failure of the pupil to dilate with atropine are unexplained.

*Pontine haemorrhage* causes severely constricted pupils. The patient is comatose and the condition is usually fatal.

*Drug induced miosis* . Instillation of miotics, e.g. pilocarpine, for the treatment of pre-existing glaucoma may confuse the diagnosis in patients presenting with neurological signs. Opiates also induce miosis.

## The afferent pupil defect and the swinging flashlight test

The resting size of the pupils is normally symmetrical. However in cases of unilateral afferent defect the affected pupil does not constrict to direct light stimulation as well as its fellow, but reacts equally on consensual testing. Unilateral or unequal afferent pupil defect is revealed by the 'swinging flashlight test'.

*The swinging flashlight test.* This simple test is best practised at first on a patient who has markedly impaired vision in one eye and after a little practice it will reveal even small degrees of afferent defect. The light from a pen torch is alternately shone from one eye to the other exposing each eye to the light for about five seconds. When the worse eye is stimulated the pupil dilates.

This is due to the pupil of the eye with reduced vision dilating following its consensual reaction. Because the consensual reaction is usually slightly less than the direct reaction a small constriction can be expected when the light is swung from eye to eye. If dilatation occurs there is an afferent defect in that eye. Sometimes it is necessary to increase the speed of the swing to demonstrate minor defects, up to one second on each eye. The importance of this test is that:

-it is *unaffected by lens opacities* which do not reduce the light entering the eye but only diffuse it,
-it is *unaffected by amblyopia* due to squint or refractive error which is mainly cortical in origin,
-it is *unaffected by refractive error,*
-it is an *objective sign of reduced vision* which is useful when dealing with possible hysterics,
-it is *very sensitive to optic nerve disease* and for example can detect visual loss in retrobulbar neuritis even before the vision is reduced below 6/6. It is however less sensitive to retinal disease and a small branch vein occlusion or macular degenerative change may not cause an afferent defect detectable by this method.

# CHAPTER 20

# CONGENITAL ABNORMALITIES AND GENETIC DISORDERS

## Molecular genetics

During the last decade there have been enormous advances in understanding the genetic basis underlying a variety of diseases. It is now possible to determine which genes are responsible for a condition and apply this information in many ways. These range from the diagnostic - determining which relatives are at increased risk of developing a tumour (eg retinoblastoma) to the therapeutic - designing new treatments based upon an understanding of how individual genes function (eg cystic fibrosis). The first step in these molecular genetics techniques is identification of the disease-causing gene. At present this is simplest for single gene disorders because polygenic conditions, such as hypertension, schizophrenia, or primary open angle glaucoma, where a number of different genes are believed to be involved, are much harder to analyse.

The main technique used, termed *linkage analysis*, involves studying the inheritance of the disease condition (or phenotype) in families where a number of relatives are affected. Evidence is sought of co-inheritance of the disease phenotype with genetic markers that are scattered, quite naturally, throughout the genome. If the gene causing the disease lies close to a marker, the gene and the marker will be inherited together more frequently than one would expect by chance alone. The degree of co-inheritance is related to the distance between the gene and the marker and gives an approximate chromosomal location for the disease gene. This position may be refined using a number of more closely spaced markers or by studying additional families. Eventually, when the genetic interval in which the gene resides has been narrowed to perhaps 1,000,000 base pairs it becomes feasible to look for sequence changes (mutations) in genes lying in this area in order to identify the disease gene. By using such techniques a large number of genes causing conditions ranging from corneal dystrophies, cataract, glaucoma to retinitis pigmentosa and other retinal dystrophies have now been identified.

A *congenital abnormality* is one that is present at birth and may be secondary to either genetic or environmental factors. Examples of environmental factors include maternal infections such as rubella, toxins, radiation or drugs. The impact these have is determined by the stage of pregnancy at which they occur. Factors acting in the first trimester of pregnancy have a much more severe effect on development compared to those occurring later. The profound effects that damage to primordial tissues induces is illustrated by neurocristopathies, a group of disorders occurring secondary to abnormal neural crest development. They cause abnormal development of the anterior portion of the eye resulting in a variety of disorders including congenital glaucoma.

Knowledge about the genetic factors responsible for congenital anomalies can be derived from noting which family members are affected. This will reveal the pattern of inheritance of this condition. Three main patterns of inheritance exist: autosomal dominant, autosomal recessive and X-linked. As disorders such as retinitis pigmentosa can be inherited in all these ways it indicates that this retinal dystrophy is in reality a group of similarly appearing but genetically distinct conditions. Some ocular disorders that are not thought to be inherited may still have a familial basis. This is illustrated by the human leucocyte antigens (HLA), some of which are associated with an increased risk of developing a particular disease (e.g. iritis in HLA-B27 positive patients). Possession of the HLA-B27 antigen increases the relative risk of iritis but its development requires the presence of an environmental stimulus in addition to the genetic component (HLA-B27). Hence ocular disease may develop due solely to genetic or environmental factors or occasionally from a combination of the two.

**Ocular malformations as the result of infection in utero**

Ocular malformations may result from intra-uterine damage as well as from genetic or chromosomal abnormalities:

-*Rubella* (p332)
-*Toxoplasmosis* (p182)
-*Cytomegalovirus disease* (pp331, 337)
-*Syphilis* (p328)

## Developmental abnormalities due to exposure to abnormal oxygen concentrations

*Retinopathy of prematurity (retrolental fibroplasia)* occurs in low birth weight infants exposed to high concentrations of inspired oxygen. The peripheral retina is vascularised at a relatively late stage of foetal development and is susceptible to damage from the high oxygen concentration. This causes constriction of the retinal vessels which subsequently respond by neovascularisation to the ischaemia caused by exposure to lower (normal) concentrations. The new vessels are fragile and leak fluid, and later whole blood, with consequent retinal separation spreading from the periphery and haemorrhages. Retinopathy of prematurity is a bilateral condition in which five stages of severity are recognised ranging from a white demarcation line in the peripheral retina, through development of neovascular proliferation to tractional retinal detachment (LCW 8.2, 8.3 p122). The majority of cases involve mild forms of the disease and these may regress without causing visual loss. Regular screening of infants at risk of developing retinopathy of prematurity (those born at less than 36 weeks gestation or weighing less than 1500 grams) will permit prompt treatment. This involves ablating the ischaemic peripheral retina and can be achieved with either cryotherapy or laser photocoagulation.

## The ocular effects of chromosomal abnormalities

The chromosomal nature of a condition may be determined by staining cells and examining the chromosomes microscopically. The appearance of the chromosomes can be recorded photographically and differences are readily detected by comparing the appearance with that of normal chromosomes.

*Down's syndrome (mongolism)* is commonly caused by an additional copy of chromosome 21 (trisomy 21) and occurs with increased frequency in mothers older than 35 years. It is believed to occur because of failure of chromosomal separation at meiosis (chromosomal non-disjunction). Other less frequent causes include translocation of part of chromosome 21 to other chromosomes or secondary to mosaicism. A variety of ocular features are found in Down's syndrome including *Brushfield's spots.* These lighter coloured areas on the irides are also present in normal individuals. Myopia, cataract, nystagmus, strabismus and keratoconus are also frequently seen.

## Genetically determined congenital malformations

*Anophthalmos* means the total absence of ocular tissues. It is a very rare condition and most suspected cases are severe forms of microphthalmos.

*Microphthalmos* (LCW 8.15 p127) is caused by a failure of normal ocular development and a spectrum of severity occurs. It results from a variety of conditions including trisomy 13 and less severe types are compatible with a reduced level of vision. Various associated findings may be present including corneal opacification, coloboma, cataract, persistent hyperplastic primary vitreous and an associated cyst (microphthalmos with cyst).

A *coloboma* in the eye results from failure of closure of the embryonic fissure. It may be partial or complete and when affecting the iris gives rise to a 'keyhole pupil' appearance (Plate 20.1 p426, LCW 8.17 p127). A coloboma may also extend posteriorly to involve the lens, lens zonule, ciliary body, retina, choroid and optic disc. In a retinal coloboma, failure of retinal pigment epithelial development results in hypoplasia of the underlying choroid and sclera. The fundal appearance is characteristic with an area of visible sclera situated inferior to the optic disc (Fig. 20.2 p427, LCW 8.18 p127). A patient with a coloboma often has reduced vision in the affected eye and the incidence of strabismus and nystagmus is increased. There is also sensitivity to light due to incomplete miosis and in the case of a retinal coloboma, to the scattering of light from the white scleral background exposed.

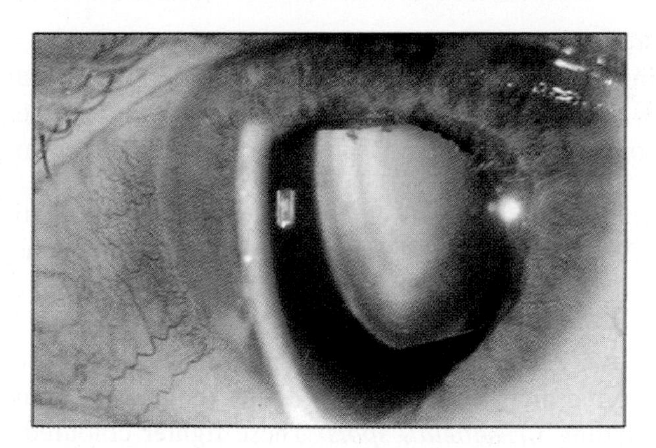

Plate 20.1 Coloboma of the iris and nuclear sclerosis as shown in optical section by a slit lamp beam.

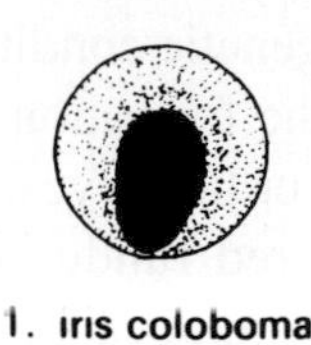

Fig. 20.2 Developmental anomalies of the eye - 1. iris coloboma and 2. choroidal-retinal coloboma.

*Anterior chamber dysgenesis, Rieger's and Axenfeld's syndromes and Peter's anomaly* (p516).

*Aniridia* is a spectrum of disorders characterised by a variable degree of iris hypoplasia. A rudimentary stump of iris is usually present and associated ocular abnormalities include corneal opacification, glaucoma, lens subluxation, together with optic nerve and macular anomalies. Cases of aniridia occurring in individuals without any similarly affected relatives (sporadic cases) are associated with renal tumours (Wilms' tumour) necessitating appropriate screening.

*Persistent hyperplastic primary vitreous* is caused by a failure of regression of the primary vitreous. It may be present at birth or shortly afterwards with a white opacity posterior to the lens. It can give rise to a white pupillary reflex (leucocoria) and may be distinguished from other causes of this condition (e.g. retinoblastoma) by features such as indrawn ciliary processes, microphthalmos and persistent hyaloid artery remnants. There is a less common posterior variant in which the unregressed tissue lies just anterior to the optic disc and which, in certain cases may cause retinal detachment.

*Persistent hyaloid artery remnants* are amongst the commonest ocular abnormalities occurring in up to 3% of healthy full term infants. If complete they extend from the optic disc to the posterior surface of the lens but frequently only remnants are present. The posterior portion may be associated with glial tissue on the disc (Bergmeister's papilla) whilst the anterior portion may be attached to a localised lens opacity adjacent to the posterior pole of the lens (Mittendorf's spot).

*Lens anomalies*

-*Lenticonus* can be inherited but appears to be a non-genetic condition in which there is a variation in the curvature of the lens and the formation of a cone at either the anterior or posterior pole of the lens of one or both eyes. On ophthalmoscopy it can be seen as a dark disc against the red fundus background. Usually it is of minor degree but occasionally it may require cataract surgery.

-*Developmental lens opacities* (p96, p100) result from incorrect formation of lens fibres and may be secondary to metabolic disorders, intrauterine infections or inherited conditions such as galactosaemia. A spectrum of severity is encountered ranging from minor lens opacities to visually significant congenital cataracts. 'Blue dot' cataracts are perhaps the most common of the minor lens opacities appearing as irregular shining blue dots when examined with the slit lamp. They are rarely symptomatic. Lamellar cataracts represent the effect of a transient insult on lens development, consisting of opacification of one or more lenticular lamellae surrounded by clear lens fibres (Fig 20.3 p428 and Plate 6.3 p96). In contrast, infections such as *rubella* contracted by the mother during the first 12 weeks of pregnancy may cause severe lens opacities as well as a variety of ocular and non-ocular defects (p332).

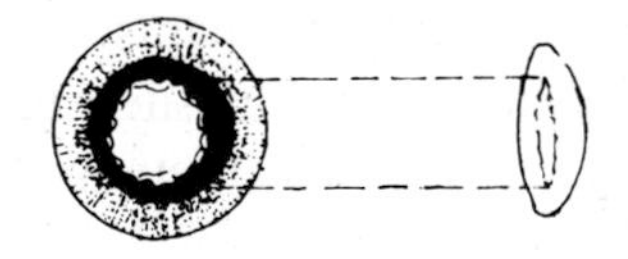

Fig. 20.3 Lamellar cataract.

-*Subluxation of the lens* may be associated with congenital anomalies of the lens zonule and in particular with conditions such as *Marfan's syndrome* (p434), *homocystinuria* (p433) and *spherophakia.*

-*Spherophakia* is a condition in which the lens is small and tends to be spherical causing myopia. It has a stretched weakened zonule often allowing subluxation. Secondary glaucoma may supervene due to pupil block by the small lens. Spherophakia may also be part of an hereditary hyperplastic mesodermal dystrophy (Marchesani's Syndrome) in which the stature and fingers are short and there is mental retardation.

*Anomaly of myelination*

-*Myelinated retinal nerve fibres.* Myelination of the optic nerve begins at the optic chiasma reaching the lamina cribrosa by term. Occasionally myelination continues beyond the disc giving rise to a localised feathery white retinal appearance (LCW 7.5 p108).

*Dermoids*

-*Dermoid cysts* are caused by sequestration of surface ectoderm occurring along lines of embryonic closure. They may enlarge with time and if the contents leak, cause a vigorous inflammatory reaction (p455).

-*Limbal dermoids* are congenital anomalies containing tissue not normally found at that location. They occur most frequently at the temporal limbus and are covered by conjunctival epithelium (Fig. 20.4 p429, LCW 8.21 p128). They can occur as part of a syndrome in which accessory auricular appendages are present (Goldenhaar's syndrome) (p476).

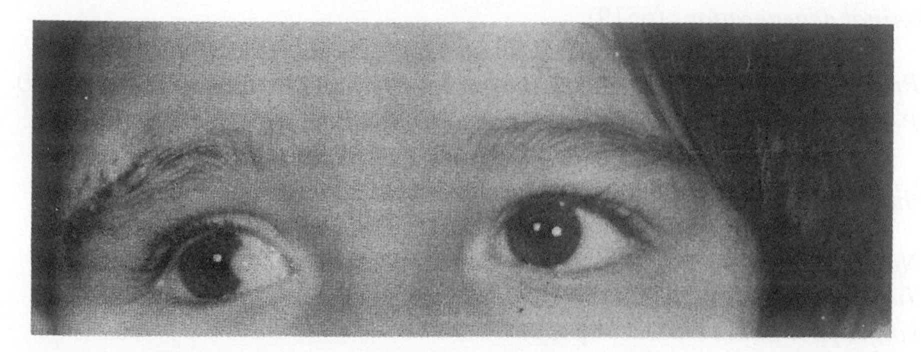

Fig. 20.4 Limbal dermoid R eye. Scar of recent excision of R. external angular dermoid cyst which in this case was also present.

-*Dermolipoma* of the orbit is described on p476.

*Lacrimal anomalies*

-*Lacrimal duct obstruction in infants* is common in the first twelve months of life. It reflects a failure of canalisation of the nasolacrimal duct and frequently resolves spontaneously but may require surgical opening (p211).

*Lid anomalies*

-*Lid coloboma (Limbal dermoid)* (p463).
-*Epicanthic folds* (pp228, 462).
-*Blepharophimosis* (p462, Fig. 22.11p462).
-*Blepharoptosis (ptosis)* (p450, Fig. 22.4 p450).

*Corneal conditions*

-*Megalo-cornea* (p477).
-*Micro-cornea* (p477).

*Developmental glaucoma (buphthalmos, hydrophthalmos)* (p572, Fig. 31.21 p573).

**Genetically determined ocular conditions**

*Corneal dystrophies* (p483).

*Retinal degenerations* (p519)

-*Primary pigmentary degenerations of the retina (retinitis pigmentosa)* (p520).
-*Primary macular degeneration* (p522).

*Phakomatoses* (p436).

-*Neurofibromatosis* (437)
-*Tuberose sclerosis* (p437)
-*von Hippel Lindau disease* (p438)
-*Sturge-Weber syndrome* (p438)
-*Coats' disease* (p438)

*Uveal degenerations* (p515)

-*Irido-corneal endothelial syndrome (ICE syndrome or essential iris atrophy.)* (p515)
-*Choroidal vascular atrophy.* (p515)
-*Choroideraemia.* (p516)
-*Gyrate atrophy.* (p516)
-*Disciform degeneration of the retina.* (p529)

*Optic atrophy*

-*Hereditary optic atrophy* (p388).
-*Leber's disease* (p384).

**Inherited disorders with systemic and ocular complications**

*Retinal pigmentary disorders* (p520 - 522)

-*Primary pigmentary degeneration (Retinitis pigmentosa).*
-*Laurence-Moon-Biedl syndrome.*
-*Usher's syndrome.*
-*Refsum's syndrome.*
-*Kearn's syndrome.*

*Disorders of lipid metabolism*

-*Cerebromacular degenerations* (retinal and central nervous system degeneration associated with lipidosis).

-*Tay-Sachs disease* is the most common of the gangliosidoses and is inherited as an autosomal recessive condition. It is caused by a deficiency of hexosaminidase A leading to an accumulation of ganglioside GM2 in neural tissue. It occurs most frequently in children of Ashkenazi Jewish extraction presenting in the first year of life with progressive neurological dysfunction. Ophthalmoscopy early in the course of the disease reveals the classic foveal cherry-red spot appearance due to ganglioside deposition in retinal ganglion cells. The lack of ganglion cells at the fovea permits visualisation of the vascular choroid which is outlined against the surrounding swollen white retina. Optic atrophy develops as the cherry-red spot appearance fades and the child usually dies prior to the age of four.

-*Niemann-Pick disease* is an autosomal recessive condition in which accumulation of sphingomyelin occurs. Several subtypes exist and like Tay-Sachs disease it is much more common in Ashkenazi Jewish children. Although the onset of symptoms occurs later than Tay-Sachs the majority of cases have an equally poor prognosis.

*Disorders of amino acid metabolism*

-*Albinism* is characterised by reduced pigmentation of the skin and eyes secondary to a deficiency of tyrosinase - the enzyme converting tyrosine to melanin. It is termed ocular albinism if only the eyes are involved and oculocutaneous if skin involvement is present. All three main inheritance patterns are present and development of more severe disease is associated with autosomal recessive inheritance. Ocular features include reduced vision, strabismus, pendular nystagmus, pale irides which transilluminate easily and a depigmented fundus. Vision is reduced due to a combination of factors including poor foveal differentiation, excessive light entry into the eye, nystagmus and amblyopia. Treatment is limited to the use of tinted spectacles or contact lenses together with correction of refractive errors and any squint that is present.

-*Alkaptonuria* is a recessively inherited disorder caused by a deficiency of homogentisic acid oxidase. This results in accumulation of homogentisic acid, a breakdown product of tyrosine, with deposition throughout the body particularly in cartilage *(ochronosis)* and connective tissue. This results in a secondary arthritis. Pigmentary deposition in the conjunctiva and episclera also occurs. Large amounts of homogentisic acid are excreted in the urine which darkens if it becomes alkaline after standing exposed to air.

-*Homocystinuria* is caused by a deficiency of cystathione synthetase, the enzyme that converts methionine to cystine. It results in an accumulation of homocysteine, some of which is excreted in the urine. Its presence in the urine may be detected by the sodium nitroprusside test helping to confirm the diagnosis. Patients with homocystinuria frequently develop progressive lenticular subluxation (partial dislocation of the lens), which characteristically occurs in an inferior direction. Occasionally the lens dislocates into the anterior chamber (Plate 20.5 p433) and can be returned to its correct position by laying the patient flat after the pupil has been dilated. Subsequent pupillary constriction will keep the lens in place until surgery can be arranged. Patients with homocystinuria bear some similarity to those with Marfan's syndrome displaying a tall stature and arachnodactyly. They differ in possessing an increased risk of thromboembolism which makes anaesthesia more hazardous. Mental retardation is also a recognised feature. Treatment with vitamin B6 and a low methionine diet improves biochemical control whilst treatment early in life may prevent or delay development of some of the complications outlined above.

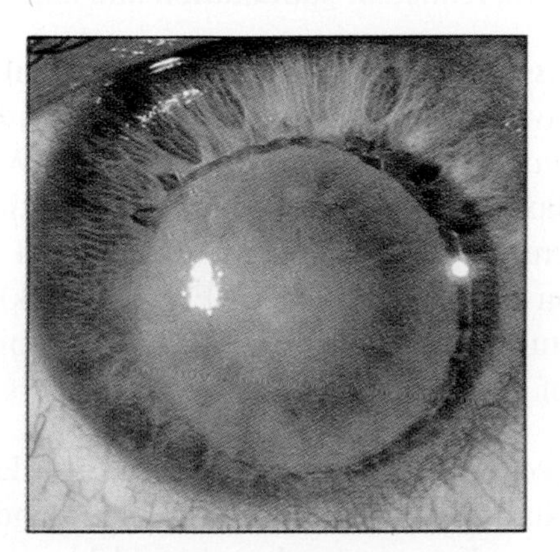

Plate 20.5 Homocystinuria - dislocated lens in anterior chamber.

-*Fanconi syndrome* is a manifestation of *cystinosis* presenting in infancy, in which renal and bone disease (rickets) also occurs. The underlying defect is one affecting lysosomal cystine transport resulting in cysteine accumulation in a variety of tissues. These fine highly refractile crystals are deposited within the corneal stroma causing a reduction in vision and severe discomfort. Oral cysteamine decreases intracellular cysteine levels and topical treatment will reduce the number of corneal cysteine crystals.

-*Lowe's syndrome* is an X-linked disorder in which ocular and renal disease occur in conjunction with mental retardation (hence the term *oculocerebrorenal syndrome*). Aminoaciduria and rickets develop secondary to the renal involvement and the most common ocular features are glaucoma and cataract.

*Disorders of connective tissue*

-*Angioid streaks* (pseudo-xanthoma elasticum, Grönblad-Strandberg syndrome) (p524, Plate 29.4 p524).

-*Ehlers-Danlos syndrome* is a rare autosomal dominant disorder. An abnormal collagen matrix is present leading to hyperextensible skin and joints. Systemic collagen defects are present whilst ocular involvement includes the presence of a blue sclera, lenticular subluxation and angioid streaks.

-*Marfan's syndrome* is caused by a mutation in the fibrillin gene on chromosome 15. Patients have a characteristic appearance with disproportionately long extremities (*arachnodactyly* or long digits) (LCW 8.19 p128), a high arched palate, joint laxity, scoliosis and ascending aortic aneurysms. As the latter may dissect, screening of the relatives of affected individuals is essential. Lens subluxation is frequently seen (Fig. 20.6 p435, LCW 8.20 p128) and other ocular features of the disorder include myopia, glaucoma and an increased incidence of retinal detachment.

-*Osteogenesis imperfecta* is an inherited collagen disorder that affects a variety of structures including bone, joints, teeth, ears and eyes. Pathological fractures, deafness secondary to otosclerosis and blue sclera all result from the presence of abnormal type 1 collagen. The sclerae appear blue due to thinning and increased transparency of the scleral collagen permitting visualisation of the underlying choroid. Keratoconus, congenital glaucoma and lens subluxation are also described.

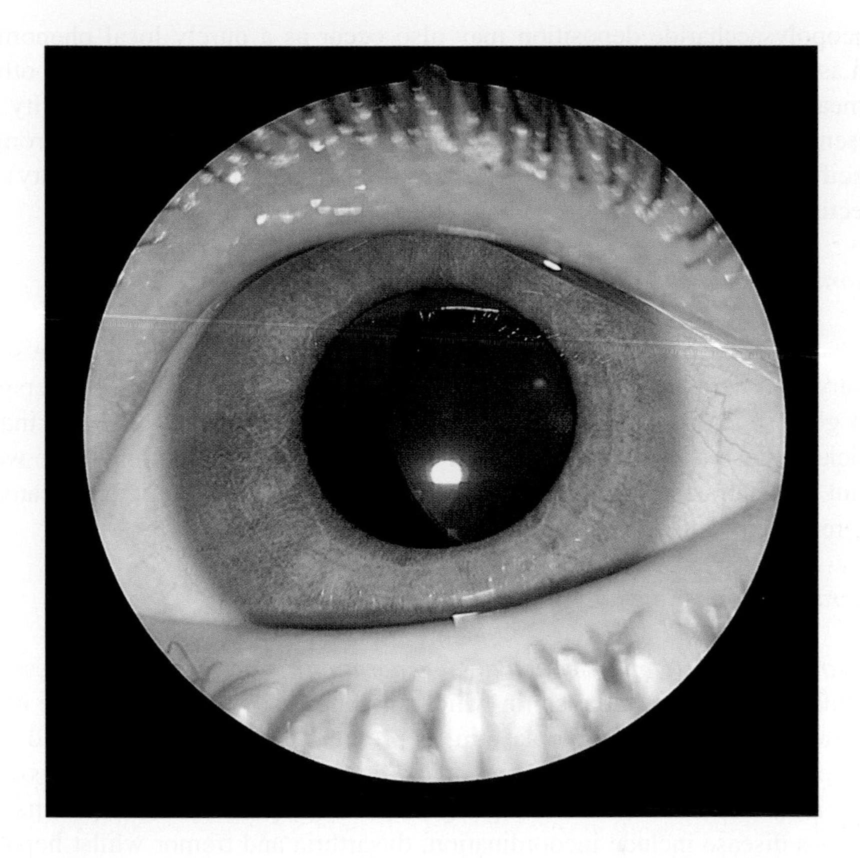

Plate 20.6 Marfan's syndrome - subluxated lens.

*Mucopolysaccharidoses*

These are a group of disorders caused by a deficiency of lysosomal enzymes involved in degrading mucopolysaccharides. Several different types exist of which *Hurler's syndrome (gargoylism)* (type I mucopolysaccharidosis) is the most common. It presents within the first twelve months of life with skeletal defects, a dysmorphic facial appearance, hepatosplenomegaly, cardiac anomalies, mental retardation and corneal clouding. The latter is caused by mucopolysaccharide deposition within the corneal stroma.

Mucopolysaccharide deposition may also occur as a purely local phenomenon as is illustrated by *corneal macular dystrophy.* This differs from the other corneal dystrophies in its inheritance (autosomal recessive) and severity. It presents within the first decade of life, progressing rapidly and causes stromal opacification. Treatment with a corneal graft (penetrating keratoplasty) is effective although the condition may recur.

*Disorder of carbohydrate metabolism*

-*Galactosaemia* is a condition caused by the accumulation of galactose secondary to a deficiency of one of the enzymes responsible for its conversion into glucose. Two main enzyme defects have been identified. Galactokinase deficiency results in the early onset of 'oil drop' cataracts in an otherwise well infant. In contrast galactose-1-phosphate uridyl transferase deficiency causes severe systemic disease together with congenital cataracts.

*Disorder of copper metabolism*

-*Wilson's disease (Hepatolenticular degeneration)* is an autosomal recessive disorder of copper metabolism in which an abnormally low level of the copper transport protein (caeruloplasmin) is present. This results in reduced biliary excretion of copper with secondary deposition in a variety of tissues especially the liver and basal ganglia. Neurological manifestations of Wilson's disease include incoordination, dysarthria and tremor whilst hepatic involvement results in hepatitis and cirrhosis. Copper deposition in the eye causes a golden brown ring situated in the peripheral cornea (Kayser-Fleischer ring). It is present in all patients with neurological features of Wilson's disease and some patients also develop a greenish lens opacity (sunflower cataract). Treatment with D-penicillamine which chelates copper is effective and reverses the ocular signs.

## Phakomatoses

The *phakomatoses* are a group of conditions in which a variety of structures derived from neuroectoderm are affected by hamartomas (congenital overgrowths of tissue which is normally found in that location). They are frequently inherited and ocular involvement is common.

-*Neurofibromatosis* exists in two forms. Type I (peripheral neurofibromatosis/von Recklinghausen's disease) is caused by a mutation on the long arm of chromosome 17. Features include café au lait spots (hyperpigmented cutaneous patches), axillary or inguinal freckling, fibroma molluscum (pedunculated pigmented nodules) and plexiform neurofibromas (enlarged peripheral nerves). Ocular involvement may include mechanical ptosis (from lid neurofibromas), proptosis, glaucoma, iris nodules, multiple retinal astrocytomas and optic nerve gliomas. Type II (central neurofibromatosis) is characterised by bilateral acoustic neuromas typically presenting in the second to third decades.

-*Tuberose sclerosis (Bourneville's disease)* is an autosomal dominantly inherited condition characterised by the triad of adenoma sebaceum (multiple facial nodules or angiofibromas) mental retardation and epilepsy. Small retinal astrocytomas (glial cell tumours) which are pale mulberry shaped masses may be present adjacent to the optic disc (Stereo Plate 20.7 p437, LCW 8.12 p126). Cerebral astrocytomas are found in a proportion of cases accounting for the presence of mental retardation and epilepsy. In the skin, adenoma sebaceum particularly affects the cheeks and forehead. (LCW 8.13 p126)

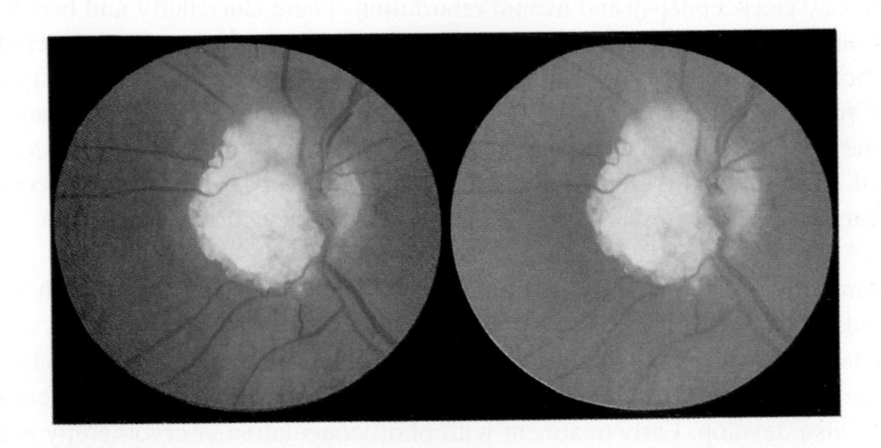

Stereo Plate 20.7 Tuberose sclerosis of the retina. (see p4)

*-von Hippel-Lindau disease* (retinal and cerebellar angiomatosis) is an autosomal dominant condition caused by inactivation of a tumour suppressor gene on the short arm of chromosome 3. As a result, patients may develop multiple hamartomas affecting not just the eye but also the brain, kidneys and adrenals. The ocular lesions are vascular retinal angiomas some 1-3 disc diameters in size. They are supplied by an enlarged tortuous feeding artery and have a similarly proportioned draining vein (LCW 8.14 p126). They may leak a lipid-rich exudate causing visual loss and this can be prevented by photocoagulation of the angioma. A proportion of patients with retinal angiomas develop tumours in the central nervous system and abdomen necessitating regular screening of both them and their relatives. These additional tumours include haemangioblastomas (affecting the cerebellum, medulla oblongata and spinal cord) as well as renal cell carcinomas and phaeochromocytomas.

*-Sturge-Weber syndrome* (encephalotrigeminal angiomatosis) is caused by an anomaly affecting the primordial vascular system at an early stage of its development from neuroectoderm. Lesions affecting the face, choroid and meninges result. The facial lesion is a unilateral angioma (naevus flammeus or port wine stain) occurring in the distribution of the first and second divisions of the trigeminal nerve. It is purple in colour and blanches under applied pressure. Similar angiomas may be present intracranially affecting the meninges and may cause epilepsy and mental retardation. These can calcify and become visible radiographically. Ocular abnormalities found in this condition include angiomatous involvement of the conjunctiva and episcleral vessels and a typical buphthalmic appearance of the iris and angle of the anterior chamber in some cases with raised intraocular pressure. This may be mild and unsuspected until adult life and then present as advanced secondary open angle glaucoma (Plate 20.8 p439, LCW 8.11 p126). There may also be choroidal angiomas.

*-Coats' disease* typically occurs unilaterally in males during late childhood. Fundal examination reveals the presence of retinal telangiectasis, aneurysmal dilatation of vessels and subretinal exudates. Progressive exudative retinal detachment occurs, resulting in visual loss and cataract. Uveitis and glaucoma may also develop. Early treatment with photocoagulation or cryotherapy may halt the progression of this condition. As Coats' disease can cause leucocoria (a white pupillary reflex) it occasionally needs to be distinguished from the other causes of this appearance including retinoblastoma.

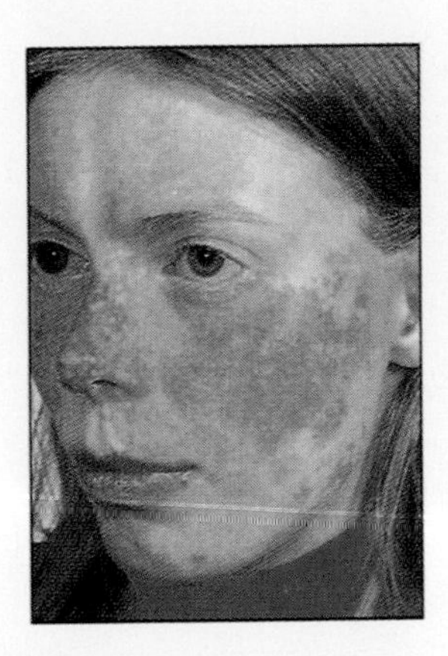

Plate 20.8 Sturge-Weber syndrome with 'adult' presentation of glaucoma.

**Developmental abnormalities of the orbit** (p543)

Congenital orbital malformations are often part of a more widespread systemic syndrome and so repair and rehabilitation must involve close co-operation between multiple disciplines. Many of these children are intellectually normal despite their physical deformity and it has been demonstrated that psychological development is enhanced by procedures which normalise physical appearance.

# CHAPTER 21

# FUNCTIONAL DISORDERS

The perceptions which comprise vision are supremely important and highly developed. The eyes are perhaps the principal media for personal communication of human emotions (as in mother and baby relationships) and the malign influences which primitive man has always felt to be around him have in many civilisations been canalised into the conception of the 'evil eye'. It is therefore to be expected that visual disturbances will accompany disordered states of mind.

*Features of functional ocular conditions*

Variability of symptoms and signs from one examination to another and lack of consistency in findings of the same type, e.g. in visual fields or visual acuities recorded at different distances are usual characteristics. Good records are essential. Psychogenic motor disturbance of the eyes are usually spastic rather than paralytic and the functions involved are only those which can be under the control of will or others which are of the character of associated movements. Miosis associated with spasm of convergence and accommodation is one example.

## Functional visual loss

*Functional visual loss* refers to a reduction in vision not fully explained by organic disease, and is the commonest functional visual disorder encountered. Previous ideas as to the underlying aetiology of such states drew heavily on the Freudian concepts of somatisation of emotional conflicts, but it is now clear that this is only true in a proportion of cases. Patients with functional visual loss represent a continuous spectrum of disease from the true hysterical conversion reaction, through the "suggestible innocent", to the deliberate malingerer. The diagnosis of functional visual loss should be established by positive means, rather than by exclusion, and it is important to bear in mind that up to 50% of cases have some concurrent ocular disease and/or psychiatric disorders.

Functional visual loss usually presents either as unilateral or bilateral *reduction in vision or blindness,* often with a *visual field defect.* Once a thorough examination has excluded relevant organic disease, bilateral feigned blindness can usually be disproved objectively by demonstrating retained optokinetic nystagmus or refixation movements with prisms placed in front of the eyes. Similar methods can be applied in cases with unilateral functional blindness if the seeing eye is totally occluded prior to testing. Less severe visual reduction, both unilateral and bilateral, can be disproved using refractive manipulations of trial lenses while the patient reads a test type, or with dissociating stereopsis testing such as the TNO chart.

Functional *visual field loss* gives visual field defects that are non physiological, often with spiral or tubular fields, which do not expand with increasing test distance. There is often an inversion of the colour field loss found in organic disease, with the red field being wider than the blue. A striking feature is that even gross field defects often do not give rise to difficulty in orientation as they would if they were organic, and watching patients moving about the examination room gives invaluable information. Central scotomas are extremely unusual as functional field defects, and should prompt a search for organic disease. However, once a diagnosis of functional visual loss has been made, direct confrontation of the patient serves little benefit, and confident reassurance is all that is usually needed.

*Blurred vision* may be merely evidence of general fatigue and it is well to avoid attributing symptoms of fatigue to a neurotic condition on the one hand, or to slight errors of refraction which are not proportionate to the degree of disability on the other. Intermittent blurring of vision may however be part of an anxiety or hysterical reaction due to relaxation of the ciliary muscle with loss of tone and accommodative power; this may be accompanied by weakness of convergence in which an unusual feature is the failure of both eyes to converge. Normally, of course, one eye continues to fix the approaching object while the other fails to follow and then diverges. Psychotropic or anti-Parkinsonian drugs may be responsible and myasthenia should be considered.

*Accommodative spasm* presents with bilateral visual reduction and headache, and is usually easily recognised because of the associated pupillary miosis and marked convergence associated with the spasm of the convergence reflex. The spasm of the ciliary muscle results in a dramatic pseudomyopic shift in the refraction of the patient, who can usually read small test type at very close range. Ocular movements if present will be jerky due to the spasm of the medial rectus muscles. Accommodative spasm can often be overcome by atropinisation and true refractive errors will need correction, but the condition is likely to recur unless the environment is modified or the patient receives psychiatric treatment. Disturbances of accommodation and convergence are found particularly in adolescent girls trying to maintain exacting standards of conduct or studying for examinations in which they have little natural interest. One-eyed adults conscientiously trying to do detailed, responsible work seem to be more liable than others to develop accommodative spasm.

*Defective night vision.*This was found to be a psychoneurotic symptom in the majority of cases of night blindness during the last war. Pigmentary degeneration of the retina, chronic glaucoma and malnutrition involving particularly Vitamin A deficiency should of course be excluded but in these cases the disorder is usually of some standing.

*Disorders of colour vision.* Inversion of colour fields is common in functional visual loss, with the red field larger than the blue. Contraction of the fields and colour amblyopia may occur. Red is usually least affected, whereas in optic nerve lesions the reverse is usual. No explanation is known for the erythropsia or pink vision following cataract extraction. It is not infrequent and soon passes off.

*Photophobia.* While there is an increased susceptibility to light in lightly pigmented persons, this symptom is frequently a neurotic symptom. Other causes of photophobia require exclusion. In corrected myopia, for example, small bright images are focused on the retina, and patients with lens or corneal opacities are often very distressed by the dazzle due to dispersion of the light which they suffer.

## Illusions and hallucinations

A hallucination is a 'mental impression of sensory vividness in the absence of an external stimulus'. Hallucinations are to be distinguished from illusions in which an external stimulus is present but wrongly interpreted.

*Visual hallucinations.* Often highly complex hallucinations may occur in psychotic states such as schizophrenia and senile dementia, but in any elderly person with poor vision, irritation of the visual pathways by ischaemic or degenerative changes may sometimes give rise to unformed hallucinations of colours or branching patterns and these may be endowed with meaning and described for example as trees with red or blue flowers. Reassurance will often make the symptoms more tolerable. Darkroom delirium with hallucinations may be induced by bandaging both eyes for several days following an ocular operation, thus causing sensory deprivation in unfamiliar surroundings. This is now usually avoided. Florid, sometimes frightening hallucinations may have a toxic basis in delirium tremens from alcohol or from drugs like mescaline or LSD.

*Muscae volitantes.* The complaint of seeing floating specks, hairs or webs is very frequent in ophthalmic practice. In myopia, or as a senile change, the vitreous gel separates, to leave strands or particles floating in fluid vacuoles. The appearance of vitreous opacities may be caused by inflammatory conditions affecting the retina and the uveal tract and by retinal haemorrhage or detachment, and these conditions require exclusion by careful examination. When they are feeling tired, anxious or introspective, those who are subject to vitreous 'floaters' know how irritating they can be, especially when the background is moderately bright and uniform. Once reassured that they are not of serious import, patients are usually unaware of them most of the time. In others, vitreous opacities may give rise to illusions of various kinds, depending on the visual imagery of the individual, e.g. flies, spiders, cobwebs, etc.

## Functional disorders involving the lids

*Hysterical ptosis* is spastic and the spasm is increased on attempted raising of the lid. The fold of the upper lid and the smoothness of the forehead demonstrate the presence of active levator palpebrae and inactive frontalis muscles. Excessive blinking is very commonly met with in children in the absence of any other condition. It indicates a degree of nervous tension, the source of which should be unobtrusively sought and the symptom itself ignored. In adults, facial tics and blepharoclonus may be distinguished from organic clonic spasms, of Parkinsonism, for example, in that a co-ordinated movement is produced.

*Idiopathic blepharospasm,* characterised by bilateral repeated forceful eyelid closure, may be so debilitating as to render the patient functionally blind. This is now thought to be an organic movement disorder, with a chemical imbalance in the eyelid movement control centres, and these patients may gain enormous benefit from injections of Botulinum toxin to the orbicularis oculi muscle. Disorders such as *hemifacial spasm* were also previously thought to have a psychogenic basis, but seventh nerve compression in the cerebello-pontine angle is now often implicated, and a neurologist's opinion should be sought in case surgical decompression of the nerve is considered necessary.

## Self inflicted injuries to the eyes

Self-inflicted injuries often spring from the same origins as the functional disorders, and are therefore included here. Conjunctivitis artefacta is usually produced by inserting some irritant substance into the lower fornix. This may be due to simple malingering in which there is a deliberate intention to deceive others, or to a mixed state of hysteromalingering in which the pretence becomes involuntary and consistent, making diagnosis difficult. Usually one eye is involved, usually the left in a right-handed person. The lesion is often localised and out of keeping with the general appearance of the eye. Self-mutilation may extend to the gouging out of one or both eyes, perhaps in an oedipal type psychotic state. In one patient, who was a prisoner, several attempts had culminated in many deep cuts being made with a safety razor blade across each cornea, the lens having been penetrated on one side. The desire not to see (or perhaps not to be seen) was so great that this end was achieved for the patient by performing tarsorrhaphies to prevent further mutilation while psychiatric treatment was attempted.

**The importance of functional disorders**

Psychiatric illness is very common, and it is not surprising that ocular involvement is frequent. Among ophthalmologists the awareness of the need to consider the part played by functional disorders of the eyes and the capacity to deal with them grows with experience, and this can make an art of the day-to-day practice of ophthalmology.

# CHAPTER 22

# EYELIDS

**The anatomy of the eyelids is described on** p25.

## Eyelid malpositions

### *Entropion*

This describes a situation in which the tarsus and lid margin rotate inward. As a result, the lashes are directed towards the ocular surface causing irritation, reflex lacrimation and recurrent conjunctivitis. The cornea is at risk of ulceration. Entropion usually affects the lower lid and is classified as congenital, involutional or cicatricial. The latter may cause misdirected eye lashes (*trichiasis*) which increases the risk of corneal ulceration as, for example, in trachoma.

*Congenital entropion* is due to hypertrophy of the anterior lamella (skin and orbicularis oculi). Mild forms are common and usually resolve with time. If the child is photophobic or the eyes appear irritated, surgical correction can be achieved by excising an ellipse of skin and orbicularis.

*Involutional entropion* is caused by laxity of the tarsus and its medial and lateral canthal tendons, laxity of the lower lid retractors and over-riding of the orbicularis. Taping the lower lid provides temporary relief while awaiting surgical correction. The simplest form of surgery consists of passing absorbable sutures from the lower fornix through the lid to emerge on the skin just below the lashes (everting sutures). The everting effect can be augmented by splitting the lid horizontally. The lid can be shortened if horizontal laxity is a factor.

*Cicatrical entropion* is caused by scarring disorders which shorten the posterior lamella. The main causes are chemical burns, chronic conjunctivitis, trachoma, ocular cicatricial pemphigoid and Stevens-Johnson syndrome. The surgical treatment is complex and may involve grafting the posterior lamella in order to lengthen it and allow the lid to rotate outwards again. Even then, the condition often recurs as the scarring progresses. (Plates 22.1, 22.2 p448 LCW 2.11 p30).

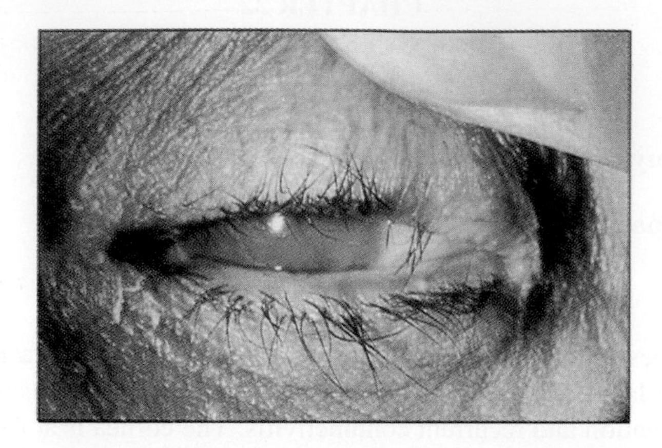

Plate 22.1 Trachoma causing cicatricial entropion of upper lid and trichiasis.

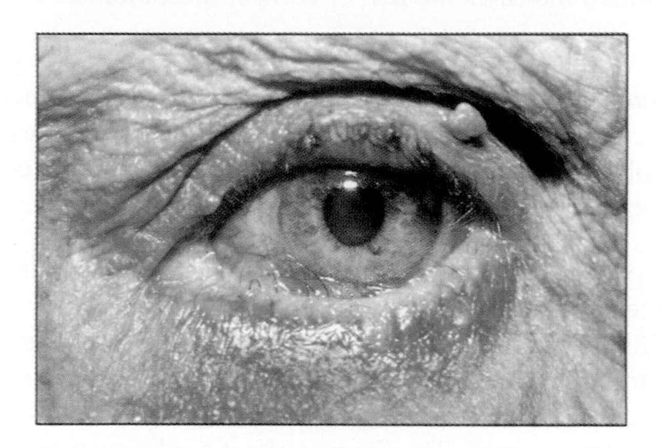

Plate 22.2 Ocular cicatricial pemphigoid - scarring of conjunctiva and cicatricial entropion.

## *Ectropion*

In this condition the tarsus and lid margin rotate outwards. The exposed conjunctiva becomes chronically inflamed and hypertrophic, the lower part of the cornea is exposed and the punctum is carried out of the tear lake, causing epiphora. Ectropion usually affects the lower lid and is classified as congenital, mechanical, involutional, cicatricial or paralytic.

*Congenital ectropion* is rare but may be seen, precipitated by attacks of crying, in babies. It may resolve spontaneously but if protracted may require surgery.

*Mechanical ectropion* is due to the weight of a lid lump. The treatment is to excise the tumour.

*Involutional ectropion* is due to horizontal laxity of the eyelid tissues. The treatment is to shorten the lid in the site of maximum laxity. In the case of medial ectropion with punctal eversion this can be combined with excision of a diamond of tarsoconjunctiva to invert the punctum.

*Cicatrical ectropion* is due to scarring of the anterior lamella due to trauma, burns or facial dermatoses. If the contracture is localised it may be corrected by a Z-plasty, but when it is generalised a skin graft is required.

*Paralytic ectropion* is due to a 7th cranial nerve palsy. If recovery is anticipated then treatment is aimed at protecting the globe. A temporary tarsorrhaphy may be required, depending on the degree of corneal exposure. A newer alternative is to drop the upper lid by paralysing the levator using botulinum toxin. If the palsy is permanent then a variety of lid-shortening procedures can be employed with or without lateral tarsorrhaphy (Fig. 22.3 p449, LCW 2.10 p30).

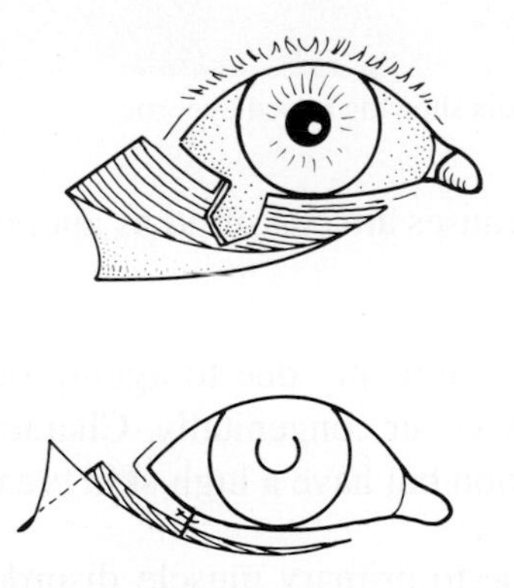

Fig. 22.3 Ectropion - correction by the Kuhnt - Szymanowski operation.

*Ptosis (Blepharoptosis)*

## *Types of ptosis*

This describes drooping of the upper eyelid, the margin of which normally lies 1-2mm below the upper limbus.

*Congenital ptosis.* If the lid covers the eye sufficiently to prevent its use, this constitutes a relative ophthalmic emergency as delay in elevating the lid results in the development of amblyopia. The most common cause of congenital ptosis is an isolated dystrophy of the levator muscle and there is sometimes a family history of the condition (Fig 22.4 p450, LCW 2.8 p30).

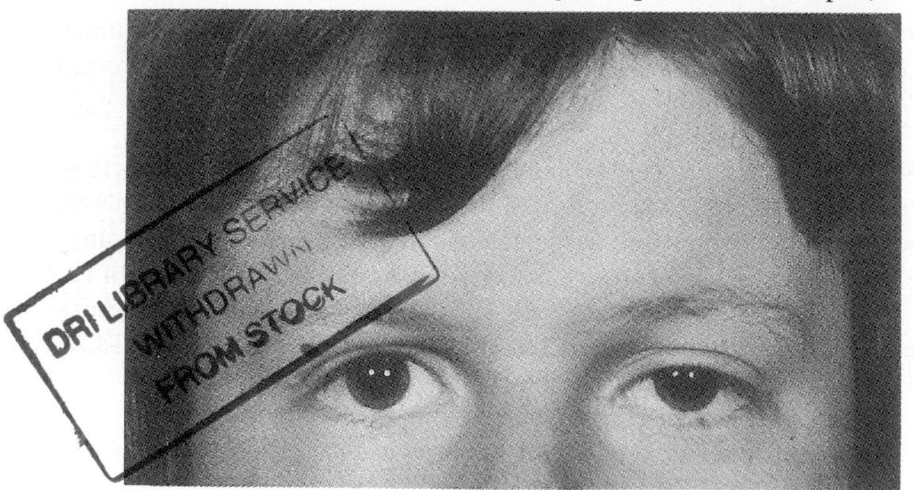



Fig. 22.4 Left congenital ptosis showing raised eyebrow.

*Acquired ptosis.* The causes are classified as aponeurotic, myogenic, neurogenic or mechanical.

*Aponeurotic ptosis* is commonly due to ageing or to repeated episodes of oedema, although it may occur congenitally. Characteristically these patients retain good levator function but have a high skin crease and a thin upper lid.

*Myogenic ptosis* is due to primary muscle disorders, such as chronic progressive external ophthalmoplegia (CPEO). *Myasthenic ptosis* is due to disordered neuromuscular transmission.

*Neurogenic ptosis* includes 3rd cranial nerve disease (Fig. 11.19 p250, LCW 2.9 p30) and Horner's syndrome, both of which may have potentially life-threatening causes, such as intracranial aneurysm or an apical lung neoplasm. In IIIrd nerve palsy there are abnormalities of ocular movement and the pupil may be dilated. In Horner's syndrome, due to sympathetic paresis, the pupil is small and there may be loss of sweating and vasomotor control on the same side of the face. A special type of neurogenic ptosis occurs congenitally due to abnormal brainstem connections. In the Marcus-Gunn jaw-winking phenomenon the ptotic lid elevates rhythmically with chewing due to synkinesis with the pterygoid muscles (Fig. 22.5 p451).

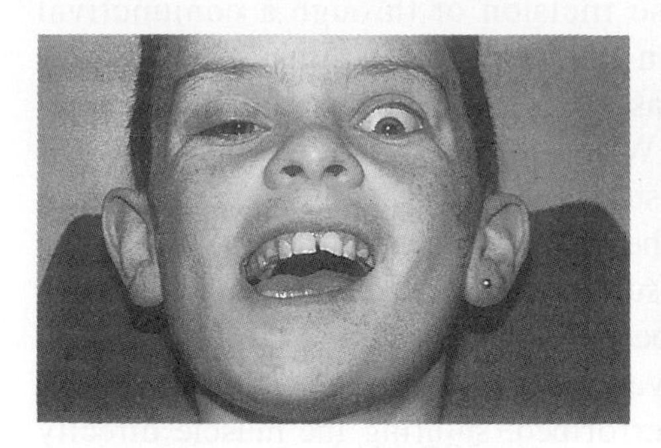
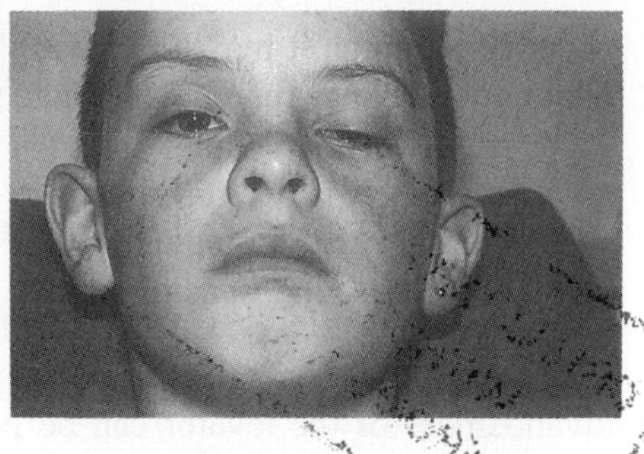

Fig. 22.5 Marcus-Gunn jaw-winking phenomenon. - 1. Mouth open. - 2. Mouth closed.

*Mechanical ptosis* is due to the weight of an upper lid tumour, such as neurofibroma.

*Treatment of ptosis*

When a congenital ptosis is occluding the visual axis treatment should be carried out without delay to prevent the development of amblyopia. Otherwise, the visual development can be monitored and surgery carried out at preschool age, when the tissues are better developed. Attention should be directed to the underlying cause in neurogenic ptosis. Treatment of myasthenia is medical, although residual ptosis in long-standing disease can be managed surgically. Surgical correction of ptosis should be undertaken with extreme caution in patients with a poor Bell's phenomenon, dry eye or impaired corneal sensation. A spectacle ptosis prop may be useful. The risk of exposure keratopathy must be explained to the patient and the possibility that the lid may have to be lowered again if exposure problems are severe must be specifically mentioned. Antibiotics and lubricants are traditionally employed post-operatively until the ocular surface becomes accustomed to the new lid height, and may be necessary life-long if the eye remains open during sleep.

In aponeurotic ptosis the aponeurosis is repaired, or advanced and sutured to the tarsal plate through a skin-crease incision or through a conjunctival approach. The choice of procedure in myogenic ptosis hinges on the pre-operative levator function. This is measured in mm between the lid positions in the extremes of down- and upgaze. Where levator function is poor, i.e. less than 5mm. it is unlikely that levator surgery will be helpful. These patients often elevate their lids by recruiting the frontalis muscle, and this effect can be augmented by performing a brow suspension procedure, tunnelling strips of inelastic material e.g. fascia lata, between the tarsus and the frontalis. In patients with moderate to good levator function, i.e. 5-15mm, graded advancement of the levator can be performed, suturing the muscle directly onto the tarsus to raise the lid and resecting the excess.

*Eyelid retraction*

This is a feature of thyroid eye disease (p306). In the early stage of hyperthyroidism lid retraction and lid lag are due to hypersensitivity to adrenergic stimulation of Müller's muscle. In established ophthalmopathy retraction is due to infiltration and fibrosis of the levator itself. Surgical correction should not be undertaken until the patient is clinically and biochemically euthyroid to avoid unnecessary anaesthetic risk. It is also prudent to wait until the lid signs have been stable for about six months to help avoid under or over-correction of the lid height. The upper lids are lowered by sectioning the levator or recessing it on adjustable sutures. The lower lids are raised by grafting a spacer material such as donor sclera between the retractor muscle and the tarsus.

## Eyelid inflammatory conditions

*Dermatitis*

The thin skin of the lids is prone to swell due to oedema in response to allergic phenomena. The eyelids may be involved in a more widespread atopic dermatitis. In this condition there is often a family history of the atopic triad - dermatitis, hay fever and asthma. Patients have high levels of circulating IgE and often have a peripheral eosinophilia. Antihistamines may be useful in controlling itch. In *allergic contact dermatitis* prior exposure to an allergen leads to a specific cell-mediated sensitisation (LCW 2.4 p28). Identification and avoidance of the sensitising substance is the key to managing these patients. Sometimes the allergen is obvious from the history but in other cases patch testing may be necessary. *Direct irritant contact dermatitis* is due to exposure to a substance which damages the normal barrier function of the epidermis. Identification of the irritant is based purely on the history and patch testing is inappropriate.

All forms of dermatitis are treated with wet dressings in the acute phase and with weak topical steroid cream (e.g. hydrocortisone 0.5%) in the chronic phase.

*Blepharitis*

This is classified as anterior or posterior inflammation of the lid margin. The junction of the anterior and posterior lamellae can be seen clinically on the lid margin as the grey line. *Anterior blepharitis* may be seborrhoeic or staphylococcal.

*Seborrhoeic blepharitis* occurs as part of the generalised condition of *seborrhoeic dermatitis,* which shows waxy crusting on the scalp, eyebrow, sternal area and skin flexures (LCW 2.1 p28). Treatment is difficult. The lids should be kept clean of crusts and if they develop frank signs of inflammation then topical steroid-antibiotic preparations may help. *They should not be used regularly without monitoring the intraocular pressure to avoid steroid glaucoma.* Daily cleaning of the lid margins on waking, with sterile water or 'baby shampoo' will allow the patient to be more comfortable but it will need to be continued.

*Staphylococcal blepharitis* is usually due to *Staphylococcus albus* or coagulase negative *Staphylococcus aureus.* The infection is based on the lash follicles.

The lid margin is erythematous and the lashes show exudate and crusts at their bases (Plate 17.4 p322, LCW 2.2 p28). Frank pus may be seen with coagulase positive organisms and in these cases trichiasis, lid ulceration and scarring are seen. Treatment is lid hygiene as described above and topical antibiotics.

*Posterior blepharitis* is characterised by telangiectasia of the lid margin, and inflammation of the meibomian glands (meibomitis). It is often associated with acne rosacea; if facial skin changes are absent then the term ocular rosacea is used. Conjunctival and corneal signs are common, in particular chronic papillary conjunctivitis and marginal keratitis (Plate 17.5 p322). Multiple or recurrent chalazia are frequent. The treatment is regular expression of the tarsal glands and long-term low-dose tetracyclines which appear to have the unique effect of favourably altering meibomian secretion.

**Cystic lesions of the eyelids**

*Stye.* This is an abscess of the lash follicle and its associated structures. It may occur in isolation or in association with anterior blepharitis (LCW 2.5 p29). Application of heat and topical antibiotic is usually sufficient to affect a cure but a systemic antibiotic may be necessary if the surrounding lid tissues begin to show signs of cellulitis.

*Chalazion* (Meibomian cyst). Retention of Meibomian secretion excites a granulomatous foreign body reaction (LCW 2.6, 2.7 p29). If infected, chalazia have a tendency to discharge through the tarsal conjunctiva. If a chalazion requires surgical drainage, the approach is always via the tarsal conjunctiva to avoid skin scarring. The posterior wall of the cyst is excised, the contents curetted out and the anterior lining destroyed by cauterisation (Fig. 22.6 p454).

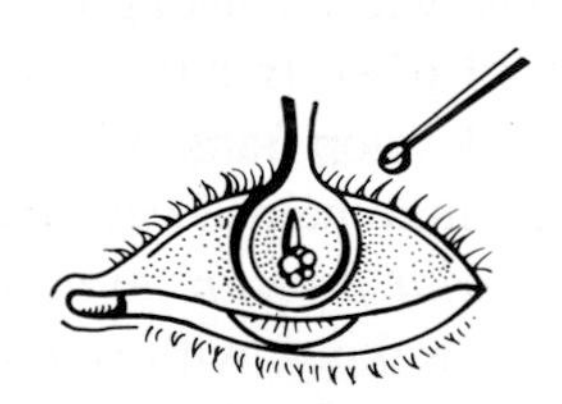

Fig. 22.6 Chalazion - incision and curettage using a ring-shaped clamp.

If the tissue is submitted for histology, a characteristic lipogranulomatous reaction is seen. Multiple or recurrent chalazia should always be sent for pathological examination to avoid missing the diagnosis of *sebaceous carcinoma,* discussed below.

*Epidermal cysts* arise from hair follicles or from traumatic implantation of epidermal elements. They are slow growing, lined by stratified squamous epithelium and keratin-filled. Rupture of the cyst provokes a brisk inflammatory reaction.

*Dermoid cysts* are due to sequestration of skin elements in embryonic lines of closure. They are often located in the upper lid at the lateral canthus and may communicate with the superotemporal quadrant of the orbit through a bony defect (dumbbell tumour) (Fig. 20.4 p429). Care is also required to distinguish a dermoid cyst from a meningococle which is usually situated on the medial side of the orbit and may be continuous with the subarachnoid space via a gap in the skull. A C.T. scan is therefore advisable before attempting excision. Microscopically, the cyst is lined by skin, including pilosebaceous units and sweat glands and contains keratin, sebum and hairs. Rupture of such a cyst provokes severe inflammation in surrounding tissues.

*Sudoriferous cysts.* These result from sweat gland obstruction. The commonest type is a cyst of Moll, which appears at the lid margin and contains watery fluid. Microscopy shows a double layer of epithelium, an inner cuboidal and an outer myoepithelial layer. Simple incision with removal of part of the anterior wall is usually effective.

## Epithelial and other lesions of the eyelid

*Basal cell papilloma.* These are well-demarcated, verrucous or velvety pigmented lesions which are common in the elderly and which are also known as seborrhoeic keratoses. The pigmentation is due to melanocytic activity in the basal layer and may be dense enough to cause diagnostic confusion with naevi.

*Squamous cell papilloma* are similar but show more pronounced keratin production and less pigmentation. It may be difficult to distinguish them from viral warts.

*Keratoacanthoma.* These occur on sun exposed areas of middle-aged and elderly patients. They grow rapidly for a few weeks to form a cratered nodule with rolled edges and a central keratinous horn. They then involute spontaneously over a period of months to leave a depressed scar. Distinction from a squamous carcinoma can be very difficult clinically and histologically but helpful features are a straight base with absence of invasion and a junction with healthy epidermis rather than with areas of actinic keratosis.

*Viral warts* (verruca vulgaris) are common in children and young adults. they are often horny or thread-like in the facial region. The causative virus is the human papilloma virus (HPV), a DNA virus of the Papovavirus group. They are differentiated from other acanthotic lesions by their vacuolated cells and the ground-glass appearance of their nuclei. They regress due to a cell mediated immune process associated with a mononuclear cell infiltrate.

*Molluscum contagiosum* (p329 and Plate 17.7 p329) present as multiple, pearly, umbilicated lesions. There is often a follicular conjunctivitis due to discharge of virus-laden material into the eye. Molluscum virus is a member of the pox virus group and the characteristic histological appearance is of cytoplasmic inclusions called molluscum bodies. The lesions are cured by expression, curettage or puncture with a sharp orange stick dipped in iodine solution. The lesions may be florid in immuno-compromised patients as in AIDS.

*Actinic keratosis* is common in fair-skinned individuals exposed to high levels of ultraviolet light. Clinically these appear as raised, scaly lesions, Histologically the epidermis shows thickening of the keratin layer (hyperkeratosis) and persistence of nuclei into this layer (parakeratosis). The underlying dermis shows elastotic degeneration. These areas of intraepithelial dysplasia are prone to develop into squamous cell carcinomas although the latent period is relatively long.

*Squamous carcinomas* present as rapidly-growing ulcerated nodules or hyperkeratotic papules. They spread locally into the orbit and by lymphatic spread to the submandibular and pre-auricular nodes. They can be graded histologically according to their degree of differentiation. The usual cause is ultraviolet light, and rarer causes are radiation, arsenic ingestion (Bowen's disease) and xeroderma pigmentosum, which is an autosomal recessive fault in DNA repair mechanisms.

*Basal cell carcinoma (Rodent ulcer)* (Plate 22.7 p457 and Fig. 22.8 p458) is the commonest malignant lesion of the lids, and occurs especially at the medial canthus. There are a variety of clinical and pathological variants. The tumour may appear as a diffuse plaque or a cyst but most often appears as an ulcerated nodule. The margins of the lesion often show a pearly, rolled-over edge with prominent telegiectases (LCW 2.14 p31). The basal cells are arranged around the periphery of the cell nests in a characteristic manner known as 'palisading'. The histological subtypes are solid, cystic, keratotic, adenoid and sclerosing. There is no difference in prognosis between the first four subtypes, all of which are curable by surgical excision (leaving a 5mm margin of healthy tissue), cryotherapy, or radiotherapy. Sclerosing tumours stimulate a prominent fibrous tissue reaction and infiltrate diffusely making the margins of the tumour difficult to define clinically. Adequate excision may require the technique of Moh's lamellar cryosurgery, which uses frozen sections to determine clearance of all the margins. These tumours have a strong tendency to infiltrate along the orbital walls, and orbital exenteration may be necessary if diagnosis is delayed.

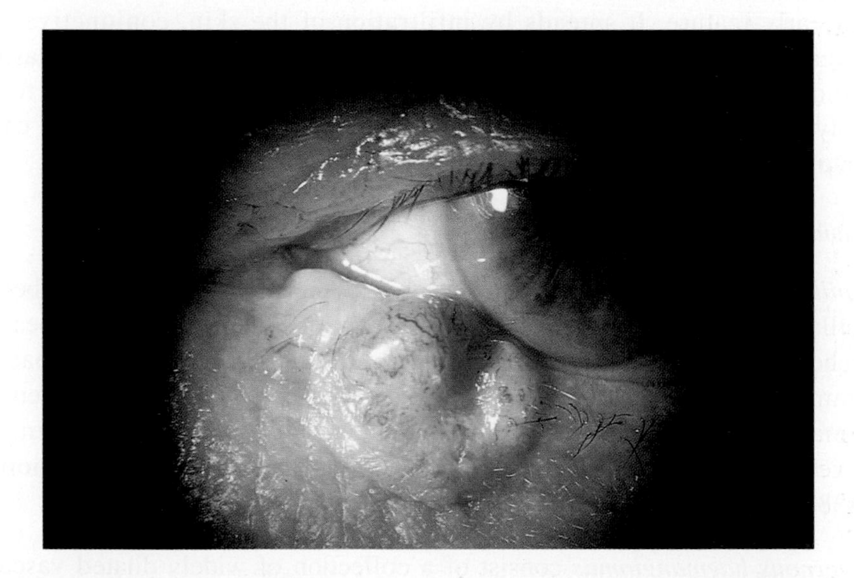

Plate 22.7 Rodent ulcer of lower lid.

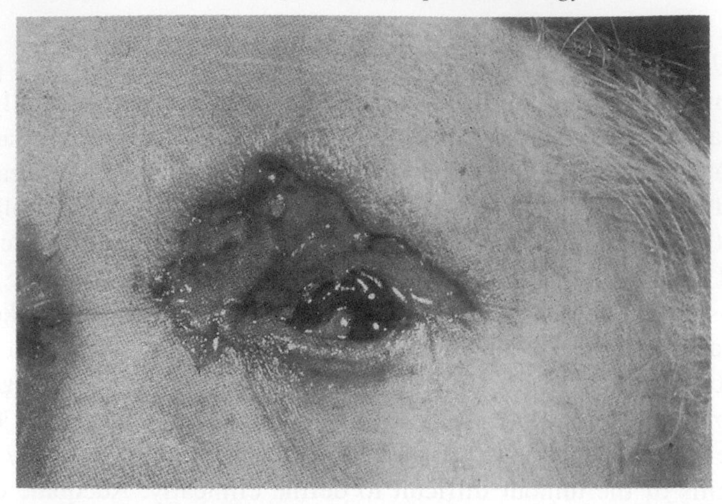

Plate 22.8 Rodent ulcer late, neglected.

*Sebaceous carcinoma* arises from the meibomian glands or the sebaceous glands of the lashes. It presents clinically as diffuse thickening of the lid, recurrent or multiple chalazia or refractory blepharitis. Lash loss is a prominent early feature. It spreads by infiltration of the skin, conjunctiva and cornea. It shows variable differentiation characterised by fine cytoplasmic lipid droplets. This is the most aggressive tumour of the eyelids, though fortunately very rare. Treatment is by wide surgical excision. In advanced cases orbital exenteration or radiotherapy may be necessary.

*Haemangioma*

*Capillary haemangiomas* are characterised by an increase in the number of capillaries in the superficial plexus of the dermis. Clinically they appear as patches of purplish discoloration of the skin, the port-wine stain, or naevus flammeus. The Stürge-Weber syndrome comprises facial capillary haemangiomas in the territory of the first and second divisions of the trigeminal nerve which may be associated with meningeal and choroidal angiomas (p438 and Plate 20.8 p439, LCW 8.22 p128).

*Cavernous haemangiomas* consist of a collection of widely-dilated vascular spaces in the dermis. They are elevated and bright red (strawberry naevus) and tend to grow with the child for the first year then involute slowly over a number of years. The importance of eyelid angiomas lies in the fact that they may cause occlusional amblyopia (as in Plate 22.9 p459). For this reason it may be necessary to treat these lesions, which in other body sites may be left to regress spontaneously. Treatment involves injecting the lesion with corticosteroids at monthly intervals until it has regressed sufficiently to clear the visual axis and allow normal visual development. This will often also require patching of the fellow eye and refractive correction of any astigmatism induced by the weight of the lesion pressing on the cornea. Excision of any residual fold of redundant skin may be undertaken for cosmetic reasons once the vascular anomaly has regressed completely, which is usually around the age of eight. Cavernous haemangiomata may rarely be associated with thrombocytopenia.

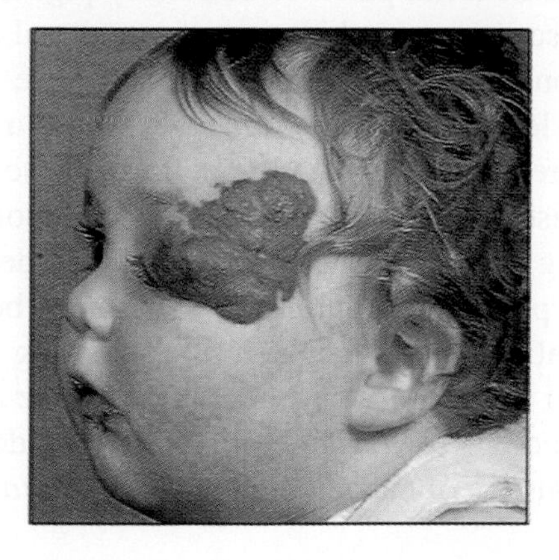

Plate 22.9 Cavernous haemangioma causing ptosis.

*Kaposi's sarcoma* is a spindle-cell vascular tumour occurring in the acquired immunodeficiency syndrome. It may present on the eyelid or the conjunctiva (p336).

*Neurofibroma* (see also p437 and p538). These are benign tumours of peripheral nerves which occur in type I neurofibromatosis (von Recklinghausen's disease). In the upper lid the tumour is often of the plexiform type and causes ptosis. Debulking and raising the affected lid presents a considerable surgical challenge. Other features of neurofibromatosis, which is an autosomal dominant condition, include café au lait spots, other cutaneous neurofibromas, axillary freckling, iris Lisch nodules, prominent corneal nerves, choroidal naevi, astrocytic hamartomas, optic nerve gliomas, absence of the greater wing of the sphenoid bone (which results in a spheno-orbital encephalocele, causing pulsatile exophthalmos) and multiple endocrine neoplasia, particularly phaeochromocytoma. Eyelid plexiform neurofibromas are associated with ipsilateral glaucoma and with hemifacial hypertrophy.

*Naevus* is a term applied to benign tumours due to proliferation of melanocytes (LCW 2.13 p31). When the clusters of melanocytes are confined to the basal layer of the epidermis the term *junctional naevus* is used. Clinically this appears as a flat patch of uniformly pigmented skin. The term *compound naevus* describes a combination of junctional activity in the basal layer and proliferation in the underlying dermis. These appear clinically as darker brown raised lesions. *Intradermal naevi* show a normal basal layer with packets of mature naevus cells in the dermis. These appear clinically as faintly pigmented raised lesions. Junctional and compound naevi are more common in the young and intradermal naevi in the elderly, so it is thought that this represents a process of continuous maturation between these forms. *Blue naevi* are palpable blue-black lesions consisting of mature spindle-shaped naevus cells in the dermis. It is thought that these are neural crest cells which have failed to complete their migration to the epidermis. In the eyelids these congenital lesions are often called the *naevus of Ota.*

*Malignant melanoma* is common in fair skinned individuals exposed to sunlight. It may arise de novo or in a pre-existing naevus. The signs that a naevus is undergoing malignant change are growth in size, change in shape, change in pigmentation, bleeding or ulceration. A melanoma often undergoes a phase of intra-epidermal growth before invading the dermis. This phase may last many years but is only a few months in a superficial spreading melanoma. In these conditions, the development of nodules correlates to deep invasion. Nodular melanoma is the most aggressive variant and often arises in healthy skin. In all forms, the prognosis relates inversely to the depth of invasion of the dermis in mm. Cryotherapy may be used in lentigo maligna. Surgical excision is the treatment of choice in more aggressive forms but chemotherapy has a developing rôle in metastatic disease. (Plate. 22.10 p461).

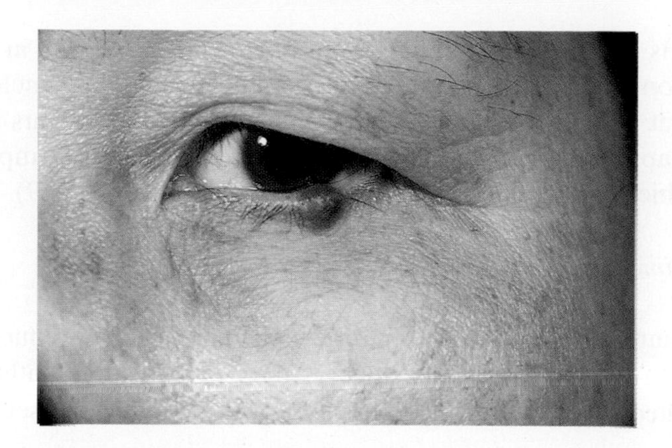

Plate 22.10 Malignant melanoma of the lid margin.

*Lipid deposition*

*Xanthelasmas* are yellow or creamy plaques of lipid deposition in the dermis of the upper and lower lids near the medial canthi (LCW 2.12 p31). Histologically they are composed of clusters of lipid-laden macrophages. The condition is usually idiopathic although it may be associated with type IIa hyperlipidaemia and therefore the serum lipids should be checked if there is a family history of hyperlipidaemia or premature cardiovascular disease. Treatment is by surgical excision, but as this may involve removal of extensive amounts of skin, laser ablation is becoming more popular.

**Eyelid lacerations** (p160)

**Congenital abnormalities**

Congenital ectropion, entropion and ptosis have been discussed above.

*Cryptophthalmos*

In this condition the eyelids fail to separate and the skin is continuous over the eyes. It is invariably associated with small malformed eyes (microphthalmos or even anophthalmos), with a large cystic space lying between the rudimentary ocular structure and the closed lids. It is frequently associated with widespread systemic malformations such as dyscephaly, syndactyly and urogenital anomalies. It is best left untreated surgically.

*Epicanthus* is a common condition. A vertical skin fold occurs in the medial canthal region tending to conceal the medial angle and the caruncle. It is normally present in Orientals and is seen in many Caucasian infants in whom it usually disappears during facial growth. It may give the impression of esotropia which can be excluded by the cover-uncover test (p229).

*Blepharophimosis* (Fig. 22.11 p462)

This is an autosomal dominant condition characterised by reduction of the palpebral aperture in all its dimensions, with epicanthus, telecanthus and ptosis. Associated features are maxillary hypoplasia, small, low-set ears and a

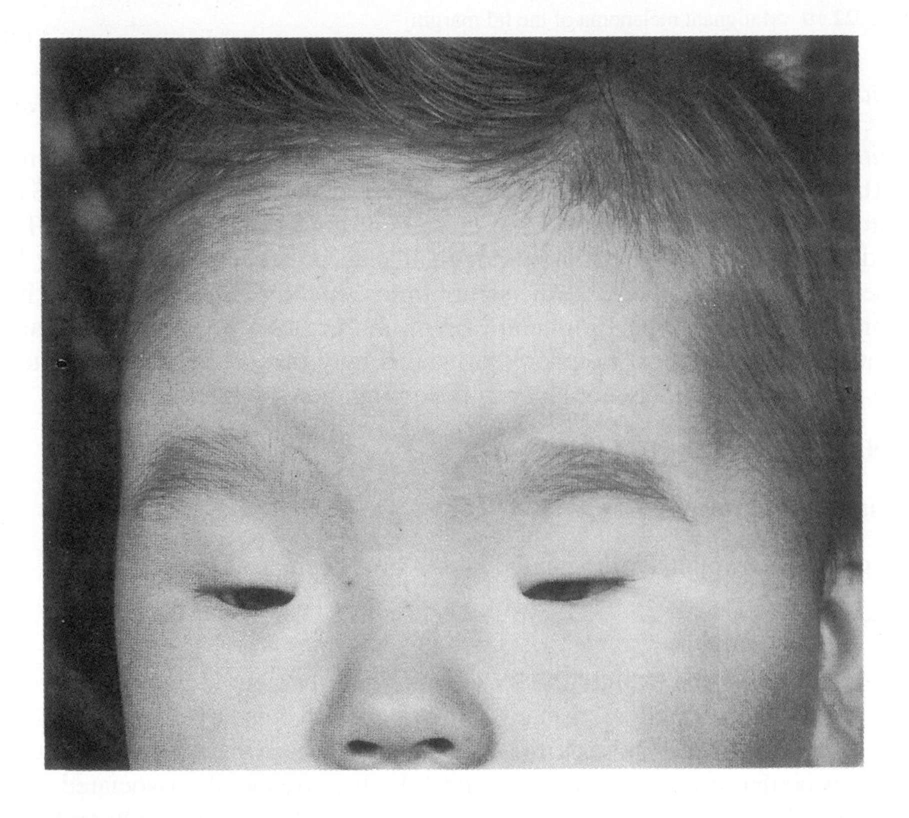

Fig. 22.11 Blepharophimosis.

fair complexion. The patients are normal intellectually and developmentally. Treatment is to correct the epicanthus and telecanthus using a Y-V or Mustardé procedure combined with plication of the medial canthal tendon, followed by ptosis repair as a second procedure. As levator function is usually poor, this normally necessitates a brow suspension procedure, so it is advisable to wait until the child is old enough to obtain fascia lata.

*Coloboma* (LCW 8.16 p127)

This is a defect in the lid margin and may be associated with a limbal dermoid. In isolation they are most common in the medial third of the upper lid. They also occur as part of the facial cleft syndromes or in the branchial arch syndromes such as Goldenhaar's or Treacher-Collins' syndromes (p546). Upper lid colobomas can be repaired directly when they are less than 4mm long; larger defects require a composite flap from the lower lid. Lower lid colobomas often require more extensive flaps and grafts for satisfactory repair.

# CHAPTER 23

# LACRIMAL APPARATUS

## Lacrimal gland swelling

When considering the possible causes of a lacrimal gland swelling it is helpful to be aware that approximately 50% of swellings are inflammatory or lymphoid and 50% are epithelial tumours. Of the primary epithelial tumours 50% are benign (pleomorphic adenomas) and 50% are malignant. Of the malignant epithelial tumours, 50% are adenoid cystic carcinomas. Of the remainder, 50% are adenocarcinomas and 50% are malignant mixed tumours.

*Dacryoadenitis.* (Plate 23.1 p465) Inflammatory causes include swelling and tenderness of the lacrimal gland, caused by viral infections such as mumps (paramyxovirus) or infectious mononucleosis (Epstein-Barr virus). Bacterial infection causing suppurative adenitis is rare but may occur in debilitated patients. Tuberculosis or sarcoidosis (p365) cause granulomatous inflammation.

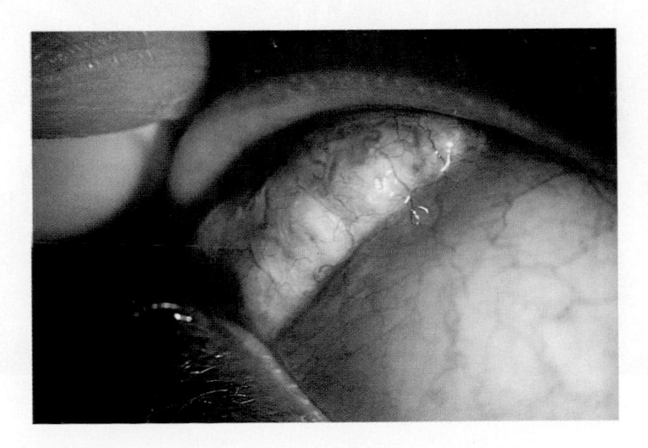

Plate 23.1 Swelling of lacrimal gland.

*Lymphoid proliferation.* The lacrimal gland contains a significant proportion of lymphocytes and may be the seat of orbital lymphoid proliferation.

*Tumours of the lacrimal gland*

*Pleomorphic adenoma.* Patients present with a long history of gland enlargement and bony erosion of the lacrimal fossa. Histologically the tumour consists of epithelial (cuboidal cell acini surrounded by myoepithelial cells) and stromal (fat, fibrous and chondroid tissue) elements. The tumour has a pseudo-capsule which contains epithelial elements presumably responsible for the recurrences which occur if removal is incomplete or the capsule is breached. Its extent can be assessed by C.T. scanning (Fig. 23.2 p466).

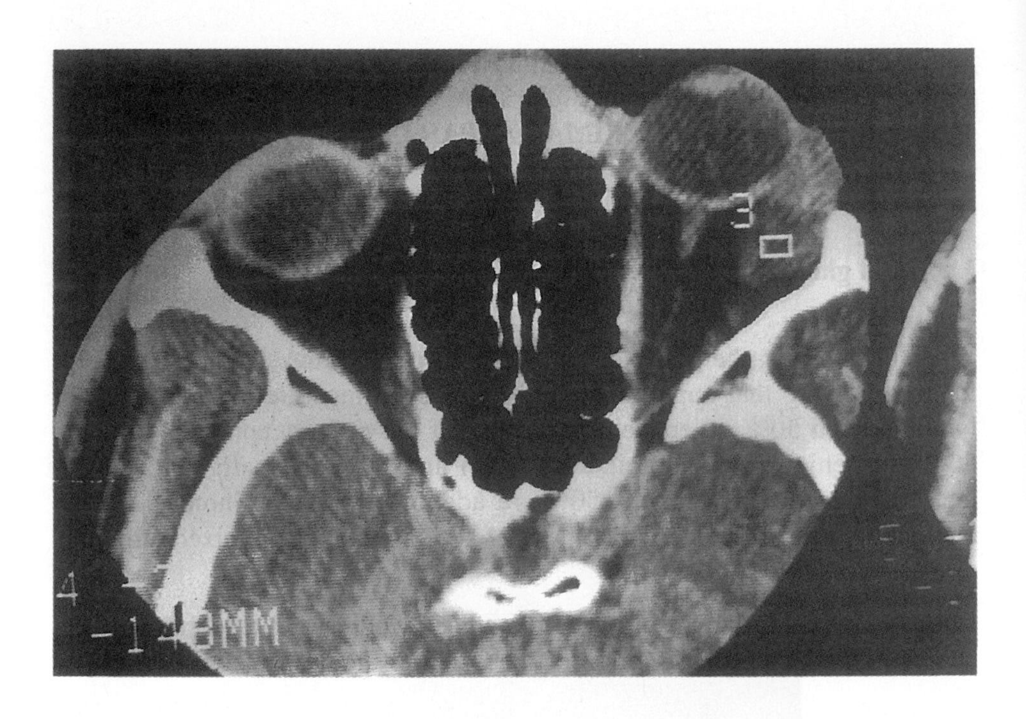

Fig. 23.2 Lacrimal gland adenoma, C.T. Scan.

*Lacrimal gland carcinoma.* Adenocarcinoma presents as a rapidly-growing mass, which is painful due to bony destruction. The tumour is radio-resistant and treatment is primarily surgical. Adenoid cystic carcinoma is the commonest and most aggressive variant. Local invasion of bone and perineural spread are seen histologically when exenteration is carried out. Malignant mixed tumour refers to carcinoma arising in a pleomorphic adenoma.

## The lacrimal drainage system

### *Congenital anomalies*

*Punctal and canalicular agenesis.* Atresia of the puncti can be relieved by perforation with a punctum seeker then dilatation. Absence of both puncta is generally accompanied by absence of all canalicular tissue and often by either partial or total absence of the sac. An exploratory dacryocystorhinostomy operation (DCR) (pp212, 213 and Figs 9.2 - 9.7 pp214 - 215) is carried out and a retrograde search made for any rudimentary canaliculi which can be marsupialised in the fornix. If none can be found then a bypass tube is inserted. Absence of a single punctum is strongly associated with nasolacrimal obstruction, which is treated by probing or if necessary by a DCR.

*Nasolacrimal obstruction.* This is usually due to non-perforation, during development, of a membrane at the lower end of the nasolacrimal duct (pp29, 49, 211), at the junction of nasal and lacrimal mucous membranes (LCW 8.9 p125).

*Amniotocele.* This is a congenital tense swelling of the sac associated with nasolacrimal obstruction. Pressure on the sac usually empties it, after which the child is managed as any nasolacrimal obstruction. If pressure fails then early probing is necessary to avoid the development of dacryocystitis.

*Supernumerary puncta.* These cause epiphora when one of the puncta is incompetent. The treatment is surgical obliteration of the canaliculus provided that the other canaliculus is normal.

*Congenital lacrimal fistula.* These usually communicate with the common canaliculus. Symptoms are due to leakage of tears from the abnormal opening, to secondary skin changes, or to infection. In symptomatic patients the duct is usually abnormal and when this is so, fistulectomy is carried out in combination with a DCR.

*Acquired conditions*

*Punctal stenosis.* Closure of the puncta often occurs in association with medial ectropion of the lower lid and the resulting metaplasia of the exposed conjunctiva. Replacement of the punctum in the tear lake by lid surgery will allow the punctum to re-open, failing which it may be marsupialised by punctoplasty (the 'three snip' procedure).

*Canaliculitis* - p212

*Dacryocystitis - acute and chronic* - p212

*Lacrimal sac tumours.* Primary tumours are rare and are usually transitional cell carcinomas or of the poorly-differentiated squamous cell type. Lymphoma or secondary spread from sinus carcinoma occur in this site (p213).

*Lacerations of the lacrimal drainage apparatus* (LCW 9.11 p136)

Considerable controversy surrounds the subject of treatment of canalicular lacerations, some authorities consider that all such injuries to the superior or inferior canaliculi should be repaired. Some studies have demonstrated that symptoms of epiphora are rare unless both canaliculi are damaged. In addition, stenosis at the site of repair is common, regardless of the surgical method used. It follows from this that in lacerations of a single canaliculus, no method of repair should be used which compromises the fellow canaliculus. In isolated canalicular laceration priority should be given to accurate repair of the lid, with good apposition to the globe, and marsupialisation of the medial end of the divided canaliculus. In bicanalicular lacerations both medial ends can be marsupialised or repair of the canaliculi can be carried out with intubation of the canaliculi and nasolacrimal duct; it must be remembered, however, that this carries the risk of iatrogenic damage to the duct.

Lacerations of the common canaliculus or lacrimal sac require a DCR and intubation as part of the primary repair (p213 and Figs. 9.2 - 9.7 pp214 - 215).

# CHAPTER 24

# THE CONJUNCTIVA

(see also p191)

## Conjunctivitis

Conjunctivitis is in general characterised by the presence of red sticky eyes due to inflammation causing dilatation of its blood vessels and stimulation of mucus production by its mucous glands (p191)

## Conjunctival Degenerations

*Conjunctival lithiasis (concretions)*

These yellowish deposits are a relatively common finding in older people especially if the conjunctiva is chronically inflamed. If present, they are usually found in palpebral conjunctiva. They are composed of finely granular material and membranous debris including phospholipid and elastin, but no phosphate or plasmin, suggesting that they occur secondary to cellular degeneration without calcification. If they ulcerate and cause irritation they can be removed using a sterile needle.

*Pinguecula*

These common yellowish interpalpebral perilimbal plaques are areas of elastoid degeneration of conjunctival substantia propria with epithelial thinning. They are rarely of more than cosmetic importance but can occasionally become inflamed (pingueculitis) or calcified (LCW 3.15 p46).

*Pterygium*

A pterygium is a triangular shaped lesion which consists of hyperplastic subepithelial bulbar conjunctival tissue (Plate 24.1 p470, LCW 3.14 p46). They are usually seen in the nasal interpalpebral region of persons who have been exposed to a hot, dry climate for a number of years. An early pterygium can be difficult to distinguish from a pingueculum (see above). The apex of the lesion involves superficial peripheral cornea and with progression into the

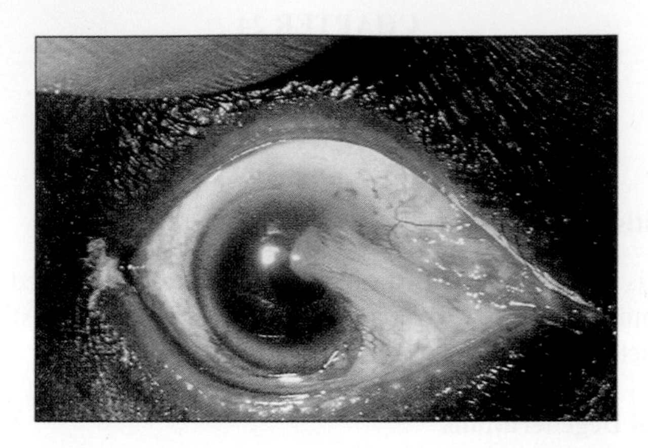

Plate 24.1 Pterygium.

cornea there is associated destruction of Bowman's membrane. Deposition of iron within the corneal epithelium may be seen at the advancing head of a pterygium (Stocker's line) and if the pterygium is prominent there may be an associated localised depressed area of corneal drying and thinning (dellen). (Plate 24.2 p470)

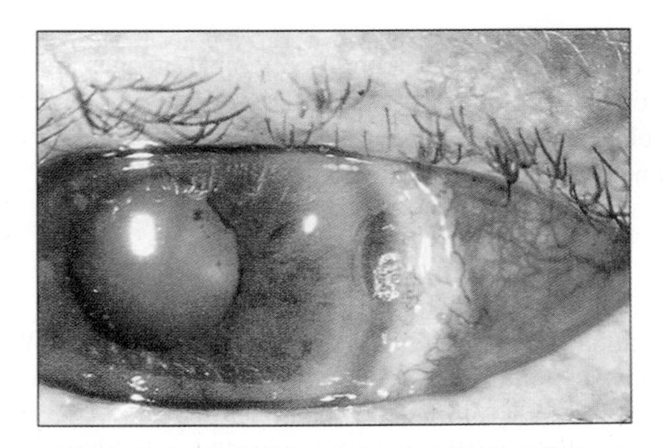

Plate 24.2 Dellen ulcer.

Surgical excision (Figs. 24.3, 24.4, 24.5 p471) is indicated if there is progression towards the visual axis or if the lesion is cosmetically unacceptable. Recurrence is common but is reduced by complete excision with beta irradiation or autoconjunctival grafting using superior bulbar conjunctiva from the same eye.

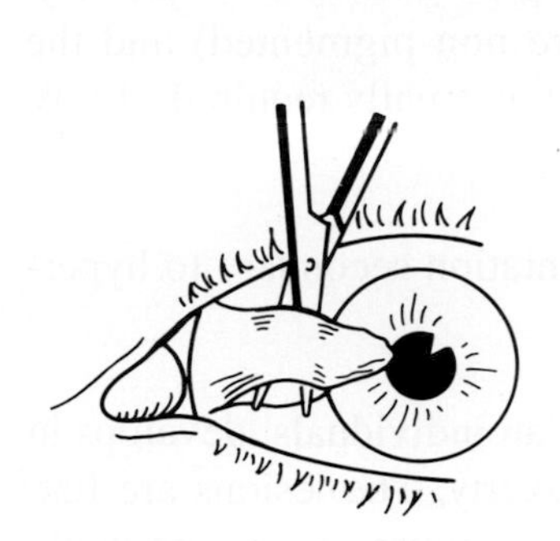

Fig. 24.3 Excision of pterygium - under mining the body.

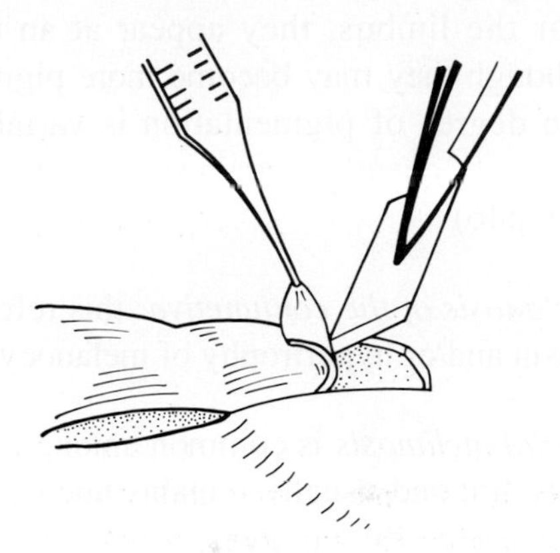

Fig. 24.4 Excision of pterygium - dissection of the head.

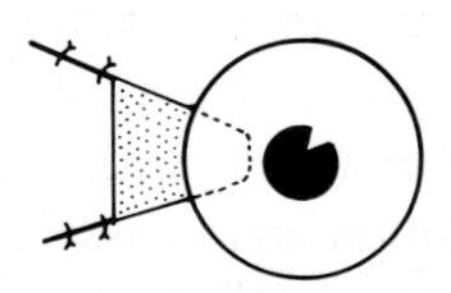

Fig. 24.5 Excision of pterygium - closure leaving bare sclera to reduce recurrence.

*Pseudopterygium*

A true pterygium is adherent to underlying tissue throughout its length, whereas a pseudopterygium is an adhesional fold of conjunctiva attached to the cornea by its apex alone. Pseudopterygia may form in the presence of a chronic peripheral corneal ulcer or other corneal damage.

## Conjunctival tumours

### *Pigmented conjunctival lesions*

*Melanotic conjunctival naevus:* Pigmented naevi are common tumours consisting of naevus cells which are similar to melanocytes. Frequently situated near the limbus, they appear at an early age, are benign and stationary, although they may become more pigmented during pregnancy or at puberty. The degree of pigmentation is variable (30% are non-pigmented) and the lesions frequently contain small cysts. Excision is not usually required. (LCW 3.16 p46)

*Melanosis of the conjunctiva:* this refers to pigmentation secondary to hyperplasia and/or hypertrophy of melanocytes .

*Racial melanosis* is common among black and Asian individuals, develops in early life and usually remains unchanged after puberty. The lesions are flat, with patchy brown areas of pigment within the epithelium, most commonly near the limbus. Excision is not required.

*Congenital melanosis* is subepithelial and may occur in two forms. Congenital melanosis oculi is an isolated lesion which characteristically appears as an area of bluish pigmentation because of its subepithelial position. A similar lesion may be associated with skin hyperpigmentation in the distribution of the first and second divisions of the ipsilateral trigeminal nerve (oculodermal melanosis or Naevus of Ota). The condition may occur with scleral or uveal hyperpigmentation and is associated with a slightly increased risk of developing choroidal or iris malignant melanoma.

*Primary acquired melanosis (PAM)* develops in adult life and is characterised by single or multiple areas of pigmentation within the conjunctiva, occasionally involving the skin of the eyelids. The lesions usually continue to enlarge and may undergo malignant transformation. Patients who have atypical PAM require life-long follow-up. Forniceal and palpebral PAM has a worse prognosis, mainly because pathological changes go unnoticed. Changes of importance include increasing thickness, changes in degree of pigmentation and haemorrhage. Treatment may be indicated for suspicious areas.

*Malignant melanoma of the conjunctiva* is a malignant tumour of melanocytes, which may be melanotic or amelanotic. Unifocal malignant melanoma in the form of a single nodule may arise in apparently normal tissue or rarely within a naevus. The prognosis following excision of these single lesions is excellent. Multifocal or diffuse malignant melanoma may arise in areas of PAM (Plate 24.6 p473) and has a worse prognosis than unifocal tumours, although it remains good if excision is early and performed before invasion of deeper structures occurs. Spread is by local invasion via lymphatics or it may be haematogenous (especially to the liver). Poor prognostic features include: thickness greater than 1-2mm, palpebral/forniceal location (partly because of delay in diagnosis), plica or caruncle location, multifocality, association with PAM, invasion of deeper structures (e.g. sclera) or spread to the ipsilateral nasal cavity, invasion of lymphatics, satellitosis, a high mitotic rate, a pure epithelioid or mixed cell type and metastatic disease. Depending on the tumour's clinical characteristics treatment may involve cryotherapy, irradiation, local excision, enucleation (Figs. 28.4 p512, 28.5 p513), regional lymph node dissection, exenteration (rarely) (Figs. 24.7, 24.8 p474) and chemotherapy for metastatic disease. The overall 5 year survival rate for favourably located tumours (bulbar/limbal conjunctiva) is about 90% and that for those at an unfavourable site (fornix/palpebral/plica/caruncle) is about 70%.

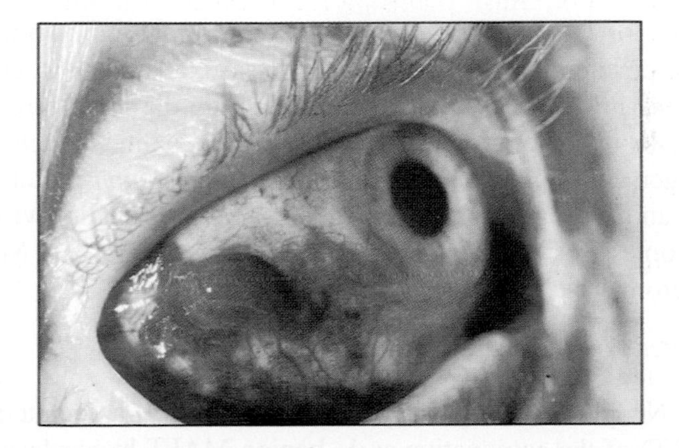

Plate 24.6 Malignant melanoma arising from Primary Aquired Melanosis of conjunctiva.

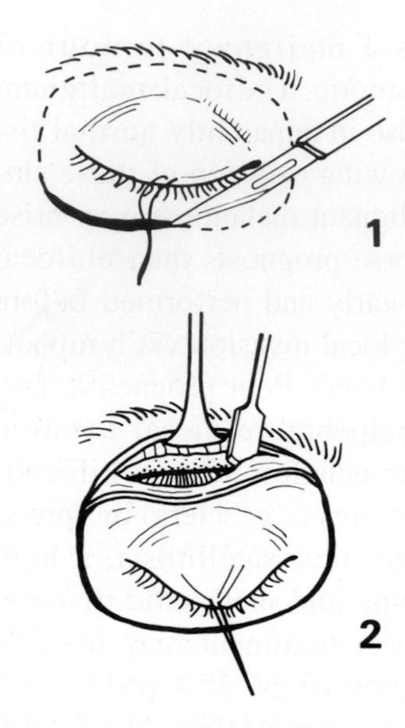


Fig. 24.7 Exenteration of the orbit
1. Skin incision
2. Periosteal reflection.

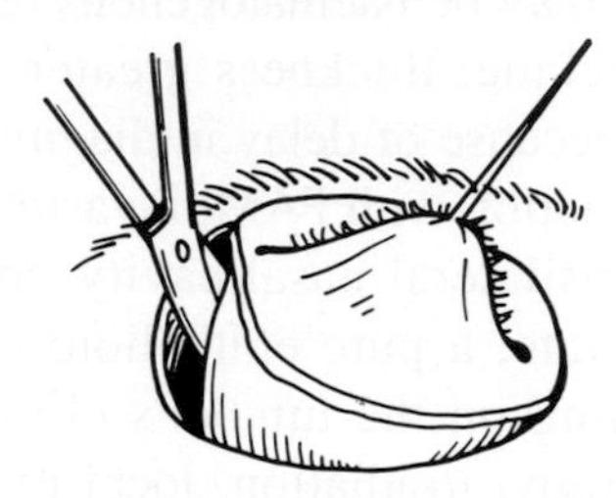

Fig. 24.8 Exenteration of the orbit
-amputation of orbital contents.

*Secondary melanocytic pigmentation* of the conjunctiva may occur with exposure, ectropion, trachoma or onchocerciasis.

*Non-melanocytic pigmentation: Endogenous pigmentation* may occur in patients who develop jaundice, Addison's disease, alkaptonuria and ochronosis. *Exogenous pigmentation* may occur following the long-term administration of topical sympathomimetic drops such as adrenaline, which can be used in the management of glaucoma *(adrenochrome deposition),* after chronic use of mascara and with argyrosis. *Pseudopigmentation* occurs with the "blue sclera syndromes" (osteogenesis imperfecta, Ehlers-Danlos and Marfan's syndrome), scleromalacia perforans and staphylomas.

*Non-pigmented conjunctival tumours*

*Lymphoma:* Neoplastic change may affect the mucosal-associated-lymphoid-tissue (MALT) of the conjunctiva as 1: benign MALT hyperplasia, 2: MALT lymphoma, or 3: non-MALT lymphoma (non-Hodgkin's). Some cases of apparent conjunctival lymphoma represent extension of an orbital lymphoma and it is not always possible to determine the true site of origin. The classical appearance is of diffuse "salmon-pink" infiltrates within forniceal and bulbar conjunctiva. However, clinical presentation is variable and the tumour is often yellowish in colour. Sometimes a lymphoma presents with large pale masses affecting the upper fornix and tarsal conjunctiva and can be confused with giant papillary conjunctivitis. In other cases the appearance is of deep orange-pink masses in the inferior fornix which can be confused with follicular conjunctivitis or Meibomian granulations. Forniceal tumours often present with lid swelling or even ptosis. MALT tumours tend to behave in a less aggressive manner and they usually remain localised in comparison with non-MALT tumours. However, the type of tumour can only be distinguished histopathologically. All conjunctival lymphocytic neoplasms reported to date are of B-cells and the presence of a clonal population of B-cells is taken as evidence of malignancy. MALT lymphomas have a good prognosis and if asymptomatic can be observed following biopsy. For symptomatic cases, radiotherapy is the treatment of choice since recurrence is common following attempted surgical excision. Non-MALT lymphomas have a poor prognosis and patients require full systemic evaluation, including bone marrow biopsy, since most have evidence of systemic spread. They are treated with chemotherapy.

*Papillomas* are benign tumours which are often pedunculated and are composed of thickened squamous epithelium. Excision biopsy may be required to confirm the diagnosis and for cosmetic reasons.

*Squamous cell carcinoma (Syn. intraepithelial carcinoma, carcinoma in-situ, Bowen's Disease)* usually affects elderly men as a slowly growing neoplasm near the inter-palpebral limbus which causes chronic conjunctivitis and often has a gelatinous or red-grey appearance. Aetiological factors include UV-radiation, cigarette smoking, petroleum exposure and human papilloma virus infection. Initially the lesion does not involve the substantia propria but it may be highly vascular and with time may invade the cornea. The treatment of choice is surgical excision, which should bc combined with a lamellar corneal graft if the cornea is involved. Some surgeons have advocated frozen section control and adjunctive cryotherapy. Radiotherapy is an alternative. Recurrence may occur even after careful excision. If surgery is not performed invasion of the substantia propria occurs *(invasive squamous cell carcinoma or epithelioma).* Eventually invasion of the globe may occur and an exenteration may be required. The prognosis for life is excellent but local disease control can be difficult.

*Kaposi's Sarcoma* (p336). Prior to the 1980's this tumour was rare but the incidence increased due to its association with AIDS. Clinical presentation is usually as a vascular, bluish-red lesion which can be either nodular or diffuse. The lesion may be pigmented. The cell of origin is thought to be the lymphatic endothelial cell. Most cases of conjunctival Kaposi's sarcoma respond well to radiotherapy.

## Conjunctival cysts and dermoids

### *Simple conjunctival cysts*

Macroscopic conjunctival cysts are almost always benign and of little consequence. A simple serous cyst may be treated by puncture with a sterile needle and removal of part of its anterior wall, but occasionally may recur. Surgical or traumatic implantation cysts have to be excised to prevent recurrence. Foreign body granulomas are often cystic, but when solid may be difficult to distinguish from more serious conditions. Excision biopsy may be required to confirm the diagnosis and to improve appearance.

### *Conjunctival, epibulbar and limbal dermoids*

A dermoid consists of congenitally displaced embryonic epithelium. A conjunctival dermoid is raised, whitish, epithelialised, may contain sebaceous glands and hair follicles and is usually situated at the limbus in the lower temporal quadrant. Excision with a lamellar corneal graft is usually very successful. Occasionally the limbal lesion is associated with a strand of tissue which extends to a lid notch. Another type of dermoid which consists mainly of fatty tissue is a *dermolipoma,* usually found as a yellow mass under the temporal conjunctiva. It may be unsightly and can be excised. It will be found to envelope the lateral rectus muscle and blend with the orbital fat. Care is required to avoid injury to the lateral rectus. In *Goldenhaar's Syndrome* (oculoauriculovertebral dysplasia) bilateral limbal dermoids are associated with pre-auricular skin tags, vertebral anomalies (fused cervical vertebrae, spina bifida, atlas occipitalisation, hemivertebrae), aural fistulae and sometimes other facial, skeletal, cardiovascular, abdominal or neuromuscular anomalies. Conjunctival dermoids should not be confused with lid or orbital dermoid cysts (p455) but they may occur together (Fig. 20.4 p429).

## CHAPTER 25

## CORNEA AND SCLERA

### Cornea

*The practical anatomy* of the cornea and sclera is described on p32.

*Congenital abnormalities*

*Megalocornea* is a bilateral enlargement of the cornea with a diameter of >14mm but without an associated rise in intraocular pressure. The entire anterior segment of the globe may be enlarged and the iris stroma atrophic but the corneal changes are often the most prominent feature. Megalocornea is a benign hereditary condition, the only problems being a large refractive error and a posterior subcapsular cataract with aging. It is essential to distinguish it from congenital glaucoma (p572) where the enlarged cornea is the result of a raised intraocular pressure. Examination under an anaesthetic is usually required.

*Microcornea* is characterised by the presence of a small cornea in an otherwise normal eye. Vision is not affected when any refractive error has been corrected. Microphthalmos by contrast involves all parts of the eye.

*Corneal inflammation (Keratitis)*

The cornea may be involved in a variety of inflammatory processes including trauma (p151) and infection and sometimes the corneal reaction may be secondary to conjunctival disease. Infection and injury of the cornea have already been discussed in Chapter 7 p151. The principal infections of the cornea are:

-*bacterial* (pp165, 319),
-*viral:*
  - herpes simplex (pp168, 330)
  -herpes zoster (pp171,330)
  -adenovirus (pp171, 200, 331)
-*fungal* (p339)
-*chlamydial:* (p338)
  - trachoma (pp173, 195)

*Non-infective corneal inflammation*

Many inflammatory conditions of the cornea are of uncertain aetiology.

*Mooren's ulcer (and Terrien's marginal degeneration)*

This ulcer is a rare, severe corneal lesion of unknown cause in patients over the age of fifty. It commences as guttering at the limbus with an overhanging edge on the central side of the ulcer, which progresses relentlessly across the whole surface of the cornea with marked thinning, irregular astigmatism and corneal opacity. There is secondary uveitis, severe pain, and vision is reduced (Plate 25.1 p478). This should be distinguished from *Terrien's marginal degeneration of the cornea* which is much more slowly progressive than Mooren's ulcer and is more common in the 20-40 year age group being usually confined to the limbal area. There is progressive thinning at the limbus which may result in ectasia and even rupture. The major problem is severe astigmatism. Terrien's marginal degeneration is included in this section because it is sometimes confused with Mooren's ulcer and treatment in both cases is as conservative as possible with protection by soft contact lenses and in the case of Mooren's ulcer the use of topical steroids. If judged necessary lamellar corneal grafts have been recommended and sealing with cyanoacrylate glue. The eyes should be protected from injury to avoid rupture of the thinned cornea.

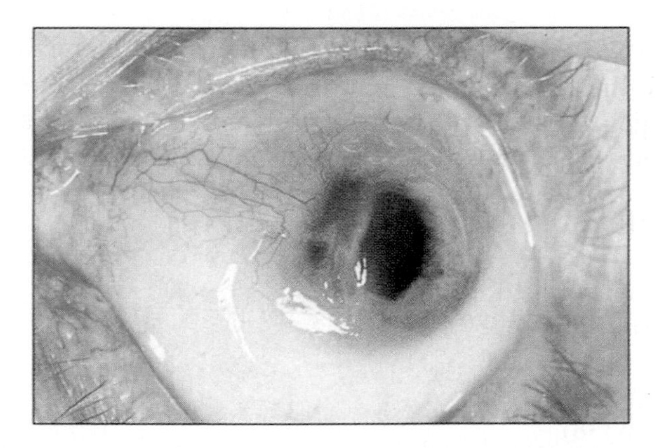

Plate 25.1 Mooren's ulcer.

*Phlyctenular Keratitis* (pp205, 324)

*Filamentary keratitis*

This is characterised by the presence of filaments of mucus attached to the corneal epithelium. The filaments may tear the epithelial attachments resulting in painful corneal epithelial erosions. Filaments are commonly seen in patients with diminished tear secretion (e.g. Sjögren's syndrome).
*Filamentary keratitis in association with kerato-conjunctivitis sicca.* The reduction of tear flow is associated with excess mucus in the pre-corneal tear film and may result in filament formation and Rose Bengal staining (Plates 10.1 p217 and 10.2 p219). The treatment for dry filamentary keratitis is adequate rehydration and this may be achieved by the frequent instillation of tear film supplements (artificial tears) (p616). This treatment alone is usually enough to prevent further filament formation. However, a small percentage of patients require mucolytic drops (e.g. acetyl cysteine 5-10% three times daily) in addition to the artificial tears (pp220, 615). Temporary occlusion of the lacrimal puncta with collagen or silicon punctal plugs may give relief from symptoms by preventing what little drainage occurs through the canaliculi in these patients. Should this prove particularly successful the puncta can be sealed permanently using cautery.

*Filamentary keratitis of Theodore.* This condition of unknown aetiology is characterised by the formation of filaments of mucus in the presence of adequate tear flow. Examination of the conjunctiva of the upper lid shows a marked papillary reaction and the bulbar conjunctiva above the limbus is hyperaemic and may stain with Rose Bengal. Treatment entails the use of mucolytic drops of adequate strength applied frequently (e.g. acetyl cysteine 10 percent four-hourly). Unfortunately, acetyl cysteine eyedrops are painful on instillation.

*Superficial punctate keratitis*

Punctate epithelial staining of the cornea occurs in a wide variety of corneal disease. The dead or degenerating epithelial cells take up Rose Bengal stain and are seen as scattered fine red spots on the surface of the cornea. Bare areas of the Bowman's layer are revealed by fluorescein stain. The distribution of the epithelial staining sometimes gives a clue to the underlying disease which may be:

-*viral infection*
-*bacterial infection*
-*keratoconjunctivitis sicca*
-*exposure to ultra-violet light* e.g. welder's flash, sun lamps
-*insensitivity* (anaesthetic cornea)
-*continued corneal exposure*

*Superior limbic keratoconjunctivitis* (Plate 25.2 p480)

This is an epithelial keratitis associated in many cases with hyperthyroidism in which the upper bulbar conjunctiva is injected and prominent at the limbus and the adjacent cornea shows punctate erosions with filaments, both of which stain with Rose Bengal. The palpebral conjunctiva in apposition to the affected area shows a papillary reaction and the conjunctival mucus is abnormal. The condition causes a burning sensation. Treatment with mucolytic acetyl cysteine may be effective in reducing the pain associated with filaments (p615).

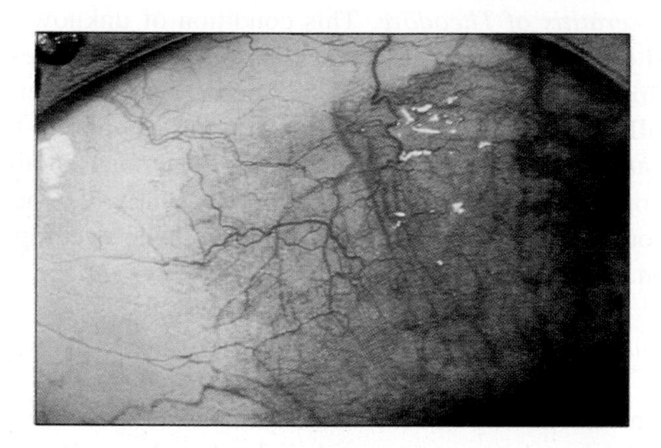

Plate 25.2 Superior limbic keratitis.

*Thygeson's superficial punctate keratitis*

The sudden onset of coarse, grey, punctate epithelial opacities, similar to those of adenoviral keratitis, which may extend to involve the superficial stroma, characterises this uncommon condition. The symptoms are mild to moderate discomfort and, if the lesions occur in the visual axis, slight impairment of vision. The eye is usually less red than might be expected from the severity of symptoms. The condition may continue for several years with periodic remission and relapses. Local steroids give symptomatic relief. The cause is unknown.

**Corneal changes secondary to conjunctival disease**

These range from punctate epithelial keratitis due to the toxic effects of bacterial conjunctivitis to the profound corneal scarring which occurs in *chronic progressive cicatrising conjunctivitis.* This includes *Stevens-Johnson syndrome*, ocular cicatricial pemphigoid (benign mucous membrane pemphigoid), dermatitis herpetiformis and drug induced or pseudopemphigoid. If the conjunctiva becomes elevated for any reason, so that the lid is lifted from contact with the peripheral cornea, the cornea dries, becomes dehydrated and thins to form a 'dimple' at this point (dellen ulcer) (Plate 24.2 p470).

*Ocular cicatricial pemphigoid* p203 *syn. (Benign mucous membrane pemphigoid)*

*The Stevens-Johnson syndrome (Erythema multiforme)* (p203 and Plate 8.4 p204, LCW 3.5 p42)

This is of acute onset and commonly affects children and young adults. It is characterised by the generalised erythematous bullous skin eruptions of erythema multiforme, with similar changes in mucous membranes. The disease may be fatal during the acute phase but those patients who recover are left with severely scarred mucous membranes, including the conjunctiva (Plate 8.4 p204). Although the disease process is no longer active after the acute episode, progressive corneal damage may result from the effects of conjunctival contracture. The Stevens Johnson syndrome is thought to be an immune vasculitis with lymphocytic and eosinophilic cellular reaction caused by antigen-antibody complexes being deposited. It can be provoked by various antigens including viruses, bacteria and drugs (sulphonamides, anticonvulsants). Treatment is protective and symptomatic.

*Pterygium* (p469 and Plate 24.1 p470)

**Abnormal pigmentation of the cornea**

The *Hudson-Stahli line* is a thin brown line running horizontally across the cornea in the interpalpebral fissure. It is comprised of iron deposited from the tear film in the basal epithelial cells and is of no clinical significance. Iron may be deposited similarly whenever the corneal surface has become irregular, for example at the head of a pterygium (Stocker's line), at the edge of a trabeculectomy bleb (Ferry's line) or following laser corneal surgery (Gartry-Kerr Muir lines).

The *Krukenberg spindle* is a vertically distributed line of uveal pigment deposited on the endothelial surface of the cornea. It rarely becomes sufficiently dense to impair vision. It raises the suspicion of both iris epithelial degeneration and the pigment dispersion syndrome which may be revealed by transillumination on the slit lamp examination. The second may be complicated by pigmentary glaucoma (p579).

*Blood staining* may occur with persistent hyphaema, particularly if associated with raised intraocular pressure. Haemosiderin is deposited in the corneal stroma giving the cornea a greenish brown colour (Plate 25.3 p482).

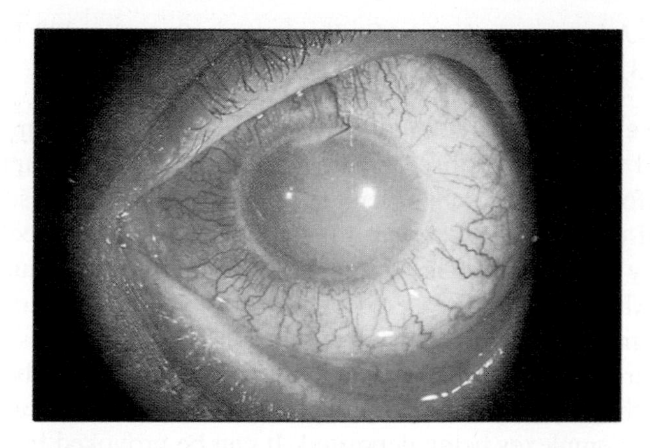

Plate 25.3 Blood staining of cornea.

A *Kayser-Fleischer ring* may occur in Wilson's disease *(hepato-lenticular degeneration)* (p436), an hereditary, inborn error of metabolism involving the absence of a protein, caeruloplasmin, responsible for the transport of copper. Copper is deposited in many tissues including the cornea, the liver and the lentiform nucleus of the brain (hence 'hepato-lenticular'). The corneal deposit is typically in a brownish green ring (or part of a ring) deposited in the peripheral cornea at the level of Descemet's membrane but ending sharply at Schwalbe's line. It is important to discover it early in the disease if possible when treatment with the chelating agent penicillamine may prevent neurological damage. Deposits also occur on the anterior lens capsule and may be associated with the development of 'sunflower' cataract.

*Other metallic deposits* such as silver and mercury may stain the cornea and these are also deposited at the level of Descemet's membrane. Silver and mercury were once commonly used as medicaments to treat a wide variety of diseases but these metallic deposits are now only seen as a rare occupational hazard.

**Corneal dystrophies**

A large number of corneal dystrophies are described. Many are excessively rare and some may be the same underlying disease with varying appearances. For a corneal condition to be considered a dystrophy it should be bilateral although one eye may be affected more than the other and most dystrophies show an hereditary pattern, the majority being autosomal dominant. They may affect any layer of the cornea. Those involving the epithelium and Bowman's membrane often result in painful ulcers due to the associated corneal epithelial erosions. The stromal dystrophies show varying degrees of corneal opacity, those in the visual axis being the most serious. Dystrophy of the endothelium results in failure of the endothelial aqueous 'pump' and development of generalised corneal oedema (p486). Only a few of the more common varieties are described:

*Anterior dystrophy:* The dominant dystrophy of Reis and Buckler is perhaps the best known. This presents in the first decade of life with episodic pain due to recurrent corneal erosions which may become extensive. Superficial threadlike corneal opacities appear at the level of Bowman's membrane with visual impairment.

*Map-dot or fingerprint dystrophy (Cogan's)* is characterised by linear or dot irregularities in the epithelium which can break down to form frank erosions.

*Stromal dystrophies:* There are three of importance. *Lattice dystrophy* (Plate 25.4 p484) is autosomal dominant and develops in the first decade of life as translucent lines, which are composed of amyloid, in the anterior stroma. The epithelium over these lines can become eroded, resulting in intermittent pain. *Granular dystrophy* (Groenouw I) is also autosomal dominant and is characterised by milk-white spots in the superficial stroma (Plate 25.5 p485, LCW 3.20 p48). The cornea in the visual axis is mainly affected. It appears in the first decade of life and is usually painless. Vision may be impaired but not seriously until late middle age. *Macular corneal dystrophy* is an autosomal recessive condition (Groenouw II) which also develops in the first decade of life. There is diffuse clouding of the cornea due to mucopolysaccharide deposition which is superficial and central initially. Eventually a full thickness opacity develops which extends to the limbus. Vision is usually markedly reduced by early middle age (p436).

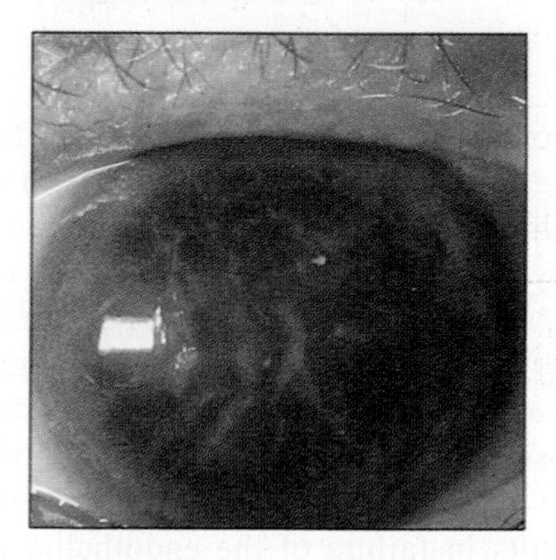

Plate 25.4 Lattice corneal dystrophy.

*Dystrophies affecting the endothelium* include the autosomal dominant condition of *Cornea guttata* in which hyaline excrescences occur in Descemet's membrane. The overlying endothelial cells often have a bronze or pigmented discoloration on slit lamp examination. The vision is not seriously affected unless Fuch's endothelial dystrophy develops. Existing endothelial impairment demands great care in avoiding further endothelial cell loss during any intraocular surgery. *Fuch's dystrophy* (LCW 3.21 p48) occurs commonly in older women and may present as cornea guttata. Increasing endothelial decompensation leads to the retention of stromal fluid, this corneal oedema causes progressive swelling and clouding of the cornea. This is at its worst on waking because no dehydration due to evaporation to the atmosphere can occur during sleep. Open angle glaucoma may be associated with Fuch's dystrophy, but whether this condition has an excess risk has ben disputed.

*Irido corneal endothelial syndrome (ICE syndrome)* p515

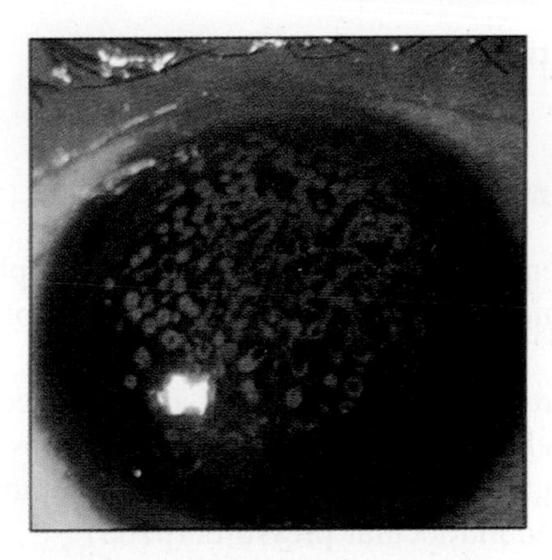

Plate 25.5 Granular corneal dystrophy. (Groenouw type 1)

*Corneal ectasia or ectatic dystrophy (keratoconus)* (LCW 3.22 p48) can be hereditary, although the genetic pattern is probably multifactorial and therefore not clear. It usually presents in the second decade of life and may cause progressive impairment of vision. Usually both eyes are affected but one more than the other. The cornea becomes progressively thinned centrally and assumes a conical shape. This may be revealed by the use of a Placido's Disc (Fig 5.4 p87). At the apex of the cone Descemet's membrane may develop splits resulting in sudden stromal oedema with pain and blurred vision *(acute hydrops)*. Subsequently, central corneal scarring results. The induced refractive error in keratoconus, which is commonly one of compound myopic irregular astigmatism, may be corrected with spectacles at first; later when it becomes impossible to correct the irregular astigmatism by spectacles a rigid gas permeable contact lens becomes necessary. In around 10 to 15% of cases of keratoconus a corneal graft is necessary due to contact lens intolerance.

## Management of corneal dystrophies

Attempts to decrease corneal oedema, such as that occurring in Fuch's dystrophy, by drops of a hypertonic solution (e.g. 5% sodium chloride) may give a little relief but, when vision is appreciably impaired by a corneal dystrophy, a corneal graft (penetrating or lamellar keratoplasty) is the only effective treatment. The result of keratoplasty in many dystrophies is very good though some types react unfavourably with recurrence in the graft. Endothelial dystrophy requires regular monitoring of the intraocular pressure.

## Miscellaneous conditions of the cornea

### *Corneal oedema*

This may be due to a number of causes:

-Endothelial decompensation (p93) as a result of dystrophies, trauma (including surgery or birth injury), and the presence of keratic precipitates in anterior uveitis.
-Raised intraocular pressure in acute closed angle glaucoma will give rise to corneal oedema, which may occur also in buphthalmos and occasionally in open angle glaucoma, due to a combination of endothelial decompensation and a chronically raised intraocular pressure. (p188)
-Corneal inflammation (Plate 25.6 p486) can result also in corneal oedema, particularly the viral inflammations, (p486) ophthalmic herpes simplex and herpes zoster ophthalmicus.

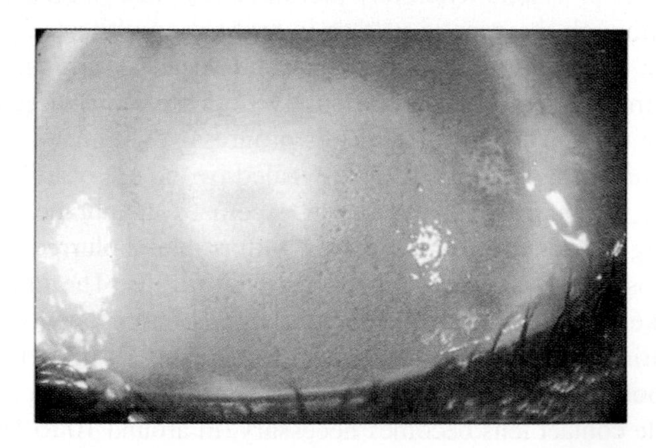

Plate 25.6 Corneal oedema and corneal scar.

*Vitamin A deficiency (keratomalacia, nutritional xerophthalmia)* (p352)

*Corneal arcus (arcus senilis, arcus juvenilis)* (Plate 7.27 p186, LCW 3.25 p49)

This is a ring shaped whitish deposition of lipid and cholesterol in the stroma of the periphery of the cornea separated from the limbus by a narrow clear zone. The arcus may at first only affect the lower and upper cornea but it becomes more complete with age. In younger patients an arcus suggests some anomaly of lipid metabolism but it does not appear to be a risk indicator for vascular disease. The central cornea remains clear and there are no symptoms, so treatment is unnecessary. It may be associated with megalocornea.

*Band-shaped keratopathy* (p92)

This is characterised by a band-shaped grey opacity extending across the eye in the interpalpebral fissure. The opacity, which represents calcium deposition at the level of Bowman's layer, starts at the nasal and temporal limbus, appearing first as a collection of whitish dots and later becoming more dense. The opacities, which can exhibit 'holes' or gaps in the otherwise confluent grey appearance, slowly spread from the limbus and visual acuity is severely impaired when the optical axis is involved. Causes of band-shaped keratopathy may be:

-*hypercalcaemia* due to hyperparathyroidism, hypervitaminosis D or sarcoidosis (p365 and Plate 18.9 p368, LCW 3.26 p49)
-degenerative changes in the cornea due to chronic iridocyclitis, e.g. as occurs in juvenile chronic arthritis and Still's disease (p359), or generalised ocular degeneration, e.g. following trauma or inoperable retinal detachment. Silicon oil within the anterior chamber following vitrectomy will result in a diffuse deposition of calcium in the cornea.

Band opacity of the cornea may be treated by reduction of serum calcium when it is due to hypercalcaemia. In chronic cases the calcareous material has been dispersed by removing the epithelium mechanically and applying a solution of the chelating agent sodium ethylene diamine tetra acetate (EDTA) to the opacity with which it forms a mush. This can then be irrigated and wiped away. More recently, the argon fluoride excimer laser has been used successfully to remove band keratopathy and treat superficial corneal pathological conditions.

*Exposure keratitis and neuroparalytic keratitis* (p174) *Amiodarone opacity* (p619)

**Corneal grafting (keratoplasty)**

A corneal graft is indicated either when the cornea becomes sufficiently scarred or distorted as to impair vision (i.e. an optical indication) or when there is a corneal perforation which requires repair or reinforcement (a tectonic graft). Before undertaking this major ocular surgery, it is important to ensure that the corneal disease is the principal cause of visual loss so that successful grafting will significantly improve vision. In general, corneal grafting is considerably more successful than other transplant operations, because of the avascularity of the tissue and the relatively low antigenicity of collagen. Antibodies reach the graft by the limbal vessels and the aqueous. When a graft is put into a vascularised cornea, there is a higher risk of rejection.

*Factors adversely affecting the fate of the graft*

-vascularised recipient cornea
-raised intraocular pressure
-poor tear film and lid abnormalities
-abnormal recipient conjunctiva
-poor quality donor material
-a previous failed graft
-an inflamed eye
-anterior segment disorganisation and collapse

*Donor material.* Fresh young adult cornea is the ideal material for a penetrating keratoplasty. Elderly corneas have a reduced endothelial cell population and are more likely therefore to develop graft oedema. The donor material should be collected as soon as possible (within four hours) and may be stored for a few hours at four degrees centigrade in suitable sterile conditions. Certain centres have the facilities for deep freezing and therefore longer storage of donor cornea. Microscopic examination of the donor cornea is mandatory to exclude unsuitable material (e.g. early Fuch's dystrophy).

*Types of graft*

*Penetrating keratoplasty.* A full thickness disc of cornea is removed from the patient (Figs. 25.7, 25.8 p489) and replaced by a disc of donor cornea (Fig. 25.9 p489). This type of graft is necessary when the stromal opacities are deep or the endothelial activity of the recipient cornea has become decompensated (Plate 25.10 p489, LCW 3.27, 3.28 p50).

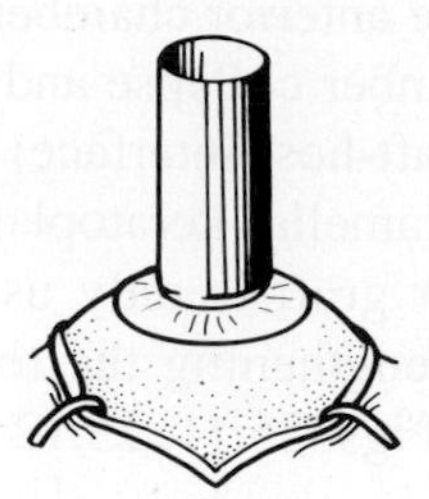

Fig. 25.7 Penetrating keratoplasty - Trephine of recipient's cornea.

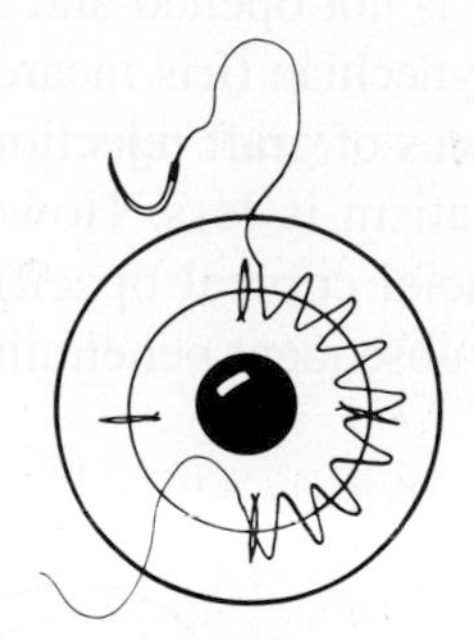

Fig. 25.8 Completion of removal of corneal disc with scissors.

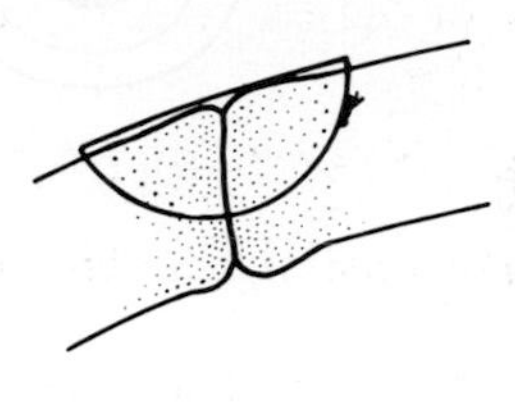

Fig. 25.9 Suturing donor graft into position.

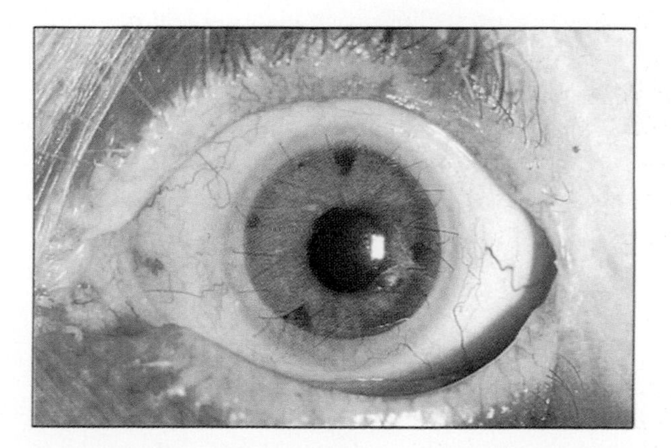

Plate 25.10 Completed penetrating corneal graft.

*Lamellar keratoplasty* is a safer operation because the anterior chamber of the recipient cornea is not opened and thus anterior chamber collapse and subsequent anterior synechiae (iris incarceration in the graft-host interface) cannot occur. The chances of graft rejection are reduced in lamellar keratoplasty and induced astigmatism is less. However the lamellar graft is only useful in cases of superficial corneal opacification or for strengthening the recipient cornea prior to subsequent penetrating keratoplasty (Figs. 25.11, 25.12 p490).

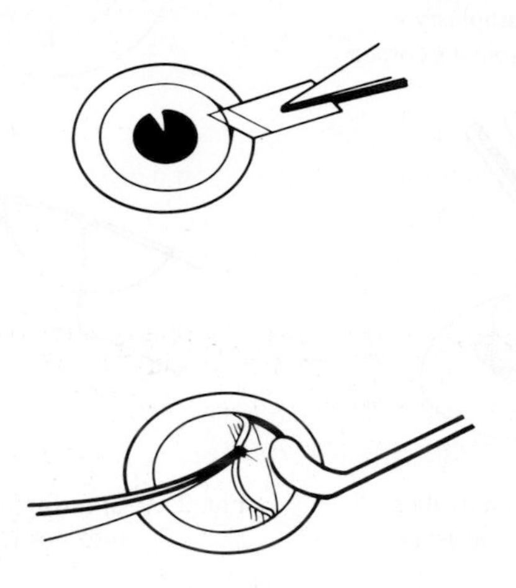

Fig. 25.11 Lamellar keratoplasty - Cleavage of lamellar graft.

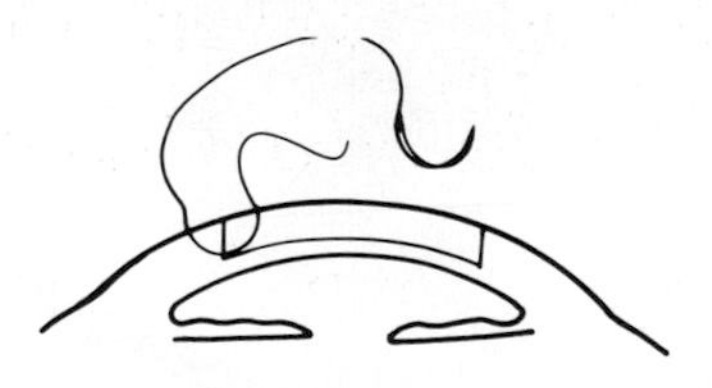

Fig. 25.12 Suturing of lamellar graft.

**The surgical and laser correction of refractive error**

As the cornea contributes about two thirds of the refractive power of the eye and is easily accessible, this has naturally been favoured as the operation site. Most operations have involved the correction of myopia but methods have also been devised to improve astigmatism, hypermetropia and presbyopia. Many techniques have been explored but chronologically the most successful have been radial keratotomy (RK) for medium myopia (-4D up to even -7D), photorefractive keratectomy (PRK) for moderately high myopia (-8 to –10 D) and laser in-situ keratomileusis (LASIK) for high myopia up to –20D (Fig. 25.13 p491). For even higher degrees, if considered essential, it would be necessary to proceed to clear lens extraction using phakoemulsification and posterior chamber intraocular lens implant, alternatively an anterior chamber phakic lens could be employed, however these intraocular methods would invite the risk of complications.

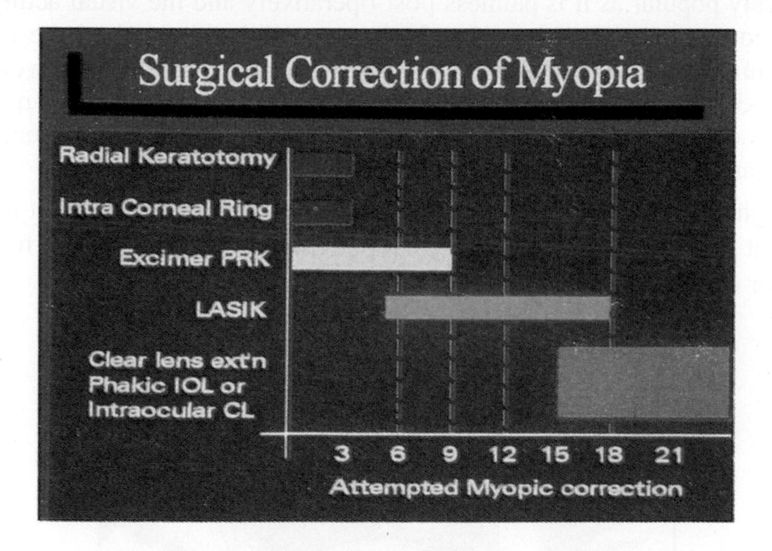


Fig. 25.13 A surgical schematic for the treatment of myopia.

*Radial Keratotomy* was introduced in the mid 1960's and consists of about 8 deep radial incisions avoiding the central zone. These led to flattening of the cornea but there was always a tendency to slow change towards hypermetropia and troublesome diurnal variation. (LCW 10.7 p147)

*Photorefractive Keratectomy* was made possible by the development of argon-fluoride excimer photo ablation (p76) which in the mid 1990's was employed to sculpt layers accurately from the central 6mm zone of the cornea to reduce its refractive power. Correction of up to 8D can be obtained. Inevitably there is some transient inflammation and foreign body sensation but this is variable. Loss of best corrected acuity, scarring and rarely infective keratitis can occur in about 2% of cases.

*Laser In-situ Keratomileusis*, which can correct up to about 20D of myopia, involves the creation of an 8mm hinged corneal flap by an automated keratome (Plate 25.14 p492) and ablation of the exposed stromal bed by the appropriate application of excimer laser. The flap is replaced and allowed to dry into place without sutures. The procedure requires skill and meticulous attention to detail. As with PRK the ablation zone must be central but haze and myopic regression complicating healing are minimised by the flap and great care is taken to avoid its damage or separation. LASIK is becoming progressively popular as it is painless post operatively and the visual acuity usually recovers to 6/12 or better in 24 hours. With this method the corneal epithelium and Bowman's layer are preserved and stability achieved after about 2-3 months, although it is reported that excimer laser ablation may disturb the orderly arrangement of corneal fibres, reducing contrast sensitivity which can impair the ability to drive vehicles at night.

Although laser or surgery is valuable for certain groups of patients, it is probable that for the immediate future most patients will continue to opt for spectacles or contact lenses.

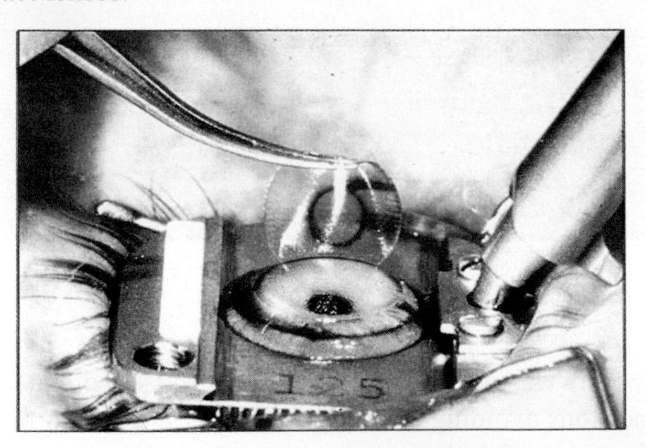

Plate 25.14 Elevation of the 160 micron thick corneal flap during a LASIK procedure. (Courtesy Dr. Stephen Slade)

## The sclera

*The practical anatomy* of the sclera is described on p34.

## Symptoms and signs of scleral disease

A *staphyloma* is a thinned area of sclera lined by uvea which has bulged outwards under the influence of intraocular pressure. The underlying uvea gives it a bluish black appearance. It may occur in high myopia or as a result of injury or discase and may be *anterior* or *posterior*. The anterior type is described as *ciliary* when the ciliary body is involved or as *intercalary* when between this zone and the corneo-scleral limbus. When the sclera is very thin it may appear uniformly light blue as in osteogenesis imperfecta (p434) or in Marfan's syndrome (p434).

*Scleral inflammation* may be superficial *(episcleritis)* or deep *(scleritis)*, anterior or posterior, and nodular or necrotising.

*Episcleritis* usually appears overlying the nasal or temporal sclera within the interpalpebral aperture and is characterised by a dusky reddish raised area. It may ache a little but usually responds well to local corticosteroids under ophthalmological supervision (Plate 25.15 p493, LCW 6.21 p99).

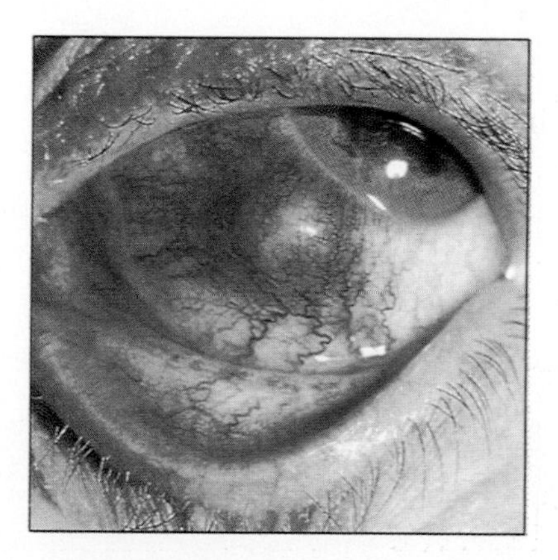

Plate 25.15 Episcleritis.

*Scleritis (diffuse and nodular)* can be serious and is frequently associated with rheumatoid arthritis (p357 and Plate 18.1 p358) although it is common in leprosy and some cases are attributed to tuberculosis or syphilis. It often affects the entire circumference of the sclera just behind the limbus *(diffuse scleritis)* (LCW 6.22 p99), with a dusky red swelling and infiltration of all the coats of the eye, and may spread to involve the cornea which also becomes opaque and vascularised like the sclera in the region affected *(sclerosing keratitis).* The inflammation is accompanied by a severe, deep and 'boring' pain. Treatment is by local and systemic steroids but even when the condition responds to treatment, thinning may lead to exposure of a large area of ciliary body or choroid (anterior staphyloma) and the eye may even perforate. Any accompanying uveitis is treated in the usual way (p181).

*Necrotising scleritis (scleromalacia perforans).* Severe rheumatoid arthritis (p358) may be complicated by scleritis with rheumatic nodules in the sclera. The sclera may be severely thinned as part of an intense inflammatory condition *(necrotising scleritis)* or it may be bilateral and progressive, sometimes with minimal inflammation (*scleromalacia perforans*) (LCW 6.23 p99). Perforation and a relentless course to blindness sometimes results. A massive granuloma of the orbit may develop from a rheumatoid nodule. The diagnosis may be clear when this is placed anteriorly but it may be difficult to distinguish from an intraocular or orbital tumour when it affects the posterior sclera.

**Injuries**

These are considered in Chapter 7, p151.

# CHAPTER 26

# LENS

**The practical anatomy and physiology of the lens is described on** p35

## Cataract

The classification, symptoms and general management of cataract have already been discussed (p94) and the correction of the high hypermetropia of the aphakic eye. (pp65,108) The complications of cataract surgery will now be considered further. Lens dislocation will also be described.

*Complications of cataract surgery*

Cataract surgery is a safe and successful procedure in the majority of patients, but complications do occur. These may be divided into operative, early post-operative and late.

*Operative complications*

*Loss of vitreous* in the extra-capsular method usually arises from faulty technique with damage to the posterior capsule and in cases with weak zonular attachments. It may also occur during intracapsular extractions especially in eyes that have been traumatised and highly myopic eyes with a fluid vitreous. It increases the risk of uveitis, cystoid macular oedema and retinal detachment due to the adhesion of vitreous to the wound edges and to the retina. These strands of vitreous exert traction on the retina and encourage the formation of peripheral retinal breaks which can then lead to detachment of the retina. Adhesions of vitreous between the iris and the wound also contract, drawing the iris up to the wound. This can cause secondary glaucoma and an up-drawn or 'peaked' pupil. Another complication of vitreous loss is vitreo-corneal contact which may cause corneal oedema. Should vitreous loss occur it is most important to remove all prolapsed vitreous to ensure that none remains incarcerated in the wound. Scrupulous attention to clearing the wound of vitreous is essential and will reduce complications appreciably.

*Haemorrhage* into the anterior chamber or occasionally into the anterior vitreous is not uncommon. Fortunately small haemorrhages usually clear spontaneously without adversely affecting the final visual result. A rare and disastrous complication is that of *expulsive haemorrhage* when acute bleeding into the choroid occurs during surgery. This blood is under high pressure and inexorably forces the contents of the eye through the incision so that all vision is lost.

*Early post-operative complications*

*Severe Infection (endophthalmitis)* is fortunately rare (approximately 1:1000 cases), although when it does occur (usually within 48 hours of surgery), prompt treatment is essential and even then there may be loss of vision or even of the eye (see p166 for treatment). Iritis of varying severity may follow cataract surgery although this is usually remarkably mild after phakoemulsification surgery. Provided infection is not the cause, this usually responds to mydriatics and local steroids.

*A transient rise in intraocular pressure* may follow an otherwise normal cataract extraction. This is usually adequately controlled with oral or intravenous acetazolamide or beta blocking drops and settles spontaneously in a few days. *Pupil block, flat anterior chamber and the aqueous misdirection syndrome ('malignant' glaucoma)* involve a much more serious rise in intraocular pressure with apposition or adhesions of the iris and the vitreous face blocking the flow of aqueous from the posterior to the anterior chamber. This predisposes to pooling of aqueous behind the peripheral vitreous, pushing it and the lens and iris forwards, so that the anterior chamber becomes flat. A vicious circle is thus set up which can only be broken by release of the trapped aqueous either by dilatation of the pupil with mydriatics or if this fails, by incision of the vitreous face or vitrectomy. Topical beta blocking drops or systemic acetazolamide reduce aqueous production. This, fortunately, is now a rare problem but was much more common when intracapsular surgery was performed routinely. One or more peripheral iridectomies, performed prior to the lens extraction, allowed aqueous to pass from the posterior to the anterior chamber, by-passing the pupil, unless the iridectomies became occluded. With developed phakoemulsification techniques such surgical manoeuvres are unnecessary.

*Iris prolapse* may occur through the incision (in the case of large incision extracapsular surgery) and is usually found in the first post-operative day though it sometimes can occur later. Coughing, straining, or a blow on the eye raises the intraocular pressure and the aqueous carries the iris in front of it as it escapes through a weak point in the incision. Iris prolapse is prevented by careful suturing and the avoidance of vitreous incarceration. Treatment involves excision of the prolapsed iris and re-suture. Associated vitreous prolapse requires vitrectomy with care to avoid any vitreous incarceration in the incision.

In *choroidal detachment* fluid may exude into the space between the choroid and sclera. It is seen as a dark balloon just behind the iris. The anterior chamber becomes very shallow. Choroidal detachment is probably the result of the sudden loss of the anterior chamber and hypotony that may occur when the eye is opened to remove the lens in some forms of cataract surgery. It usually resolves spontaneously without influencing the final result, although in marked cases the persistent shallow anterior chamber may lead to iritis and raised intraocular pressure, making necessary drainage of the sub-choroidal fluid. The use of small corneal or scleral incisions and visco-elastic fluids to maintain the anterior chamber during the operation may also help to prevent this happening and also protect the lining tissues from damage.

*Macular oedema* may follow cataract extraction and it may be a common finding in the first few days after operation. The oedema usually resolves spontaneously. Occasionally persistent cystic macular oedema may adversely affect the final visual result (p527), sometimes due to incisional vitreous incarceration.

*Late complications*

*Retinal detachment* can be precipitated by cataract extraction and if so, it usually occurs within 12 months of surgery. Vitreous loss and incarceration at the time of operation increases the risk of retinal detachment. Unlike phakic detachments it may be difficult to find the break or breaks in the retina but most aphakic detachments respond to conventional detachment surgery.

*Glaucoma in aphakia.* Raised intraocular pressure may occur some years after cataract extraction. These patients have a number of possible causes for the embarrassed aqueous outflow. Some are due to chronic simple glaucoma unrelated to the aphakia and others to vitreous in the anterior chamber making the free flow of aqueous through the trabeculum more difficult. It may also be secondary to peripheral anterior synechiae often following a flat anterior chamber in the immediate post-operative period. Each case is treated according to its cause.

*Corneal oedema.* Injury to the corneal endothelium will impair the endothelial pump (pp33, 34) and cause corneal oedema (pp484 - 486) following cataract extraction. It may result from vitreo-corneal contact. The corneal oedema may eventually necessitate corneal grafting. Patients with Fuch's endothelial dystrophy are particularly at risk as they already have a compromised endothelium.

*Posterior capsule opacification.* In an estimated 30% of patients, the posterior capsule thickens and opacifies in the first 18 months following extracapsular surgery (including phakoemulsification). If this causes significant deterioration in vision, or marked glare and scatter of light (disability glare), an aperture can be made in the posterior capsule without surgery, using the YAG laser. This is a relatively rapid and painless out-patient procedure.

*Decision making in cataract treatment*

It is not possible to generalise about the 'best' operation, because in addition to the age of the patients and their ability to co-operate, the choice of operation and anaesthetic depends so much on the operating conditions and equipment, the expertise of the surgeons, anaesthetists, nurses and other assistants and the opportunities for follow up. The decisions can only be made by those immediately responsible for the welfare of the people under their care. For instance, where the demand is great and resources sparse, a more rapid and cheaper intracapsular operation with the spectacle correction of aphakia might be judged to restore sight to more people than expensive methods which when successful give an improved result and in some patients the simple extracapsular technique may be safer. The aim, of course, is still to make the best universally available.

**Dislocation of the lens**

Total dislocation of the lens *(luxation)* or partial dislocation *(subluxation)* may occur as the result of trauma or, occasionally, due to an inherited disorder of the lens zonule.

*Traumatic dislocation* (p153)

Total dislocation of the lens following trauma is rare but partial dislocation is not. Secondary rise in intraocular pressure may occur (LCW 4.17 p63). A problem arises when the subluxated lens becomes cataractous and requires removal. Vitreous is frequently found in the anterior chamber and as the trauma tends to occur in younger people, the remaining firm zonular attachments of the lens as well as vitreo-lens attachments increase the risk of vitreous loss. Such operations are best carried out by surgeons experienced in vitreo-retinal techniques.

*Spontaneous dislocation*

-*homocystinuria* (see p433, Plate 20.5 p433)

-*Marfan's syndrome* (see p434)

## CHAPTER 27

## VITREOUS

The practical anatomy and physiology of the vitreous has been described on p36 and its development on p46.

### Pathological changes of the vitreous

### Vitreous collapse (posterior vitreous detachment)

The firm gel structure of the vitreous changes with age to allow liquefaction of the central vitreous. This occurs earlier in myopia or in the presence of any type of retinopathy or inflammatory process. As the central vitreous liquifies and the central collagen framework collapses, the vitreous cortex peels away from the internal limiting membrane of the retina (Fig. 27.1 p501). The symptoms of vitreous collapse are flashing lights due to traction on the retina as the cortex pulls away from the internal limiting membrane and the sudden appearance of black spots representing opacities on the posterior surface of the vitreous, now visible as they hover in front of the retina. In addition condensations of the vitreous gel cause strand-like and ill-defined floaters.

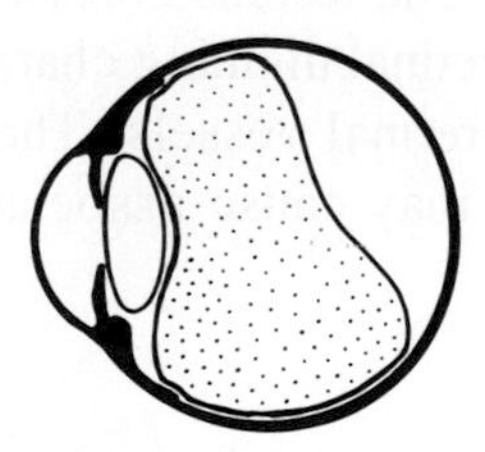

Fig. 27.1 Posterior detachment of the vitreous.

Vitreous collapse may be seen clinically by the appearance of an optically clear zone in front of the retina; the former attachment of the vitreous to the optic disc margin is seen as a ring-shaped opacity floating in the vitreous cavity.

*The relationship between vitreous collapse and retinal detachment*

Posterior vitreous detachment is common and has been estimated to have occurred in 75% of the population over 65 years of age. Therefore in the majority of patients vitreous detachment, although often associated with disturbing symptoms, is essentially a benign condition. However in a few cases vitreous detachment may be followed by retinal detachment. In these patients there are abnormally strong areas of vitreo-retinal adhesion so that when the cortex pulls away from the retina, tears occur at the sites of firm adhesion. A retinal tear allows fluid to pass from the vitreous into the potential space between the pigment epithelium and the neural layer of the retina, the plane in which 'retinal detachment', which is more correctly described as a 'separation', takes place (pp113, 525).

*Vitreo-retinal changes associated with lattice degeneration*

Lattice degeneration (p520, Plate 29.1 p520) is one of the more common peripheral degenerative conditions which may predispose to retinal detachment, hence its importance. The lesions consist of well demarcated, circumferentially oriented areas of retinal thinning characterised by variable pigmentation and white sclerosed retinal vessels. The vitreous is attached to the edges of these lesions and may cause associated tears when the vitreous detaches.

*Pre-retinal membranes*

Pre-retinal membrane formation may occur in diabetes, and following retinal vein occlusion or trauma, due to cell migration across the surface of the retina and these cells form a membrane lying on its surface. Such membranes may form firm attachments to different parts of the retina and as these membranes contract they cause wrinkling of the retinal surface. When distortion occurs at the macula considerable visual loss occurs (cellophane maculopathy). The track made by a foreign body passing through the eye often subsequently fibroses, causing a traction band. This traction band may cause elevation of the retina and late retinal detachment.

*Proliferative vitreo-retinopathy*

In the presence of a full thickness retinal break, cells of the pigment epithelium may migrate into the space behind the detached vitreous gel, and in the presence of a detached retina these cells are particularly liable to cause fibrous membranes, either on the surface of the retina or vitreous. The contraction of these membranes causes the retina to become immobile and thrown into fixed folds. This type of retinal detachment does not respond satisfactorily to conventional surgery.

*Vitreous haemorrhage*

This has been considered on p111.

*Asteroid hyalosis*

Calcium soaps occasionally form in the vitreous. These are seen by the examiner as multiple whitish floating spheres in the vitreous. Although obvious to the examiner they sometimes are not noticed by the patient and though they may cause slight blurring of vision they are of no clinical consequence. They occur unilaterally in the eyes of older patients.

*Synchysis scintillans*

This rare bilateral condition is characterised by the presence of tiny yellowish-white floating crystals of cholesterol in the vitreous. It tends to occur in eyes that have had uveitis or vitreous haemorrhage and is a degenerative change. The crystals settle in the lower part of the vitreous and may be stirred up by ocular movements.

## Vitreous surgery

Vitrectomy is performed via the anterior or posterior segment using specially designed cutting instruments which remove the vitreous without causing traction on the retina.

### *Anterior vitrectomy*

This becomes necessary when vitreous loss occurs during cataract surgery, or the posterior lens capsule is ruptured. The aim is to avoid incarceration of vitreous in the cataract wound which will lead to an irritable eye and the risk of retinal detachment or macular oedema. Long standing vitreous touch may decompensate the corneal endothelium after intracapsular cataract surgery and anterior vitrectomy may prevent or reduce corneal oedema. Pupil block after cataract surgery, causing angle closure glaucoma can be relieved by anterior vitrectomy.

*Lensectomy* (p105) is removal of the lens with a vitrectomy instrument, and this is a useful technique for some types of congenital cataract, traumatic cataracts where there is perforation of the posterior capsule, and cataracts secondary to uveitis. During this operation the posterior capsule and anterior vitreous is removed.

Anterior vitrectomy inevitably is part of the removal of pupil membranes in aphakic patients; which are usually the result of trauma.

### *Pars plana vitrectomy*

This technique, in which the vitrectomy instrument is introduced into the eye via a stab incision through the pars plana, is used to remove opaque vitreous or during complex retinal detachment procedures or for penetrating injuries and the removal of intraocular foreign bodies.

Vitreous haemorrhage (p112) is the usual opacity requiring removal by vitrectomy and is most commonly seen in proliferative retinopathy . If possible surgery should be delayed for four to six months in case spontaneous clearing occurs. Retinal tears are an important cause of vitreous haemorrhage and if suspected early vitrectomy should be considered. Amyloidosis of the vitreous is a rare opacity which responds well to vitrectomy. Vitrectomy is performed during detachment surgery to allow space for internal tamponade of the retina by gas or silicone oil or for the manipulation of the retina with heavy liquids. It is also necessary for the dissection of retinal membranes or the release of vitreo-retinal traction. The removal of intraocular foreign bodies requires vitrectomy if retinal detachment is to be avoided. Foreign body or instrument tracts in the vitreous inevitably cause fibrosis and retinal traction which may lead to retinal tears and detachment.

# CHAPTER 28

# UVEA

The important conditions affecting the uveal tract are uveitis, injuries, serous choroidal detachment, tumours, developmental abnormalities and degenerations.

**Uveitis** (p176).

**Injuries** (p151).

## Serous choroidal detachment

A choroidal detachment appears as a smooth convex bullous elevation frequently in the upper peripheral fundus. The commonest symptom is reduction in visual acuity. Field defects are often absent as are flashing lights and floaters because the vitreous is not involved. Sometimes the condition can be difficult to distinguish from a retinal detachment. Retinal detachments stop at the ora serrata whereas choroidal detachments include the ciliary body and this can be seen on ultrasonography. These detachments are due to the collection of fluid in the supra-choroidal space. Normally there is a flow of fluid from the vitreous, across the coats of the eye to the orbit and from there into the lymphatics. The retina poses no significant resistance to flow and the retinal pigment epithelium (RPE) actively pumps fluid across it, so there is no hydrostatic pressure drop across this either. The driving force for the flow of fluid across the sclera is the hydrostatic pressure drop generated by the intraocular pressure. Fluid will collect in the suprachoroidal space if there is a drop in the intraocular pressure (hypotony) such as may occur after trauma or excessive drainage following trabeculectomy. (This may also occur in posterior scleritis due to resistance to fluid flow across the sclera). It is often associated with a shallow anterior chamber. A leaking wound can be detected by Seidel's test (p156, Plate 7.5 p157) which can then be sealed if necessary. The treatment is that of the underlying cause. Most cases settle spontaneously.

## Tumours of the uvea

There are essentially only two primary tumours which arise in the uvea: those arising from melanocytes and haemangiomas. Metastatic deposits may form in the uveal tract. Other tumours can arise in the uveal tract such as leiomyomas and neurofibromas but they are extremely rare and the correct diagnosis is almost never made or only following enucleation for a suspected melanoma. It is difficult to biopsy lesions in the uveal tract due to the risk of damage to adjacent structures but this is an option in cases of great clinical uncertainty.

### *Posterior malignant uveal melanocytic lesions*

*Choroidal malignant melanomas* are solid lesions which arise in the choroid (Stereo Plate 28.1 p508). They have two growth patterns. Over 95% grow in a nodular manner and the remainder diffusely. The latter have a poor prognosis mainly due to delay in diagnosis. If the central retina is affected the patient may complain of a visual disturbance at an early stage such as flickers of light or distortion. The diagnosis is made by indirect ophthalmoscopy and ultrasonography. Fluorescein angiography is of limited help (Stereo Plate 28.2 p509). Current enucleation series show that the correct diagnosis is made on clinical features in almost all cases. Pigmentation is not required for the diagnosis of melanoma but they are usually pigmented and will throw a dark shadow on scleral transillumination dimming the red reflex normally seen in the pupil. The other clinical features depend upon the size of the tumour. Small growths usually present either close to the fovea due to

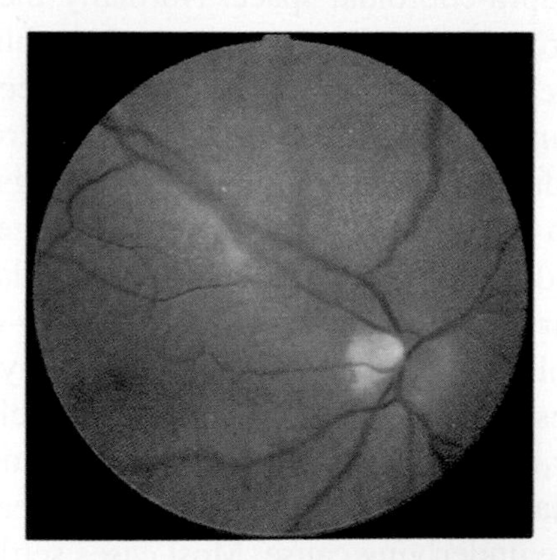

Stereo Plate 28.1 Early choroidal malignant melanoma. (see p4)

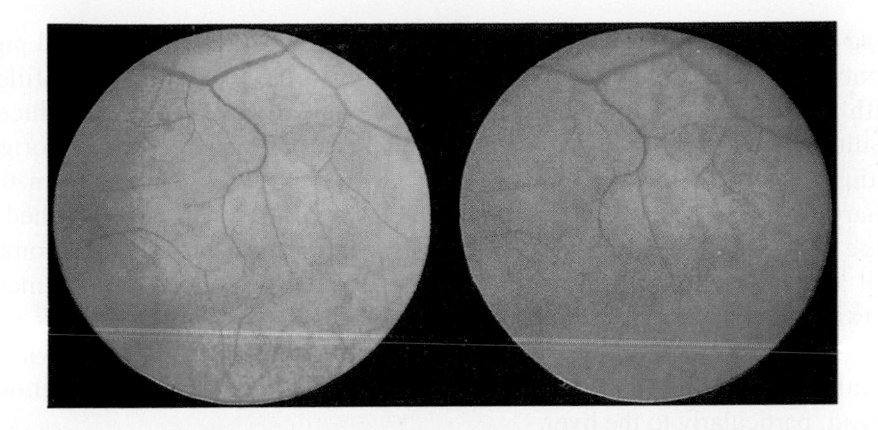

Stereo Plate 28.2 Early choroidal malignant melanoma. Fluorescein angiography of Stereo Plate 28.1 (see p4)

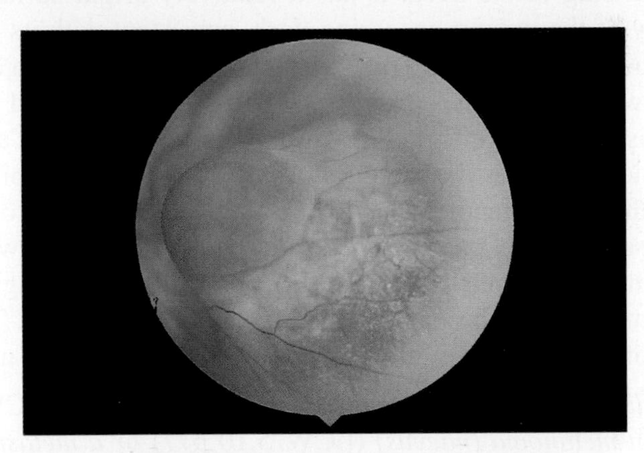

Plate 28.3 Advanced choroidal malignant melanoma invading through Bruch's membrane (collar stud appearance).

reduced visual acuity at an early stage, or close to the optic disc where they can interfere with vision or be noticed by chance on a routine eye examination. The features suggesting a diagnosis of malignant melanoma are an irregular outline, the presence of lipofuschin which is a bright orange pigment present on the surface of the tumour arising from degenerating retinal epithelial cells and the presence of subretinal fluid or changes in the surrounding retinal pigment epithelium suggestive of the previous formation of sub-retinal fluid. As the tumours grow, they will usually invade through Bruch's membrane and adopt a 'collar-stud' appearance (Plate 28.3 p509). Such a tumour does not have lipofuschin pigment on its surface having grown through the retinal pigment epithelium. Continued growth will result in the globe becoming filled with tumour. At some point, extensive sub-retinal fluid will be produced resulting in a serous retinal detachment and when this happens tumours originating away from the fovea or optic disc, especially those arising from the ciliary body, will then give rise to symptoms and often will have reached a large size with a resulting poor prognosis. Malignant choroidal melanomas will also invade into the orbit. The sclera forms a natural barrier to tumour growth and invasion tends to be along the channels transmitting nerves and blood vessels. However, despite their relatively low grade local behaviour, uveal melanomas carry a 40% ten year mortality due to haematogenous spread, particularly to the liver.

The single most important investigation is ultrasonography, which reveals a solid lesion arising in the uvea. It has a relatively bright surface reflection with the body of the tumour appearing dark, consistent with its homogenous texture and there is typically scleral excavation. It is also able to detect a 'collar stud' growth pattern, the presence of sub-retinal fluid and extra-scleral extension. Histologically the grading system employed is the Callendar classification which classifies the tumours into Spindle A cells, Spindle B cells, Mixed epithelioid and spindle cells and Epithelioid cells, the prognosis worsening progressively from one category to the next. A problem with this classification is that the Spindle A cell category includes both benign and malignant lesions and it is not possible to distinguish between them. The best single measure of stage is the height of the tumour as measured by ultrasound.

*Differential diagnosis of a malignant choroidal melanoma* (LCW 5.11 p77) *from a benign melanoma (naevus)* (LCW 5.10 p77) *or a metastatic deposit.* There are two major clinical problems encountered in their diagnosis: 1. to distinguish between an amelanotic melanoma and a metastatic deposit and 2. in the case of a small lesion, is to distinguish a malignant melanoma from a benign choroidal naevus.

1. *Differentiation of a malignant melanoma from secondary malignant disease of the uvea*

The uvea is a relatively common site for secondary tumours (surprisingly so considering its relatively small tissue volume). Frequently they arise from lung primaries in men and breast primaries in women but almost any tumour can metastasise to the eye. The commonest intraocular site for metastasis is the posterior uvea. It is important to examine the fellow eye because secondary tumours are bilateral in one third of cases. Not only does this help to distinguish secondary deposits from an amelanotic melanoma but also because in the case of secondaries, the aim of treatment is not to cure the patient (which unfortunately is not usually possible) but to give effective palliation so that sight is preserved for the patient's lifetime. The clinical picture can closely resemble that for an amelanotic melanoma and the distinction can be difficult if the patient is not previously known to have a primary neoplasm elsewhere and no other primary can be readily found by physical examination and simple investigations such as chest X-ray. There are three approaches which can be helpful. A short period of observation is the simplest. Secondary tumours often show rapid growth with observable enlargement in under three weeks whereas melanomas are slowly growing tumours in which it is rare to see much change in any period less than three months. An immunoscintigraphy scan using a technetium labelled anti-melanoma antibody (225.28S) which is extremely specific for melanoma is unfortunately not very sensitive being positive in only 70% of cases of melanoma. Thus a negative result gives limited help. Biopsy can be very useful in this situation as needle biopsy usually yields enough material to make the distinction between a primary melanoma and a secondary deposit. Secondary carcinomas usually respond well to external beam radiotherapy.

2. *Differentiation of a malignant melanoma from a choroidal naevus*

Choroidal naevi (LCW 3.16 p46) are common and found in 5% of eyes at post mortem. The distinction from a small malignant melanoma is mainly on size. Lesions less than 2mm in height, as measured by ultrasound, are usually benign while those greater than 3mm are usually malignant. Other features that suggest malignancy are the presence of the orange lipofuschin pigment, subretinal fluid, an irregular outline and the absence of drusen (drusen are white surface deposits and suggest a long standing lesion). There are times when a clear distinction is not possible on clinical grounds and only a working diagnosis of 'suspicious naevus' is possible . Biopsy is not helpful as there are no established histological criteria to distinguish benign from low grade malignant melanocytes and, in addition, small lesions are technically difficult to biopsy. A characteristic feature of malignant lesions is growth, so that careful observation is then required which may need to be extended for several years. Observed growth being an indication for treatment.

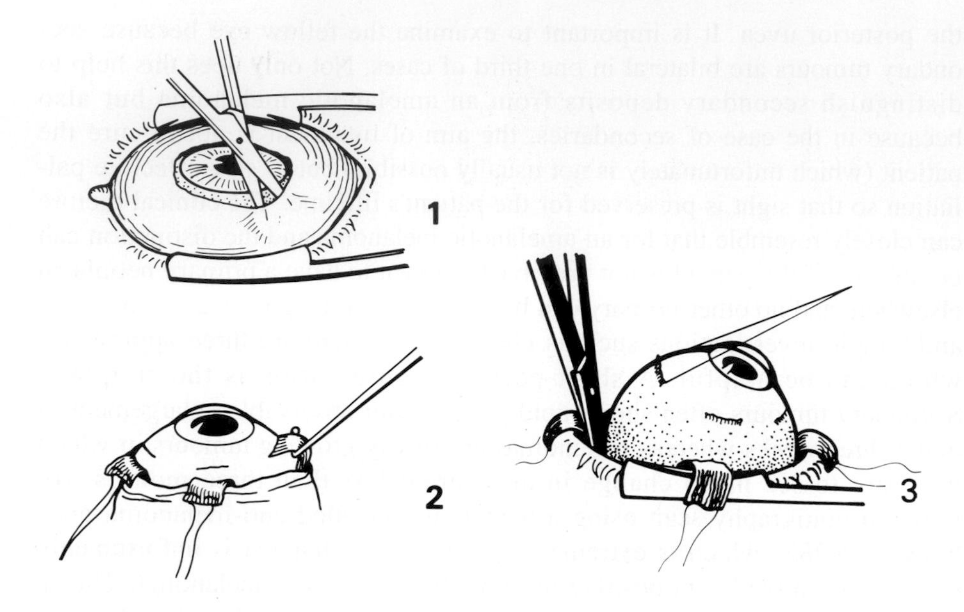


Fig. 28.4 Enucleation of the eye 1. conjunctival incision 2. division of muscles 3. division of the optic nerve.

*Treatment of malignant choroidal melanoma.* Small tumours can be treated by radiotherapy. Malignant melanomas are relatively radio resistant and the eye is radio-sensitive. Therefore two techniques are employed to treat the tumour while limiting the total dose received by the eye. One technique is to use a small radioactive plaque which can be sewn to the sclera overlying the tumour (brachytherapy) and the other uses charged particles, either protons or helium nucleii which have special properties allowing very precise localisation of the administered radiotherapy. The choice between these two forms of radiotherapy is still being evaluated. Large tumours require a high dose of radiation which gives rise to a number of side effects and often results in a blind, painful and unsightly eye, so that large tumours are probably still best treated by removal of the eye (enucleation) (Figs 28.4 p512, 28.5 p513). Another option which is rarely performed is local resection of the tumours. This is a technically demanding procedure which requires profound hypotensive anaesthesia and it is not clear if the results are any better than those for radiotherapy. All these approaches aim to eradicate the primary tumour but

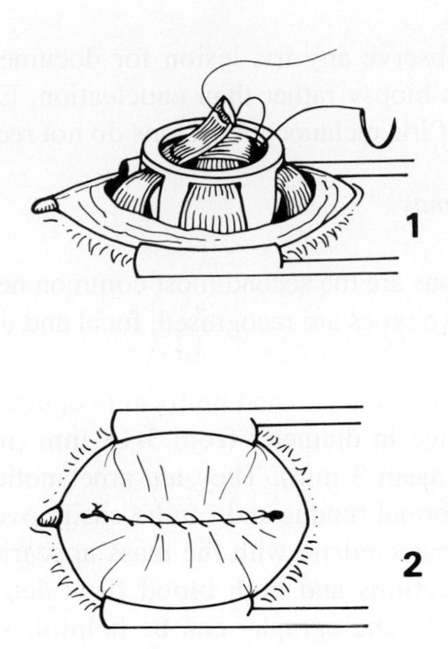


Fig. 28.5 Enucleation of the eye 1. orbital implant insertion 2. conjunctival closure.

no difference in survival (correcting for tumour size) has been demonstrated between these treatment options. There is currently no satisfactory treatment of metastatic disease arising from a primary choroidal malignant melanoma and at this stage it carries a very poor prognosis (median survival from diagnosis of metastasis is about six months).

*Iris melanocytic lesions*

These are described separately from posterior uveal lesions as their behaviour is surprisingly different. Their prognosis is much better, only 4% undergoing metastatic spread. This is probably because they present much earlier in their natural history, being clearly visible and because they are of the relatively benign spindle cell type. It is also clear that most melanocytic lesions of the iris are naevi and are not malignant melanomas but the distinction is however extremely difficult on clinical grounds. Occasionally, the iris lesion can represent the 'tip of an iceberg' and is an anterior extension of a large ciliary body melanoma (which carries a very different and much worse prognosis) and for this reason, it is important to perform ultrasonography and examine the fundus to exclude a posterior melanoma.

Current policy is to observe any iris lesion for documented growth before proceeding to excision biopsy rather than enucleation. Even if incompletely excised, the majority of iris melanocytic lesions do not recur.

*Choroidal Haemangiomas*

Choroidal haemangiomas are the second most common neoplasm of the uvea. They are benign and two types are recognised, focal and diffuse.

*Focal haemangiomas* are unilateral and unifocal neoplasms, with no systemic associations. They range in diameter from 3-18 mm (mean 7 mm) and in height from 1-7 mm (mean 3 mm). They are amelanotic masses often of a colour similar to the normal fundus and can be easily overlooked. The ultrasound appearance is characteristic with the mass appearing very bright with multiple internal reflections and high blood flow detectable on Doppler examination. Fluorescein angiography can be helpful, showing rapid early filling of the mass which is often followed by leakage. Their major complication is leakage of sub-retinal fluid causing a localised retinal detachment. They only require treatment if they are persistently leaking fluid, which often is only transitory or intermittent, so that a period of observation is justifiable but prolonged leakage over several months can lead to irreversible damage to the fovea (with macular hole formation). A laser grid treatment to the surface of the tumour can be helpful in reducing or stopping leakage. These tumours are extremely radiosensitive and refractory cases often respond well to a relatively low dose of radiation which is best given in a similar manner as for melanoma either by radioactive plaque or by particle beams, thereby limiting the dose to the rest of the eye.

*Diffuse choroidal haemangiomas* occur in patients with the Sturge-Weber Syndrome (p438). They can be subtle and they are best detected by noting the difference in the colour of the red reflex. The diffuse haemangioma has a red rather than orange-red reflex. They can give rise to a chronic serous retinal detachment which if untreated can lead to rubeotic glaucoma. The aim of treatment is to flatten the retina and so prevent rubeotic glaucoma. Two approaches have been described. The sub-retinal fluid can be drained surgically and the retina flattened by the injection of gas into the vitreous cavity, followed by laser treatment to the surface of the tumour. Alternatively they can be treated by radiotherapy because like their focal counterpart, they are radiosensitive. An external beam approach is used in view of their diffuse nature, if possible sparing the lens from the radiation field. The relative merits of either approach have yet to be fully assessed.

**Genetically determined congenital malformations of the uvea** (p426)

**Degenerations of the uvea**

*Iris atrophy* is a natural aging process but may follow inflammation or any cause of anterior segment ischaemia such as an extensive retinal detachment or squint operation. There is scattering of pigment granules revealing the pigment epithelial layer and the brown sphincter muscle. Miosis and rigidity are also common. In some cases a cleft may form between anterior and posterior stromal layers, *iridoschisis*. Progressively the anterior stroma splits into fibres containing a blood vessel core with their ends floating freely in the aqueous.

*Irido - Corneal Endothelial Syndrome (ICE Syndrome)* includes a spectrum of ocular disorders which particularly affect white middle aged females. It may appear unilateral but the other eye usually shows slight changes of the same type. It does not appear to be familial. The basic defect is an abnormal corneal endothelium, which gives a 'beaten silver' appearance. By proliferation it forms a membrane of cells which creeps on to the iris across the drainage angle. This membrane then contracts, causing distortion of the iris creating holes in it, ectropion of uveal pigment and patchy angle closure. A mild form is known as *Chandler's Syndrome* and a more severe type as the *Cogan Reese Iris Naevus Syndrome,* which although benign, has sometimes been misdiagnosed as malignant melanoma due to the presence of pigmented iris nodules. Corneal oedema due to endothelial dysfunction may occur in the absence of raised intraocular pressure which, if present, would of course aggravate it. If the pressure is raised, standard glaucoma treatment is given. The condition used to be called *Essential Iris Atrophy* before its corneal endothelial nature was revealed.

In *choroidal vascular atrophy* the choriocapillaris may atrophy as a genetically determined abiotrophy (autosomal dominant or recessive) or secondary to local changes in the choroid or to degenerative changes in the adjacent retina.

The larger choroidal vessels become exposed to ophthalmoscopic view and they appear outlined in connective tissue. *Although the name choroidal sclerosis has been applied to this condition, fluorescein angiography demonstrates a normal lumen in these vessels.* There is visual loss depending on the distribution of the disturbance and poor dark adaptation especially in the genetic cases.

*Choroideremia* is a rare X chromosome-linked condition of choroidal atrophy with secondary retinal atrophy. The affected homozygous male suffers from night blindness in early life and vision is typically lost in middle age. Clinically it commences with 'pepper and salt' pigmentation which then progresses to complete atrophy with a featureless white appearance. The process starts in the periphery and moves centrally with the macula spared until late in the disease.

*Gyrate atrophy* is an autosomal recessive disorder due to a loss of the enzyme ornithine aminotransferase and the characteristic biochemical finding is hyperornithinaemia in the blood. The disease is rare and starts in the first decade of life and progresses giving 6/60 vision or worse by the age of 40 years. The symptoms are night blindness, myopia, constricted visual fields and reduction in visual acuity. The appearances are sharply defined areas of chorio-retinal atrophy with scalloped borders. Treatment is an argenine free diet.

*Disciform degeneration of the retina:* p529

**Anterior segment dysgenesis** p427

*Rieger's* and *Axenfeld's syndromes* are examples of this condition and have a dominant transmission. There is an abnormally prominent Schwalbe's line and posterior embryotoxon with a peripheral corneal opacity. Strands or sheets of atrophic iris may be attached to Schwalbe's line and iris holes may form. The angle of the anterior chamber is often grossly abnormal and a raised intraocular pressure may be found in many cases. If this occurs at an early stage buphthalmos may result. Dental anomalies and hypoplasia of the maxilla may be present giving a characteristic flattened facial appearance. *Peter's anomaly* is a rare autosomal recessive condition in which there may be strands of iris attached to the cornea and lens associated with lens and vitreo-retinal abnormalities.

## CHAPTER 29

## THE RETINA

### Assessment of retinal function

-*visual acuity* (pp77, 273)
-*dark adaptation* (p263)
-*colour vision* (p265)
-*contrast sensitivity* (p269)
-*motion sensitivity* (p268)
-*visual field examination* (pp 82, 273)
-*E R G* (p285)
-*E O G* (p285)
-*fluorescein angiography:* the retinal circulation can be studied by fluorescein angiography both anatomically and functionally and this has greatly helped our understanding of retinal vascular disease. The technique is briefly described on p289.

### Disorders of the retinal vascular system

These are considered in general in the section on cardiovascular conditions (p289), also *retinal arterial occlusion* (p118) and *retinal venous occlusion* (p121)

The disorders may be:

-*degenerative* (arteriosclerosis) (p293)
-*hypertensive* (p294)
-*diabetic* (pp290, 313)
-associated with *haemorrhage* and *blood disorders* (p297)
-*phakomatous* congenital overgrowth of retinal vascular and other tissues normally found in that location (hamartomas) (p436)
-*inflammatory* (retinal vasculitis) (p518)

### Inflammatory (retinal vasculitis)

Retinal vasculitis may occur as part of sarcoidosis (p365), systemic lupus erythematosis (p363), polyarteritis nodosa (p364), Behçet's disease (p360) and Wegeners granulomatosis (p368). Many cases were formerly attributed to tuberculosis, but most commonly the cause is not known (see Eales' disease below). It seems possible in these cases, as has been demonstrated in lupus erythematosis and polyarteritis nodosa, that the vasculitis is a result of immune complex deposition. Temporal arteritis can cause symptoms resembling central retinal artery occlusion or branch occlusion and in some cases the central artery is affected, but usually this condition, when it involves the eye, causes occlusion of the posterior ciliary vessels supplying the optic disc resulting in an anterior optic atrophy.

*Idiopathic vasculitis retinae* (including Eales' disease) (Plate 6.21 p112). In the early phase of inflammation, perivascular whitish exudates may be seen around the vessels and extending into the vitreous. The veins are particularly affected and show irregular tortuosity and variations in the calibre of the blood column. This results in slowing of the blood flow and varying degrees of obstruction and ischaemia which act as a stimulus to new vessel formation. The patient complains of seeing floating specks and of blurred vision. This stage then merges with the chronic phase in which the inflammatory foci become less active or inactive and in which the fragile neovascular capillary twigs, many of which have grown forwards feebly supported along a detached posterior vitreous face, are liable to rupture. Haemorrhage into the subhyaloid space or the vitreous is followed by organisation into fibrous bands causing retinal distortion and sometimes a retinal tear and detachment with field loss spreading from the periphery. A haemorrhage into the vitreous causes the symptoms of a sudden appearance of black dots in the field of vision, which may be described as looking like a 'fall of soot', and very blurred vision, but the sight may recover completely on absorption of the blood. There is, however always the liability to recurrence with slower clearing of the media. The condition is also liable, as in diabetic retinopathy and retinal venous occlusion, to lead to new vessel formation in the iris and in the angle of the anterior chamber with the subsequent closure of the angle by contraction of the connective tissue accompanying the vessels and an intractable secondary rubeotic glaucoma (pp316, 581). Treatment is unsatisfactory but steroids have been used in an attempt to suppress the inflammatory process in the early phase. However, laser treatment should certainly be used if proliferative retinopathy develops, in order to prevent vitreous haemorrhage.

*Retinal infections*

-*rubella* (p332)
-*toxoplasmosis* (p182)
-*syphilis* (p326)
-*tuberculosis* (p323)
-*cytomegalovirus* (pp331, 337)
-*toxocara* (p347)
-*A.I.D.S.* (p333)

*Retinal disorders associated with haemorrhage and blood diseases*

-*severe blood loss* (p297)
-*sickle cell disease* and *thalassaemia* (p297)
-*polycythaemia* (p300)
-*severe anaemia,* whether acute or chronic, of which pernicious anaemia is an example (p300)
-*leukaemia* (p301)
-*multiple myeloma* or *macroglobulinaemia* (p302)

## Retinal degenerations

Degeneration will in any case occur with advancing years but it may be a genetically determined abiotrophy, with an onset at any age characteristic of the particular disorder. Secondary retinal degenerative changes may also follow trauma, inflammation, vascular disturbances and retinal detachment. The essential defect may be in the metabolism of the cells of the pigment epithelium or in the rods and cones themselves as the 'dry' type of age related macular degeneration (ARM, dry type), but similar changes may occur in them due to circulatory changes, especially subretinal neovascular membrane formation leading to the separation of layers by clear fluid or haemorrhage (ARM, wet type). Suspected neovascular formation requires early fluorescein and sometimes indocyanine angiography, because photocoagulation of well defined membranes reduces the risk of severe visual loss from extra-foveal lesions and even foveal lesions may respond to photo-dynamic therapy (PDT) using wavelengths specific to a photo-sensitising dye eg.verteporphin and may prove effective with minimal retinal damage and this combined with surgical translocation is being pioneered.

Common modes of degeneration include the cobblestone, lattice and cystic as well as myopic central types (p64, LCW 5.13 p78, 5.16 p79) and a most important retinal degeneration, the hereditary primary pigmentary degeneration of the retina (retinitis pigmentosa).

-in *cobblestone degeneration* there are small patches of retinal atrophy with pigment at the edges. It is of little significance.

-*lattice degeneration* (Plate 29.1 p520) occurs in the equatorial region with thin branching white lines and pigment changes. The vitreous is attached to the edges of the degenerate area but is liquefied over it. It is more common in myopic eyes and predisposes to retinal holes and detachment. Prophylactic treatment by cryotherapy or laser is sometimes indicated.

-*cystic (or cystoid) degeneration* (p527)

***-primary pigmentary degeneration of the retina (retinitis pigmentosa).*** The retinal pigment epithelium is sometimes referred to as the tapetum and hereditary pigmentary degenerations are called *tapetoretinal degenerations.* The most important of these is primary pigmentary degeneration of the retina or 'retinitis pigmentosa'. This is a bilateral hereditary affection predominantly of the rods and the pigment epithelium although the cones are also involved. The mode of inheritance in any particular pedigree may be autosomal recessive, or autosomal dominant or X chromosome linked.

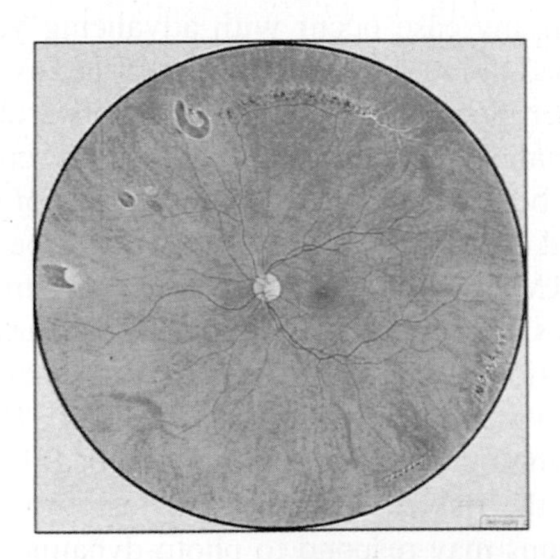

Plate 29.1 Lattice retinal degeneration.

*Symptoms and signs* (Plate 19.16 p389) The earliest symptom, usually in youth, is difficulty in seeing in dim light ('night blindness'), followed by the development of an annular scotoma in the mid-periphery of the field of vision which enlarges until only a small central area of field remains. This may be retained for years with a visual acuity of 6/9 or 6/12, but eventually central vision may be lost in late middle age. The sight may also become blurred by a posterior subcapsular cataract. Ophthalmoscopically the most striking features are extreme narrowing of the retinal vessels, pigmentary deposits in the mid-periphery of the fundus and pallor of the optic disc, which is often described as yellowish or 'waxy'. The pigmentation may be of the characteristic 'bone corpuscle' type, but may also be in the form of fine lines or speckling, and it does not usually involve the macular region. The retinal pigment epithelium has a diffusely thinned appearance revealing a patchy view of the choroidal circulation. Pathologically, degeneration of the rod and cone layer of the retina and of the pigment epithelium is demonstrable with spread of the released pigment into the retina.

*Variants of primary pigmentary degeneration of the retina and the differential diagnosis*

Occasionally in childhood there may be impaired vision with little apparent retinal disturbance, the characteristic signs appearing in adolescence *(Leber's congenital amaurosis).* Abnormal macular appearances in childhood may also be followed by a generalised tapetoretinal degeneration. For prognostic reasons it is important to carry out electro-diagnostic tests to distinguish these patients from children with heredomacular degeneration in whom very useful peripheral vision will be retained indefinitely. In the adult too there may sometimes be little retinal pigment disturbance *(retinitis pigmentosa sine pigmento),* or instead there may be white spots scattered over the fundus *(fundus albipunctatus, retinitis punctata albescens).* Some of the latter patients may have a better visual prognosis than those with retinitis pigmentosa and others may not progress at all. It should be remembered that retinitis pigmentosa is sometimes difficult to distinguish ophthalmoscopically from scars of diffuse choroidoretinitis although in the latter the pigment patches tend to be in a coarser pattern and less uniform in distribution. The visual field defects are also less characteristic.

Occasionally variants of pigmentary degeneration of the retina may be associated with other abnormalities such as deafness and, in the *Laurence-Moon-Biedl syndrome,* with polydactyly and hypogenitalism. In *Usher's syndrome* there may also be congenital deafness and dumbness and in *Refsum's syndrome* chronic polyneuritis leading to an enlargement of peripheral nerves, ataxia and deafness. This is due to accumulation of a fatty acid, phytanic acid, due to enzyme deficiency. *Kearn's syndrome,* in which there are pigmentary retinal changes in association with ocular myopathy, may also be associated with conduction defects in heart muscle. Allied conditions are: *neuroepithelial dysgenesis* in which there is no retinal pigmentary disturbance. A normal EOG indicates that the pigment epithelium is not affected although the ERG is abnormal. The visual defect is similar to but not as severe as that found in primary pigmentary degeneration of the retina. *Oguchi's disease,* is a rare congenital and inherited condition in which there are few rods, many cones and a defect of the pigment epithelium. The main symptom is extreme night blindness.

## Primary macular degenerations

These are genetically determined. A juvenile type *(Stargardt)* and an adult type *(Behr)* have been described. The group also includes vitelline dystrophy *(Best)*, fundus flavimaculatus and drusen.

*Vitelliform degeneration* (Plate 29.2 p523) (vitelliruptive degeneration, Best's disease) is an autosomal dominant disorder. The onset is at birth or shortly afterwards. It is so named because there is an orange deposit like the yolk of an egg in the macular region. The EOG is abnormal and dark adaptation is impaired. Although the lesion appears to be confined to the macular region, there is general impairment of the pigment epithelium. The sight is normal until, after a variable interval, there is dispersal of the yellowish material, the 'scrambled egg' stage, followed by scarring and pigmentary disturbance with loss of central vision.

*Fundus flavimaculatus* consists of yellowish round or usually fishtail shaped flecks in the central fundus associated at some stage with macular degeneration. Ill defined hyperfluorescence is seen at the site of old lesions on fluorescein angiography. It is a genetic disorder transmitted as an autosomal recessive character.

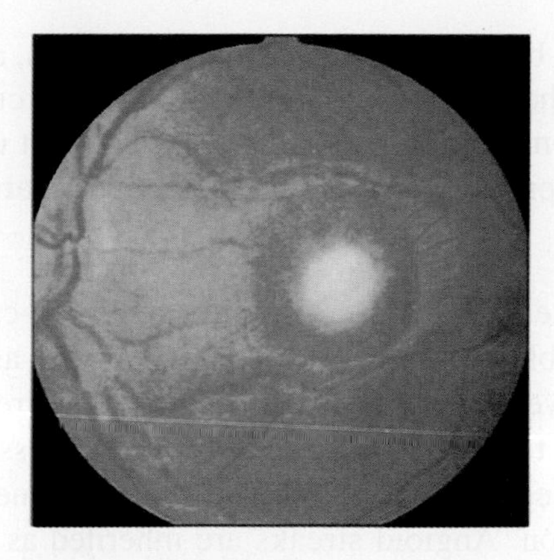

Plate 29.2 Vitelliform degeneration of the retina.

*Juvenile (Stargardt)* (LCW 5.14 p78) and *adult (Behr)* types of primary macular degeneration show central pigmentary change with a variable degree of loss of central vision. Fluorescein angiography shows punctate areas of fluorescence where the choroidal fluorescence is seen through the damaged retinal layers.

*Drusen (colloid or hyaline bodies)* (Plate 29.3 p523, LCW 5.15 p78) may be a genetically determined abnormality of Bruch's membrane and possibly the pigment epithelium, mainly affecting the posterior pole. They can also occur

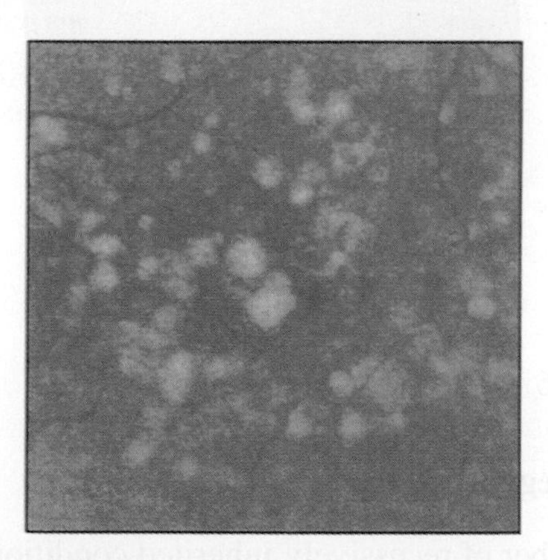

Plate 29.3 Colloid bodies - drusen.

as a senile change or be associated with angioid streaks, and may occur at the optic disc (p376). They appear as yellowish white circular dots of variable size which may become confluent. Central vision is not usually affected until a late stage. Fluorescein angiography shows them as sharply demarcated fluorescent areas.

*Angioid streaks* (Plate 29.4 p524) Angioid streaks, so-called from their fancied resemblance to blood vessels, show histologically as ruptures in Bruch's membrane. They have the appearance of dull reddish irregular and frequently broad streaks which tend to be distributed more or less radially around the optic disc. Macular degeneration usually develops, frequently in the form of a disciform degeneration. Angioid streaks are inherited as an autosomal recessive abnormality and are seen particularly in association with the yellowish skin lesions on the neck of pseudo-xanthoma elasticum (of Grönblad and Strandberg), in Ehlers-Danlos syndrome, sickle cell anaemia and with osteitis deformans of Paget).

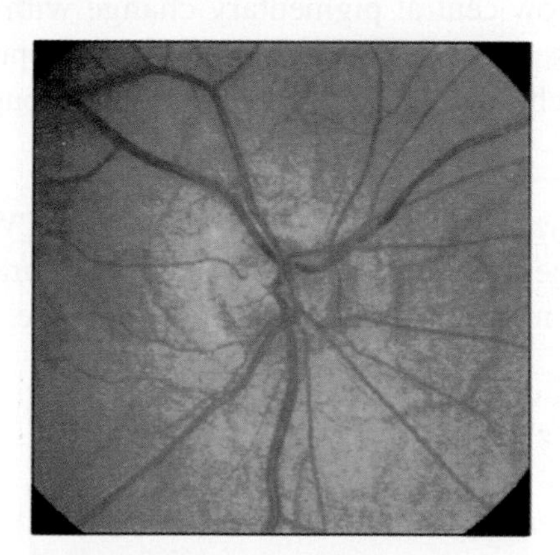

Plate 29.4 Angioid streaks.

**Phakomatoses** (p436 )

**Cerebro-macular degenerations**

These include a number of recessively inherited conditions involving the ganglion cells of both retina and brain, two of which are genetically determined deficiencies of the enzymes concerned in lipid metabolism. The latter are:

-*Gangliosidosis (Tay-Sach's disease)* (p431)
-*Sphingomyelin lipidosis (Niemann-Pick disease)* (p432)

**Fluid separation of retinal and choroidal layers**

This group of conditions frequently results from degenerative change in the retina and choroid and the various types are conveniently considered here. From a practical point of view they are probably best described as they present clinically:

*Retinal (separation) detachment*

-with retinal hole or tear (rhegmatogenous serous retinal detachment)
-without retinal hole
 -exudative detachment
  inflammatory
  neoplastic
 -traction detachment
 -schisis
 -cystoid oedema of the macula
-associated with central choroido-retinal degenerations
 -central serous retinopathy
 -pigment epithelial detachment
 -disciform degeneration of the retina
-*choroidal detachment*
 -with inflammation or trauma
 -with hypotony

**Retinal (separation) detachment (serous retinal detachment)**

*Retinal detachment with a retinal hole or tear (rhegmatogenous)*

Although correctly a separation of the retinal layers, the term 'detachment' is usual and this will be used here. In this case the fluid lies in the space between the neural retina and the retinal pigment epithelium, i.e. between the layers of the secondary embryonic optic vesicle. The hole, which may result from degenerative change in the retina, is usually associated with *vitreous traction* or with trauma. It allows the passage of fluid from the vitreous into the intraretinal space. The neural layer of the retina then progressively separates under the influence of eye movements and gravity, like a tear in wallpaper in a flooded room might allow water to seep in behind the paper which gradually falls away from the wall. Severe ocular contusion may be followed by retinal tears particularly in the peripheral retina at the ora serrata, a *retinal dialysis*. Sometimes these can be very large and take the form of *giant retinal tears* with the retina folded on itself.

*Clinical aspects.* The signs, symptoms, treatment and differential diagnosis have been considered under the important causes of 'painless loss of vision' (p113).

*Retinal detachment without a retinal hole*

This may be exudative or due to traction and between the two embryonic retinal layers or within the neural layer.

*Exudative retinal detachment* may be associated with *uveal inflammation* which may be in the central fundus, as is frequently seen in general uveitis, or in the periphery as an annular retinal detachment, due to choroidal inflammation which may complicate scleritis. The fluid may also arise in association with a *choroidal neoplasm* where fluid escaping from the disordered tumour circulation frequently passes into the intraretinal space and gravitates to the lower part of the fundus (p509). The treatment is that of the primary condition.

*Traction retinal detachment* by vitreous bands may occur in proliferative diabetic retinopathy (p315 and Plate 16.12 p317) or vasculitis retinae. Treatment by vitrectomy and membrane peeling have improved the prognosis in the treatment of traction detachments.

*Senile retinoschisis* is a form of retinal separation which begins as a degenerative cystic change in the periphery of the retina situated in the plexiform layer and most commonly in the superior temporal quadrant, leading to a collection of fluid within the neural retina which may be shallow or increase to form a large cyst. It is often bilateral and symmetrical, the inner layer does not show undulations and the fluid does not shift with alteration in the position of the head. The retina of the inner leaf is thin and may in time develop holes. There is an absolute field defect corresponding to the area of schisis because of the irreversible separation of the neural connections within the retina. Treatment is conservative with serial visual field tests and photography to monitor progress. If, however, holes form in the *outer* leaf leading then to retinal detachment or if the macular area is threatened by the advancing schisis, surgical treatment will become necessary, but is often difficult.

*Juvenile retinoschisis* is genetically determined and foveal cysts are formed, sometimes associated with peripheral schisis as well.

Retinal separation also occurs in *cystic (cystoid) macular degeneration* or *oedema,* when fluid containing cavities form within the retina in a radiating pattern around the fovea. Fluorescein angiography reveals the rosette pattern of leakage into the retina centered on the fovea. It may result from senile degeneration or subjacent uveitis, from any vascular accident or retinopathy involving retinal oedema, from trauma (*commotio retinae*) following cataract extraction or drainage operation for glaucoma and the use of adrenaline or prostaglandin drops in aphakic eyes which in this case may be reversible on cessation of medication. In some cases a bright red circular *macular cyst or hole* may occur which normally remains unchanged although it causes reduced central vision (Plate 29.5 p527).

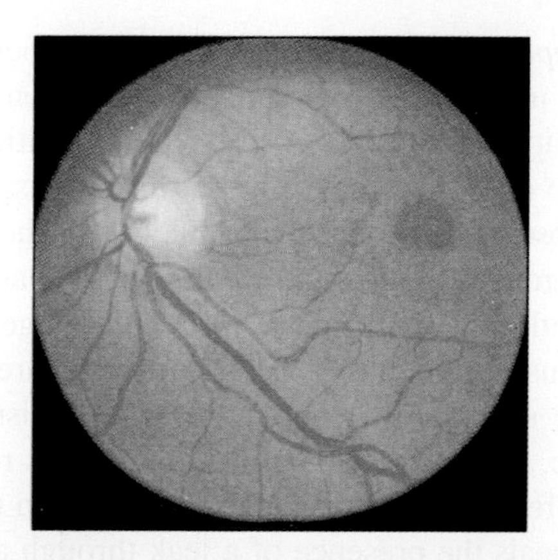

Plate 29.5 Macular hole surrounded by a zone of intra - or sub - retinal fluid.

*Retinal separation associated with central choroido-retinal degeneration*

The layers to be considered are shown in Fig. 29.6 p528. Impairment of function may occur in the membrane of Bruch and in the pigment epithelium. It is usually not possible to say whether this is a specific functional change or degeneration of these tissues or whether it is brought about by degeneration or other disturbance in the choriocapillaris.

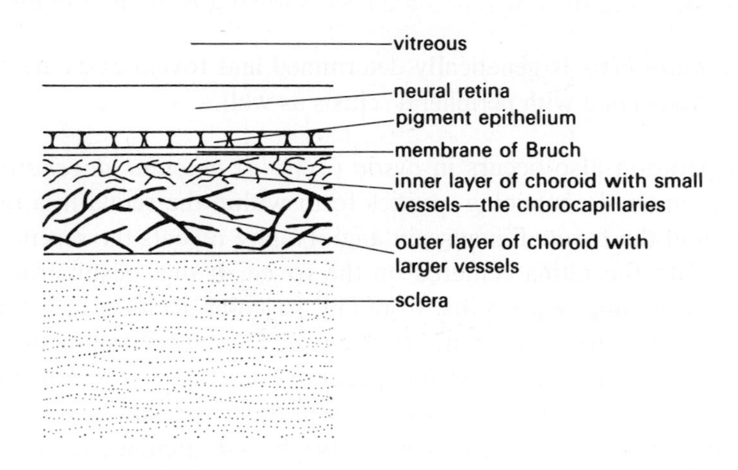


Fig. 29.6 Choroido-retinal layers.

*Central serous retinopathy.* The pigment epithelium is believed to pump fluid actively through the membrane of Bruch for absorption by the choriocapillaris. Focal abnormality in these two layers, possibly initiated by lipid deposition in the membrane of Bruch, may impede this process with the accumulation of fluid and subsequent neovascular changes in the choriocapillaris. In the case of central serous retinopathy the fluid accumulates in the intraretinal space. It predominantly occurs in men (90%) in the age group 20-45 years. The precipitating cause is unknown. The symptoms are blurring of vision, micropsia due to the separation of retinal receptors, distortion, and delayed recovery from dazzle. About 80% undergo spontaneous recovery over a period of 6-12 weeks, a few being left with some distortion (1 in 20). If fluorescein angiography reveals the presence of a leak through an impaired focus of the pigment epithelium away from the macula, an attempt may be made to seal the leak by the application of laser. A condition which may cause some difficulty in differential diagnosis is the central retinal accumulation of fluid which may be associated with a *craterlike pit in the optic disc.* Absorption is usual but some central visual impairment will remain (Plate 29.7 p529, LCW 5.18 p79).

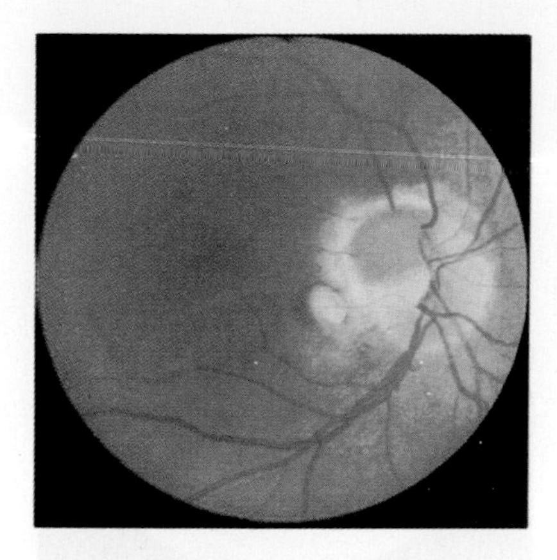

Plate 29.7 Central serous sub - or intra - retinal fluid associated with a crater hole/cyst at the optic disc.

*Age related macular degeneration (ARM wet type)*

*Pigment epithelial detachment* is due to the accumulation of fluid between the pigment epithelium and the membrane of Bruch. Some fluid may also accumulate in the intra-retinal space. Drusen or focal degenerative change in the membrane of Bruch may be the initial change. Ophthalmoscopically it is seen as a well defined dome shaped swelling with an even fluorescence on angiography. It may disappear spontaneously. Unfortunately laser therapy has hitherto been shown to be ineffective or even harmful.

In *disciform degeneration of the retina* blood vessels grow through the membrane of Bruch. They may arise in relation to drusen of the membrane of Bruch or in a pre-existing pigment epithelial detachment. They can form an extensive network and may be associated with fluid transudation causing elevation of the retina and typically with haemorrhage either outside the pigment epithelium or inside it in the intra-retinal space (Plate 29.8 p530 and Fig. 29.9 p530, LCW 5.17 p79).

Clinically it is seen as a greyish central swelling with haemorrhages and exudates surrounding it. It has to be distinguished from the greyish swelling due either to malignant melanoma or secondary choroidal neoplasm affecting the central fundus, or to toxocara.

*Treatment* (p519)

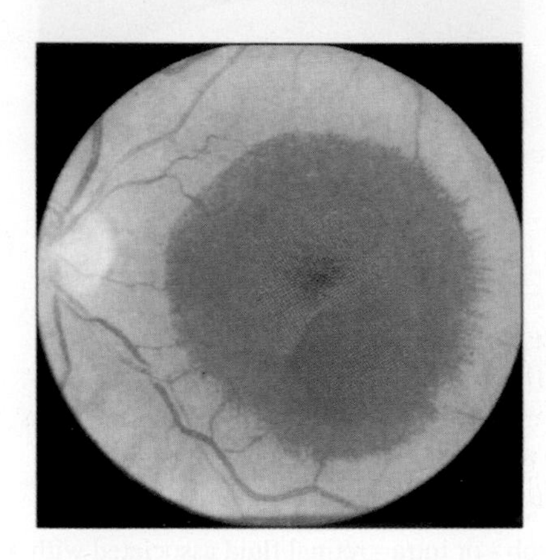

Plate. 29.8 Disciform degeneration of the retina.

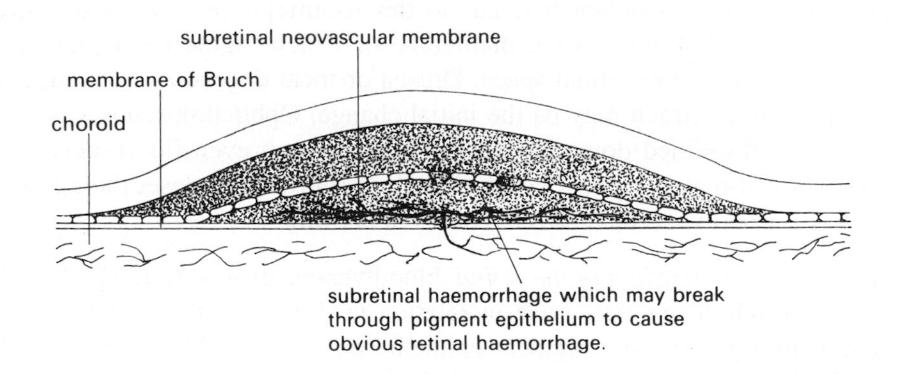


Fig. 29.9 Disciform degeneration of the retina.

**Choroidal detachments** (p507)

**Retinal injuries** (p153)

Retinal effects of ocular contusion (p153) or perforating injury (p157)
Retinal *macular oedema, macular cyst, macular hole* (p527)
Retinal effects of *radiant energy* (p163)
Indirect effect of *fat embolism* of the retinal artery due to fractures of the long bones

**Toxic disorders of retina**

Chloroquine, phenothiazines, epinephrine, quinine (Chapter 34 p617).

**Retinal neoplasms (retinoblastoma)**

The most important neoplasm of the retina is retinoblastoma. Retinal tumours and vascular malformations may also be seen in the phakomatoses (p436) and retinal astrocytomas and tumours of the ciliary epithelium (dictyoma, medullo-epithelioma) may rarely occur.

*Retinoblastoma* is a malignant tumour which arises in the nuclear layers of the retina from multiple foci (Plate 29.10 p532, LCW 8.1(b) p121). It is bilateral in a third of patients and these are almost always hereditary cases transmitted as an autosomal dominant with variable penetrance. Unilateral cases usually represent a fresh mutation but about one in ten may be a unilateral expression of the hereditary disease. Individuals who survive the bilateral disease will themselves transmit it as a dominant characteristic, but in those who survive the unilateral disease and have no family history there is a 5% chance of transmitting the disease. When the non-inherited type occurs in a child of normal parents there is about a 5% chance of a further child also being affected.

Clinically retinoblastoma commences, usually undetected in the first eye, as a whitish swelling in the fundus which may have new vessels and haemorrhages on its surface and may spread in or behind the neural part of the retina to cause a retinal separation or may grow forwards as white masses into the vitreous (LCW 8.1(a) p121). Calcification is common and may show on CT scanning and help in the differential diagnosis. Tumour cells may float in the vitreous or aqueous and give the appearance of uveitis. Secondary glaucoma or uveitis may complicate the picture. Untreated the neoplasm spreads up the

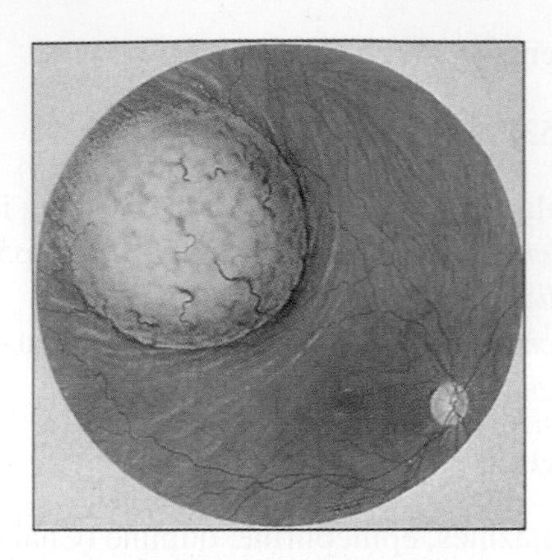

Plate 29.10 Retinoblastoma.

optic nerve or eventually into the orbit. It usually does not form metastases except in advanced cases. The presentation of retinoblastoma may be as a white reflection seen in the pupil of the affected eye, the 'amaurotic cats eye' appearance, or the eye may appear to have uveitis or juvenile glaucoma. In many cases poor vision may precipitate a squint and the tumour is then found on investigation of the squint (LCW 8.5 p123). When diagnosed, the other eye is examined with full mydriasis under a general anaesthetic and careful ophthalmoscopy is necessary at two month intervals until late childhood.

*Treatment.* When an eye is first diagnosed as having retinoblastoma the condition is often well advanced so that enucleation including as much optic nerve as possible is the safest measure. If tumour cells are found extending down the optic nerve the orbit is irradiated. If the condition is bilateral and it is considered justifiable to retain the less affected eye, the lesion in this eye is treated with local irradiation. This work is best carried out at subspecialist centres and careful surveillance by a variety of regimes has greatly improved the prognosis.

*Differential diagnosis.* So important is the differential diagnosis that all other conditions with signs which may be confused with those of retinoblastoma have been designated as 'pseudoretinoblastoma'. These include retinopathy of prematurity, persistent hyperplastic vitreous, persistent vascular sheath of the lens, endophthalmitis, and Coat's disease.

# CHAPTER 30

# THE ORBIT

## Evaluation of orbital disease

*Orbital position and contours* can be assessed by noting the general mid-facial shape and by palpating the orbital rims. The orbit may be smaller on one side in cases of hemifacial microsomia, or after childhood radiotherapy. The orbital walls may be displaced in fracture cases, when steps in the bony contour may also be felt. Local or generalised enlargement on one side may be seen in childhood orbital tumours. Hypertelorism is an increase in the normal distance between the medial orbital walls and is best measured on X-ray.

*Position of the globes within the orbits:* anteroposterior (axial) displacement is detected by examining the patient from above and noting the position of the corneal apex in relation to the supraorbital ridge, or from below in relation to the malar eminences. It is quantified with the exophthalmometer, which uses two mirrors mounted at 45° to compare simultaneously the position of the apices in relation to the lateral orbital walls (Fig. 16.5 p310). The globe is displaced anteriorly (exophthalmos, proptosis) in congenitally shallow orbits, in orbital tumours (LCW 2.15 p32), haemorrhage and thyroid associated disease and in 3rd cranial nerve palsy. It is displaced posteriorly (enophthalmos) in fractures of the orbital walls. Horizontal displacement is noted by measuring the position of the corneal reflexes from the midline with a ruler, and vertical displacement can be estimated by measuring their position downwards from a second ruler held horizontally at the brows.

*Functional integrity of the globes:* examination of the visual acuity and pupillary reactions is mandatory because many forms of orbital disease interfere with optic nerve function; visual fields and colour vision are also useful in monitoring this. The state of the cornea and whether the epithelium stains with fluorescein must be noted, as one of the most devastating complications of exophthalmos is exposure keratopathy. The fundus should be examined for optic disc changes and for choroidal folds.

*Ocular movements* are often affected in orbital disease either due to involvement of the muscles or 3rd, 4th or 6th cranial nerves in the orbital apex. Orthoptic methods of recording eye movements (such as the Hess test) are useful for subsequent comparison (p243).

*Fifth nerve function:* assessment of infraorbital nerve sensation is particularly useful in fractures involving the orbital floor.

*Radiological assessment*

*Plain X-rays:* the Caldwell view is a PA view and is useful as a general survey of the orbital bony features. The Waters view is taken with the patient's head tilted backwards in order to see the orbital floor. X-rays may reveal increased or reduced orbital size, localised indentation or erosion in tumours, or hyperostosis in meningiomas, Paget's disease, osteoblastic metastases, fibrous dysplasia or osteitis. Intraorbital calcification is seen in optic nerve meningioma and as phleboliths in venous malformations.

*Computerised tomographic (CT) scanning:* this is the most useful investigation in orbital disorders as it provides information about the globes, nerves, extraocular muscles, fat and bony walls. The use of intravenous contrast media provides information on the vasculature and may enhance the appearance of tumours. The relationship to the intracranial structures, particularly the cavernous sinuses, and to the paranasal sinuses is well demonstrated.

*Angiography:* the use of orbital arteriography and venography has declined due to the development of CT scanning, but angiography still has a place in the diagnosis of vascular malformations.

*Magnetic resonance imaging (MRI) scanning:* this is occasionally useful particularly in small soft tissue lesions but is limited by the fact that bone does not produce a signal.

*Ultrasonography:* orbital B-scans use a 18,000 Hz probe to produce a two dimensional image. CT scanning has largely superceded ultrasound in orbital diagnosis but it may be useful occasionally when CT is not available, particularly with cystic lesions. The development of colour Doppler ultrasonography may have an increasing rôle in the investigation of vascular tumours.

**Thyroid associated disease** (Chapter 16 p305)

**Orbital inflammatory conditions**

*Orbital cellulitis* presents as redness, swelling and tenderness in the periocular tissues (LCW 2.17 p32). In preseptal cellulitis the infection is confined to the eyelids, and is usually a complication of a localised infective focus such as a chalazion. In true orbital cellulitis the structures behind the orbital septum are involved, producing the 'orbital apex syndrome' of proptosis, visual loss, external and internal ophthalmoplegia, ptosis and sensory loss in the distribution of the 1st and 2nd divisions of the 5th cranial nerve (p380 and Figs. 2.12 p18 and 2.11 p15). The infection is often derived from the paranasal sinuses, but may be from haematogenous spread or from an infected globe (endophthalmitis). Once established, orbital cellulitis may be complicated by subperiosteal abscess formation or spread via the orbital veins to produce cavernous sinus thrombosis or brain abscess. The causative organisms are generally staphylococci, streptococci or haemophilus. Treatment is with high dose parenteral antibiotics while closely monitoring the patient for complications. Chronic orbital infection due to tuberculosis or syphilis is now very rare. Fungal infection may complicate sinus disease (aspergillosis) or diabetes (mucormycosis).

*Orbital pseudotumour* is an idiopathic, non-granulomatous inflammation of the orbital soft tissues. It presents acutely, mimicking orbital cellulitis, or more slowly, mimicking thyroid - associated disease. It may be unilateral or bilateral. Middle - aged females are most commonly affected. Pain is a prominent symptom; other features are lid swelling, chemosis, proptosis and restrictive ocular myopathy. Histologically it consists of a mixed inflammatory cell infiltrate consisting of neutrophil and eosinophil polymorphs, plasma cells and lymphocytes. With disease progression there is a prominent fibroblastic reaction involving orbital fat and extraocular muscle. A similar process may affect the orbital apex (Tolosa - Hunt syndrome) causing early visual loss. Treatment involves steroids, immunosuppressives and radiotherapy. Palliative orbital decompression may have a rôle in selected cases.

*Wegener's granulomatosis:* (p368)

**Lymphoma** (p302)

Orbital lymphoid tumours occur predominantly in the elderly and tend to present as proptosis of slow onset without pain or bony destruction. The mass is generally in the superior orbit and about half have their primary site in the lacrimal gland. The mass may present subconjunctivally as a salmon patch in a similar manner to primary conjunctival disease (Plate 23.1 p465).

There are several problems in classifying orbital lymphomas pathologically. As there are no lymph nodes in the orbit, the disease is by definition extranodal, so the features of nodal disruption cannot be used to assess if a lymphocytic infiltrate is reactive or malignant. There is a spectrum of cytological features from mature, well differentiated lymphocytes forming follicles (benign reactive lymphoid hyperplasia) at one end and atypical lymphocytes without follicular architecture at the other (diffuse lymphoma). Unfortunately, the majority of lesions are 'grey - zone' lymphomas which are diffuse and lacking in germinal centres but which are composed of well - differentiated lymphocytes without features of cellular atypia. These are termed diffuse or atypical lymphoid hyperplasia.

In an effort to decide whether the infiltrate is benign or malignant, immunohistochemistry is widely used. The proportions of T and B cells can be determined, and it has been found that when T cells predominate the infiltrate is generally benign whereas when B cells predominate it tends to be malignant. It can also be determined, by studying the immunoglobulin light chains, whether the tumour is mono- or polyclonal, the former indicating malignancy. It remains controversial whether polyclonality always indicates that the lesion is benign, because some authors have reported that some of these lesions are associated with systemic disease.

The diagnosis of an orbital lymphoid infiltrate as malignant lymphoma by these means does not necessarily carry a bad prognosis as many of these lesions are slowly-growing. The majority of these patients are clinically at an early stage (1E) and only a quarter are ever associated with extraorbital disease; this latter is probably a distinct group with secondary orbital involvement in a disseminated lymphoma. The 5 year survival rates for the early stage patients, including those with bilateral disease is 85%, whereas in those with extraorbital disease it is nearer 20%.

It follows that thorough clinical staging is mandatory in these patients, who have a good prognosis if they are at an early stage at presentation and six months later, regardless of histological or 'immuno' features of malignancy.

The treatment of localised disease is radiotherapy; this is combined with chemotherapy in disseminated lymphoma. For other tumours arising in the lacrimal gland see p466 and Fig 23.2 p466.

**Orbital mass lesions**

*Mucocele*

This is a cystic mass derived from the mucosa of the paranasal sinuses. Histologically it is lined by pseudostratified columnar epithelium and has a fibrous wall. It usually occurs in older patients with a history of sinus disease but may occur in the young with allergic disorders or after sino-orbital trauma.

*Dermoid* (pp429, 455, 476)

*Dermolipoma* (p476)

*Vascular tumours*

*Capillary haemangioma* occurs in the anterior orbit, presents in infancy and undergoes a period of growth followed by regression in childhood. It may be associated with a strawberry naevus of the lids and can be treated with steroid injections if it is threatening vision. *Cavernous haemangioma* is a relatively common orbital lesion and occurs as a unilateral intraconal mass in adults. It grows progressively, compressing the optic nerve and causing proptosis, and has to be surgically excised. Pathological examination shows widely dilated vascular channels enclosed in a fibrous capsule. Occasionally it presents explosively as an acute orbital haemorrhage. *Haemangiopericytoma* and *haemangioendothelioma* are rare in the orbit. *Lymphangioma* is also a rare tumour but is important as a cause of 'spontaneous' orbital haemorrhage. Chronic haemorrhage gives rise to the macroscopic appearance of 'chocolate cysts'.

*Neural tumours*

*Neurofibroma and neurolemmoma (Schwannoma)* (pp437, 459) are tumours of the fibrocytes of the perineurium and the Schwann cells of the nerve. They usually occur in multiple foci associated with the other manifestations of von Recklinghausen's disease but occasionally occur singly in otherwise healthy patients. Neurofibromas are non-encapsulated whereas neurilemmomas have a fibrous capsule. Thc former have a rather uniform fibrous morphology, often requiring special stains to show the neural elements, whereas the latter have contrasting cellular and hypocellular areas and pallisaded nuclei known as Verocay bodies.

*Meningioma* (p380) occurs mainly in adults in their middle years and affects females more commonly. It arises from the optic nerve sheath and grows slowly, compressing the nerve and causing visual loss before it enlarges sufficiently to cause proptosis. Compression causes the signs of optic disc swelling, optic atrophy and finally optico-ciliary shunt vessels. CT scanning shows a well defined enlargement of the nerve and there may be flecks of calcification. Treatment is surgical removal of the nerve and histology reveals a syncytial pattern with prominent Psammoma bodies.

*Glioma:* this tumour occurs in children, approximately half of whom have neurofibromatosis. The optic canal is enlarged early in the course of the disease. The growth rate of optic nerve gliomas is low and the usual treatment is excision of the nerve. The morphological features are astrocytic proliferation, destruction of myelin sheaths and myxoid degeneration. The proximal end of the nerve is examined for clearance as the tumour can grow along the nerve to the chiasm and into the contralateral side. The condition carries a good prognosis although a recurrence rate of 5% has been estimated. An alternative mode of therapy is radiation, which is particularly useful in chiasmal tumours.

Adult optic nerve gliomas are rare and carry a lethal prognosis as they behave as glioblastoma multiforme, and frequently invade intracranially.

*Soft tissue tumours*

*Rhabdomyosarcoma* is the most common orbital primary tumour of childhood. They are most often situated in the superonasal quadrant and present as lid swelling (which may mimic a chalazion) or proptosis. Biopsy reveals a mass of eosinophilic cells and cross-striations which are more readily apparent in well differentiated tumours. In the commonest, most poorly differentiated (embryonal) variety myosin filaments may only be revealed on electron microscopy. The prognosis is poor, as the tumour metastasises widely, but is improving with advances in radiotherapy and chemotherapy. *Lipomas* are hamartomatous tumours and *liposarcomas* are malignant tumours derived from orbital lipocytes. *Fibrous histiocytoma* is the commonest orbital mesenchymal tumour in adults. It tends to present in the superonasal quadrant of the orbit and causes non-axial proptosis. There is a spectrum from small, well differentiated, well circumscribed lesions to a widely infiltrative, pleomorphic variety (malignant fibrous histiocytoma). Wide surgical excision is recommended but this condition carries a poor prognosis due to a high local recurrence rate.

*Osteoid tumours*

*Osteoma* is a benign bony outgrowth from the orbital wall. Wide excision is recommended as local recurrence is common. *Osteosarcoma* is a highly malignant tumour which may occur in isolation or in association with the retinoblastoma gene.

*Tumours invading the orbit by local spread*

These may originate in the paranasal sinuses, most commonly the maxillary, when infraorbital nerve anaesthesia may be a useful sign. Skin tumours such as basal cell carcinomas invade along the orbital walls from the lids and neglected primary intraocular tumours, retinoblastomas and melanomas, through the sclera. Intracranial tumours, most often meningiomas (p380, Figs. 19.11, 19.12 p382) invade through the superior orbital fissure or optic canal and nasopharyngeal tumours, which may be unsuspected in the early stages, through the inferior orbital fissure.

### *Orbital metastases*

These occur most often in previously diagnosed tumours but are the presenting feature of systemic disease in about 40% of cases. In decending order of frequency they originate from breast, lung, prostate, melanoma, gastrointestinal tract or kidney and involve bone (especially prostate), orbital fat (especially breast) and extraocular muscles (especially melanoma). Metastatic scirrhous carcinoma of the breast may cause enophthalmos. Cytology of specimens from fine-needle aspiration has a rôle in the diagnosis of these lesions but biopsy may be necessary when a dry tap is obtained or when cytological features are not diagnostic. Radiotherapy is often a success in relieving symptoms such as pain and preserving vision, but overall survival is very poor.

## **Orbital trauma** (LCW 9.21, 9.22 p139)

Priority must be given to life support and diagnosis of the full extent of bodily injury. The appropriate specialists should work as a team from the outset. Orbital fractures are frequently associated with injury to the globe, canthal tendons and lacrimal apparatus. The optic nerve, infraorbital nerve and nerves to the extraocular muscles may be involved in apical fractures.

### *External fractures*

*Zygomatic (malar) fractures* occur with blows to the cheek and are common in assaults and as sporting injuries. The fractures occur at the frontozygomatic and zygomaticotemporal sutures and sometimes also in the zygomatic arch (tripod fracture). The clinical signs are periorbital and intraoral haematoma, flattening of the cheek and infraorbital nerve anaesthesia. The bone frequently displaces downwards and backwards with a medial hinge effect and the globe may be displaced due to involvement of the suspensory ligament. Treatment is moving from blind elevation techniques through the temporal approach towards direct exposure and internal fixation. Transient diplopia is common after reduction of these fractures but usually settles with time.

*Naso-orbital fractures.* These result from direct trauma in the region of the nasal bridge. The nasal bones and frontal processes of the maxillae are fractured and the internal nasal structures driven backwards. The nasal bridge appears flattened and disruption of the medial canthal tendons gives rise to telecanthus. Complications include anterior fossa dural tear and CSF leak, damage to the globe and laceration of the lacrimal sac and canaliculi. Treatment is moving from blind intranasal elevation of fragments to open reduction, which has the advantage of allowing simultaneous dacryocystorhinostomy and transnasal wiring techniques to correct lacrimal injuries and telecanthus.

*Midfacial fractures* are classified according to the degree of injury which depends on the force applied. Midfacial fractures have a characteristic clinical appearance with severe oedema, a 'dish face' deformity and underbite malocclusion of the teeth. Movement of the maxillary segment can be elicited by palpation. Treatment of the less severe fractures is by closed reduction and intermaxillary fixation. The more severe ones are treated by open reduction methods and transnasal wiring to correct the associated telecanthus.

*Internal (blow-out) fractures* (p154, Fig. 30.1 p542)

The orbital contents are separated from the paranasal sinuses by thin plates of bone. Force transmitted to the eye causes a shockwave of increased intraorbital pressure, which results in fracture of the weakest points, namely the lamina papyracea of the ethmoid bone and the posteromedial orbital floor. The clinical features are periorbital haematoma, conjunctival haematoma, axial displacement and vertical displacement of the globe, abnormal eye movements and infraorbital nerve anaesthesia. Axial displacement usually takes the form of enophthalmos due to prolapse of orbital contents. Normally, Lockwood's ligament prevents downwards displacement of the globe, but this may occur in large floor defects due to rupture or stretching of the ligament. Exophthalmos may occasionally be seen, due to orbital haemorrhage. A variety of mechanisms are invoked to explain the motility abnormalities, including entrapment of muscle or connective tissue septa in the fracture site, haemorrhage into the posteroinferior fat, contusion injury to the muscle or intramuscular haematoma, nerve injury or Volkmann's ischaemic contracture (LCW 9.23 p139).

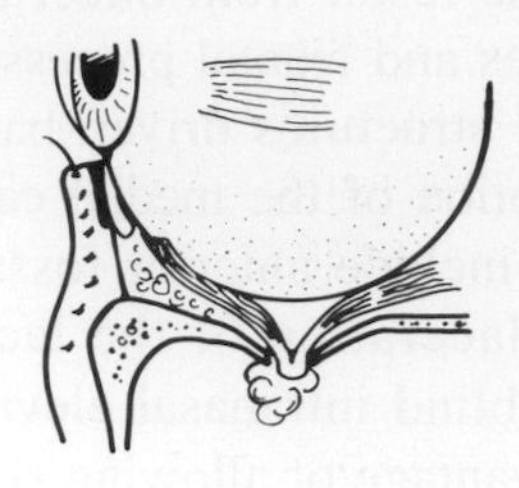

Fig. 30.1 Blow-out fracture of orbital floor. Inferior rectus incarceration.

*Investigations.* Plain X-rays may show a fracture line and opacification of the roof of the antrum (hanging drop sign) usually due to prolapse of orbital contents, but sometimes due to haemorrhage or oedema in the sinus mucosa, CT scans may be of some use in predicting outcome and defining those cases which will benefit from early surgical intervention.

*Complications.* The rate of ocular injury is quoted at between 15-30%. This varies from inconsequential signs such as subconjunctival haemorrhage to serious injuries with varying degrees of long-term visual dysfunction such as hyphaema, angle recession, dislocated lens, cataract, vitreous haemorrhage, commotio retinae, choroidal rupture, macular scarring or even rupture of the globe. Hypo-aesthesia of the infraorbital nerve has a tendency to spontaneous recovery due to axonal regrowth. Orbital emphysema is due to air from the paranasal sinuses being forced into the soft tissues at the time of injury or subsequently. It resolves spontaneously but may be complicated by infection or, rarely, by optic nerve compression.

*Treatment.* Any associated injury to the globe must be dealt with prior to bony manipulations. Controversy exists as to the selection of cases for early surgical repair and its timing. A reasonable approach is to re-assess the patient at two weeks post-injury and operate on those patients with cosmetically unacceptable enophthalmos or non-resolving, handicapping diplopia of a type which is likely to respond to floor repair, namely that which is due to restriction rather than paresis, as demonstrated by either a positive forced duction test or CT evidence of tissue entrapment.

Orbital floor repair is carried out by a transconjunctival or subciliary dissection to the orbital rim, elevation of the periosteum of the orbital floor, reduction of the prolapsed orbital tissues, and repair of the bony defect. Care must be taken not to damage the optic nerve during repair and the visual function must be closely monitored post operatively.

## Developmental abnormalities of the orbit

The craniofacial malformations encountered can be broadly divided into four categories. These may well overlap and representative examples only are given to illustrate the broad principles of surgical rehabilitation.

-Craniofacial synostoses
-Clefting syndromes
-Branchial arch syndromes
-Aplasia or hypoplasia of the brain or eye

### *Craniofacial synostoses*

This is a group of disorders characterised by premature fusion of the sutures. This results in reduced bone growth in the direction perpendicular to the synostosis and compensatory deformities of the skull and face. Raised intracranial pressure is a common feature. The ocular manifestations are secondary to disordered orbital development. A variety of syndromes have been described. The commonest to involve the ophthalmologist are Crouzon's and Apert's syndromes.

*Crouzon's syndrome* consists of premature craniosynostosis, midfacial hypoplasia and exophthalmos (Fig. 30.2 p544). There is a variable head shape depending on the sequence of closure. Hypoplasia of the frontal, sphenoid, ethmoid and maxillary bones results in a characteristic frog-like facies with relative prognathia and dental malocclusion. Exophthalmos results from very shallow orbits and leads to problems with corneal exposure and even luxation of the globes beyond the eyelids. Optic nerve compression and divergent squint are common features.

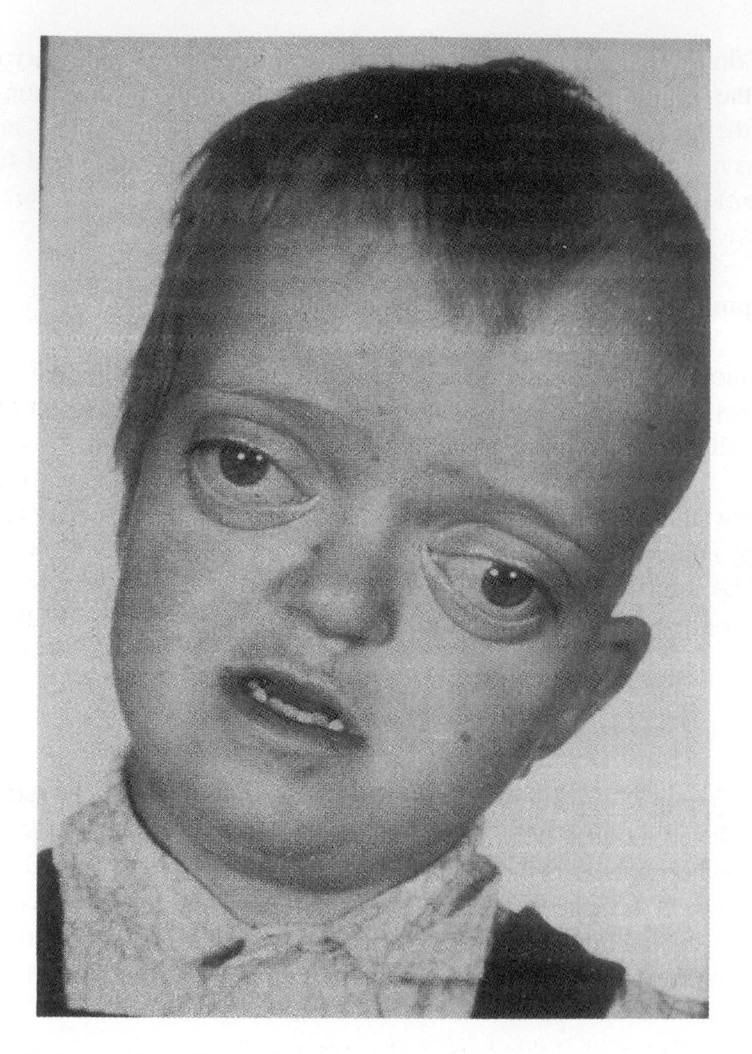

Fig. 30.2 Shallow orbits in Crouzon's disease causing the appearance of proptosis.

*Apert's syndrome* produces craniofacial abnormalities similar to those of Crouzon's, but with prominent abnormalities of the teeth, palate and ear. These patients also have syndactyly of the hands and feet involving the 2nd 3rd and 4th digits, and may have associated cardiac and gastrointestinal anomalies. The ophthalmological manifestations generally parallel those of Crouzon's syndrome.

Surgical management of craniosynostosis generally consists of fracture, mobilisation and advancement of the anterior two-thirds of the orbit with bone grafting into the spaces thus created. Particular attention must be paid to treatment of corneal exposure and amblyopia from the earliest stages of presentation. Strabismus surgery is generally deferred until the orbital abnormailties have been corrected.

*Clefting syndromes*

The aetiology of clefts remains controversial. There are two main theories, the first being failure of fusion of normal facial processes and the second being the failure of neural crest cell influence (neurocristopathy). A variety of influences have been implicated, including ionising radiation, influenza virus, anticonvulsants or tranquillisers. A distinct aetiological entity is the amniotic band syndrome. In this condition there is amniotic rupture leading to oligohydramnios and the formation of bands of amnion, which may be swallowed by the foetus and cut into the developing face. Concurrent constriction or compression may produce intrauterine amputations or deformities such as club foot.

Facial clefts are classified by a *Tessier number*. This assigns each cleft a number 0-14 in a circle running clockwise around the right eye and anticlockwise around the left from 0 in the lower facial midline to 14 in the upper facial midline. They are responsible for a variety of clinical presentations e.g. hypertelorism, colobomatous malformations and medial canthal dystopia and nasolacrimal obstruction, for example median facial clefts 0,1,13,14 present with hypertelorism. Surgical repair is best carried out by a craniofacial team of specialists and the staged repair should deal as a priority with functions such as feeding or corneal protection. Later, procedures may be necessary to improve appearance or deal with epiphora.

*Branchial arch syndromes*

The branchial arches are a series of 5 or 6 paired structures on the ventrolateral aspect of the embryonic head present by the fourth week of gestation. They correspond phylogenetically to the site of the gill arches. The branchial arch syndromes are disorders of development of the structures derived from the first and second arches. The first arch consists of the mandibular process (Meckel's cartilage) and the maxillary process. The second arch produces the facial muscles and the main components of the developing ear.

*Treacher-Collins-Franceschetti syndrome (mandibulofacial dysostosis)* is often called the first arch syndrome although structures derived from both the first and second arches are involved. The main features are hypoplasia of the maxillae and mandible, beak nose, malformed ears, conductive deafness and dental anomalies. The ocular features include an antimongoloid slant and a characteristic lower lid coloboma which occurs at the junction of the medial two-thirds and lateral third of the lid, the medial portion being devoid of lashes. The lower puncta may be absent (p467).

*Goldenhaar's syndrome* consists of ocular dermoids or dermolipomas in association with abnormal ears, pre-auricular tags, hypoplasia of the mandibular ramus and vertebral anomalies. Lid colobomas and strabismus are common. Cardiac and pulmonary anomalies are sometimes associated.

# CHAPTER 31

# DISORDERS ASSOCIATED WITH THE LEVEL OF INTRAOCULAR PRESSURE

## Relationship between glaucoma and intraocular pressure

The intraocular pressure (IOP) and its variations in health depend on several factors but basically upon the balance between the rate of aqueous production and the rate of its drainage back into the venous circulation. It is usually found in health to be between 10 and 20 mm Hg., the eye thereby having sufficient rigidity to function as an optical instrument. The mean intraocular pressure in a white population aged over 40 years is 15.5 mm Hg. with a standard deviation of 2.5 and it is usually accepted that the upper limit of 'normal' intraocular pressure is 21 mm Hg., but this is a statistical concept. It is important to realise (as it has not been in the past) that for the individual a pressure moderately in excess of 21 mm Hg. will not necessarily cause injury to the eye, nor will eyes with pressures of 21 mm Hg. and below necessarily be free of the chronic type of optic nerve and other damage usually associated with moderately higher levels of intraocular pressure viz. glaucoma.

*Chronic glaucomatous damage to the eye in primary open angle glaucoma* depends partly on the level of the intraocular pressure and partly on the viability of the ocular tissues. (p129) The result is damage to the optic nerve in the region of the optic disc and lamina cribrosa, although more diffuse deterioration may occur due to retinal ganglion cell death. ***Thus there is no actual 'normal' intraocular pressure level and the term only indicates an arbitrary but increasing statistical degree of risk to the retinal ganglion cells and their axons in the optic nerve when the intraocular pressure rises above about 21 mm Hg.*** The clinical convenience of recognising this degree of risk is that it indicates patients requiring further investigation on grounds of pressure alone. There is, however, epidemiological evidence that the relative risk of intraocular pressure causing damage to non-compromised optic nerves only becomes significant at levels from about 26 mm Hg although optic nerves already damaged for this and other reasons are partially, sometimes fully, protected by its reduction to a level found appropriate to the individual.

The *aqueous humour* originates from the epithelium covering the ciliary body, the area of which is greatly increased by the presence of about eighty radial villous ciliary processes which are heavily vascularised. The ciliary epithelium consists of two layers of ectodermal cells representing the outer pigment layer and the inner neural layer of the optic vesicle. From the posterior chamber the aqueous passes into the anterior chamber. From here it returns to the venous circulation mainly through the pores of the corneo-scleral trabeculum in the antero-lateral wall of the angle of the anterior chamber (Fig. 6.39 p131) and through the endothelium of the canal of Schlemm, from the lumen of which it drains to collector channels which either empty directly into the scleral venous network or continue to the surface as aqueous veins where they can be seen joining the episcleral veins. This *trabecular outflow* comprises about 85% of the total aqueous flow. Some aqueous also drains from the anterior chamber directly to blood vessels in the uvea and sclera and through the sclera to the orbital vessels. The *uveo-scleral flow* is about 15% of the total. Both outflows are variable.

The *blood-aqueous barrier.* The tissues separating the blood from the aqueous humour form the blood-aqueous barrier. The aqueous is similar in many ways to plasma from which most of the protein has been removed. Its composition differs, however, from a filtrate or dialysate, implying that it is a secretion requiring work to be done by the cells of the ciliary epithelium. Aqueous production is also subject to such factors as the capillary blood pressure in the ciliary body and iris and the osmotic pressure of the plasma; if there is a change in plasma osmolarity, adjustment rapidly takes place. An increase in the osmotic pressure of the plasma especially if brought about by a substance which does not penetrate into the aqueous (e.g. mannitol) or only with difficulty (e.g. urea) will therefore rapidly reduce the intraocular pressure. Aqueous is also in equilibrium with fluid in the vitreous gel. The concentrations of the various constituents of aqueous are not only dependent on the composition of the secreted aqueous but vary in different parts of the aqueous because they are modified by the metabolism of the adjacent tissues, particularly the retina, lens, iris and cornea.

*The blood supply of the optic nerve head* (Fig. 19.3 p373). The blood supply of the ocular portion of the optic nerve is mainly derived from the posterior ciliary branches of the ophthalmic artery. These supply twigs directly to join the arterial network at the nerve head. There they are joined by arterial twigs from the choroid, which is also supplied by the posterior ciliary arteries and by small branches from the pial vessels surrounding the optic nerve. Recurrent branches from the main divisions of the central retinal artery just beyond the optic disc have been said to join the network and the central retinal artery may also supply some twigs to the centre of the nerve. The venous drainage is to the central retinal vein, which joins the ophthalmic veins and to choroidal veins.

**'The glaucomas'**

'The glaucomas' is the collective term for a number of conditions which are basically different from each other, in which the optic nerve sustains damage in the region of the optic disc and lamina cribrosa resulting in characteristic changes in the optic disc and the visual fields. The intraocular pressure is a risk factor of varying importance in the different types of glaucoma.

*An outline of the practical essentials of primary open angle glaucoma (POAG),* which is a chronic condition, has been given on p129 and of *primary angle closure glaucoma (PACG)* (sometimes called primary closed angle glaucoma), which is usually of acute or sub-acute onset, on p186. These and other types will now be considered further.

**Concepts basic to the classification of the glaucomas**

It is customary to classify glaucoma as primary or secondary.

*Secondary glaucoma* is usually considered to be present when a pressure of more than 21 mm Hg. is accompanied by an ocular disturbance which can reasonably be expected to have caused the raised pressure. It is convenient to recognise an arbitrary level of pressure in these cases because the direct effects of the disturbance, which may be traumatic, inflammatory, degenerative, vascular or toxic in nature, sometimes makes it difficult to assess the extent to which visual impairment may be related to the level of intraocular pressure. If the condition occurs *in utero,* secondary congenital glaucoma will result. The causes of secondary glaucoma are considered more fully on p578.

In *primary glaucoma* there is no evidence of an ocular or general cause of secondary glaucoma, many cases result from an inherited abnormal mutation and others from a new mutation during meiosis, producing the same or a similar phenotypic change in an individual patient, which subsequently may be passed on to descendants.

*A consideration of the pathogenesis of different types of primary glaucoma*

*Primary developmental glaucoma* (p572) is due to ocular developmental anomalies. If these are not present, primary glaucoma is usually divided into angle closure or open angle types.

***Primary angle closure glaucoma (PACG)*** (usually acute or subacute) causes optic nerve damage due to raised intraocular pressure. In the *acute* type (pp186, 574) when the flow of aqueous is suddenly stopped by closure of the drainage angle, the vascular circulation cannot adapt to the sudden and marked rise in intraocular pressure. A shallow anterior chamber is frequently found in small hypermetropic eyes which are then predisposed to angle closure. The factors which precipitate closure in the predisposed patient are not fully understood but the circumstances surrounding the event which occurs against a background of steady lens growth, seem to point to a vascular change involving vasodilatation of the uveal tract which swells, increases vitreous hydration and tends to push the lens forward and simultaneously thickens the iris, aggravated by mid-dilatation of the pupil. Older middle aged women are particularly susceptible, possibly because they are more liable to vasomotor instability. It is associated with anxiety and has been shown to be more likely to occur in weather which shows large variations in barometric pressure which accompanies thunder storms and in many people is associated with a feeling of headache and congestion. (LCW 4.12 p62)

Once the process is initiated the sequence of events tends to perpetuate and aggravate it. The vascular circulation cannot adapt to the acute and marked rise in intraocular pressure. The veins leaving the eye, the central retinal vein at the disc and the vortex vein, have at their exits the lowest pressure of all the intraocular vessels and are consequently the first to become compressed at their points of exit; the pressure then rises in the venous end of the capillary network. The blood flow is slowed and the capillary pressure rises, which, with anoxia of the vessel walls, leads to increased capillary permeability and oedema in the tissues of the uveal tract, retina, and optic nerve head and the escape of protein and cells into the aqueous. If the condition is not relieved, patchy infarction results in necrosis and eventually in atrophy. The vessels supplying blood or draining it from the eye also become dilated due to nervous or chemical influences. Here too capillary permeability increases and causes orbital and conjunctival oedema. Focal areas of necrosis may appear as grey spots in the subcapsular epithelium of the lens *(glaukomfleken)*. The corneal endothelium normally controls the degree of hydration of the stroma by actively transmitting fluid from the cornea into the aqueous. This process is resisted by the suddenly raised intraocular pressure, resulting in corneal oedema and the formation of droplets under the epithelium. The escape of protein into the aqueous from dilated anoxic capillaries is followed by the formation of fibrinous adhesions across the already closed angle. The adhesions are known as *peripheral anterior synechiae* and may eventually prevent all or part of the angle opening even if the pressure is relieved.

*Primary chronic angle closure glaucoma* may arise in the presence of a narrow anterior chamber angle from the apposition of the peripheral iris to the back of the cornea, with closure of the narrow angle by slowly progressive anterior synechiae and cause effects resembling those of primary open angle glaucoma which require similar treatment but the response to trabeculectomy is less favourable with an increased risk of flat anterior chamber and its sequelae (p581).

*Plateau iris* - is an uncommon condition due to the anterior position of the ciliary body causing it to press forwards the peripheral iris which may lead to chronic angle closure or be revealed by the unexpected recurrence of acute angle closure despite laser peripheral iridotomy for this condition in a younger patient, in the presence of an only mildly shallow anterior chamber. Indentation gonioscopy may fail to open the angle but cause a backward bowing of the iris midperiphery. Ultrasonography may reveal the underlying cause. Mydriasis can increase the angle closure to a marked extent in this condition. Topical pilocarpine and latanoprost may help to control the pressure but a trabeculectomy may be necessary, despite the risk of aqueous misdirection due to ciliary block (malignant glaucoma) (pp496, 581).

***Primary open angle glaucoma*** *(POAG).* This is a chronic optic neuropathy and this particular type of optic nerve damage (as has been indicated above p547) can have a variable relationship to intraocular pressure. It is usually associated with an intraocular pressure more or less constantly raised above the statistically normal range *(chronic simple glaucoma)* but, in many of the cases running a very chronic course, the intraocular pressure may be either at or below the statistical norm (these are described as *low tension glaucoma (LTG)* or *'normal' pressure (or tension) glaucoma (NPG or NTG).* Those POAG eyes with pressures above 'normal' are often in the mildly raised zone. It is clear that in all these relatively low tension glaucoma patients the optic nerve is unduly vulnerable to intraocular pressure. On the other hand there are a considerable number of people who have intraocular pressure levels appreciably raised above the 'normal' range who have no signs of optic nerve impairment over long periods of observation, nevertheless they are more liable to develop it than those with normal pressures. Such people have optic nerves which are able to resist the adverse effect of higher than average pressures for a variable time, a condition designated *ocular hypertension (OH),* which by definition is not glaucoma, nevertheless, they are much more likely to develop it than those with 'normal' pressures.

In *primary open angle glaucoma,* the intraocular pressure rises due to increased resistance to the passage of aqueous, the site of which may be the deep trabecular drainage meshwork, the canal of Schlemm, its efferents or the episcleral veins. The pressure rise is usually intermittent at first with increased diurnal variation and gradually the general level of pressure increases to exceed the critical pressure at which glaucomatous optic nerve head damage occurs in the particular patient, revealed by the insidious advance of scotomas in the arcuate areas of the field of vision and the progressively pale and cupped optic disc of glaucomatous optic atrophy.

***It is thus probably best to regard the anterior optic nerve damage which characterises primary open angle glaucoma as "a POAG complex" resulting from the interplay of two main variables:1. the intraocular pressure and 2. an abnormality of the retina and optic nerve. Together these tend to result in damage particularly overt in the region of the optic disc and lamina cribrosa.*** The second being due to inadequate blood supply or genetic quality or to distortion of structure or other genetically determined reactions leading to ganglion cell death by apoptosis (p553) all of which may themselves also be affected by the level of the intraocular pressure. This would also increase optic nerve vulnerability to whatever level of intraocular pressure is present. Ischaemia would tend to cause reduced glucose and oxygen supply, resulting in inadequate ATP, while distortion of the lamina cribrosa might well narrow and kink both the nerve fibres and the capillaries which nourish them. Both factors would lead to interruption of rapid axoplasmic flow locally in optic nerve fibres. This in turn blocks the transport of trophic signals from the synapse of the axon and causes ganglion cell death in the retina (see also nitric oxide etc. p553). Primary open angle glaucoma thus can be revealed as a complex of chronic simple glaucoma and low tension glaucoma, recognising at the same time the existence of a third entity, ocular hypertension, which does not satisfy the definition of primary open angle

| Sub-division of POAG complex | Glaucomatous anterior optic neuropathy | Intraocular pressure raised above statistical 'normal' level |
|---|---|---|
| Chronic simple glaucoma | + | + |
| Low tension glaucoma | + | - |
| Ocular hypertension | - | + |

Fig. 31.1 The primary open angle glaucomatous complex. This is how it can be arbitrarily compartmentalised for convenience of clinical description, but it represents the inter action of two variables, one continuous (IOP) and the other (optic nerve state) with many possible components.

glaucoma because there is no detectable optic nerve damage; nevertheless there is in these patients an appreciable chance of developing it, such patients being one type of 'glaucoma suspect' (Fig. 31.1 p552).

A note on *primary chronic angle closure glaucoma (mixed glaucoma).* Primary angle closure glaucoma and primary open angle glaucoma are genetically, pathologically and clinically two separate diseases with a common factor of uncompensated intraocular pressure. The tendency towards these two conditions may of course co-exist and cases of mixed glaucoma may be seen when closure of even one sector of the angle of the anterior chamber may lead to a raised intraocular pressure because the outflow through the remaining drainage mechanism is of a low order. This may be due to trabecular damage resulting from previous subacute angle closure episodes so that there may also be an element of secondary glaucoma. The closure of the angle may slowly advance with progression of signs and symptoms. For the purposes of classification such patients are best regarded as primary chronic closed angle glaucoma, but it is essential to appreciate their dual or even triple nature for a clear understanding of the situation.

*Nitric oxide, endothelin and glutamate*

There is evidence that both the ganglion cell loss and the obstruction to aqueous flow in primary open angle glaucoma may, in addition to the direct effect of the level of the intraocular pressure, involve the interactions of ill understood complex systems initiated by the release of vasoactive factors by the vascular endothelial cells, in particular a peptide, endothelin and various nitric oxide synthases. *This is here briefly outlined to indicate the complexity of problems which underlie some of the changes in primary open angle glaucoma,* whether the intraocular pressure is above or below statistical 'normal' values.

*Ganglion cell death. Apoptosis*

The irreversible visual field loss in primary open angle glaucoma is a manifestation of ganglion cell death without evoking any sign of inflammation i.e. by apoptosis. This is a genetically programmed physiological process to allow cell replacement, believed to involve glutamate, which is a neuro-transmitter found in some retinal neurones. If small amounts are released extracellularly in normal cell replacement, they are taken up by Müller cells and cause no trouble. If neurones are exposed to damaging factors such as a raised intraocular pressure or ischaemia, they may leak excessive glutamate extracellularly and by the glutamate activation of an aspartate membrane receptor N-methyl-D aspartate

(NMDA) lead to an influx of calcium into certain susceptible macroganglion cells of the retina. This stimulates neuronal nitric oxide synthases (NOS) within them to produce nitric oxide as well as a free radical superoxide which then react to form a very toxic peroxynitrite which causes apoptotic cell death. Glutamate is reported to be increased in the vitreous of glaucomatous eyes.

*The regulation of intraocular pressure*

Endothelin is a biologically active peptide produced by vascular endothelial cells. It is strongly vasoconstrictive which is balanced by the vasodilator effect of nitric oxide produced by nitric oxide synthases. This balance may involve the inherently contractile elements in the trabecular meshwork and so influence the passage of aqueous through it and thereby the intraocular pressure.

*The regulation of optic disc blood supply*

A similar regulatory system involves the increased endothelin which has been detected in the aqueous of patients with primary open angle glaucoma. Their ocular tissues also showed a marked decrease in nitric oxide synthase, reducing the vasodilator element of the system, allowing vasoconstriction to cause ischaemia of the ocular tissues including those at the optic nerve head.

*The control of vasospastic reactions*

Some vascular smooth muscle cells are also stimulated to contract by acetyl choline leading to vasoconstriction. This may be regulated by the vasodilator reaction of rapidly diffused nitric oxide from the adjacent endothelial cells which are also being stimulated by acetyl choline. If the endothelial cells are damaged so that less nitric oxide is produced, this may reduce the vasodilator element and release from control a local vasospastic reaction which may contribute to an impaired blood supply to the optic nerve and retina. A tendency to vasospasm has been shown to be a risk factor for primary open angle glaucoma in some patients.

*Neuroprotection*

Thus in several ways the knowledge of these systems which appear to play a part in primary open angle glaucoma is likely to unfold gradually to reveal sub-types of the disorder characterised by a variety of factors which operate to varying extents in different patients. Molecular genetic studies are also making it possible to relate phenotypes to genotypes and to identify the presence and nature of abnormal proteins and enzyme systems. Research on many fronts thus brings closer the time when in addition to controlling intraocular pressure, new and fundamental treatments will be found for these other elements of the primary open angle glaucoma complex, foreshadowed perhaps by the neuroprotective effect of aminoguanosine and of betaxolol and sympathetic alpha 2 receptor blockers in experimental nerve injury.

## Diagnostic techniques

The essential examinations for the assessment of glaucoma are:

-the *measurement of intraocular pressure* (tonometry),
-*optic disc assessment* by ophthalmoscopy including slit lamp examination using the non contact +90 or +78D lens or the fundus contact lens and more recently the laser scanning ophthalmoscope,
-*visual field examination* (perimetry),
-the *examination of the angle of the anterior chamber* (gonioscopy).

In addition, evidence is accumulating that impairment of certain aspects of visual function may antedate conventional perimetric criteria of glaucomatous change in the optic nerve. Tests for such impairment may prove to be valuable in providing an opportunity for earlier recognition and treatment of patients with primary open angle glaucoma, especially those with borderline intraocular pressures or other risk factors. Examples are blue on yellow perimetry, motion detection and contrast sensitivity impairment as revealed by psychophysical tests such as those for combined spatial contrast and frequency doubling temporal contrast (FDP p272) or electrical recordings from eye and brain. These are considered in Chapter 13, p263. Intensive research to substantiate these findings and to devise simple reliable tests and equipment is in progress.

### *Tonometry*

Palpation: a very approximate assessment of the intraocular pressure can be obtained by palpation and judging the indentability of the eye by rocking between the index finger of each hand applied to the globe through the lids while the patient looks downwards. This is *not* a reliable method of estimating intraocular pressure but is justified if the examiner is experienced and in the absence of special equipment.

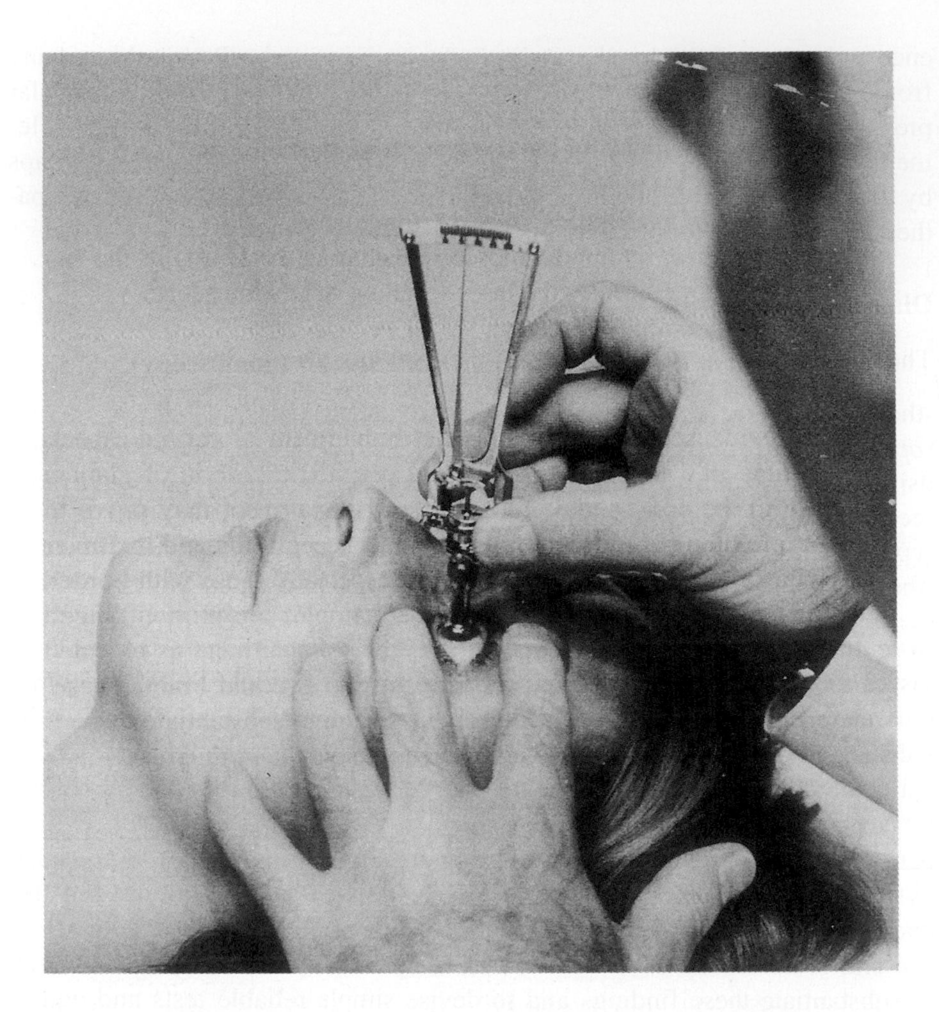

Fig. 31.2 The Schiötz indentation tonometer.

*Applanation tonometry* by the Goldmann tonometer (this is the standard method) (Fig. 31.3 p557, LCW 1.32 p20). Using a spring balance the pressure required to flatten a standard area of the anaesthetised cornea is estimated by an optical method (Figs. 31.4, 31.5 p558). The tear film is stained with fluorescein and viewed with blue light through a plastic applanation head containing a prism splitting the image. The circular meniscus of contact is seen through the prism as two yellow semicircles. The observer varies the pressure of applanation by a spring balance until the semicircles touch as shown in Fig. 31.5 p558. The balance is calibrated to record directly in mm Hg. of intraocular pressure. It is less liable to error than the Schiötz tonometer. The applanation tonometer is normally attached to a slit lamp microscope. A similar portable instrument is available for use by the general practitioner or optometrist and for the dextrous, the necessary expertise is not difficult to acquire (Fig. 31.6 p559). To prevent cross infection great care is essential to ensure the sterility of tonometer heads and gonioprisms. Disposable plastic tonometer heads are now being widely used.

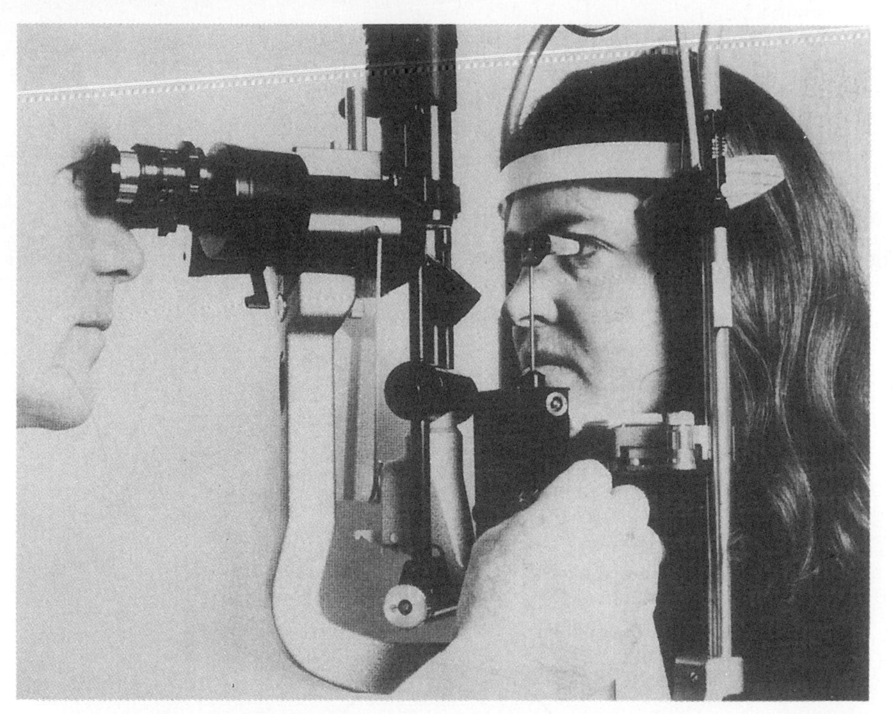

Fig. 31.3 Measurement of the intraocular pressure using the Goldmann applanation tonometer.

*Non-contact tonometry.* An applanation method using a standardised puff of air to flatten the cornea (non-contact tonometry) is being used for screening and phasing tests in some centres and has the advantage that no topical anaesthetic or risk of corneal abrasion is involved. It has been shown, when well calibrated, to give reasonably reliable results.

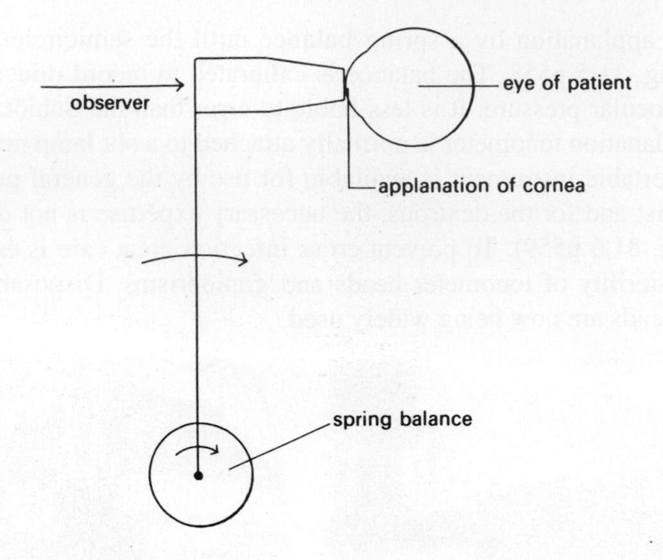


Fig. 31.4 Principle of the Goldmann applanation tonometer.

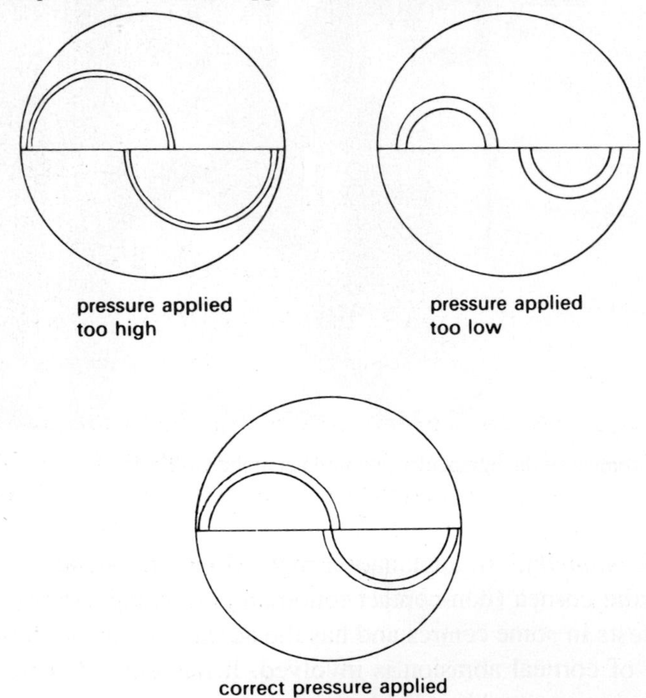


Fig. 31.5 Optical principle of the Goldmann applanation tonometer.

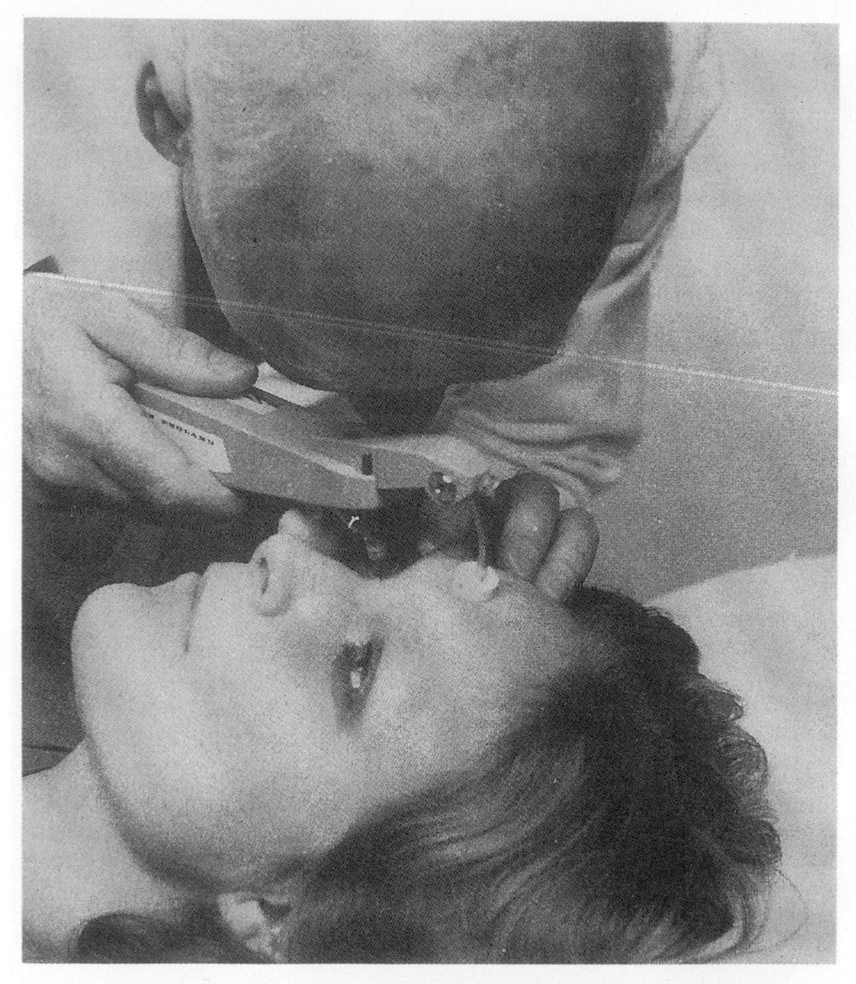

Fig 31.6 Measurement of intraocular pressure using a Perkins portable applanation tonometer.

*Tonography* is a method which aims to assess the aqueous outflow capacity of the drainage mechanism when a standard pressure is applied to the eye. It is carried out by a recording electronic tonometer (Fig. 31.7 p560) and has been considered to be useful in detecting a predisposition to glaucoma, in detecting early cases and as a guide to the most effective form of therapy, nevertheless, due to variability of results and doubt as to its theoretical validity, tonographic information is now considered less significant than was the case a few years ago for individual examinations, although statistically it is still thought to be of value when studying groups of patients.

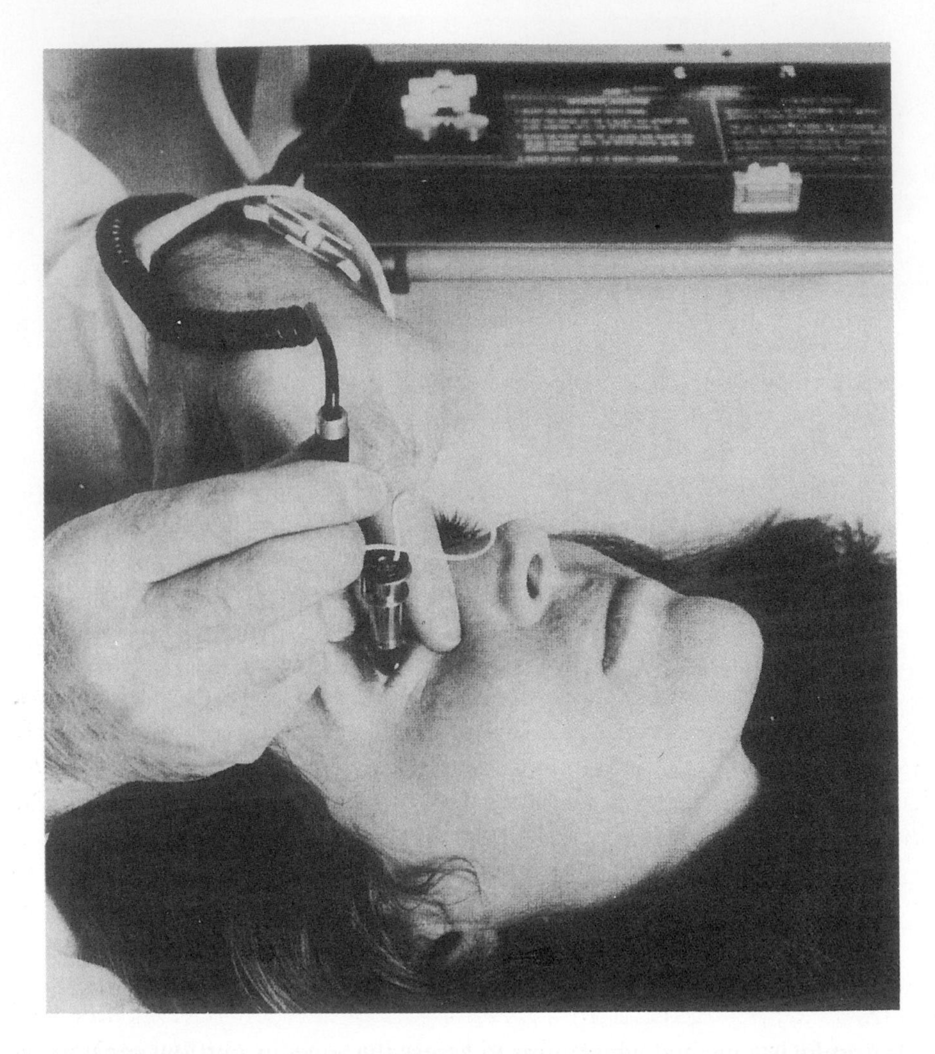

Fig. 31.7 Tonometry with the electronic tonographer.

*Ophthalmoscopy of the optic nerve head* (Stereo plates 31.8 p561, 31.9, 31.10 p562, 31.11 p563, LCW 4.9, 4.10. 4.11 p61). We have seen that optic atrophy may result from a raised intraocular pressure. Characteristically there is atrophy of both the optic nerve fibres and the supporting glial framework so that the normal physiological cup widens and may even become excavated with overhanging edges.

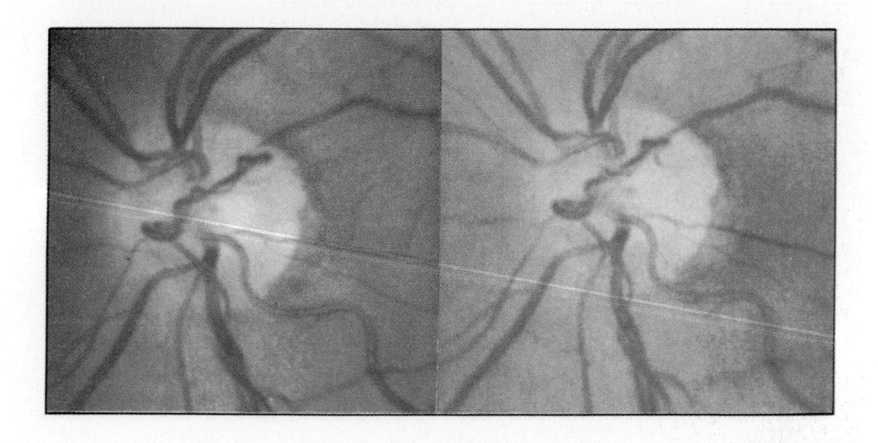

Stereo Plate 31.8 Physiological cupping of the optic disc (coincidental convoluted vessel at disc). (see p4).

A very pale area in the centre of the disc is of no significance because the white fibrous tissue of the lamina cribrosa may normally be visible here. In addition where the nerve fibres entering the disc lie over the glial tissue bordering the optic nerve there may be a narrow paler ring, the border zone. The ring of tissue between the border and the indefinite boundary of the central cup, where the nerve fibres are changing direction and running posteriorly to be in the axis of the optic nerve, is the neuroretinal rim. *This is the only part of the disc which allows a reliable assessment of pallor.* If there is atrophy of the nerve fibres and glial tissue there will be a sharp edge between the neuroretinal rim and the border zone and although this may be a glaucomatous feature it is frequently mistakenly thought to indicate the limits of the diameter of the cup when recording the ratio of the diameters of the cup and disc (C/D ratio). This may lead to grossly variable interpretation of this ratio. It is noteworthy that the outer edge of the neuro-retinal rim may occasionally be sharp, especially in extreme old age, without glaucomatous field defects, although in age there would be a general reduction of sensitivity. The optic disc in myopia tends to be paler than usual and may have a white crescent due to an exposed area of sclera around it. The optic disc may be excavated in severe cases of glaucoma so that while the retinal blood vessels can be seen in the

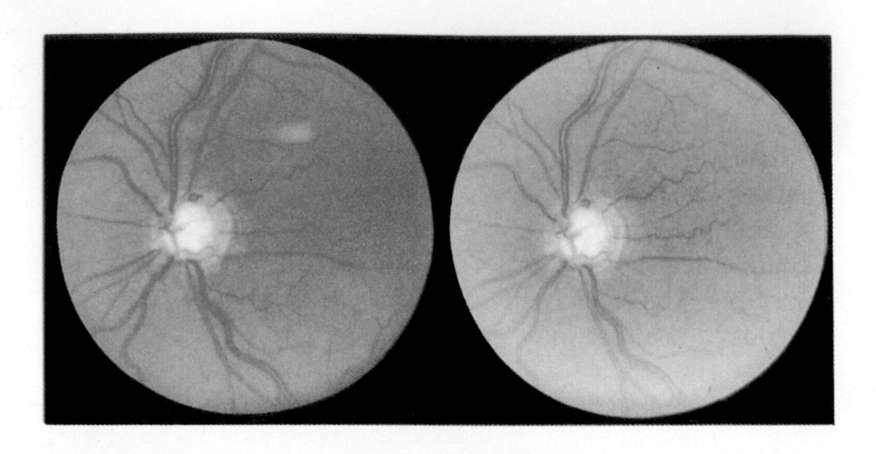

Stereo Plate 31.9 Glaucomatous cupping and atrophy of the optic disc. (see p4)

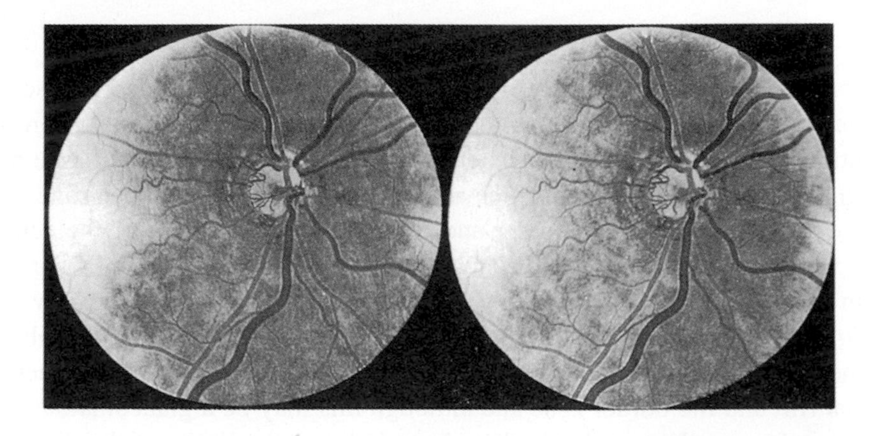

Stereo Plate 31.10 Fluorescein angiography of Stereo Plate 31.9 (see p4)

depths of the cup they are out of sight until they reappear at the edge of the cup (as in Plate 31.12 p563). Nevertheless overhanging cup margins may sometimes be a physiological variation. Other factors suggesting that atrophy of an optic disc is glaucomatous are (Fig. 31.11 p563):

-*Ovality* and *notching* in which the neuro-retinal rim is especially narrowed or pale in one part of its circumference, usually in the vertical axis and more commonly below (Plate 31.13 p564). Hence it is the vertical cup/disc ratio which is usually measured, especially as the entry of the retinal blood vessels at the nasal side of the disc makes accurate horizontal measurement difficult. An arcuate scotoma will frequently be found to correspond to a notch in the neuro-retinal rim.

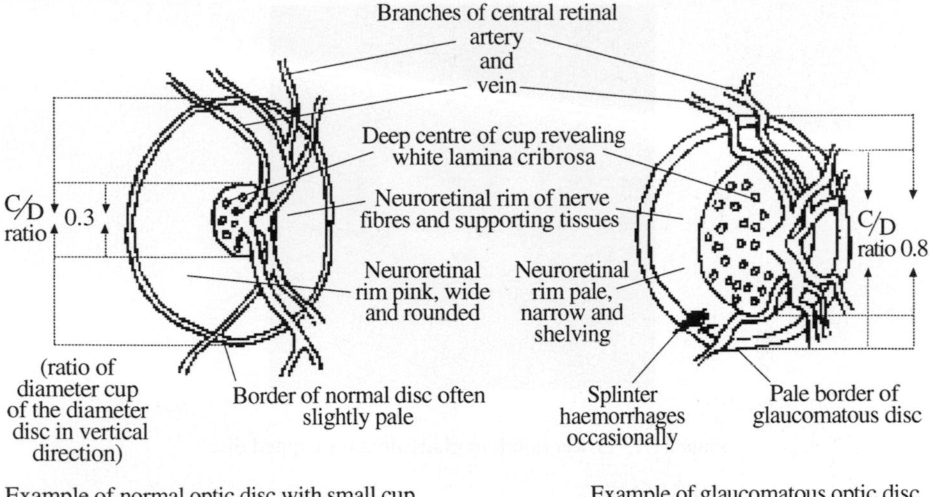


Fig. 31.11 The normal and glaucomatous optic disc features contrasted.

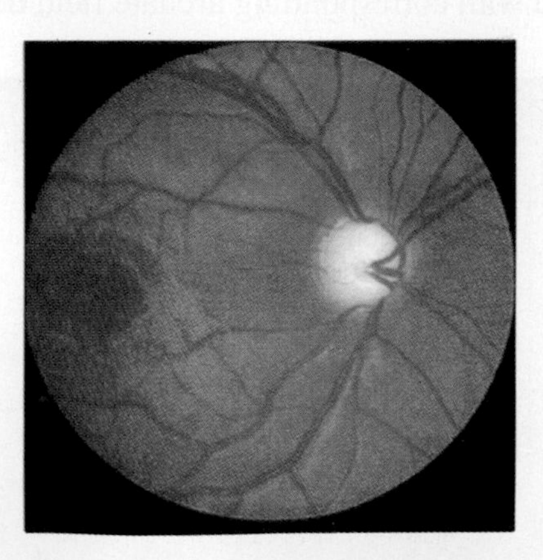

Plate 31.12 Buphthalmos. Advanced optic atrophy and cupping with excavation and retinal vessels reappearing at edge of cup.

*-Asymmetry of cup/disc (C/D) ratios:* the physiological optic cups usually have the same diameter in each eye. A cup/disc ratio difference between the eyes in excess of 0.2 is suggestive of glaucomatous cupping. The vertical C/D ratio is the more reliable measurement and should be specified.

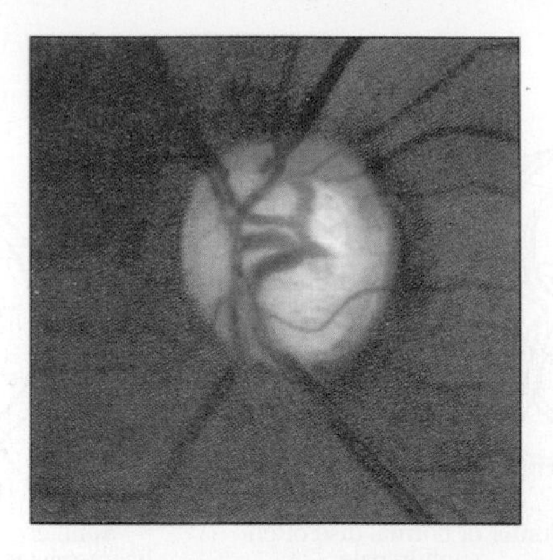

Plate 31.13 Lower notch in glaucomatous cupped disc.

*-Splinter haemorrhages,* (Plate 31.14(a) p564) small linear radial haemorrhages at the border of the disc, indicate a sector of disc ischaemia and are sometimes associated with corresponding arcuate field defects.

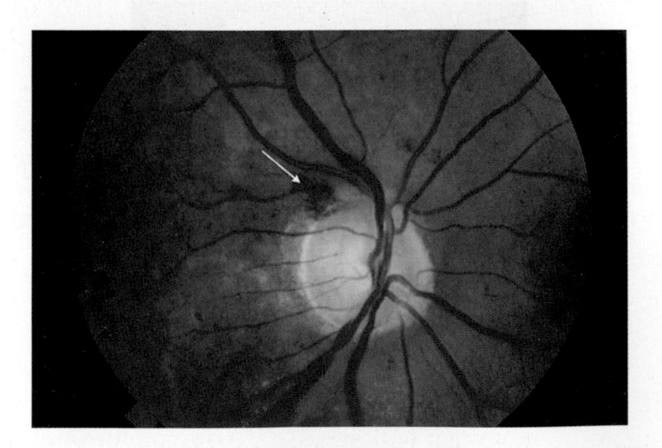

Plate 31.14 (a) Disc haemorrhage.

- *Retinal nerve fibre layer defects.* The layer is thickest in the peripapillary area, the fibres from peripheral ganglion cells are situated deeply to those from more central ones and also occupy the periphery of the nerve at the disc. Routine ophthalmoscopy will reveal the normal opaque silvery appearance of the fibres, especially in a green light. Dark gaps where the fibres are missing correspond well with arcuate scotomas. The arrow in Fig. 31.14 (b) p 565 demonstrates this well with the suggestion of a splinter haemorrhage just above the sector where the fibres would have entered the disc.

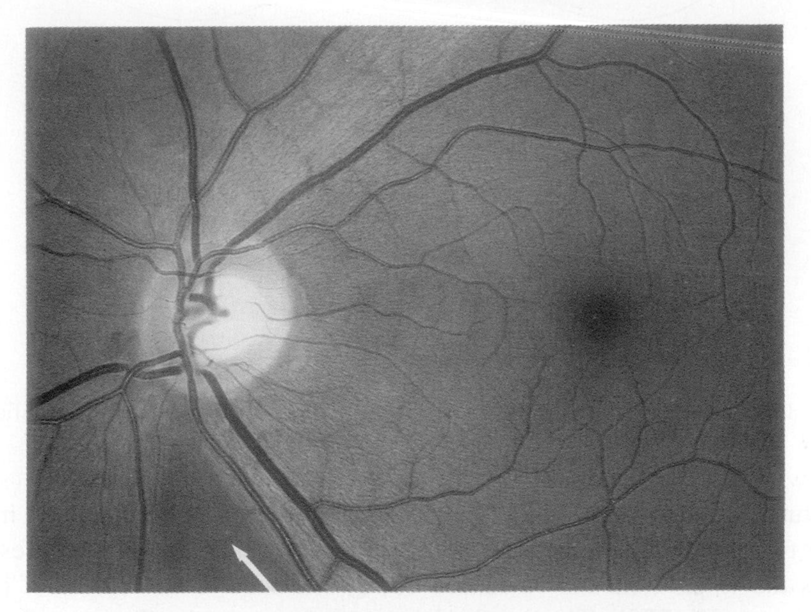

Fig. 31.14 (b) Retinal nerve fibre defect (as indicated by arrow)

*Visual field examination (perimetry)* and the optic disc appearance are the main methods of assessing glaucomatous injury to the optic nerve both for diagnosis and follow up. Visual field defects (scotomas) of nerve fibre bundle type (either arcuate or much less commonly in a temporal wedge distribution) are sought and their progress studied (Fig. 31.15 p 566). While the pattern of nerve fibre bundle loss is characteristic, numerical records of threshold sensitivity at many points in the field have revealed that an overall loss of sensitivity may also be a feature of glaucomatous field loss. In the untreated or inadequately managed patient the scotomas will steadily progress until only a tubular field of central vision remains, usually on and below the horizontal meridian.

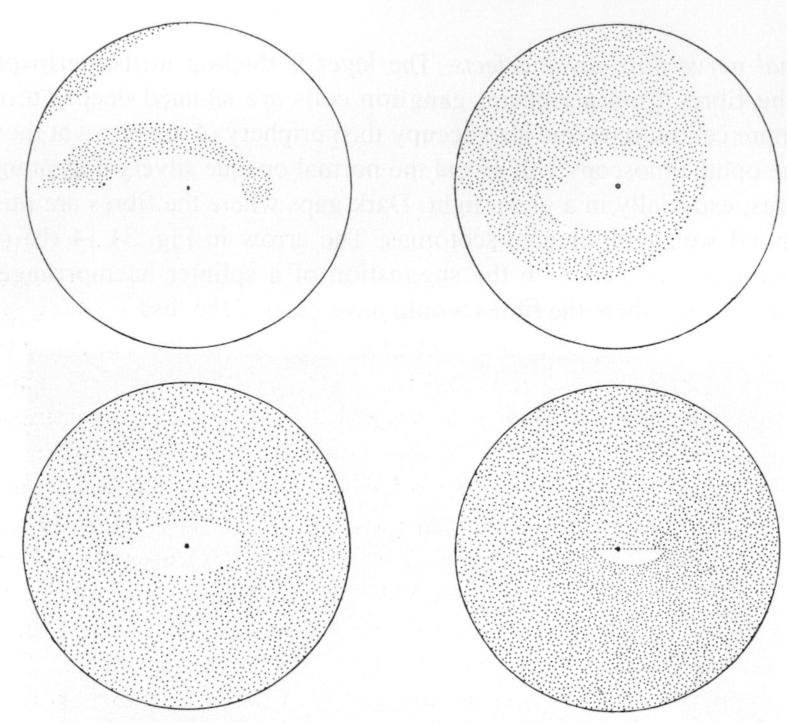

Fig. 31.15 Progressive glaucomatous arcuate visual field loss right eye, the upper hemifield usually being affected earlier and more severely.

Eventually central vision will be lost but a small eccentric residual field in the lower temporal region may persist until the loss of perception of light. Tests to assess the visual field can be carried out by methods of all degrees of complexity (pp83, 273). The original instrument was the Tangent Screen (Bjerrum) (Fig. 13.9 p276) succeded by the Goldmann perimeter (Fig. 13.8 p275). The Friedmann Visual Field Analyser and the Henson Central Field Analyser 3200 (Fig. 13.10 p277) semi-automated perimeters with numerical field assessment, have been found particularly useful in surveys to detect chronic glaucoma in large groups of people. Ideally where circumstances allow, a fully automated instrument such as the Octopus (Fig. 13.11(a) p278 ) or the Zeiss-Humphrey perimeter (Fig. 13.11(b) p279) with a computerised strategy gives reliable results more independent of the skill of the perimetrist. The Zeiss-Humphrey F.D.P. perimeter is designed for early detection (Fig. 13.7 p272) by using responses to spatial and temporal (frequency doubling) contrast sensitivity testing. (p272)

*Gonioscopy.* It is not possible to see into the angle of the anterior chamber by looking obliquely through the cornea because rays of light from this region are subject to total internal reflection and fail to pass from the cornea to air (p52 and Fig 4.3 p53). If however, a material of suitably higher refractive index is applied to the cornea, e.g. a plastic contact lens, the rays of light can escape and the angle be visualised. Such a device is called a gonioscope and allows a division of patients into those with wide and those with narrow or even closed angles. Where there has been an angle closure episode or coincidental uveitis leaving adhesions between iris and cornea, these peripheral anterior synechiae can also be seen. Figs. 31.16 - 31.19 pp567, 568 , show the Goldmann 3 mirror lens which allows which allows by rotation a gonioscopic slit lamp view with one mirror, the others and the centre give fundus views. This lens is easy to use but requires a layer of fluid between the lens and the cornea. The Zeiss 4 mirror gonioscopy lens is directly applied to the cornea. The whole 360° of the angle can be viewed with little manipulation. Pressure on one side of this lens, indentation, will displace fluid in the anterior chamber to open the opposite part of the angle and help distinguish between peripheral synechiae and mere apposition of iris to the trabeculum. The surface view of the angle as seen through the gonioscope is shown in Fig 6.39 p131. The structures visible and the degree of iris convexity allow a classification of angle widths and the risk of angle closure.

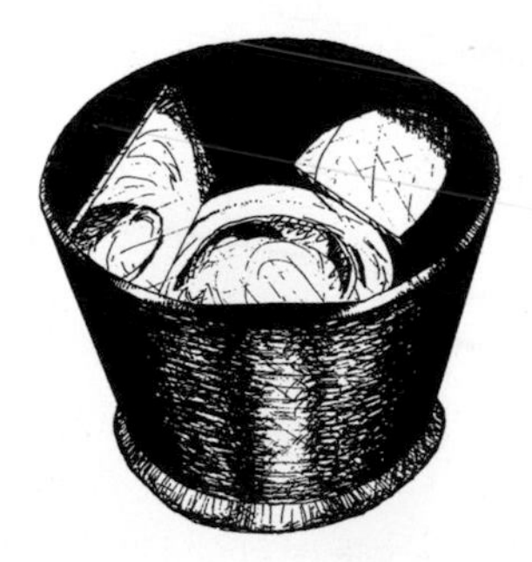

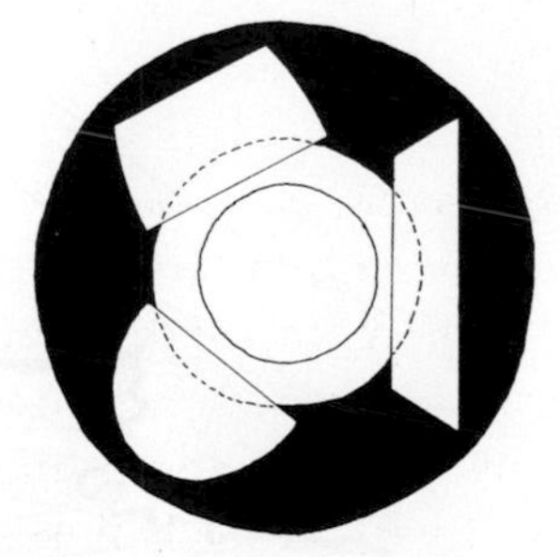

Fig. 31.16 Goldmann 3 mirror gonioscopy lens.

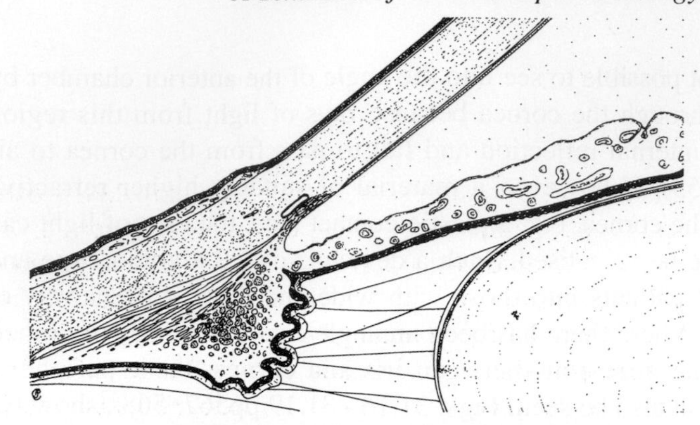

Fig. 31.17 An open anterior chamber angle of intermediate width.

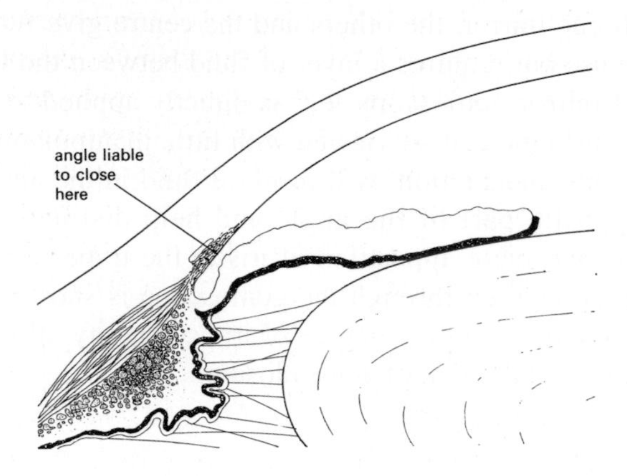


Fig. 31.18 A narrow anterior chamber angle with liability to close.

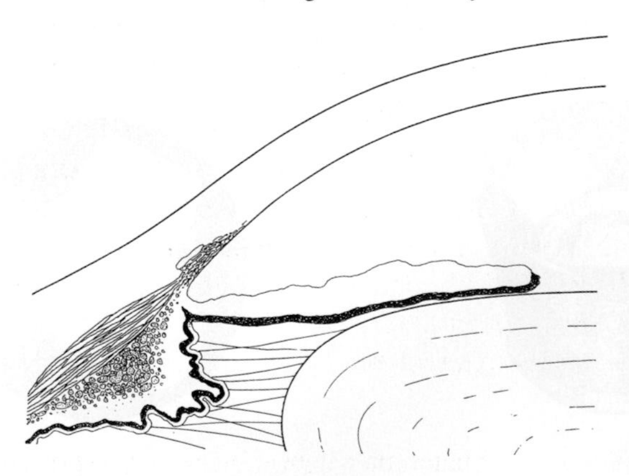

Fig. 31.19 A widely open anterior chamber angle.

## Clinical aspects of glaucoma

### *Epidemiology and screening*

Glaucoma of all types is estimated to affect 2% of the white population over the age of 40 years and suspect cases and treatworthy ocular hypertension patients increase the numbers involved to about 3%. Primary open angle glaucoma affects more than half of these. There are *racial differences* in the relative proportion of the different types of glaucoma. In the white population of Illinois U.S.A. for example, open angle glaucoma has been found in 52% of all types of glaucoma, closed angle glaucoma in 14%, secondary glaucoma in 20% and other types in 13%. In the black population of Illinois the corresponding figures were 73%. 5%, 11% and 11%. Comparable figures have been found elsewhere in the West but in S.E. Asia and among Eskimos the proportion of closed angle glaucoma is reported to be appreciably higher and of primary open angle glaucoma, less. In Baltimore U.S.A. in a large epidemiologically controlled survey it was reported that the overall prevalence in whites over the age of 40 of primary open angle glaucoma was 1.7% and in blacks 5.6%. In St Lucia, West Indies in a black population the prevalence ranged between 8.8% and 14.7% depending on the stringency of the criteria for diagnosis.

There are also *age differences.* The prevalence of primary open angle glaucoma increases with age from 35 years and steeply after 60 years. Certain autosomal dominantly inherited juvenile-onset types may however have an earlier age of onset even before the teens. The prevalence at different ages in eight large surveys of white populations is summarised in Fig. 31.20 p570. There are some differences in the criteria for diagnosis and selection but the general picture is similar in each survey. Composite prevalences for quinquennial age groups were calculated by curve fitting.

*Blindness due to glaucoma* was found in the Model Reporting Area of the U.S.A. to be 0.016% of the whole population covered. It was the third most important cause of blindness in the white population but the leading cause in blacks in whom the risk of blindness was eight times that of whites. The Baltimore survey found that bilateral blindness from primary open angle glaucoma was six times as frequent among blacks as among whites and began ten years earlier.

The *relative sex incidence* of glaucoma has been variously reported and remains equivocal, although there is some evidence of an increased risk in males.

**Per cent of population with POAG**

| Age (years) | Ferndale (Wales) 1966 | Framingham (USA) 1977 | Rotterdam (Neth.) 1994 | Age (years) | Baltimore (Whites) (USA) 1991 | Roscommon (Eire) 1993 | Blue Mountains (Australia) 1991 | Casteldaccia (Sicily) 1995 | Age (years) | Beaver Dam (USA) 1992 | Composite Prevalence Tuck & Crick 1998 | Composite Implied Incidence Tuck & Crick 1998 |
|---|---|---|---|---|---|---|---|---|---|---|---|---|
| 40-44 | 0 | | | | | | | | | | 0.12 | 0.01 |
| | | | | 40-49 | 0.18 | | | 0.36 | | | | |
| 45-49 | 0 | | | | | | | | | | 0.21 | 0.02 |
| | | | | | | | | | 43-54 | 0.99 | | |
| 50-54 | 0.25 | | | | | | | | | | 0.35 | 0.04 |
| | | | | 50-59 | 0.32 | 0.72 | 0.29 | 0.34 | | | | |
| 55-59 | 0.84 | 0.48 | 0.2 | | | | (48-59 yrs) | | | | 0.54 | 0.06 |
| | | | | | | | | | 55-64 | 1.29 | | |
| 60-64 | 0.44 | 0.92 | 0.2 | | | | | | | | 0.97 | 0.09 |
| | | | | 60-69 | 0.77 | 1.76 | | 1.36 | | | | |
| 65-69 | 1.01 | 0.92 | 0.9 | | | | 1.07 | | | | 1.51 | 0.13 |
| | | | | | | | | | 65-74 | 2.65 | | |
| 70-74 | 1.13 | 1.70 | 1.8 | | | | | | | | 2.25 | 0.16 |
| | | | | 70-79 | 2.85 | 3.20 | | | | | | |
| 75-79 | | 1.94 | 1.6 | | | | 4.17 | 3.65 (70+yrs) | | | 3.12 | 0.18 |
| 80-84 | | 2.92 | 3.1 | | | | | | 75+ | 4.71 | 4.01 | |
| | | | | 80+ | 1.94 | 3.05 | | | | | | |
| 85-89 | | | 3.3 | | | | 8.17 | | | | 4.79 | |
| All ages in survey | 0.43 | 1.15 | 1.11 | | 1.10 | 1.88 | 2.38 | 1.22 | | 2.11 | 1.23 | |
| Total number: examined | 4608 | 2352 | 3062 | | 2913 | 2186 | 3654 | 1062 | | 4926 | | |
| POAGs | 20 | 27 | 34 | | 32 | 41 | 87 | 13 | | 104 | | |
| % of newly detected cases with normal intraocular pressure at initial screen | 50 | n.a. | 39 | | 49 | 62 | 74 | 33 | | ca. 33 | | |
| Summary of main screening tests | Virtually all subjects were tested by ophthalmoscopy and tonometry at the first stage in every survey. Any history of glaucoma was also ascertained. The % tested initially by perimetry varied, however, as below. | | | | | | | | | | | |
| | 33 | 0 | 100 | | 98 | 56 | 89 | 0 | | 95 | | |
| Main diagnostic criteria* | F+D(S) | F | F+D(M) or ›21mmHg+F | | F or D(S) ** | F+D or ›30mmHg +(ForD(S)+V) | F+D(S) | F+D(M) or ›20mmHg+F | | F+D(S) or ›21mmHg +(ForD(S)) | | |

* = glaucomatous visual field defect. D = optic disc defect, but where this was reported more precisely the following is added: (S )= severe, typified by cup disc ratio of 0.6 ; or (M) = mild, typified by CD ratio 0.4. V = visual acuity 6/60. IOP is expressed as mm Hg. The precise definition of each individual criterion varied as between the different surveys.
** For patients unable to complete a satisfactory visual field test.

Fig. 31.20 The prevalence of primary open angle glaucoma (POAG) in predominantly white populations by age group, according to eight population surveys and Composite quinquennial calculated prevalence obtained by curve fitting. Implied incidence per annum by age group also calculated up to 80 yrs. From Tuck MW, Crick RP, 'The age distribution of primary open angle glaucoma' Ophthalmic Epidemiology 5 (4): 173-183 (1998) in which decennial and other age ranges are also included. Aeolus Press. Reproduced by permission.

*Genetic aspects of primary open angle glaucoma*

Surveys report a median prevalence of 6% of relatives of primary open angle glaucoma sufferers compared with a normal prevalence of 1.2%. A decade of molecular genetic studies has revealed the DNA loci of 8 chromosomes for primary open angle glaucoma. The myocillin TIGR (trabecular meshwork inducible glucocorticoid repsonse) gene on chromosome 1 at 1q21-q31 has been studied in many populations and 35 mutations reported. An abnormal protein which becomes insoluble and accumulates in the trabecular cells possibly contributing to impaired aqueous drainage. Its characteristic phenotype is found in about 4% of primary open angle glaucoma patients and is a severe early onset type often requiring surgery to control intraocular pressure.

Mutations of chromosome 10 at 10p14 in a gene OPTN coding for the protein optineurin have also been identified causing adult onset primary open angle glaucoma with low or mildly raised intraocular pressure, up to 26 mm Hg. This involves 17% of this type of glaucoma in 54 unrelated families in the UK, USA and Canada with the same intraocular pressure range. Evidence suggests that normally, optineurin protein may be neuroprotective against cell apoptosis in the optic nerve and retina and that the abnormal proteins produced by this or other mutated genes allow this nerve damage characteristic of primary open angle glaucoma, particularly of the type in which the intraocular pressure is below about 26 mm Hg. Molecular genetics may well allow a more informed classification of glaucoma, thus preventing sight loss, not only by the early detection of many of those at risk, but eventually by achieving a fundamental cure for some of the factors causing the disease.

*Screening for primary open angle glaucoma - Risk factors*

It is clear that the absence of early symptoms in primary open angle glaucoma means that great damage can be done to the optic nerve without anyone being aware of it. Screening by regular examination of optic disc appearance, of intraocular pressure and of the visual fields in persons most likely to be affected is clearly desirable i.e. those 1. over 40 years of age (35 years of age if other risk factors are present) 2. with a family history of glaucoma in a near blood relative 3. of African origin 4. with moderate or high myopia 5. with diabetes mellitus (the risk of primary open angle glaucoma in this condition has not been substantiated in some population studies although other types of glaucoma may complicate diabetes).

The need for all three tests is not so much that one is better than the others but that they all have deficiencies and if employed together sensitivity can be improved without undue loss of specificity. Tests which will reveal glaucomatous optic nerve damage before visual field or ophthalmoscopic evidence is available are being developed but are not yet either sufficiently validated for adoption (p268) or independent of opacities in the ocular media.(p272) In due course DNA screening may well reveal many of those at risk, even early low tension patients.

## Signs, symptoms and treatment of glaucoma

### Developmental glaucoma *(buphthalmos, hydrophthalmos, congenital glaucoma)*

*Primary.* We have seen (p49) that the anterior chamber develops between the sixth and ninth month of foetal life as a chink in the mesoderm between the iris root and the trabeculum. If this mesoderm does not entirely regress, an impervious layer may remain bridging the angle between the iris and cornea which impedes access of aqueous to the trabeculum.

*Genetic aspects.* This is an autosomal recessive disorder. Loci of genes for this condition have been identified at 2p21 and 1p36 and the mutations discovered in the human cytochrome (CYPIB) gene on chromosome 2 prevent the function of the CYPIBI protein which is expressed in the trabecular meshwork and may affect its development.

*Secondary developmental glaucoma.* Various causes of secondary glaucoma may operate in the developing eye and cause an enlarged globe. Prominent among these are uveitis (due for example to maternal rubella (p332), retinoblastoma (p531), retinopathy of prematurity (p425) and trauma. Such changes may also complicate uveal neurofibromatosis (p437), the Sturge Weber syndrome (p438), Marfan's syndrome (p434) and anterior segment dysgenesis (pp427, 516).

*Symptoms and signs*

Although buphthalmos is a relatively rare form of glaucoma it is important to be alert to the possibility of congenital glaucoma in any child who has large corneae or high myopia (Plate 31.21 p573, LCW 8.7 p124). Suggestive

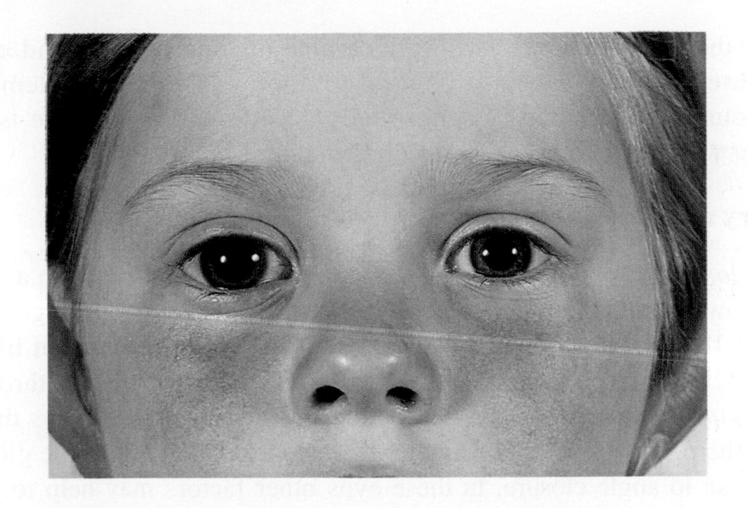

Fig. 31.21 Buphthalmos more marked in the right eye. (note watery meniscus of the tear film at the lower lid margin of the right eye.

symptoms are burying the head in the pillow, watering and eye rubbing. Enlargement is due to stretching of the young globe by increased pressure giving rise to the term buphthalmos or 'ox eye'. This may be present at birth or gradually become noticeable. The coats of the eye are thinned, the sclera appears bluish and in the cornea grey lines may appear due to splits in Descemet's membrane. These splits and impairment of the endothelium lead to corneal haze due to oedema, cause photophobia and tend to prevent a view of the optic disc which may be cupped and atrophic from an early stage (Plate 31.12 p563). The pupil may be dilated and the conjunctiva reddened with some ciliary injection. Large corneae may also be found in the inherited condition megalocornea, which is relatively benign. In this condition the intraocular pressure is normal and a corneal arcus may be present (p477).

*Treatment*

It must be emphasised that a good visual result is the goal, rather than just normalisation of intraocular pressure. The prevention of amblyopia by occlusion, especially in unilateral or unequal cases is most important, but it is often difficult to achieve. Primary cases are usually treated without delay by incision of the abnormal tissue covering the trabeculum (goniotomy) and if this fails, fistulising procedures are carried out using measures to minimise scarring of the bleb. Treatment with pilocarpine or timolol drops and carbonic anhydrase inhibitors may also be used to improve the situation temporarily before surgery. In secondary congenital glaucoma the basic cause is treated together with similar measures to reduce the intraocular pressure.

**Primary angle closure glaucoma** (p186)

*Angle closure,* as described above, is liable to occur in an eye with a markedly shallow anterior chamber and a very narrow angle as seen by gonioscopy (Fig. 31.18 p568 and Plate 7.27 p186). The lens grows throughout life and a thick anteriorly placed lens makes it difficult for aqueous to pass through the pupil *(pupil block),* so that the iris balloons forwards and narrows the angle still further. A small corneal diameter and short axial length of the globe also predispose to angle closure. In these eyes other factors may help to precipitate closure. These include:
- *mid-dilatation of the pupil* due to sitting in dim lights or following the use of mydriatics,
- *variation in aqueous production and vitreous volume,*
- *hydration* and thus swelling of the lens as a pre-cataract change or *slackness of the suspensory ligament,*
- *forward movement of the lens* on accommodation or following the use of miotics which also cause contraction of the ciliary muscle, or the prone position in eyes with a slack lens suspensory ligament.
- *congestion of the uvea,* pushing the lens forwards and tending to make the iris swell.

All these factors tend to contribute to 'pupil block' which implies that there is increased difficulty for aqueous to pass between the iris and lens in the region of the pupil. This in turn causes a bowing forward of the iris peripherally which further narrows the angle of the anterior chamber and may precipitate angle closure. A flaccid iris is more likely to bulge forward whereas a more rigid iris may tend to protect against angle closure attacks.

The frequent coincidence of acute angle closure glaucoma complicating crises of other types especially in anxious middle aged women, suggests that nervous and endocrine influences contribute significantly to the attacks, possibly aggravating vasomotor instability leading to congestion in the uveal tract and narrowing of the angle as well as by causing dilatation of the pupil.

*Effects of angle closure* (p550)

*Symptoms, signs and treatment of primary acute angle closure glaucoma* (p186)

*Provocative tests for angle closure*

Some patients who have narrow anterior chamber angles give a history suggesting previous sub-acute episodes of raised pressure. Such patients may be given provocative tests which aim to recreate the conditions under which the angle may close and may help to indicate the degree of liability to angle closure by measuring the pressure before and after the test. A rise of more than about 7 mm Hg. is considered a positive result. The validity of these tests is still being assessed but a negative result is generally unreliable. The tests which should be very closely supervised, include:

-the *dark room test:* the patient sits in the dark for one hour, so that the pupils are dilated but is kept awake to avoid sleep miosis,
-the *mydriatic test:* a drop of phenylephrine 10% or one of cyclopentolate 1% is instilled,
-the *pilocarpine-phenylephrine test:* in which the iris is stretched over the convex surface of the lens thus precipitating pupil block in predisposed eyes,
-the *reading test:* a rise in pressure may be found in susceptible patients after reading small print for a hour due to displacement forwards of the lens in accommodation,
-the *prone test:* the patient lies prone and in some patients with a slack zonule the lens may fall forward enough to close the angle of the anterior chamber in parts and cause a rise in intraocular pressure.

The prone and dark room tests may be combined.

**Primary open angle glaucoma** (Plate 31.9 p562)

This accounts for about 70% of *primary* glaucoma in a white population and has a familial tendency, children and siblings of patients with primary open angle glaucoma having about six times the normal liability to develop it. There is also evidence that patients with this condition and their unaffected close relatives are predisposed to a rise of intraocular pressure following steroid medication especially in the form of local drops or ointments. In primary open angle glaucoma, while there may conceivably sometimes be an element of aqueous hypersecretion, an increase in intraocular pressure is almost always caused by obstruction of the outflow of aqueous due probably to changes in the cells of the trabeculum and particularly in the pericanalicular tissue. Variations in the resistance in veins draining the aqueous veins from the canal of Schlemm may also play a part. The intraocular pressure may rise slowly to approach or exceed the perfusion pressure of the blood vessels supplying the various eye tissues which, therefore, suffer ischaemic changes; this is particularly important when it affects the optic nerve head.

*Clinical aspects* (p129)

*Types of primary open angle glaucoma*

*Chronic simple glaucoma* is the term used for types of primary open angle glaucoma with an intraocular pressure continuously or intermittently in excess of 21 mm Hg. and in the absence of any developmental abnormality. This condition accounts for the majority of cases of glaucoma in Western countries and Africa. It should also be noted that two conditions which in the past were classed as primary open angle glaucoma, *pigmentary glaucoma* and *pseudocapsular glaucoma,* are now considered to be forms of secondary glaucoma although their management is similar to that of chronic simple glaucoma (p137), but see p579.

While increased intraocular pressure remains the most important risk factor for primary open angle glaucoma, there is, as discussed on p552, increasing recognition of the role of other causes of this type of optic neuropathy, including the disturbance of local enzyme systems leading to apoptosis of the retinal neurones particularly the ganglion cells. Some of the enzyme systems are themselves influenced by abnormal intraocular pressure but they may well operate in cases of *low tension glaucoma.* It is important to appreciate that intraocular pressures vary by about 5 or 6 mm Hg., so that many patients thought to be in the 'low tension' glaucoma category with a pressure below 22 mm Hg. may well have a pressure of 22 mm Hg. or more on subsequent examinations hence the unreliability of diagnostic decisions based on isolated tonometry readings which are fairly near the cut off point. Thus ***there is no "normal" intraocular pressure level*** and below 22 mmHg is merely a guide to the probability of glaucoma and nothing more. It was noteworthy that in the Baltimore Survey, on initial screening 55% of those judged to have primary open angle glaucoma had intraocular pressures below that level and appeared to justify a classification of low tension glaucoma but on further testing the proportion of low tension glaucomas fell to 16%. Nevertheless this is still an important group for which explanations are required.

*Principles to be considered in treatment*

As we have seen, the intraocular pressure may remain at a higher than average level for many years, perhaps indefinitely, without causing optic nerve damage *(ocular hypertension),* but nevertheless such eyes are more likely to suffer such damage over a period of years, than those with pressures of or around average levels. The presence of other risk factors such as a family history might influence the ophthalmologist to give treatment to ocular hypertensives to lower the pressure, others may prefer to review the condition regularly for the first signs of visual function loss, although it is necessary to realise that it has been demonstrated that *considerable loss of optic nerve fibres has occurred before either the visual fields show defects or the optic disc has glaucomatous features using routine methods of examination.* The corollary is that the average range of intraocular pressures may be injurious to an optic nerve made vulnerable by other factors, for example an impaired blood supply. When treating these patients the ophthalmologist aims to reduce materially the intraocular pressure in primary open angle glaucoma or low tension glaucoma, below whatever pressure was present when the optic nerve damage took place. Reducing it just to below 21 mm Hg. will be inadequate in very many cases, because half the patients with chronic simple glaucoma (i.e. intraocular pressure more than 21 mm Hg., continuously or intermittently, with evidence of optic nerve damage) have only an intraocular pressure of 22-25 mm Hg. Therefore the level of pressure at which to aim (the target pressure) in these cases might be anticipated to be not more than about 16 mm Hg. and preferably about 12 mm Hg. The adequacy of the treatment will be reflected in a stabilisation of the visual field as shown by several examinations performed at intervals under identical conditions.

*Possible causes of optic nerve vulnerability in low tension glaucoma* include cardiovascular abnormalities such as episodes of blood loss, anaemia or abnormal blood viscosity, cardiac arrhythmias, carotid stenosis and a vasospastic tendency as in migraine or Raynaud's disease. With improved Doppler and other techniques to measure blood flow, vascular aspects of glaucoma are the subject of intense study. Weakness of the supporting tissues of the disc and lamina cribrosa leading to distortion and strangulation of the nerve fibres and their blood supply has also been considered to play a part as may other unsuspected genetically conditioned factors.

**Treatment:** is described on p137

In the majority of cases of primary open angle glaucoma treatment is given to reduce the intraocular pressure, if possible to 10-12 mm Hg.. Following trabeculectomy, pressure reduction may not be maintained due to bleb fibrosis in some patient groups. The application of antimitotic agents e.g. 5-fluorouracil or mitomycin C, to the exposed Tenons capsule after reflection of the conjunctival flap, followed by irrigation will prevent this cause of operative failure in many of these cases but great care is required to avoid side effects.

## Secondary glaucoma

This may be said to be present when an intraocular pressure cut off point of more than 21 mm Hg is found in the presence of an ocular disturbance which can reasonably be expected to lead to a raised pressure. There are many causes which are summarised in the table, but a few deserve special mention (Fig. 31.22 p578):

Fig. 31.22 Types of secondary glaucoma.

| • Pigmentary | | •Toxic | steroids, siderosis, ammonia |
|---|---|---|---|
| •Pseudocapsular | | | |
| •Uveitic | acute, granulomatous, lens induced uveitis, heterochromic cyclitis. | •Heterochromic cyclitis | |
| •Lenticular | intumescence, subluxation, phakolysis. | •Essential iris atrophy | |
| •Post operative | due to aqueous misdirection, ('malignant'). | •Traumatic | epithelial ingrowths, injury to the trabeculum and recession of the anterior chamber angle, intraocular haemorrhage, displacement of the lens. |
| •Rubeotic | diabetes, retinal vein occlusion, vasculitis retinae. | | |
| •Raised orbital pressure | either directly or by raised venous pressure. | •Intraocular neoplasm | |
| •Ghost cell | long standing vitreous haemorrhages with ghost cells | | |

*Pigmentary glaucoma:* This condition arises in a small proportion of cases of the pigment dispersion syndrome which mainly affects middle aged white myopic males. It is thought to be due to the rubbing by a peripherally concave iris against the fibres of the lens zonule which causes shedding of pigment granules and their dispersion by the aqueous. They become engulfed by the endothelial cells of the surrounding tissues. On the back of the cornea this forms a vertical line of pigment (the Krukenberg spindle) and in the trabecular spaces the endothelial cells, having phagocytosed the pigment are thought to migrate, denuding the trabecular sheets which collapse and stenose so that the flow of aqueous is obstructed and causes a rise in intraocular pressure. Peripheral iris thinning is revealed by transillumination on slit lamp examination. These cases are treated in the same way as primary open angle glaucoma but although miotics would help reduce the iris concavity, they are thought to be contraindicated by increasing the existing risk of retinal detachment in the pigment dispersion syndrome. The iris concavity is said in some cases to be reduced by peripheral laser iridotomy by equalising pressure in the anterior and posterior chambers, the apposition of iris to lens otherwise acting in these eyes as a non-return valve on blinking or eye movement.



*Pseudocapsular glaucoma* may arise in about 20% of cases of the pseudoexfoliation syndrome which was formerly thought to be caused by exfoliation of the lens capsule. The prevalence of this condition varies widely in different geographical locations and increases after the age of 60 years. White flakes are found on and in the cells of tissues lining the anterior chamber. The trabeculum is pigmented and this pigmentation can extend forward on to the corneal endothelium. In the iris, flakes are also present in blood vessel walls, appearing to have occluded some vessels and the pupillary border is often atrophic. Characteristically the deposits are rubbed off the front of the lens in a ring shaped area by movements of the iris where it touches the lens near the pupil margin. Patients who also have cataracts may present difficulty in cataract surgery due to bonding of the posterior surface of the iris to the preequatorial lens capsule. The untreated intraocular pressure has been found at diagnosis to be significantly higher than in primary open angle glaucoma and the lowering of pressure by trabeculectomy to be more effective. This has led some ophthalmologists to recommend surgery earlier in these cases, which otherwise are treated in the same way as primary open angle glaucoma.

*Secondary glaucoma due to anterior uveitis*

This is the most important cause of secondary glaucoma. A raised pressure may occur during a phase of acute iridocyclitis due to obstruction to the drainage of aqueous by protein and cells. This is usually controlled by acetazolamide and improves with treatment of the uveitis. The granulomatous uveitis of sarcoidosis, where nodules frequently occupy the angle of the anterior chamber, is also liable to be complicated by a raised pressure, somctimes even when the inflammatory signs are mild. If inadequately treated, the nodules fibrose and chronic angle closure glaucoma may require a fistulising operation. During the course of uveitis, posterior synechiae are liable to form near the pupil margin and sometimes fibrinous exudates organise across the pupil. Aqueous is then partially or totally prevented from passing through the pupil and the iris becomes bowed forwards (iris bombé) narrowing the drainage angle so that in either case the pressure rises. Energetic treatment of the uveitis will usually prevent synechiae and secondary glaucoma (p181) and established adhesions can if necessary be surgically treated by removal of a segment of iris after separation of adhesions with an iris repositor through a peripheral iridectomy.

Sometimes marked intermittent rises of tension occur in the presence of minimal signs of uveitis *(glaucomato-cyclitic crises - the Posner-Schlossman syndrome).* It appears that in these cases the inflammation affects particularly the trabecular region. Release of prostaglandins may play a part in acute uveitic intraocular pressure rises. The prognosis is usually favourable providing that the individual episodes are energetically treated. Even if recurrent, they tend to become less frequent and may cease altogether.

*Heterochromic cyclitis* is a type of uveal change causing degeneration and depigmentation of the iris and cataract, but freedom from posterior adhesions of iris to lens. Heterochromic cyclitis is believed to be associated with disturbance of the sympathetic nerve supply of the blood vessels of the iris, although it may occasionally be difficult to distinguish it from chronic uveitis with secondary atrophy of the iris or with cataract. Secondary glaucoma may occur.

*Lenticular causes of secondary glaucoma* (p96)

These include intumescence, phakolytic glaucoma and lens induced uveitis.

*Flat anterior chamber and the aqueous misdirection syndrome ('malignant glaucoma')* (p496 )

A type of glaucoma which may follow operations is associated with a 'flat' anterior chamber, so-called 'malignant' glaucoma. Following intracapsular cataract operations this may be due to a vitreous plug sealing the gap between the iris and vitreous and leading to an accumulation of aqueous behind it (aqueous misdirection) or in fistulising operations an excessive freedom of drainage may allow the lens to come forward with the same effect. Plateau iris increases this risk p551 This can usually be prevented by accurate suturing and ensuring that the intraocular pressure is brought to a low level for some days pre-operatively, thus avoiding sudden decompression of the anterior chamber. Treatment is by mydriatics to allow fluid to come forwards and by carbonic anhydrase inhibitors and osmotic agents. Occasionally surgical methods to incise the vitreous face, or after drainage operations to seal the excessive drainage combined with reformation of the anterior chamber may be required. In phakic cases lens extraction may be necessary.

*Rubeotic glaucoma*

This is due to neovascularisation of the iris (rubeosis iridis) and may occur particularly in diabetes (p316 and Plate 16.11 p316 , LCW 4.15 p63) and following central retinal vein occlusion (p123 and Plate 6.33 p122) or vasculitis. It can sometimes be prevented by photocoagulation of ischaemic areas of the retina (p316). When neovascularisation occurs, the shrinkage of the connective tissue associated with new vessels draws the iris across the angle of the anterior chamber which it closes. Intractable rubeotic glaucoma follows, often requiring enucleation eventually, although occasionally a procedure to reduce production of aqueous by exposing the ciliary body to the effect of a *diode laser probe* applied intraocularly or to the sclera at points over its circumference (p143) or the similar external use of a cryoprobe which was previously employed. *Cyclocoagulation* is often successful in reducing otherwise uncontrollable intraocular pressure, but it may need to be repeated and hypotony and corneal and retinal complications may occur, so caution should be observed in patients with useful vision. Sometimes a drainage procedure may be successful when employing a plastic drainage tube of the Molteno type.

*Increased orbital pressure or orbital venous pressure*

Endocrine exophthalmos, orbital tumour or carotico-cavernous fistula will raise the intraocular pressure, directly or by increasing the episcleral venous pressure. This must be considered in the management of these conditions and tonometry performed. Fistulising operations in eyes with suspected raised episcleral venous pressure require caution due to the risk of massive choroidal detachment and flat anterior chamber.

*Toxic effects causing secondary glaucoma*

*Corticosteroids:* the toxic effects of various agents will cause a rise in intraocular pressure. The most important are corticosteroids to which a small proportion of patients, especially those with open angle glaucoma themselves or with a family history, will react quite severely. Continued local treatment with steroid drops is particularly liable to precipitate this rise in pressure leading to optic atrophy and field loss which may be advanced before this complication is suspected. Steroids of higher potency are more likely to cause this effect, the mechanism of which is uncertain. *Steroid drops should therefore be avoided if possible in predisposed patients and never prescribed for regular use in chronic relatively less important conditions such as blepharitis. All patients who require prolonged treatment with steroids should have tonometry at intervals.*

*Alkali burns:* severe kerato-uveitis from alkali burns may result from eye contact with ammonia or lime and is liable to cause secondary glaucoma.

*Siderosis.* Intraocular injuries involving retained ferrous foreign bodies may be followed by the toxic effects of dissolved iron salts (siderosis) on the trabeculum (p160) with consequent rise in intraocular pressure.

*Irido-corneal endothelial (ICE) syndrome (Essential iris atrophy)* (p515)

This characteristically affects mainly one eye in white middle aged females. An abnormal corneal endothelium spreads over the angle of the anterior chamber and on to the iris. By its contraction, holes may form in the iris and broad peripheral anterior synechiae may block the angle of the anterior chamber. It may be genetically determined but this is not certain.

*Trauma*

This may be followed by an immediate rise in intraocular pressure which may be associated with prostaglandin release and which settles but may be complicated by haemorrhage blocking the angle, by direct injury to the trabeculum and the root of the iris and by rupture of the lens suspensory ligament with displacement of the lens. If there is also rupture of the lens capsule, this will allow access of aqueous to the lens fibres which swell and fill the anterior chamber with soft lens matter which is phagocytosed by leucocytes which block the trabeculum with a rise in intraocular pressure (phakolytic glaucoma) (Plate 7.8 p159). The lens matter may also cause lens induced uveitis to aggravate the situation. Some cases of subluxation may be controlled by medical means but removal of the lens may be necessary although difficult. Occasionally corneal or conjunctival epithelium may be displaced into the wound in perforating injuries or operations, followed by epithelialisation of the anterior chamber and late intractable glaucoma.

*Ghost cell glaucoma.* Trauma, or retinal disease can cause haemorrhage which enters the vitreous cavity. Gradually fresh pliable biconcave erythrocytes are transformed into brownish, fairly rigid spheres which may remain for some time in the vitreous until the hyaloid membrane is disrupted for some reason and they gain entrance to the aqueous. They are only able to pass with difficulty through the trabecular meshwork pores and accumulate there causing obstruction and a temporary but sometimes marked rise in intraocular pressure. In large numbers they may form a 'striped' pseudohypopyon with a layer of red blood cells superimposed. The condition may be controlled usually by standard antiglaucomatous medication but in some cases irrigation or vitrectomy may become necessary.

*Intraocular neoplasms*

Intraocular tumours of all types may give rise to a variety of secondary glaucomas.

**Low intraocular pressure**

The intraocular pressure may become constantly reduced below about 5 mm Hg., a condition known as *hypotony.* It is usually due to poor aqueous secretion resulting from atrophy of the ciliary body following uveitis. There is reduced nutrition and respiration of the intraocular tissues. Cataract, wrinkling of Descemet's membrane with corneal opacities and chalky corneal deposits follow. The circulation in the retina and choroid may fail to adjust to the low pressure leading to retinal oedema, swelling of the optic disc and degenerative changes. Similar results may follow hypotony caused by injury or by excessive drainage from fistulising operations for glaucoma, when choroidal detachment may be an early feature. It is possible to attempt to seal the fistula in these cases but otherwise no treatment is possible except that for uveitis. Persistent variable pain may sometimes justify enucleation and, despite initial reluctance, the patient will appreciate not only the freedom from pain but the improved appearance resulting from a carefully designed prosthesis.

# CHAPTER 32

# THERAPEUTICS PART 1

## A. Routes of administration

One of the main problems in the drug treatment of ocular disease is the difficulty of achieving a sufficient quantity of drug at the desired site of action. Direct application is easy and usually effective for treating conjunctival disease, but the cornea acts as a barrier to drugs penetrating into the eye, and the tight junctions of the capillaries of the iris and retina effectively form an impermeable blood/eye barrier preventing passage of drugs from the blood into the aqueous and vitreous. Another factor which must be considered is the rate of removal from the eye of any drug that does actually penetrate into the aqueous or vitreous, because although inflammation may reduce the barrier to penetration of the drug into the eye, the associated hyperaemia will also speed the removal of the drug from the eye. The routes of administration are local and systemic.

### Local application

Local application of drugs for the treatment of superficial eye disease, e.g. conjunctivitis, is a very satisfactory route of administration. When the desired site of action of the drug is inside the eye, then the problem of the ocular barrier arises.

*The corneal barrier.* For practical purposes the cornea can be considered to consist of three layers. The outer and inner layers (the epithelium and the endothelium) prevent water soluble agents, e.g. ionised molecules, passing into the eye, but permit the passage of lipid soluble agents, whereas the corneal stroma resists the passage of lipid soluble agents but freely allows the passage of those which are water soluble. Theoretically the ideal drug is therefore capable of existing in two forms: a non-ionised fat soluble form for penetration through the corneal epithelium and endothelium and a water soluble form for passage through the corneal stroma. Drugs with this dual ability are usually capable of changing from lipid solubility to water solubility by ionisation.

*The scleral barrier.* The sclera, unlike the cornea, does not act as a differential solubility barrier and it is relatively porous. If this fact alone were considered then drugs applied directly to the scleral surface (by subconjunctival injection) should pass rapidly into the eye. However, this is not the case. There is a unidirectional flow across the sclera from the inside to the outside of the eye. The intraocular pressure may be partially responsible for this.

*Methods by which drugs may be locally applied to the eye:*

-*direct application* to the corneal surface: drops, ointment, soft lenses, constant release membranes (Ocuserts), iontophoresis.
-*sub-conjunctival injection.*
-*peribulbar, retrobulbar or orbital floor injection .*
-*direct injection into the aqueous or vitreous.*

*Application to the corneal surface.* The drug must fulfil the necessary criteria for passing the corneal barrier for it to penetrate into the eye. The effectiveness of the corneal barrier may be considerably reduced by inflammation or removal of the corneal epithelium, and increasing the concentration of the topical agent or prolonging its corneal contact time may also affect the ocular penetration of the drug. When drops are used much is lost because they are washed away by the tears. Viscous and ointment preparations of drops including oil suspensions and methyl cellulose solutions prolong contact time, as do soft lenses and 'ocular inserts' which contain a measured amount of the required drug which is slowly released over a number of days. This reduces the total quantity of drug given, as well as reducing the unwanted frequency of regular local applications. Iontophoresis is the technique of applying a small electric current to the eye to enhance penetration of ionisable drugs into the anterior chamber. It is rarely used in clinical practice.

*Subconjunctival injection.* Subconjunctival injection may be used to achieve a high concentration of drug in the anterior chamber. Antibiotics, steroids and mydriatics may all be given by this route. Subconjunctival injections are painful and carry the risk of globe perforation when given by inexperienced personnel. It is for these reasons that they are best only given by ophthalmologists. They are used for the more severe cases of acute inflammation or infection of the anterior segment. The eye is anaesthetised with local anaesthetic drops, and the appropriate drug injected using a long narrow tuberculin syringe instead of a normal syringe, which makes a much smoother injection possible. The method of absorption of the subconjunctivally injected drug is not clearly understood

*Sub-Tenon, peribulbar, retrobulbar and orbital floor injection.* Drugs can be delivered by sub-Tenon injection or to the back of the orbit directly and this is often used to achieve local anaesthesia in ocular surgery, the former being much safer. Steroids may be injected by these routes to reduce optic nerve or posterior pole inflammation, and depot injections of orbital floor steroid can give prolonged periods of treatment.

*Direct injection into the eye.* Drugs are often introduced directly into the eye during ocular surgery, e.g. acetyl-choline with mannitol (Miochol). Care must he taken that the concentration of the drug, the vehicle and the type of preservative is suitable. Antibiotics and steroids may be injected directly into the aqueous and vitreous in cases of endophthalmitis, and carefully calculated volumes and doses are essential, as direct retinal toxicity may be a problem.

## Systemic administration

General rules for systemic drug administration apply, but the effective blood/aqueous and blood/vitreous barrier means that intraocular levels of systemically administered drugs are usually much lower than in the serum. Drugs that are strongly protein bound are less permeable, and in general low molecular weight agents cross the barrier more easily. Many antibiotics including systemic penicillins do not usually achieve therapeutic vitreous levels, although the blood eye barrier is certainly less impermeable in cases of intraocular inflammation. Newer agents including the quinolone ciprofloxacin and the thienamycin imipenem do reach bactericidal vitreous levels. Oral steroids are used in treating severe ocular inflammatory disease, and may be extremely effective, but close monitoring is required because of the systemic complications associated with the high doses often required.

## B. Cycloplegics and Mydriatics

The mydriatics may be divided into two groups: the parasympatholytics, which block the action of the parasympathetic nervous system and the sympathomimetics.

### 1. Agents acting on the parasympathetic nervous system

Acetylcholine is the neuro-humoral transmitter at the receptor site of the parasympathetic nervous system and all the drugs in this group act by competitive blockade of acetylcholine for uptake at the postganglionic muscarinic receptors. This results in mydriasis due to the unopposed action of the dilator pupillae (which is under sympathetic control) and cycloplegia. A higher dose of a parasympatholytic agent is required to achieve cycloplegia than is required for mydriasis. This is significant for cycloplegic refraction because full mydriasis need not indicate complete cycloplegia.

*Atropine.* Atropine is the most powerful long-acting cycloplegic used in ophthalmology. It is an alkaloid and is used in its water-soluble form (atropine sulphate), and it is available as both drops and ointment at 0.5% or 1% strength. It may also be given by subconjunctival injection. Atropine is used both for its cycloplegic and mydriatic effects, a single drop of atropine 1% giving maximal mydriasis in about forty minutes and partial cycloplegia in about three hours. The effects of a single dose may last between three and seven days. Atropine may be indicated in severe anterior segment inflammation as it is a potent long lasting mydriatic, reducing the risk of posterior synechiae, and because its effect on the ciliary muscle reduces the ciliary spasm. It is sometimes combined with a sympathomimetic agent for maximal effect e.g. as a subconjunctival injection, Mydricaine No.2 (Martindale) (atropine sulphate 1mgm, procaine hydrochloride 6mgm, adrenaline acid tartrate 216 micrograms, sodium metabisulphite 300 micrograms, sodium chloride 300 micrograms, water for injection to 0.3ml). Atropine may be useful for cycloplegic refractions, especially in children with esotropia where undercorrected hypermetropia is suspected, and may also be used to induce a pharmacological form of occlusion in the treatment of amblyopia, instilling it into the dominant eye to reduce the vision on that side in order to encourage the patient to use the amblyopic eye. This 'atropine occlusion' is in general only effective in treating mild amblyopia. Atropine is sometimes necessary to help children adjust to a recently prescribed high hypermetropic correction, and may also be of benefit in treating accommodative spasm (p443).

*Contraindications and side effects.* Atropine is contraindicated in an eye with a shallow anterior chamber where there is a risk of precipitating angle closure glaucoma. Local allergy to atropine is not uncommon, with excessive stinging on instilling the drops and the development of a red, excoriated, itchy rash on the skin of the lids. Systemic absorption may cause signs of **atropine poisoning** with tachycardia, tremor, delirium and hot dry skin. Young children and the elderly are particularly vulnerable, and significant systemic absorption may be reduced by using atropine ointment instead of drops. However, in children cyclopentolate 0.1% or 0.5% drops are considered to be safer.

*Homatropine.* This is a semi-synthetic alkaloid prepared from atropine. It is used in its water soluble form, homatropine hydrobromide, and is commonly used as 1% or 2% eye drops. It acts more quickly than atropine, mydriasis usually being complete within forty minutes and cycloplegia after one hour of instillation. The duration of action is shorter than atropine and recovery occurs in about twenty-four hours. The newer mydriatics have largely superseded homatropine. It must not be used in cases of potential angle closure glaucoma.

*Cyclopentolate.* This is a very effective and short acting synthetic mydriatic and cycloplegic agent. Maximal mydriasis occurs in about thirty minutes but can occur as soon as fifteen minutes. Cycloplegia is usually present in forty minutes. However, to ensure complete cycloplegia it is suggested that two applications of cyclopentolate are used with a ten minute interval between the applications. Recovery of accommodation usually occurs between four and twelve hours, but may occasionally take longer. Cyclopentolate is usually used as drops of 1% or 0.5% strength, though 0.1% strength drops are available. Its main uses are for cycloplegic refraction and to dilate the pupil for fundus examination. Cyclopentolate (often in conjunction with a sympathomimetic agent, e.g. phenylephrine) may be used to dilate the pupil in cases of anterior segment inflammation or prior to intraocular surgical procedures. Local allergic reactions are rare.

*Tropicamide.* Tropicamide is another rapidly acting mydriatic and cycloplegic. It acts faster than cyclopentolate and its duration of action is shorter but it only induces partial cycloplegia. On account of its short latency, brief duration and effective mydriasis, tropicamide in 0.5% solution is perhaps the best mydriatic for the usual fundus examination.

## 2. Agents acting on the sympathetic nervous system

The peripheral neuro-humoral transmitter of the sympathetic system is *noradrenaline.* The effect of noradrenaline and pharmacological sympathomimetic agents is mediated by at least two classes of receptors, alpha and beta receptors. Alpha stimulation results in pupil dilatation, together with generalised vasoconstriction except in skeletal muscle. Beta receptor stimulation does not affect the pupil, but does cause an increase in aqueous production, as well as dilatation of the bronchioles and increased heart rate.

There are three possible mechanisms by which sympathomimetic agents can stimulate sympathetic receptors:

1. direct stimulation of the postganglionic receptors (e.g. phenylephrine and adrenaline).
2. indirectly by stimulating noradrenaline release from the preganglionic nerves (e.g. ephedrine).
3. impairing re-uptake of released noradrenaline, effectively increasing its effective concentration (e.g. cocaine).

The use of sympathomimetics as mydriatics is really confined to the production of short term mydriasis for fundus examination or to augment mydriasis by an atropine-like agent.

*Phenylephrine.* Phenylephrine hydrochloride is the most frequently used sympathomimetic agent in ophthalmology. It is an alpha stimulant and does not have beta stimulating properties. It acts directly on the alpha receptors of the dilator pupillae causing pupil dilatation. Phenylephrine 10% is used as a weak mydriatic for fundus examination as it is short acting and without cycloplegic effect (which therefore makes it unsuitable for refraction). It is often used in combination with parasympatholytic drugs (e.g. atropine, cyclopentolate) to dilate the pupil initially in iritis. Local side effects include pigment release in the anterior chamber and corneal clouding, and systemic absorption may give toxic effects, with tachycardia, arrhythmias and hypertension. Its action can be reversed by thymoxamine 0.1% which is an alpha adrenergic blocking agent

*Adrenaline.* Adrenaline (Epinephrine) (see also p418) has both alpha and beta receptor stimulatory effects but has an inconstant mydriatic effect. The principal use of adrenaline and its prodrug dipivefrin (Propine) is in the treatment of open angle glaucoma, and the pharmacology of adrenaline is also considered in the section on the therapeutics of glaucoma (p138). A valuable preparation for achieving maximum mydriasis in cases of marked anterior uveitis is Mydricaine No2, used as a subconjunctival injection. It contains adrenaline acid tartrate 216 micrograms, atropine sulphate 1mgm, procaine hydrochloride 6mgms, boric acid 5mgms, sodium metabisulphite 300 micrograms, sodium chloride 300 micrograms and water for injection to 0.3ml. A weak solution of adrenaline (1 in 10000) may be added to anterior chamber infusions to enhance mydriasis during cataract surgery. It is also combined with zinc sulphate to act as a mild conjunctival vasoconstrictor.

*Cocaine.* This is an alkaloid and, used as cocaine hydrochloride 2% or 4%, acts as a mydriatic by inhibiting the action of amine oxidase and preventing the re-uptake of noradrenaline by the pre-synaptic sympathetic nerve endings, thereby prolonging the effect of noradrenaline at the receptor sites (p420). It is toxic to the cells of the corneal epithelium and this effect may be used to advantage, in that the damage to the epithelium allows a greater penetration of drugs through the cornea. The major use of cocaine is as a local anaesthetic (p612), but it is also useful in the interpretation of Horner's syndrome, where the pupil fails to dilate following instillation of 4% cocaine drops if the third order sympathetic neurone is affected (p420).

*Sympathetic beta blocking agents* (p139)

## C. Miotics

Miotic agents constrict the pupil, and their principal use is in the treatment of both open (p137) and closed angle glaucoma (p188). Miotics may also be used in the treatment of convergence excess esotropias (p226).

*Mechanism of action of the miotics*

Pupillary constriction is under the control of the parasympathetic nervous system, with acetylcholine released from the terminal fibres of the parasympathetic nervous system stimulating the smooth muscle fibres of the sphincter pupillae inducing miosis. The acetylcholine is then destroyed by the enzyme cholinesterase. Drugs which potentiate acetylcholine or mimic its action will therefore cause miosis. The miotics that enhance parasympathetic activity fall into two groups: 1. *acetylcholine analogues* acting directly on the receptors of the muscle fibres of the sphincter pupillae. 2. *anti-cholinesterases* blocking the activity of cholinesterase and allowing the build up of natural acetylcholine thus causing miosis indirectly.

### I. Acetylcholine analogues

These drugs mimic the effects of acetylcholine but are more stable than acetylcholine and have a much longer duration of effect.

*Pilocarpine.* Pilocarpine is an alkaloid, being derived from the leaves of *Pilocarpus microphyllus.* It is used in its salt form, pilocarpine hydrochloride, as this is freely water soluble. The concentrations available range from 0.5% to 4%. The effect of a single drop takes 30 minutes to take effect, and lasts approximately six hours, so to maintain control of intraocular pressure four applications per day are usually required. The strength and frequency of the instillation depends on the nature and severity of the condition under treatment. Pilocarpine acts on the parasympathetic receptor sites of smooth muscle in a similar manner to acetylcholine *(muscarinic action).* However it has little effect on the receptor sites of skeletal muscle which normally would react to acetylcholine *(nicotinic action).* Pilocarpine causes miosis and contraction of the ciliary muscle. It is relatively ineffective in overcoming the mydriasis produced by atropine, because atropine competes more effectively for the same receptor sites as pilocarpine, and in addition the duration of the effect of atropine is also considerably longer than that of pilocarpine. Pilocarpine is usually the miotic of choice for the control of both closed and open angle glaucoma (pp188, 137).

*Side effects of pilocarpine.* Spasm of the iris and ciliary muscle produced by pilocarpine often causes ocular pain and brow ache. However, these symptoms usually disappear if the drug is used consistently for a few days. The miosis greatly reduces the amount of light entering the eye and many patients complain of darkening of their vision. Ciliary muscle contraction also causes transient myopia following administration. Most patients adjust to these symptoms, but patients with central lens opacities not only suffer more from the loss of light, but also notice a fall in visual acuity. These side effects can be reduced by the use of slow release lamellae (Ocuserts) or a viscous vehicle (Pilogel).

Long-term therapy with any strong miotic will produce an atonic pupil with posterior synechiae, and iris cysts and lens opacities may also occur. Retinal detachments have been reported, and long term of use of pilocarpine may cause cicatricial conjunctival changes which may adversely effect the outcome of drainage surgery for glaucoma.

Systemic absorption in sufficient quantity causes the symptoms of parasympathetic over-activity including bradycardia and increased peristalsis in the gut. Severe systemic side effects (including collapse) have been reported following the intensive use of pilocarpine drops, particularly in the treatment of acute closed angle glaucoma in elderly patients.

*Carbachol.* Carbachol has properties similar to pilocarpine and causes miosis by a direct effect on the sphincter pupillae. It is used as a treatment for glaucoma, but is usually reserved for those patients who are allergic to other miotics. It can be used in concentrations of 0.75% to 3%.

*Metacholine.* Metacholine chloride (Mecholyl) is rarely used as a treatment for glaucoma. The principal use is as a diagnostic agent for Holmes-Adie pupil (p417).

*Miochol.* Acetylcholine chloride with mannitol (Miochol) is a synthetic acetylcholine, and is as readily destroyed by cholinesterase as endogenous acetylcholine and is therefore valueless as a conventional miotic. However, if 1% acetylcholine chloride is combined with 5% mannitol and injected directly into the anterior chamber, effective miosis occurs very rapidly. This combination can be used to produce miosis if required during anterior segment surgery. It is an unstable preparation and must be made up at the time of the operation.

## 2. Anticholinesterase miotics

The anticholinesterase agents may be reversible or irreversible.

### *Reversible anticholinesterases*

*Physostigmine (eserine).* This is now little used. It is an alkaloid derived from the Calabar bean. It acts by combining with and inactivating endogenous cholinesterase, allowing the build-up of acetylcholine at the motor end plate. Slow hydrolysis separates the physostigmine from the cholinesterase within 12 hours and the cholinesterase is reactivated (the reversible effect). Physostigmine used to be a well recognised agent for the treatment of both open and closed angle glaucoma. However, pilocarpine is preferred to eserine which has significant side effects. Physostigmine causes ciliary spasm, ocular pain and marked conjunctival hyperaemia. Lid twitching may occur due to the nicotinic effect of physostigmine on the orbicularis oculi, as well as generalised twitching and paraesthesia. Systemic absorption can lead to toxicity, with nausea, bradycardia, hypotension, vomiting and coma. Physostigmine is used at 0.25% strength and may be combined with pilocarpine 4% with which it is synergistic to produce miosis in refractory cases with heavily pigmented irides. Physostigmine causes miosis in about 30 minutes and lasts 12 hours. If the drops are exposed to air and light they turn pink due to the formation of irritating rubeserine. When this happens the drops should be discarded.

*Irreversible anticholinesterases.*

These drops combine with cholinesterase and permanently inactivate it so that cholinesterase activity is prevented until new cholinesterase is synthesised.

*Demecarium bromide (Tosmilen)* is a very powerful miotic synthesised from neostigmine, the effect of one drop may last three to ten days. Its use has been discontinued because of pain due to ciliary spasm.

*Echothiophate iodide (Phospholine iodide).* This is derived from the highly toxic alkyl phosphate group of compounds and is available as 0.03% drops. The two principal uses of echothiophate iodide were in the treatment of resistant open angle glaucoma (but this is now discontinued) and in the management of some convergence excess esotropias (see p226). The side effects are extreme miosis and ciliary spasm with pain and myopia. In addition prolonged treatment may result in iris cyst formation and even vacuoles in the lens of the eye. Medication requires careful monitoring but the lowest concentration may be free of these complications. The addition of low concentration phenylephrine drops to this miotic treatment reduces the incidence of iris cysts.

**Echothiophate iodide (Phospholine Iodide) and general anaesthesia**

Systemic absorption of echothiophate iodide depletes the circulating pseudocholinesterase. This enzyme destroys the depolarising agent succinyl choline which is used to facilitate intubation during the induction of anaesthesia. The effect of succinyl choline is limited normally to a few minutes but if pseudocholinesterase has been depleted then total paralysis may last several hours. The echothiophate iodide drops should therefore be stopped some weeks before a planned anaesthetic or, if this is not possible, the anaesthetist must be warned that the patient is receiving an anticholinesterase. *The patient or parents should also be informed in case a general anaesthetic should be necessary for an unrelated problem.*

## D. Carbonic anhydrase inhibitors

*Acetazolamide (Diamox)* (pp138, 188). Acetazolamide is a carbonic anhydrase inhibitor derived from the sulphonamide group. It was originally developed as a diuretic but now it is chiefly used as a potent agent for the reduction of intraocular pressure. Carbonic anhydrase is an enzyme found in most parts of the body where active transport systems are employed in the production of fluids, e.g. ciliary epithelium, stomach mucosa (hydrochloric acid secretion) and the ventricles (CSF secretion), and it was originally thought that acetazolamide reduced the production of aqueous by blocking this enzyme. This is now thought to be an oversimplification and the mechanism of intraocular pressure reduction is not fully understood. It is an effective agent for rapid reduction in intraocular pressure, and may be given orally or intravenously. Its effect lasts for 6 to 8 hours, but longer acting sustained release formulations are also available.

The *side effects* of carbonic anhydrase inhibitor therapy by mouth or injection may be troublesome. A tingling sensation in the hands and feet is common, but usually only inconvenient. Many patients complain of indigestion, nausea and general malaise, and renal stones or crystalluria may occur. Acetazolamide may cause an increase in urinary potassium with consequent systemic hypokalemia, and serum potassium levels should be monitored with prolonged use. Severe metabolic acidosis can occur, particularly in diabetics, and rarer problems including blood dyscrasias and Stevens Johnson syndrome have also been described. Both systemic and topical use of carbonic anhydrase inhibitors are contraindicated in pregnancy and breastfeeding.

*Dichlorphenamide (Daranide, Oratrol).* This is an alternative to acetazolamide. It is also a carbonic anhydrase inhibitor which has similar properties to Diamox but sometimes due to idiosyncrasy one drug is tolerated when the other is not.

*Dorzolamide (Trusopt)* (p138) For many years there has been an intensive search for a form of sulphonamide carbonic anhydrase inhibitor which could be used as an eye drop to lower the intraocular pressure by reducing aqueous production, yet be free of most of the side effects which follow the use of acetazolamide systemically. This was achieved in 1995 in the form of dorzolamide which is a potent inhibitor of carbonic anhydrase II. It was subsequently approved for general use as 2% eye drops. There is some individual

variation but in a representative study of the treatment of primary open angle glaucoma, when used three times daily, dorzolamide drops caused a mean peak reduction of intraocular pressure monitored over a year of 6mm Hg (23%), the initial figure being 27 mm Hg. This compares closely with the effect of timolol eye drops administered twice daily. In another study, dorzolamide 2% eye drops used twice daily as additional drops, further reduced the mean peak intraocular pressure of 25.5mm Hg achieved by timolol 0.5% drops twice daily, by a further 20% to 20 mm HG.
Dorzolamide may cause local sensitivity reactions in patients who react to sulphonamides. It may also result in a bitter taste due to drainage by the tear duct and some stinging and irritation of the lids and conjuntiva but in general, has been well tolerated and clinically free of the systemic effects of oral carbonic anhydrase inhibitors.

*Brinzolamide (Azopt)* (p139) is a carbonic anhydrase inhibitor available in suspension as a 1% eye drop used twice daily as an adjunct to beta blocking therapy or twice or, if necessary, three times daily as monotherapy. It is claimed to be as effective as dorzolamide but with less superficial discomfort

*Timolol and dorzolamide as a combination drop (Cosopt)* Timoptol 0.5% plus dorzolamide 2% has been formulated as a combined drop known as Cosopt. Increased aqueous flow resulting from the combination has been confirmed by fluorophotometry. The reduction of intraocular pressure produced by the twice daily use of Cosopt is approximately equivalent to that resulting from the twice daily application of both drops separately and would thus favourably affect compliance. In addition to the timolol, topical dorzolamide has also been shown to increase blood flow velocity in the epipapillary capillaries which may improve the circulation of the optic nerve at the disc.

## E. Osmotic agents

Osmotic agents increase the osmotic pressure in the blood, thereby encouraging the flow of water from the eye into the blood. They are used for rapid reduction in the intraocular pressure, e.g. in the early management of acute glaucoma or to reduce the pressure in an eye immediately prior to surgery. The use of oral glycerol and intravenous mannitol in glaucoma is described on p188. The associated dehydration and reduction in volume of the vitreous may be helpful in reducing the risk of vitreous loss in susceptible patients during cataract surgery. Its action is illustrated in Fig. 32.1 p598.

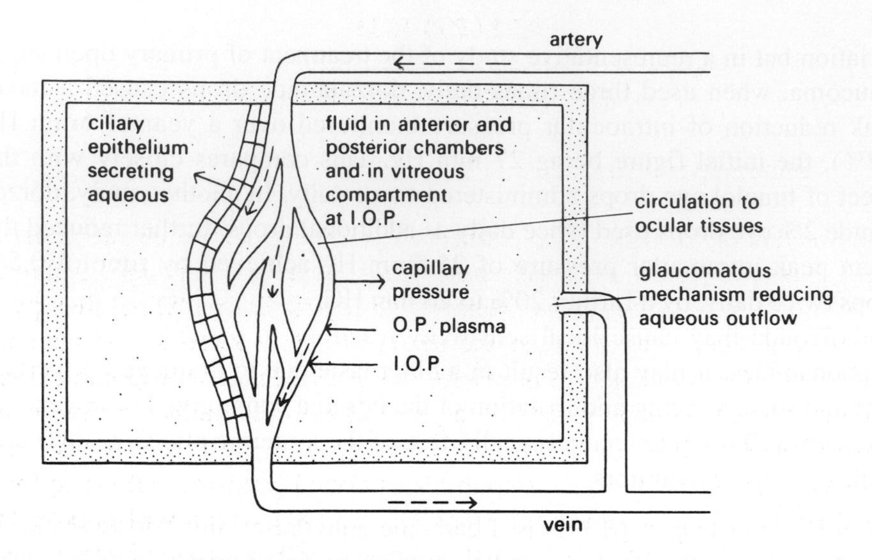


Fig. 32.1 Mode of action of osmotic agents.

*Glycerol* is an oral osmotic agent and is given at a dose of 1.0 to 1.5 g/kg body weight. The unpleasantly sweet taste may be made more tolerable by mixing it with an equal quantity of iced lime juice, but it is often necessary to prevent the associated nausea by prior treatment with a suitable anti emetic, e.g. metoclopramide (Maxolon). The maximum effect of glycerol occurs about 45 minutes after ingestion and lasts about four hours.

*Mannitol* may be given intravenously if a rapid fall in intraocular pressure is required. The effect on intraocular pressure is seen after 30 minutes and is maximal at one hour. The dose of mannitol is up to 2 g/kg body weight, usually 150 to 300 ml of 20% solution is given. Solid crystals of mannitol may form in the 20% solution so the mannitol should be infused with a drip set containing micro filters. Mannitol causes a marked diuresis so that patients under general anaesthesia may sometimes require catheterisation and acute retention of urine may rarely occur in patients with prostatic hypertrophy.

*Side Effects.* All parenteral osmotic agents used to reduce the intraocular pressure may induce profound hypotension, nausea, vomiting and confusion. Congestive cardiac failure, pulmonary oedema and diabetic hyperglycaemia are also possible, and patients should be adequately monitored.

# CHAPTER 33

# THERAPEUTICS PART 2

## A. Antibacterial agents

Many factors must be considered when choosing an appropriate antibacterial agent. Firstly a diagnosis is made based upon symptoms and signs followed by an informed guess as to the offending organism. Before 'blind' therapy is commenced, conjunctival swabs and scrapes, corneal scrapes, and aqueous and vitreous taps are taken as necessary and placed in an appropriate culture medium as well as being placed on a glass slide for immediate staining and microscopic examination. Identification and sensitivity testing may take several days and so therapy based upon the informed guess but allowing for a broad cover is commenced immediately. The route which will deliver the most drug to the site of infection but will cause the least toxic effects is chosen. For ocular surface infections such as conjunctivitis and keratitis topical therapy will usually suffice, however in cases of endopthalmitis both systemic, and even intravitreal administration may be required to achieve adequate drug levels in the posterior segment.

### *Chloramphenicol*

Chloramphenicol 0.5% drops and 1% ointment is a widely used ophthalmic broad spectrum antibiotic. It inhibits protein synthesis and has a bacteriostatic effect against a wide range of Gram negative and Gram positive bacteria although it has poor activity against *Pseudomonas aeruginosa*. Chloramphenicol is highly lipid soluble, penetrating the cornea and achieving high aqueous concentrations with topical therapy. Topical chloramphenicol is used for treating bacterial conjunctivitis and blepharitis, as a prophylactic antibiotic following ocular surgery and immediately following penetrating ocular trauma. Topical hypersensitivity can occur. Due to its toxic effects *systemic* chloramphenicol therapy is reserved for life threatening CNS infection. The first of these is a reversible bone marrow depression that can occur in a dose-related fashion. Secondly a more serious irreversible idiosyncratic aplastic anaemia with an incidence of 1:25-50,000 has been suspected to occur weeks or months following the cessation of therapy but the evidence for this is controversial. This risk is still theoretically present with topical therapy. Systemic chloramphenicol is also reported to be associated with an irreversible optic neuropathy.

*Fusidic acid*

Fusidic acid 1% (Fucithalmic) twice daily is a useful alternative to chloramphenicol. It can be used for external eye infections being active against a wide variety of Gram positive organisms, particularly *Staphylococcus aureus,* although less effective against haemophilus and streptococci. It is formulated as a gel which liquifies on contact with the eye and is particularly useful in children.

*Penicillins*

Penicillins are bactericidal and act by inhibiting the linking of polysaccharide chains in the bacterial cell wall. The older penicillins e.g. penicillin G (benzylpenicillin) either intravenously or intramuscularly and penicillin V(phenoxymethylpenicillin) by mouth are still used in the treatment of preseptal cellulitis and orbital cellulitis because they are so effective against Gram positive streptococci. Resistance by beta lactamase producing staphylococci is common and cover using a newer penicillin such as flucloxacillin is necessary for these organisms.

Gonococcal ophthalmia neonatorum in the absence of keratitis can be treated systemically with penicillin G (50,000 units/kg/day) in two divided doses over seven days. Where there is keratitis, fortified penicillin G drops (5,000 units/ml) can be used hourly. Penicillinase producing strains of gonococci can be treated effectively with a single dose of Ceftriaxone 125 mg intramuscularly.

Hypersensitivity to penicillins occurs in up to 10% of the population.These reactions may range from a mild skin rash to severe anaphylaxis. There is a cross reactivity of about 5-10% with the cephalosporins although there is generally not a problem in using cephalosporins in patients with an equivocal or mild penicillin allergy.

*Cephalosporins*

Cephalosporins act similarly to penicillins inhibiting cell wall synthesis but also altering cell membranes resulting in cell lysis. Generally these agents are effective against Gram positive lactamase producers such as *Staphylococcus aureus,* however with cases of lactamase resistance third generation agents such as cefotaxime and ceftazidime should be used. In the treatment of suppurative keratitis the combination of a fortified cephalosporin such as cefuroxime drops 5% in combination with fortified gentamicin drops 1.5% administered half hourly is the initial treatment of choice. These fortified drops are not commercially available and need to be made up by the hospital pharmacist or from parenteral preparations. Cefuroxime drops 5% have a superior effect on Gram positive bacteria to ceftazidime drops 5% which are also effective against *Pseudomonas aeruginosa.* Cephalosporins are frequently injected sub-conjunctivally at the end of surgery and systemically by intravenous or intramuscular injection are a useful alternative for treating pre-septal and orbital cellulitis in patients with penicillin allergy. Caution is still required in patients with penicillin sensitivity or renal impairment.

*Aminoglycosides*

Aminoglycosides inhibit bacterial protein synthesis by binding to the 30S subunit of the bacterial ribosome. Aminoglycosides are particularly effective in the treatment of Gram negative infections such as *Pseudomonas aeruginosa,* and proteus and serratia species but also have activity against staphylococci and *Haemophilus influenzae.*

Fortifed gentamicin drops 1.5% (not commercially available) in combination with a fortified cephalosporin is used in the treatment of suppurative keratitis. Gentamicin drops 0.3% and ointment 0.3% may be used in the treatment of a variety of bacterial infections of the external eye and adnexae. Gentamicin 40mg in 1ml injected subconjunctivally is frequently used for prophylaxis following cataract surgery, but it causes irritation and surgeons may now prefer cephalosporins.

Neomycin 0.5% drops or 0.5% ointment is similarly effective, although allergic rections are fairly common. Neomycin may be used in combination with a steroid for its combined antibacterial and anti-inflammmatory effects. For the treatment of endophthalmitis amikacin in combination with vancomycin may be injected into the vitreous. Gentamicin may be used intravenously for the treatment of endophthalmitis, however it has largely been superceded by the fluoroquinolones due to its potentially toxic effects. Systemic aminoglycosides may result in vestibular and auditory dysfunction due to 8th nerve damage and may cause acute renal tubular necrosis. Plasma levels should be carefully monitored.

*Fluoroquinolones*

Although structurally related to the older nalidixic acid, fluoroquinolones such as ciprofloxacin, norfloxacin and ofloxacin were introduced more recently. They act by inhibiting DNA gyrase which prevents DNA replication, recombination and repair. They are bactericidal as well as bacteristatic because organisms which are not killed initially will not begin to grow for up to six hours after exposure. This post antibiotic effect is of particular relevance to topical therapy where peak concentrations may fluctuate considerably. The fluoroquinolones are effective against aminoglycoside and lactamase resistant bacteria and have a wide range against Gram positive and Gram negative organisms including *Staphylococcus aureus, Pseudomonas aeruginosa,* and *Serratia marscecens,* as well as neisseria and chlamydia species. Activity against streptococci is relatively poor. Resistance to the fluoroquinolones is rare.

Norfloxacin 0.3% and ofloxacin 0.3% drops can be used as single agents in the treatment of microbial keratitis. Fortified preparations are not required due to excellent corneal penetration and there are minimal toxic effects when used intensively. Topical fluoroquinolones can be used in conjunction with systemic ciprofloxacin for the treatment of endopthalmitis. Systemic ciprofloxacin 250-750mg by mouth twice daily and 200mg intravenously twice daily produces a relatively low incidence of side effects which include gastrointestinal symptoms, dizzyness and headaches. Hypersensitivity reactions are rare. *In juvenile animals quinolones produce an erosive arthropathy and therefore these agents should be used with great caution in children if clinically essential.*

*Tetracyclines*

The tetracyclines act by binding to the 30S subunit of the bacterial ribosome which inhibits protein synthesis and exerts a bacteriostatic effect. The tetracyclines are broad spectrum antibiotics whose value has decreased due to increasing bacterial resistance, however they remain the antibiotic of choice in the treatment of chlamydial infections. For the treatment of adult trachoma, doxycycline 100mg (a long acting preparation) once weekly for three weeks repeated six months later is effective. Adult inclusion conjunctivitis can effectively be treated with doxycycline 100mg daily for two weeks. Chlamydial ophthalmia neonatorium should be treated with erythromycin 50mg per kilogram per day in divided doses. *Systemic tetracyclines should be avoided in children or pregnant women due to their deposition in teeth causing permanent staining and dental hypoplasia. Increased cerebrospinal fluid pressure with optic disc swelling has also been reported in babies and adults.*

Systemic tetracyclines are useful in the treatment of mixed blepharitis, rosacea blepharitis and acne vulgaris. It is thought that as well as antibacterial actions, they alter the fatty acid content of the meibomian and sebaceous gland secretions as well as having direct anti-inflammatory effects. Tetracycline 250mg by mouth twice daily, oxytetracycline 250mg by mouth twice daily or doxycycline 50mg by mouth once daily for up to three months is effective. Systemic tetracyclines form a non-absorbable salt with calcium ions and patients on this therapy should not take dairy products or antacids for up to three hours after each dose. They are also contraindicated in renal impairment, except doxycycline and in hepatic impairment. Topical tetracycline (Achromycin 1% ointment) three times daily is sometimes used in conjunction with systemic tetracyclines for the treatment of chlamydial and infective external diseases.

## B. Antiviral agents

Whilst there has been rapid progress in the development of antibacterial drugs there has been decidedly less success in producing effective antiviral agents. Viruses are obligate intracellular organisms which, in order to replicate, have to take over the metabolic machinery of the host cell. Any effective antiviral agent has to be selectively toxic to the virus and not to its host. The herpes viruses are the main group amongst which such effective selective toxicity has been found. The herpes viruses include, herpes simplex virus (HSV), varicella zoster virus (VZV), Epstein-Barr virus (EBV) and cytomegalovirus (CMV).

*Acycloguanosine (aciclovir)* 3% eye ointment, 5% cream, and 200mg tablets (Zovirax) is generally the most effective and least toxic of the antiviral agents. It is active against HSV and VZV. Only the herpes infected cell phosphorylates aciclovir by thymidine kinase and it is the action of the phosphorylated aciclovir against herpes-specific DNA polymerase which results in DNA chain termination. Thymidine kinase does not phosphorylate aciclovir in uninfected cells making this agent selectively toxic to the viral and not the host cell DNA.

Aciclovir eye ointment applied five times daily is probably the most effective pharmaceutical agent for treating HSV epithelial keratitis. When there is failure of epithelial healing, resistance should be considered and an alternative topical antiviral agent can be used. In combination with topical steroid, aciclovir ointment is effective in the treatment of HSV stromal keratitis and HSV disciform keratitis, although the epithelium should be healed before steroids are commenced.

Oral aciclovir 200mg five times daily is also effective in the treatment of HSV epithelial keratitis and disciform keratitis. The role of oral aciclovir in the treatment of herpes zoster ophthalmicus (HZO) is controversial, but it may hasten the resolution of signs and symptoms and reduces the incidence and severity of secondary ocular inflammatory disease, especially if treatment is commenced within 72 hours after the onset of the skin lesions. Oral aciclovir should always be used in immunocompromised and AIDS patients with HZO. A dosage of 800mg five times daily is required to be effective against VZV. Topical ointment is also applied to the skin during the vesicular stage. Topical lubricants are generally sufficient for the treatment of epithelial disease but there still remains considerable debate about the most effective topical therapy for established keratitis and scleritis. If it is impossible to exclude HSV as the cause of a keratitis and especially if topical steroids are to be used the addition of aciclovir would be prudent. Intravenous aciclovir has a role in the treatment of acute retinal necrosis syndrome. Aciclovir is the least toxic topical antiviral agent but may produce superficial punctate keratitis and conjunctivitis. Orally administered aciclovir may cause nausea, vomiting, vertigo and a rash. An alternative to aciclovir is valaciclovir tablets 500 mgm. This is rapidly converted to aciclovir and valine in the liver. Better bioavailability affords a 500 mgm twice daily dosage in HSV infections and 100 mgm three times daily in HZO infection.

*Trifluorothymidine (F3T or Trifluoridine)* 1% drops is an analogue of thymidine and is an effective inhibitor of thymidine synthetase inhibiting DNA synthesis in both virus infected and normal cells. It is only useful in the treatment of HSV. It is similar to idoxuridine but is more effective in the treatment of epithelial keratitis and less toxic to the cornea. It is used five times daily. Being more lipid soluble than idoxuridine it penetrates the cornea and is more effective in the treatment of HSV stromal keratitis. Adverse effects include superficial punctate keratitis, conjunctivitis and punctal fibrosis. Prolonged treatment may result in irreversible corneal scarring.

*Idoxuridine (IDU)* 0.5% eye ointment resembles thymidine and is phosphorylated by viral and host cells, subsequently being incorporated into viral and host cell DNA. The result is antiviral activity with considerable host cell cytotoxicity. Applied every four hours, this agent is effective in treating HSV epithelial keratitis, however due to poor corneal penetration it is unsuitable for the treatment of stromal disease. It is a useful agent when there is a resistance to other antiviral agents. Adverse effects include superficial punctate keratitis, slowed epithelial healing, stromal scarring and punctal fibrosis.

*Ganciclovir sodium* is effective against CMV and is of particular use in treating CMV retinitis in AIDS patients. It is a synthetic analogue of 2'-deoxyguanosine which is phosphoryated preferentially by CMV infected cells. Ganciclovir triphosphate competitively inhibits viral DNA polymerases and is also directly incorporated into viral DNA, resulting in termination of viral DNA elongation. The drug is administered intravenously over one hour at a dose of 5mg/kg every 12 hours. The drug is very toxic, frequently inducing a granulocytopenia and thrombocytopenia which is usually reversible. Excretion is via the kidneys, hence a need for adequate hydration during therapy. A reduced dosage should be used in patients with renal impairment. Ganciclovir is a potent carcinogen and mutagen. Frequent blood counts and creatinine tests are required during therapy. Oral guanciclovir will be available soon.

*Foscarnet sodium (Foscavir)* exerts its antiviral activity by a selective inhibition at the pyrophosphate binding site on virus specific DNA polymerases at concentrations that do not affect host cell DNA polymerases. Foscarnet is indicated for the treatment of CMV retinitis in AIDS patients who have a resistance or intolerance to guanciclovir. The drug is administered intravenously at doses of 20-200mg daily according to renal function. The major adverse effect of foscarnet is renal impairment which is generally reversible. Dosage schedules should relate to serum creatinine levels both during and after induction therapy. Anaemia, granulocytopenia and electrolyte (calcium, magnesium, potassium and phosphorous) disturbances may occur resulting in seizures. Frequent blood counts, electrolyte and serum creatinine tests are required during therapy.

### C. Antifungal preparations (p340)

### D. Antiprotozoal agents (p341et seq)

### E. Anti-inflammatory agents

#### *Corticosteroids*

The corticosteroids are secreted by the adrenal cortex under negative feedback control from the hypothalamus and anterior pituitary gland. They fall into three categories: 1. Glucocorticoids 2. Mineralocorticoids 3. Sex hormones. Only the glucocorticoids are of importance in ophthalmic therapeutics. They have a powerful anti-inflammatory and immunosuppressive action. It is these properties which have made them such important agents in the treatment of many ocular and systemic diseases. Unfortunately they have a large number of unwanted effects.

Corticosteroids, often shortened to 'steroids' in general use, are cholesterol based molecules which exert their effects upon the cell via steroid receptors in the cytoplasm. Activation results in altered levels of mRNA production which regulates protein synthesis. This causes the diverse physiological and pharmacologic effects seen following steroid administration. The anti-inflammatory effects of the corticosteroids are non specific and they inhibit inflammation without regard to cause. In general, corticosteroids appear to be more effective in acute rather than chronic conditions. The beneficial effects of these agents on inflammation are numerous and include:

-a reduction in capillary permeability and cellular exudation,
-the inhibition of mast cells, basophils and neutrophils by stabilising membranes,
-the inhibition of fibroblasts,
-the suppression of lymphocyte proliferation,
-the inhibition of phospholipase A production, reducing prostaglandin and leukotriene synthesis,
-the inhibition of the cell mediated response,
-the inhibition of collagenase activity.

The use of steroids to reduce inflammation in the presence of infection is a double-edged sword. On the one hand suppression of inflammation minimises tissue damage whereas immunosuppression reduces resistance to infection.

*Local Steroids*

*Topical.* The specific indications for local steroids are discussed under the various chapter headings. The following are the commonly used topical steroid drops:
-prednisolone acetate 1% (Pred Forte Drops),
-rimexolone 1% suspension (Vexol),
-dexamethasone alcohol 0.1% (Maxidex Drops),
-betamethasone sodium phosphate 0.1% (Betnesol Drops and Ointment)
-fluoromethalone alcohol 0.1% (FML Drops),
-prednisolone sodium phosphate 0.5% (Predsol Drops).

These drugs are listed in a decreasing order of potency as measured by their ability to reduce inflammation at the corneal surface. Their penetration into the eye varies. For example FML is potent at the corneal surface, but its penetration is poor, making it unsuitable for the treatment of anterior uveitis. The frequency of administration depends upon the degree of inflammation. For example in severe anterior uveitis or corneal graft rejection Maxidex drops can be used every 30 minutes but as the inflammation resolves the frequency is tailed off over several weeks to prevent rebound effect. Predsol drops can be diluted all the way down to 0.001%. This is especially useful in the prevention of recurrent herpetic keratitis, where patients may remain on once daily drops at this dilution for many months with minimal adverse effects.

Local and systemic steroid treatment is associated with the formation of posterior subcapsular cataract. This effect appears to be related to the dose and duration of therapy. Local and systemic steroids can in susceptible subjects, produce raised intraocular pressure and glaucoma. Patients with primary open angle glaucoma or first degree relatives of these patients are particularly at risk. Myopes may be similarly affected. FML tends to produce this effect *less* due to its poor absorption into the anterior chamber, making it a suitable drop for ocular surface disease requiring prolonged treatment but rimexolone is reported to cause less rise in susceptible patients without loss of potency. Injection of rimexolone is not advised. Steroids can delay healing due to their inhibitory effects on collagenase and keratocyte activity. It is for this reason that they should only be used in the initial treatment of alkali burns and subsequently withdrawn to prevent a delay in healing.

Whilst reducing inflammation, steroids produce immunosuppression, thus masking the signs of infection. The use of steroids and antimicrobials together in cases such as herpetic keratitis and endophthalmitis is a fine balancing act playing off anti-inflammatory effects against immunosupressive effects. In general steroids should not be prescribed unless there is confidence that infection is covered by an appropriate anti-microbial agent and even then regular review is mandatory. ***Red eye conditions should never be treated with a steroid until the diagnosis has been confirmed. The treatment of herpes keratitis with steroids alone can have a disastrous effect upon vision.***

*Local injection.* Periocular steroids can be administered by subconjunctival, sub-Tenon's, or retrobulbar injection. Betnesol (betamethasone sodium phosphate) 4mg or Depo-Medrone (methylprednisolone acetate) 40mg are the most commonly used injections. Depo-Medrone has the advantage of being a slow release preparation being delivered from the site of injection for over a week. Injected steroids are particularly useful in the treatment of posterior segment inflammations, where topical administation fails to deliver enough steroid to the site of inflammation. These agents are frequently used to suppress inflammation following anterior segment surgery.

*Systemic steroids*

Inflammation of the posterior segment, optic nerve, or orbit usually requires the systemic administration of steroids. Because systemic adverse effects (although not unknown with topical therapy) are more likely, dosage should be individualised for each patient and, when long-term therapy is necessary, the lowest possible dose to control the disease should be given.

*Oral*

Prednisolone tablets (1-80mg daily). This is the most commonly used oral steroid. Generalised systemic effects following systemic administration include:

-immune suppression,
-Cushing's syndrome,
-peptic ulceration and perforation,
-hypertension and congestive cardiac failure,
-hyperglycaemia and ketoacidosis,
-euphoria and dysphoria,
-suppression of the pituitary axis resulting in growth failure and amenorrhoea. Addisonian crisis may occur following brisk withdrawal and especially in children, even slow withdrawal may cause papilloedema and post papilloedema optic atrophy (p384)
-impaired wound healing and subcutaneous tissue atrophy.

Patients should be monitored at regular intervals with urinalysis, blood pressure and weight measurements. Patients at risk or with a previous history of tuberculosis should be screened with a chest X-ray and commenced on antituberculous treatment if necessary. Diabetics should be monitored very closely as insulin requirements will increase and non insulin dependent patients may require insulin during steroid therapy. Patients with a history of peptic ulceration should receive $H_2$ antagonists concurrently.

*Intravenous.* In certain acute sight threatenening conditions - such as optic nerve compression due to orbital pseudotumour and thyroid ophthalmopathy, or in the acute stages of temporal arteritis, especially with sight loss already in one eye, the swift administration of intravenous steroid may be useful in preventing any or further visual loss before adequate plasma levels are achieved by the oral route. Hydrocortisone 500mg and methylprednisolone lg may be given by slow intravenous infusion.

*Non-steroidal anti-inflammation agents (NSAIDS)*

These drugs which have analgesic, antipyretic and anti-inflammatory effects are available for topical and systemic administration. The mechanism of action of NSAIDS involves the inhibition of cyclo-oxygenase. This enzyme is important in the synthesis of prostaglandins from their precursor arachidonic acid. The prostaglandins act as mediators of inflammation in ocular structures. They can cause vasodilatation of ocular blood vessels, disrupt the blood-aqueous barrier and induce miosis.

*Systemic.* Indomethacin 50-200mg by mouth daily in divided doses and flurbiprofen (Froben) 150-300mg by mouth daily in divided doses are effective in the treatment of anterior non-necrotising and posterior scleritis. In recurrent and unresponsive episcleritis, indomethacin 50mg by mouth three times daily may be helpful. Both agents can be given as rectal suppositories. They can also be used for analgesia and the reduction of inflammation following vitreoretinal surgery. NSAIDS may result in peptic ulceration and should be discontinued if there is gastro-intestinal discomfort but administration of H2 antagonists may permit recommencement. NSAIDS are contraindicated in patients with a previous history of hypersensitivity to aspirin or other NSAIDS. They may also result in deterioration of renal function especially in those with pre-existing renal disease.

*Topical* NSAIDS are of use in maintaining pupillary dilatation during cataract surgery, and reducing post operative inflammation. They are useful agents for reducing pain and inflammation following excimer laser keratectomy. Diclofenac sodium 0.1% (Voltarol Ophtha) is used four times daily following cataract surgery for the reduction of inflammation and can be continued for up to two weeks . Flurbiprofen sodium 0.03% (Ocufen) can be given half hourly for two hours prior to surgery to maintain pupillary dilatation. The most frequent adverse effect is a transient stinging and mild epithelial keratitis. Wound healing may be delayed with these agents. Ocufen is contraindicated in epithelial herpes simplex keratitis.

## F. Local anaesthetic agents

Local anaesthesia of the eye may be achieved by topical application of anaesthetic drops or by infiltration of the sensory nerves with anaesthetic solutions. Increasing volumes of day case surgery have occurred partly as a result of improved local anaesthetic techniques, and also due to increased acceptability to both surgeon and patient alike. Local anaesthetics avoid the risks of general anaesthesia and offer the patient a relatively pleasant post operative period with little nausea and prolonged analgesia.

Local anaesthetic agents work because they prevent the generation and conduction of nerve impulses at the nodes of Ranvier by reducing sodium permeability. Pain is the first sensory modality to be affected followed by temperature, touch and proprioception. The motor nerves are the last to be affected.

### Topical anaesthesia

The three most commonly used topical anaesthetic agents are oxybuprocaine (Benoxinate 0.4%), proxymetacaine (0phthaine 0.5%) and amethocaine (Tetracaine 0.25-1%).

Topical agents may be used for temporary anaesthesia of the corneal surface in outpatient procedures such as tonometry, gonioscopy and fundus contact lens biomicroscopy. Minor surgical procedures such as suture and foreign body removal, as well as major procedures such as cataract surgery, can be performed under local anaesthesia. Careful patient selection is essential when planning major surgical procedures under topical anaesthesia.

From clinical experience proxymetacaine is the least potent and amethocaine the most potent drop. The onset of action of all these drops is from 10-60 seconds and duration of action is 15-20 minutes. The depth of anaesthesia also depends upon the dose, frequency and duration of use. Patients should be warned not to rub their eyes after the use of these agents and, if necessary, they can be supplied with a temporary protective shield. Contact lenses should not be reinserted for at least an hour. All of these agents delay epithelial healing and therefore should not be used for long-term analgesia. Abuse may result in epithelial defects which may become infected and could ultimately result in endophthalmitis.

Irritation and stinging are common to all these drops, but would appear to occur least with Ophthaine which are therefore the drops of choice in outpatient procedures and for use with children. They are all toxic to the corneal epithelium resulting in a superficial punctate keratitis and superficial corneal haze. With amethocaine the surface reaction can be more marked, resulting in sloughing of the epithelium, corneal oedema and corneal filaments. Rarely amethocaine causes a contact dermatitis. Systemic reactions are rare, but amethocaine in high doses may cause vasovagal reactions and should be avoided in frail elderly patients with cardiac dysrhythmias. Despite these adverse effects amethocaine remains the drop of choice for the more painful surgical procedures.

Cocaine can be used as a 4% solution, a 10% solution and a 25% paste. This natural narcotic alkaloid with sympathetic activity has largely been superseded by the synthetic agents due to its toxic effects. It does however have the advantage of being a potent local anaesthetic as well as producing mydriasis and local vasospasm. This enhances visualisation of the eye as well as producing a bloodless field. Cocaine drops 4% are useful in the diagnosis of Horner's syndrome (p420).

Cocaine 25% paste soaked into a nasal pack is useful for dacryocystorhinostomy, producing additional anaesthesia and reducing local bleeding. The drop is toxic to the corneal epithelium and stroma. When absorbed systemically cocaine may produce hyperactivity, headache, irregular tachycardia, delirium, seizures and even death. The cardiac effects can be reversed with beta blockers. Cocaine is a drug of abuse and is also a controlled drug.

*Injectable anaesthetic agents*

Anaesthesia of the globe and surrounding ocular structures can be achieved with locally injected agents. These may be injected:

- behind the globe inside the muscle cone (retrobulbar),
- outside the cone and alongside the globe (peribulbar),
- beneath the Tenon's capsule alongside the globe (sub-Tenon)
- beneath the conjunctiva (sub-conjunctival/anterior sub-Tenon)

Retrobulbar, sub-Tenon and peribulbar techniques produce globe akinesia whereas sub-conjunctival injections do not. Supplementary facial nerve blocks will be required for lid akinesia. The addition of adrenaline 1:100,000 or 1:200,000 in conjunction with local anaesthetics promotes local haemostasis, decreases systemic absorption and prolongs the duration of action.

The addition of hyaluronidase (Hyalase) results in the hydrolysis of hyaluronic acid in orbital connective tissue increasing tissue permeability and hence dispersion of the anaesthetic agents. Systemic absorption is also increased. Any injection behind the globe may produce retrobulbar haemorrhage which can be sight threatening. Any injectable technique using a sharp needle may result in globe perforation, particularly in the case of large myopic eyes. The retrobulbar approach may rarely result in bulbar paralysis due to sub-arachnoid injection. An anaesthetist should always be in the vicinity when using these techniques because the surgeon cannot safely take up the role of anaesthetist should anaesthetic complications occur.

The two most commonly used local anaesthetic agents are lidocaine (lignocaine, Xylocaine) 2% and 4% and bupivacaine (Marcaine) 0.25%, 0.5% and 0.75%. Both agents act by altering nerve cell membrane permeability to sodium. The agents are amides which differ in their speed of onset and duration of action. The action of lignocaine commences after 30-60 seconds and lasts 30-60 minutes, whereas bupivacaine starts after 5 minutes and lasts 8-12 hours. The duration of action is considerably increased with the addition of adrenaline as above. Frequently a combination of the two agents with hyalase and adrenaline is used. Marcaine has the additional advantage of providing prolonged post operative analgesia.

When absorbed systemically the local anaesthetic agents can have central nervous system effects including nausea, vomiting, drowsiness, convulsions and coma. Cardiac effects include bradycardia, tachyaryhthmias, asystole and hypotension. These agents should be used with caution in patients with known cardiac dysrythmias or cardiomyopathy. Acute hypersensitivity reactions and anaphylaxis occur rarely. Both agents may produce temporary myotoxicity.

The total dose of local anaesthetic should he calculated, especially when high volumes are used. For example the maximum recommended adult dose of lignocaine when used as a local anaesthetic agent is 200mg. This dose is contained in 5mls of a 4% solution.

## G. Other therapeutic and diagnostic agents

### *Staining agents*

*Sodium fluorescein* is a water soluble yellow dye, which fluoresces green in blue light, and is the commonest staining agent used in ophthalmology. It can be used locally or given intravenously. It may be instilled into the conjunctival sac by touching the conjunctiva with sterile filter papers impregnated with fluorescein, or with a drop from single dose containers of a 1% or 2% solution. (Multidose containers of fluorescein are not used because of the risk of bacterial contamination, particularly with *Pseudomonas aeruginosa)*. Local applications of fluorescein are used to demonstrate ulceration or breaks in the corneal epithelium (e.g. abrasions) (Plate 7.4 p155). Fluorescein will not stain intact epithelium, but if the epithelium is deficient the stroma is stained. This stain appears bright green when viewed with a blue light. Fluorescein may also be used to demonstrate and localise aqueous leakage. The patient should be asked to look down whilst a drop of 2% fluorescein solution is applied to the upper bulbar conjunctiva and allowed to flood down over the cornea. The appearance of the concentrated fluorescein film will be yellow, but at the site of the aqueous leakage, dilution by the aqueous turns the film bright green as it runs down the surface of the eye (Seidel's test) (Plate 7.5 p157). Fluorescein is also used for applanation tonometry and for the fitting of hard contact lenses. It must not be allowed to come into contact with soft contact lenses as it causes irreversible staining.

Intravenous fluorescein (5ml of a 20% solution injected into an antecubital vein) may be used to investigate diseases of the choroid, retina, optic nerve and iris by fluorescein angiography (p289) It is common for patients to experience a transient nausea and dizziness during the injection, but these symptoms rapidly pass. Because the dye is spread around all vessels, patients take on a yellow hue for several hours after angiography. The dye is excreted via the kidneys and the urine is stained yellow for up to 24 hours. Rarely acute anaphylaxis may occur during fluorescein angiography. It is necessary to have full resuscitation facilities close at hand.

*Indocyanin green* is a tricarbocyanin dye which is administered intravenously. Being 98% bound to blood proteins there is minimal extravasation from the choroidal capillaries, making it ideal for examining the choroidal circulation under infra-red conditions. Anaphylaxis may occur in patients sensitive to iodine.

*Rose Bengal* 1% aqueous solution is an iodine derivative of fluorescein. This dye unlike fluorescein, will stain hyalinised devitalised cells of the corneal epithelium and conjunctiva. It also stains mucus and keratin. There need not necessarily be a breakdown of the epithelium for staining to occur. This stain is used to reveal devitalised cells and filaments in kerato-conjunctivitis sicca and herpes simplex keratitis (Plate 10.2 p219). It gives rise to discomfort, especially if tear secretion is deficient.

*Mucolytics*

*Acetyl cysteine* is principally used for its mucolytic properties. It is available as eye drops at concentrations of 5%, 10% and 20%. The 20% solution, although the most effective, has the disadvantage of causing stinging on instillation, but it is relatively non-toxic. Drops are used 2-4 times daily. Mucus filaments occur in kerato-conjunctivitis sicca and in superior limbic keratitis (Plate 10.1 p217). The mucus associated with kerato-conjunctivitis sicca can usually be reduced by adequate treatment with artificial tears, but sometimes the addition of acetyl cysteine is helpful (p220). Although the 5% solution is relatively free of discomfort, it is also relatively ineffective, but as the patient may not tolerate the discomfort caused by the 20% solution, it may be of use. The treatment of superior limbic keratitis (p480) requires the 20% solution of acetyl cysteine and as the condition itself is so uncomfortable the patient will usually tolerate treatment.

*Citrate and ascorbate*

The use of systemic ascorbic acid and topical ascorbate and citrate has improved the prognosis where corneal damage has occurred as a result of chemical burns. Topical enzyme inhibitors are now believed to have little value in preventing or controlling corneal melting.

Ascorbic acid (Vitamin C) 500mg by mouth three times daily in combination with drops of sodium citrate 6.5% and potassium ascorbate 10% applied 1-2 hourly speed epithelial healing. These drops are not commercially available and need to be made up by the pharmacist. Their main side effect is intense stinging.

**Artificial tears and ocular lubricants**

The three main groups of artificial tears for the treatment of kerato-conjunctivitis sicca (pp217, 479) are:

-cellulose derivatives containing hypromellose,
-polyvinyl alcohol,
-mucomimetics containing hypromellose and dextran.

Whilst they all have theoretical advantages, in practice the patient selects the most suitable and least irritating preparation by trial and error. In keratoconjunctivitis sicca the frequency of administration is governed by patient comfort. The frequent use of artificial tears may cause preservative toxicity in about 9% of patients. Consideration should be given to preservative free solutions in patients showing signs of toxicity. The commercial names of the main groups are listed:

| | |
|---|---|
| -Cellulose derivatives: | BJ6, Isopto Plain, Isopto Alkaline |
| -Polyvinyl alcohol: | Liquifilm Tears, SNO Tears, Hypotears |
| -Polyacrylic acid: | Viscotears |
| -Mucomimetics: | Tears Naturale |

*Lubricating ointment* is particularly useful at night for the prevention of recurrent abrasions. Examples include Simple Eye Ointment and Super Soft Paraffin (SSP).

## Chapter 34

# IATROGENIC DISORDERS

## Systemic toxicity of ocular therapeutic agents

Physicians are trained to consider the possibility of drug induced toxicity and will expect side effects from orally or parenterally administered drugs. What is not so readily appreciated, is that drugs administered in the form of eye drops may also have systemic side effects due to the high concentrations required to penetrate the corneal barrier. An appreciable portion of the drop is systemically absorbed either by the conjunctiva or by the nasal mucous membrane via the naso-lacrimal duct, thereby avoiding first pass metabolism. Some of the drug may also be absorbed by the gastrointestinal system.

Systemic toxicity is particularly likely to occur following 'intensive treatment' as in the case of the treatment of acute glaucoma where a large quantity of the drug is delivered to the eye and a large proportion is inevitably absorbed.

Strategies aimed at decreasing the systemic absorption of topically applied drugs include punctal occlusion and eyelid closure, reducing the instilled volume, increasing the viscosity of the topical vehicle, altering lyophilicity and pH and using controlled release preparations. The majority of serious systemic side effects result from the use of drugs designed to alter pupil size and ciliary muscle contraction or tone by their action on the autonomic nervous system.

*The parasympathomimetic agents* (p592)

These are used for the reduction of intraocular pressure. Both the acetylcholine analogues and the cholinesterase inhibitors increase the parasympathetic action in the eye inducing miosis. Their systemic absorption in sufficient quantity will cause generalised parasympathetic overaction characterised by nausea, vomiting, abdominal cramp, diarrhoea, bronchiolar spasm, pulmonary oedema, bradycardia, hypotension and collapse. The irreversible anticholinesterases such as echothiophate iodide will also deplete circulating pseudo-cholinesterase, resulting in apnoea following the use of succinyl choline during anaesthesia. This effect can last for up to six weeks after the cessation of therapy.

*The parasympatholytic agents* (p588)

These are used as cycloplegics and mydriatics. Significant systemic absorption of these agents will inhibit the parasympathetic nervous system. The signs of systemic toxicity include tachycardia, dry mouth and dry skin. Central nervous system stimulation may occur with these drugs, particularly atropine. The symptoms include ataxia, dysarthria, hallucinations and confusion. Older patients are particularly prone to confusion even on low doses of atropine drops.

*The sympathetic beta-blockers.* (p139)

These agents are used in the treatment of glaucoma and ocular hypertension. Absorption via the naso-lacrimal duct and the avoidance of first pass metabolism can result in the therapeutic plasma levels of their systemically administered counterparts. All the beta-blockers can reduce heart rate and should therefore be avoided in patients with heart block or cardiac failure. All beta-blockers can result in bronchospasm in patients with reversible airways disease, although a cardioselective beta-blocker such as betaxolol hydrochloride (Betoptic) tends to have less of an adverse effect. There is also evidence that non cardioselective beta-blockers such as timolol maleate (Timoptol) by absorption produce bronchospasm in a significant proportion of elderly patients without a history of airways disease, sometimes being an unsuspected cause of shortness of breath, they may also reduce exercise tolerance even when there are no respiratory symptoms and rarely in insulin dependant diabetics they may mask the symptoms of incipient hypoglycaemia with the potential for serious results. Betablockers are thought to reduce libido and are suspected of causing impotence.

*The sympathomimetic agents* (p590)

These are used for pupil dilatation and the reduction of intraocular pressure. Phenylephrine has purely alpha stimulating properties, consequently its major effect after systemic absorption is to raise the blood pressure. Adrenaline possesses both alpha and beta stimulating properties and after significant systemic absorption may induce tachycardia, raised blood pressure and sweating. Occasionally cardiac arrhythmias may occur, ranging from extrasystoles to ventricular fibrillation and cardiac arrest. Its use is contraindicated in those with narrow anterior chamber angles as it may occasionally cause angle closure.

## Drugs inducing ocular adverse reactions

Frequently cases are reported of ocular problems occurring in patients on a wide variety of drugs. It is often extremely difficult to decide if the drug is responsible or if the problem is due to coincidentally developing ocular disease. The difficulties are increased when the response to the drug appears idiosyncratic and not dose related. These difficulties are further compounded when only a small number of people on the drug are affected. Any reaction to a drug should be noted and reported. There should be particular interest in the drug history when a patient presents to an eye department with an unusual ocular problem. A few examples of ocular adverse reactions to drugs are described below.

### *Amiodarone*

Amiodarone is used in the treatment of atrial and ventricular arrythmias. It is also used in the treatment of the Wolff-Parkinson-White syndrome. Deposition of the drug in the cornea and lens is dose and duration dependent. Initially there is a faint pigmented horizontal line in the epithelium which progresses to form a whorl like pattern involving the visual axis. With high doses amiodarone may produce anterior lens opacities which are subcapsular and pigmented. Virtually all patients on amiodarone therapy for over three months will show a mild degree of keratopathy. With the cessation of treatment the keratopathy resolves over 6-18 months. Symptoms are rare although the lenticular changes and the more severe forms of keratopathy may result in haloes and glare. Visual acuity is rarely affected although a toxic optic neuropathy has been recorded.

*Chlorpromazine*

Chlorpromazine is a phenothiazine used in the treatment of psychiatric disorders. It is deposited in the ocular tissues rather than producing toxic effects. Deposition tends to be related to dose and duration of therapy. Changes are rarely seen with total doses of less than 500g. Lens changes include fine anterior pigmentary deposits which progress to form a white or brownish anterior stellate pattern and finally an opaque central mass. Corneal involvement follows lens changes with the appearance of white and pigmented deposits on the endothelium of the interpalpebral area. The deposits generally remain after the cessation of therapy. These ocular changes rarely reduce visual acuity but may result in glare, haloes and misty vision. The anticholinergic effects of chlorpromazine may result in difficulties with accommodation. This might increase the risk of angle closure glaucoma in patients with narrow angles and shallow anterior chambers. Chlorpromazine can produce a slatey blue discoloration of the interpalpebral bulbar conjunctiva, sclera and exposed skin. Rarely, pigmentary changes may be seen in the retina.

*Thioridazine ( Melleril)*

The anticholinergic properties of the anti-psychotic drug thioridazine may produce accommodative difficulties similar to those of chlorpromazine, however its retinal toxicity is more important. These changes result in reduced visual acuity, altered colour vision, paracentral or ring scotomas and impaired dark adaptation. Thioridazine binds to melanin pigment in the uveal tract resulting in damage to the choriocapillaris which results in the retinal pigment epithelium initially showing pigment clumping in the periphery which progresses towards the posterior pole where macular oedema may occur. These changes are dose-dependent and are generally reversible in the early stages. The retinopathy is common when large doses are used, but if the daily dose is kept below 800mg these changes are rare.

*Chloroquine and hydroxychloroquine*

Chloroquine and hydroxychloroquine are used in the treatment of rheumatoid arthritis, discoid and systemic lupus erythematosis and other collagen diseases. Both chloroquine and hydroxychloroquine can produce a keratopathy although this is less likely with hydroxychloroquine. Initially there are punctate deposits in the epithelium with whorled greenish yellow lines appearing in the central cornea later. These findings do not appear to be dose dependent. The opacities rarely reduce visual acuity but may result in haloes and glare. They disappear after discontinuation of therapy.

Accommodative insufficiency is a common side effect of chloroquine therapy. Chloroquine and hydroxychloroquine can produce a retinopathy although the risk and severity is far greater with chloroquine. In the case of chloroquine the changes may be irreversible and may progress after the cessation of therapy. The effects appear to be dose and age related with retinopathy rarely occurring before a total dosage of 100g (equivalent to 250mg daily for one year) in the case of chloroquine. Visual changes in the form of relative central scotomas may manifest themselves before the appearance of clinical signs and it is at this stage that the maculopathy is generally reversible. Initially there is fine hyperpigmentation of the macula followed by the formation of concentric rings of hyperpigmentation with an area of hypopigmentation in between (bull's eye maculopathy). By this stage there are dense central scotomas, a reduction in visual acuity and often a reduction in colour perception. Hydroxychloroquine maculopathy is mild and does not affect visual acuity. Chloroquine may result in changes not unlike retinitis pigmentosa, but the hyperpigmentation is not associated with the retinal veins and there are minimal changes in dark-adaptation. Patients receiving chloroquine should have baseline evaluations including visual acuity, colour vision, field analysis and fundus examinations prior to the commencement of therapy. Subsequently evaluations should be repeated at six monthly intervals.

*Quinine*

Originally an anti-malarial drug, quinine is also used in the treatment of leg cramps. Quinine can reduce visual acuity bilaterally to 'no perception of light' and this may be irreversible. These effects can be idiosyncratic or dose related, being more common with daily doses above 4g. Quinine appears to be toxic to photoreceptors and ganglion cells, and often the initial fundal appearances are normal, but subsequently there may be attenuation of the retinal arterioles and retinal pallor and oedema, with a cherry-red spot at the macula, followed by optic atrophy

*Tamoxifen (Nolvadex)*

Tamoxifen is an anti-oestrogen used in the treatment of breast carcinoma. Toxic retinal effects generally occur with doses above 90mg daily. Patients may notice reduced visual acuity and central field defects. White and yellow lesions (which are the products of axonal degeneration) are concentrated at the macula with or without the presence of oedema. Usually these changes are irreversible.

*Digoxin and digitoxin*

The cardiac glycosides digoxin and digitoxin are used in the treatment of congestive cardiac failure and atrial fibrillation. The toxic effects of these drugs result in cone dysfunction producing visual symptoms but not signs. The most common symptom is distortion of vision where objects appear to be covered in snow or frosted, but there may also be dyschromatopsia and a general dimming of vision. These effects are dose related and are helpful when trying to adjust dosage levels. Most patients on therapeutic levels of these drugs have measurable but asymptomatic changes in colour vision.

*Ethambutol*

Ethambutol is used in the treatment of tuberculosis. The most common toxic effect is retrobulbar neuritis. This typically affects central nerve fibres resulting in red-green and blue-yellow colour vision defects, central and paracentral scotomas and loss of visual acuity. Generally, if colour vision alone is affected the changes resolve quickly with discontinuation of therapy. Field defects and loss of visual acuity may take many months to recover. Rarely there may be retinal changes including hyperaemia and swelling of the disc, flame shaped haemorrhages and macular oedema. These changes may be followed by optic atrophy. Toxicity is dose-related and rarely occurs when the dosage is maintained at 15-20mg/kg. Patients should have baseline evaluations of acuity, fields, colour vision and fundal appearance before treatment is commenced. When the dose is kept at 15-20mg/kg follow up should be at 3-6 month intervals. Patients with renal failure do not excrete the drug effectively and should be monitored more closely.

*Phenytoin and Vigabatrin*

Phenytoin is used in the treatment and prevention of epilepsy. Fine nystagmus may occur with therapeutic levels, but coarse downbeat, horizontal or vertical nystagmus occurs with toxic levels. Paralysis of the extraocular muscles with diplopia may occur. The ocular side effects are reversible Another antiepileptic drug vigabatrin may cause visual field loss which is sometimes permanent.

*Drugs affecting intraocular pressure*

Systemic drugs can both lower and raise intraocular pressure. Acute angle closure may be precipitated by drugs which dilate the pupil in patients with very narrow angles. Drugs which have such effects include anticholinergics, sympathomimetics, antihistamines, phenothiazines and tricyclic antidepressants. Patients deemed to have angles at high risk of closure may be advised to have prophylactic laser iridotomy performed and will subsequently be at less risk of this complication. Corticosteroids administered both topically and systemically have the potential to raise intraocular pressure in some people. All those on long term steroid treatment should have their intraocular pressures tested at intervals (p582). Other adverse effects of corticosteroids have been considered on pages 168 and 606.

Systemic drugs may reduce intraocular pressure. Such drugs include ß-adrenergic blockers, alcohol and cardiac glycosides. Even when not used therapeutically for the treatment of glaucoma, they may reduce intraocular pressure sufficiently to cause confusion in its diagnosis. Systemic ß-blockers may cause significant reductions in optic nerve perfusion, especially in patients with advanced disc damage. Collaboration between physicians and ophthalmologists is required in this area.

# Index
## (Medications classified under Therapeutic Agents)